The Discrepancy Method

The discrepancy method is the glue that binds randomness and complexity. Its aim is to resolve one of the burning issues in theoretical computer science: Does randomization really help? The discrepancy method map offers an answer through the study of distribution irregularities, also known as discrepancy theory. By drawing upon a wealth of mathematical techniques, this approach has met with considerable success; in particular, it has caused a mini-revolution in computational geometry.

This books tells the story of the discrepancy method in a few short, independent vignettes. It is an eclectic tale that features, among other topics, communication complexity, pseudorandomness, rapidly mixing Markov chains, points on the sphere via modular forms, derandomization, convex hulls, Voronoi diagrams, linear programming, geometric sampling, VC-dimension theory, minimum spanning trees, linear circuit complexity, and multidimensional searching.

The mathematical treatment is thorough and self-contained. In particular, background material in discrepancy theory is supplied as needed. Thus, the book should appeal to students and researchers in computer science, pure and applied mathematics, operations research, engineering, statistics, and computational sciences.

Bernard Chazelle is Professor of Computer Science at Princeton University and a Fellow of the NEC Research Institute. He is an ACM Fellow and a former Guggenheim Fellow. He currently serves on the editorial boards of eight scientific journals. Founder of the Princeton-Area Center for Theory (PACT) and former co-Director of DIMACS, he is also President of the Scientific Council for Computer Science at Ecole Normale Supérieure in Paris.

T0276217

The Discrepancy Method

Randomness and Complexity

BERNARD CHAZELLE

Princeton University

CAMBRIDGE
UNIVERSITY PRESS

CAMBRIDGE UNIVERSITY PRESS
Cambridge, New York, Melbourne, Madrid, Cape Town, Singapore,
São Paulo, Delhi, Dubai, Tokyo, Mexico City

Cambridge University Press
The Edinburgh Building, Cambridge CB2 8RU, UK

Published in the United States of America by Cambridge University Press, New York

www.cambridge.org
Information on this title: www.cambridge.org/9780521003575

First published 2000
First paperback edition 2001

A catalogue record for this publication is available from the British Library

ISBN 978-0-521-77093-4 Hardback
ISBN 978-0-521-00357-5 Paperback

Contents

Preface

iscrepancy theory grew out of a question posed by van der Corput in 1935: How uniform can an infinite sequence of numbers in $[0,1]$ be? To give meaning to this question, we may ask how fast the function

$$D(n) = \sup_{0 \leq x \leq 1} \left| \, |S_n \cap [0,x]| - nx \, \right|$$

grows with n, where S_n consists of the first n elements in the sequence. If the sequence were uniform—whatever we really mean by this—we would expect $D(n)$ to grow rather slowly, if at all. Indeed, there are known sequences for which $D(n) = O(\log n)$. Surprisingly, a theorem of Schmidt says that this is essentially optimal: $D(n)$ can never be in $o(\log n)$.

Schmidt's result can be viewed as a limitation on how well a certain discrete distribution, $x \mapsto |S_n \cap [0,x]|$, can simulate a continuous one, $x \mapsto nx$. In other words, a certain amount of *discrepancy* between the two distributions is unavoidable. Naturally, countless variants of this problem can be formulated. Their collective body forms the subject matter of discrepancy theory.

Intellectual curiosity aside, why should a computer scientist care? For an answer, go back to the first sentence of the previous paragraph, and replace the words "discrete" by "polynomial" and "continuous" by "exponential." The resulting sentence talks about efficiently computable distributions simulating intractable ones. By a wonderful coincidence this is the driving issue behind a central complexity theory question: Is randomization necessary? There is plenty of evidence to suggest that it is. Probabilistic algorithms are usually shorter, simpler, faster than their deterministic counterparts.

But outside of specialized problem models[1] there is no proof that probabilistic algorithms are computationally more powerful. For example, randomization allows us to test whether a number is prime in polynomial time. That we cannot replicate the feat deterministically now does not mean that we won't be able to some day. Whether random bits are truly needed or not is one of the major open problems in complexity theory today.

To understand why discrepancy theory addresses this question head-on, we must go back to the characteristic feature of a probabilistic algorithm: access to a sequence of perfectly random bits. What would happen if, instead, the algorithm were presented with merely a *pseudorandom* string of bits or, to take the idea to its limit, one that is computed deterministically? Should performance necessarily suffer? Intuitively, it would seem that if the perfectly random sequence could be approximated well enough by one that was only pseudorandom, the algorithm might be fooled into behaving the same. Indeed, unless $P = NP$, polynomial algorithms—the only ones practitioners care about—are not expected to be too good at telling apart random and pseudorandom.

In particular, suppose that one could replace an exponential-size probability space (common ones are typically *that* large) by one of polynomial size, without the algorithm realizing the subterfuge. Then, obviously, no loss of efficiency could occur. Not surprisingly, simulating a complicated (read: intractable) probabilistic distribution by a simple (read: polynomial) one is grist for discrepancy theory's mill; any student of Monte Carlo techniques for numerical integration knows that.

Complexity theory adds a new twist, however. Discrepancy no longer has to do with the accuracy of the output—as it does in numerical integration—but only with the time it takes to produce it exactly. (What would be the meaning of an approximate answer to the question: Is n prime?) This different outlook has given rise to the *Discrepancy Method.*[2] Discrepancy theory is blessed with many powerful tools and techniques developed since the nineteenth century. The discrepancy method bridges these tools with the new, vibrant field of complexity theory and algorithm design. It has been the force behind major recent developments in areas as diverse as probabilistic algorithms, derandomization, communication complexity,

---

[1] For example, secret key selection in a public-key cryptosystem, Byzantine agreement, oracle-based convex-body volume estimation, or primality testing without the extended Riemann hypothesis.

[2] The word "method" is to be understood here less as a particular proof technique (cf. the probabilistic method) than as a spotlight on the common core of a large and varied set of problems.

searching, machine learning, pseudorandomness, computational geometry, optimization, computer graphics, and mathematical finance.

The story of the discrepancy method is rich and far-reaching. This book tells fragments of it by means of specific examples, including both upper bounds (algorithm design) and lower bounds (complexity theory). The discrepancy method counts as one of the great achievements of theoretical computer science and one of its most compelling stories. But there is an added pleasure: Much of the math in the story is of great beauty. How can one resist such gems as the Alexander-Stolarsky formula or Roth's method of orthogonal functions or the Fourier transform method of Beck and Montgomery? That, in addition, such techniques can be used to prove lower bounds in complexity theory is nothing short of wondrous.

The fundamentals of discrepancy theory are presented in the first three chapters. The presentation privileges techniques over results: The aim is to introduce the main tools of the trade and use specific problems merely as a vehicle for reaching that objective. This is *not* a book on discrepancy theory. The reason we address the subject in the first place is to build a pool of techniques for us to tap into in subsequent chapters. There, we address a variety of topics, such as communication complexity, pseudorandomness, rapidly mixing Markov chains, sampling, linear programming, circuit complexity, geometry, searching, linear selection, and matroid optimization. All subjects are presented as reasonably short, independent vignettes. Three exceptions to this rule: Points on a sphere, convex hulls, and minimum spanning trees (Chapters 2, 7, 11, respectively) require a longer, technically more demanding treatment.

This book, like most, has a finite number of pages; a fact that did not always strike the author as self-evident. The main casualty is a plethora of serious omissions. Truly, what sort of book can call itself "the discrepancy method" while overlooking such perennial users of the method as computational learning theory, approximate counting, volume estimation, one-way functions, bin packing, computational finance, etc? Fortunately, excellent texts on some of these topics already exist, eg, Biggs and Anthony [46], Goldreich [143], Luby [204], Motwani and Raghavan [236], Sinclair [289], Traub and Werschulz [309]. Since Aardenne-Ehrenfest's proof [1] of van der Corput's conjecture in 1945, discrepancy theory has grown into a rich, mature subject. For a thorough, expert treatment of the field, the reader hungry for more will turn to Beck and Chen [37], Drmota and Tichy [111], or Matoušek [219]. None of these references address the discrepancy method, however.

Derandomization figures prominently in this book. Soberer minds might smell a whiff of perversion. Random bits help to make algorithms simple, practical, and (virtually) infallible. Why would anyone want to derandomize them? There are two good reasons: One is that deterministic low-discrepancy constructions are often better than randomized ones. The other reason is, very simply, to understand. For all its wonders, randomness sometimes is the tree that hides the forest. For example, expanders, convex hulls, and minimum spanning trees all lend themselves to simple, elegant probabilistic treatments. The randomized approach is, in fact, so powerful that, left to its own devices, it sheds scant light on these problems. In particular, it barely hints at their stunningly rich and beautiful structures. Forgoing randomization forces us to "look under the hood." I hope the reader will find the sight impressive and the exploration rewarding.

Who Should Read this Book?

To anyone who is curious about algorithms, complexity, and their relation to classical mathematics, this book has a story to tell. Complexity theory is one of the genuinely modern sciences to emerge in the twentieth century. While structural questions such as P vs. NP remain the tall mountains to climb, the art of designing polynomial-time algorithms and analyzing them mathematically has achieved an impressive level of maturity, resulting in a number of truly astonishing results. In its own modest way, this book tries to document this statement.

This book is for everyone with a taste for theoretical computer science. Let it be said, however, that college-level knowledge of mathematics and algorithms will make for smoother reading. Theoretical computer scientists are said to be mathematicians in a hurry. Regardless of the stereotypes, this book should please both camps, beginning at the senior undergraduate level. Some chapters can be read quickly while others simply cannot. (Which is which is for the reader to decide.) Ample background material has been supplied as needed. A fine thought, I know... I remember a textbook going to great lengths to explain that a compact surface of genus one *really* is like a donut, and then moving on to quote the Riemann-Roch theorem without a word of explanation. I tried to avoid this misordering of priorities, but to try does not always mean to succeed.

Judging from its table of contents, this book promises all the cohesion of a rummage sale. What could possibly justify putting under the same roof modular forms, minimum spanning trees, Voronoi diagrams, expanders, and linear circuits? Whether we are playing with the symmetries of the hyperbolic plane to draw dots on a beach ball, or designing error-prone priority queues to find minimum spanning trees, or calling on VC-dimension theory to compute Voronoi diagrams, or strolling down Cayley graphs to recycle random bits, or surfing wavelets to build high-spectrum matrices, we are always trying to achieve the same effect, albeit for different purposes: It is either to coax a representative sample out of a complex structure or, as in the last example, to show that this can't be done well. The discrepancy method encapsulates this idea and builds the tools to make it happen.

A few sections are marked with an asterisk; this indicates that they provide the broader mathematical context within which the material is to be understood. Reading them can be illuminating, even a lot of fun (I hope), but it is not essential. Keep in mind, however, that much of the added value of a book over, say, a collection of research articles often resides in those less than indispensable passages. To make this volume as self-contained as possible and at the same time enable different levels of reading, copious amounts of background material have been included in footnotes and appendices.

There is a natural temptation for any book on advanced topics to handle complicated arguments with quick brushstrokes and let the readers figure out the messy details by themselves. I have resisted this temptation. My rule has been to provide complete proofs for pretty much everything discussed in this book outside of appendices. (Of course, like any good rule, this one has exceptions, too.)

Notation and Terminology

All logarithms are to the base two, unless indicated otherwise. The expressions $f = O(g)$, $f = O_d(g)$, $f \ll g$, $f \ll_d g$, $g = \Omega(f)$, $g = \Omega_d(f)$, $g \gg f$, and $g \gg_d f$ all mean the same thing, ie,[3] $f \leq Cg + C'$, for two positive constants C, C' and all values of the variables. The subscript d indicates that the constant C depends on a parameter d. So, for example, one might write $O(x)^d = O_d(x^d)$. If both $f \ll g$ and $g \ll f$, we use the notation $f \approx g$ or $f = \Theta(g)$; if $f(x)/g(x)$ tends to 0 as $x \to \infty$, then we write $f = o(g)$. The expression, "for n sufficiently large," is a handy way of saying that n should exceed a constant large enough to satisfy the various inequalities in which n appears.

---

[3]The reader wondering what happened to the dots should blame no one (eg, the copy editor) but the author (ie, me) for his inability to appreciate the wisdom of unabbreviating abbreviations with extra dots. Exampli gratia, why e.g. and not eg?

The repeated use of expressions such as "it is easy to" or "obviously" can be exasperating. It should not. Think of them as punctuation with meaning. For instance, to the uninitiated, hearing that Voronoi diagrams are convex hulls in disguise, or that rational elliptic curves are modular, are two equally intimidating statements. All four concepts have rigorously nothing to do with one another. Each one is deep and central. Therefore, to prove these statements must be very hard. Wrong. In this book, the first statement might open with "It is easy to show that...," while the words "exceedingly difficult" might accompany the second: a useful thing to know. The word "obviously" means that the reader should be able to supply a proof within seconds. If he cannot, then obviously he is fully entitled to... feel very bad about himself.

Acknowledgments

I have taught most of the material in this book at one point or another. I thank the students in my *Advanced Probabilistic Techniques Seminar* at Princeton University for the comments and feedback they have given me over the years. Naturally, they should thank me too, for they surely know what finer human beings they are now. Well... apparently one didn't quite see it that way. The student dropped out because, as he so tactfully put it: "My life is already so full of discrepancies I don't see the need to add any more to it."

For sharing with me their thoughts and insights about the matters discussed in this book, and in many cases, reading and commenting on earlier versions of the manuscript, I am indebted to Pankaj Agarwal, Noga Alon, József Beck, Jin-Yi Cai, Bernadette Charron, Ketan Dalal, Faith Fich, Joel Friedman, Monika Rauch Henzinger, Rick Kenyon, Joe Kilian, Laci Lovász, Jirka Matoušek, Hugh Montgomery, Joel Spencer, Ken Steiglitz, Endre Szemerédi, Emo Welzl, Andy Yao, and David Zuckerman.

I am particularly grateful to Jean-Benoît Bost and Peter Sarnak for kindly reading the arithmetic section of Chapter 2, pointing out misstatements, suggesting additions and improvements and, in the process, helping me make some sense of the spectacular beauty of modular forms. Needless to say, all remaining errors in this book are solely mine.

I thank Lauren Cowles of Cambridge University Press for her helpful guidance throughout the publishing process, and Elise Oranges for her expert copyediting. Many thanks to Daniel Huson for letting me use his pictures of hyperbolic tilings. The cover of this book is based on an image produced by one of his computer programs. (Similar tilings are used in several places in the book.)

As a Professor of Computer Science at Princeton University and a Fellow of the NEC Research Institute, I am deeply grateful to both institutions for the privilege of working in such wonderful environments. This book was completed during a sabbatical year in France in 1998–99. I wish to thank Ecole Polytechnique and INRIA for co-hosting my visit and, as always, Ecole Normale Supérieure for its kind hospitality. What was that lovely story again about American expats in Paris drinking wine and writing books...?

This work was supported in part by NSF Grant CCR-93-01254, NSF Grant CCR-96-23768, NSF Grant CCR-96-43913, NSF Grant CCR-97-31535, ARO Grant DAAH04-96-1-0181, ARO Grant DAAD19-99-1-0205, NSF Grant CCR-91-999 (DIMACS), and NEC Research Institute.

You might read this book, but chances are you won't hear it. That's a pity because the music that infused its writing was the best part. In fact, I have sometimes wondered whether working on this volume was not just an excuse for listening to my jazz heroes endlessly while masquerading as a would-be author. Duke, Bean, Prez, Bird, Lady Day, Monk, Dexter, Miles, Trane, Clifford, Ella, Rahsaan: In invisible ink, these giants grace every page of this book. The sweet irony is, you won't find a bit of discrepancy in their blues.

Of all the sentences in this book none was easier to write and sweeter to read again than this one: To my wife, Celia, and my children, Damien and Anna, this book is dedicated with love.

Paris, August 1999   BERNARD CHAZELLE

Paperback Edition

Additions were made to Chapters 1 and 6. In particular, a discussion of the trace bound and its applications was added to the original text.

Princeton, April 2001   BERNARD CHAZELLE

1

Combinatorial Discrepancy

et $(V, \mathcal{S})$ be a set system, where $V = \{v_1, \ldots, v_n\}$ is the ground set and $\mathcal{S} = \{S_1, \ldots, S_m\}$, with $S_i \subseteq V$. (Such a combinatorial structure is often called a *hypergraph*.) We wish to color the elements of V red and blue so that, within each S_i, no color outnumbers the other one by too much. To make this notion precise, we introduce a function χ mapping each $v_j \in V$ to a "color" in $\{-1, 1\}$, and we define the *discrepancy* of the set S_i to be

$$\chi(S_i) = \sum_{v_j \in S_i} \chi(v_j).$$

The maximum value of $|\chi(S_i)|$, over all $S_i \in \mathcal{S}$, is called the discrepancy of the set system (under the given coloring). When no particular coloring is understood, the *discrepancy* of the set system, denoted by $D_\infty(\mathcal{S})$, refers to its minimum discrepancy over all possible colorings.[1]

This type of discrepancy is called *combinatorial* or, more evocatively, *red-blue*. By contrast with some of the discrepancies discussed in subsequent chapters, which involve both continuous and discrete distributions, the red-blue discrepancy compares two discrete distributions. Both types are intimately linked, however, and techniques for red-blue discrepancy often extend effortlessly to the continuous case.

Discrepancy has been defined in the worst-case sense, ie, in the L^∞ norm. This is intuitively appealing but difficult to manipulate algebraically. The L^2 norm provides a friendlier environment, so we define

$$D_2(\mathcal{S}) \stackrel{\text{def}}{=} \min_\chi \sqrt{\chi(S_1)^2 + \cdots + \chi(S_m)^2} \,,$$

---

[1] For technical convenience, we use absolute values for the discrepancy of set systems but not when referring to the discrepancy of a particular subset.

over all colorings $\chi : V \mapsto \{-1, 1\}$. This suggests an algebraic characterization of the discrepancy using matrices. Let A be the *incidence matrix* of the set system $(V, \mathcal{S})$; this is the matrix whose n columns are indexed by the elements of V and whose m rows are the characteristic vectors of the sets S_i, so that A_{ij} is 1 if $v_j \in S_i$ and 0 otherwise. The discrepancy of the set system, also denoted by $D_\infty(A)$, can be expressed as the L^∞ norm of a column vector:

$$D_\infty(A) = \min_{x \in \{-1,1\}^n} \|Ax\|_\infty.$$

Similarly,

$$D_2(A) = \min_{x \in \{-1,1\}^n} \|Ax\|_2.$$

Here is an overview of this chapter:

- In §1.1 we show that, in the absence of any special assumptions on the set system, a random coloring is nearly optimal. It ensures a discrepancy on the order of $\sqrt{n \log(2m)}$. We give several methods for computing such a coloring deterministically and, in the process, introduce a general derandomization technique.

- We show in §1.2 that if the number of sets in $\mathcal{S}$ is small enough, eg, $O(n)$, then the discrepancy can be kept in $O(\sqrt{n})$. (The bound is proven to be optimal in §1.5.) This gives us the opportunity to introduce the powerful *entropy method* of discrepancy theory.

- In §1.3 we establish the classical Beck-Fiala theorem, which says that if no element belongs to more than a constant number of sets, then the discrepancy can be kept constant.

- We discuss the case of *range spaces* in §1.4. These are well-structured set systems of central importance in discrete and computational geometry. We derive several results that form the foundation of our treatment of geometric sampling in Chapter 4.

- In §1.5 we describe several methods for deriving lower bounds on the discrepancy of set systems. All of them have to do with the spectrum of $A^T A$. The simplest one relates the discrepancy to the smallest eigenvalue. We apply this *eigenvalue bound* to derive a classical theorem of Roth on the discrepancy of arithmetic progressions. This result is optimal, but in general the eigenvalue bound is weak because it does not exploit the fact that the coloring x is a vector with ± 1 coordinates. To do that, we introduce the notion of *hereditary discrepancy* and show how determinants can be used to prove lower bounds. We give an application to set systems formed by points and halfplanes. Finally, we derive the powerful *trace bound*, which allows us to avoid determinants and eigenvalues altogether and prove tight lower bounds in a surprisingly simple manner. We give two examples: points in lines, and points in higher-dimensional boxes.

1.1 Greedy Methods

Given a set system $(V, \mathcal{S})$, with $|V| = n$ and $|\mathcal{S}| = m$, pick a random coloring χ, meaning that for each v_j, the "color" $\chi(v_j)$ is chosen randomly, uniformly, and independently, in $\{-1, 1\}$. We say that S_i is *bad* if $|\chi(S_i)| > \sqrt{2|S_i|\ln(2m)}$. By Chernoff's bound,[2] we immediately derive

$$\text{Prob}[\, S_i \text{ is bad} \,] < \frac{1}{m} \,;$$

therefore, with nonzero probability, no S_i is bad.

Theorem 1.1 *The discrepancy of a set system $(V, \mathcal{S})$ does not exceed $\sqrt{2n\ln(2m)}$, where $|V| = n$ and $|\mathcal{S}| = m$. This is achieved by a random coloring.*

Let us slightly relax the bound and say that S_i is bad if

$$|\chi(S_i)| > \sqrt{3|S_i|\ln(2m)}.$$

Then, by Chernoff's bound, the probability that no S_i is bad exceeds $1 - 1/\sqrt{m}$. Note that if the first coloring we try fails, we should keep on trying. The probability of being still unsuccessful after k attempts is only $O(1/m^{k/2})$.

The Method of Conditional Expectations

We now describe a general technique for derandomizing the probabilistic coloring algorithm, ie, transforming it into one that does the same thing without using random bits.

The idea is to assign $\chi(v_1)$, $\chi(v_2)$, etc, in that order, without ever backtracking. Let $B = \sum_{i=1}^{m} B_i$, where B_i is the indicator variable equal to 1

---

[2]See Lemma A.5.

if S_i is bad and 0 otherwise. We know that

$$\mathbf{EB} = \sum_{i=1}^{m} \mathbf{EB}_i = \sum_{i=1}^{m} \mathrm{Prob}[\, S_i \text{ is bad }] < \frac{1}{\sqrt{m}} . \tag{1.1}$$

Let $\varepsilon_1 = \pm 1$ be such that

$$\mathbf{E}[\, B \mid \chi(v_1) = \varepsilon_1\,] \leq \mathbf{E}[\, B \mid \chi(v_1) = -\varepsilon_1\,].$$

We have

$$\mathbf{EB} = \mathbf{E}_{\chi(v_1)} \, \mathbf{E}[\, B \mid \chi(v_1))\,] \geq \mathbf{E}[\, B \mid \chi(v_1) = \varepsilon_1\,]. \tag{1.2}$$

In general, let $\varepsilon_k \in \{-1, 1\}$ minimize the function of x,

$$\mathbf{E}[\, B \mid \chi(v_1) = \varepsilon_1, \ldots, \chi(v_{k-1}) = \varepsilon_{k-1}, \chi(v_k) = x\,].$$

Note that

$$\mathbf{E}[\, B \mid \chi(v_1) = \varepsilon_1, \ldots, \chi(v_{k-1}) = \varepsilon_{k-1}\,]$$
$$= \mathbf{E}_{\chi(v_k)} \, \mathbf{E}[\, B \mid \chi(v_1) = \varepsilon_1, \ldots, \chi(v_{k-1}) = \varepsilon_{k-1}, \chi(v_k)\,]$$
$$\geq \mathbf{E}[\, B \mid \chi(v_1) = \varepsilon_1, \ldots, \chi(v_k) = \varepsilon_k\,].$$

It follows from (1.2) that

$$\mathbf{EB} \geq \mathbf{E}[\, B \mid \chi(v_1) = \varepsilon_1, \ldots, \chi(v_k) = \varepsilon_k\,].$$

At $k = n$, no randomness is left, so from (1.1),

$$\frac{1}{\sqrt{m}} > \mathbf{EB} \geq \mathbf{E}[\, B \mid \chi(v_1) = \varepsilon_1, \ldots, \chi(v_n) = \varepsilon_n\,].$$

The right-hand side denotes the number of bad S_i's in the final coloring, which, being less than one, is therefore zero. Thus, the assignment $\chi(v_i) = \varepsilon_i$ $(1 \leq i \leq n)$ guarantees that each S_i satisfies

$$|\chi(S_i)| \leq \sqrt{3|S_i| \ln(2m)}.$$

The entire procedure can be carried out in polynomial time. Indeed, there are n basic coloring steps, and each of them involves the calculation of two conditional expectations of the form

$$\mathbf{E}[\, B \mid \chi(v_1) = \varepsilon_1, \ldots, \chi(v_k) = \varepsilon_k\,].$$

Each such conditional expectation is a sum of m terms of the form

$$\mathrm{Prob}[\, |\chi(S_i)| > \sqrt{3|S_i| \ln(2m)} \mid \chi(v_1) = \varepsilon_1, \ldots, \chi(v_k) = \varepsilon_k\,],$$

each of which is a sum of at most $2n$ probabilities from the binomial distribution[3] $B(|S_i \cap \{v_{k+1}, \ldots, v_n\}|, 1/2)$. Note that the difficulty of computing huge binomial coefficients is easily circumvented. Using the bound of $1/\sqrt{m}$ in (1.1) provides some slack that allows us to perform calculations with relative error $1/(nm)^{O(1)}$. This, in turn, makes a computer word size of $O(\log(n+m))$ sufficient.

The algorithm we just described is an instance of a very general derandomization technique. We will encounter it again in Chapter 7. It is *greedy* in that it follows a locally optimal strategy, but the cost function encodes information about the future. Intuitively, the idea is to keep, at all times, the relative density of good events reachable from the current state bounded from below. As long as these densities—or good enough approximations thereof—can be computed effectively, the method yields a polynomial algorithm for reaching a good event.

A steep trail might be followed by a flat portion. So, to ensure a fast descent, at each fork the skier opts for the trail whose average slope over all descents from that trail is maximum in absolute value. The decision is not a local one.

The Hyperbolic Cosine Algorithm

A somewhat simpler approach is to choose a cost function based on the partial discrepancies incurred up to the current point in time. Suppose that $\chi(v_1), \ldots, \chi(v_k)$ have already been assigned. For each S_i, let $p_{i,k}$ (resp. $m_{i,k}$) be the number of $v_j \in S_i$ $(j \le k)$, such that $\chi(v_j) = 1$ (resp. $\chi(v_j) = -1$). For $1 \le k \le n$, we define $H(k) = \sum_{1 \le i \le m} H(i,k)$, where[4]

$$H(i,k) = \cosh(\alpha(p_{i,k} - m_{i,k}))$$

and $\alpha = \sqrt{2\ln(2m)/n}$. Note that $p_{i,k} - m_{i,k}$ is precisely the "current" discrepancy $\chi(S_i)$. The strategy is to choose the assignment of $\chi(v_{k+1}) = \pm 1$ that produces the smaller value of $H(k+1)$. If $v_{k+1} \notin S_i$, then obviously $H(i, k+1) = H(i,k)$. Otherwise, by elementary properties of the hyperbolic cosine, the two possible values of $H(i, k+1)$ average to exactly $H(i,k)\cosh(\alpha)$. It follows that the two values of $H(k+1)$ corresponding

---

[3]See Appendix A.
[4]Recall that $\cosh x = (e^x + e^{-x})/2$.

to $\chi(v_{k+1}) = \pm 1$ average to at most $H(k)\cosh(\alpha)$, which, by taking Taylor expansions, is easily shown to be less than $H(k)e^{\alpha^2/2}$. This remains true if we extend $H(k)$ to the case $k = 0$ by setting $H(0) = m$. It follows that $H(k) < me^{k\alpha^2/2}$, for $k > 0$, and hence,

$$\tfrac{1}{2} e^{\alpha\max_i |\chi(S_i)|} < \cosh(\alpha\max_i |\chi(S_i)|) \le H(n) < me^{n\alpha^2/2}.$$

Thus, the discrepancy of the set system is at most $\sqrt{2n\ln(2m)}$.

Remark: The weight function $H(k)$ takes into account only the discrepancies of the prefixes of sets in S examined so far, and not the respective sizes of these prefixes. So, a small prefix may end with a discrepancy similar to that of a large prefix. For example, if all of the sets have linear size except for one of them that is very small and is spread evenly among $v_1, \ldots, v_n$, then $H(k)$ will not be influenced much by the small set, and an adversary can easily drive up its discrepancy as high as linear in its actual size. Our next discussion corrects this undesirable feature.

The Unbiased Greedy Algorithm

In the event that some sets of S might be small, it is desirable to have the stronger inequality, $|\chi(S_i)| \le \sqrt{2|S_i|\ln(2m)}$, for each $S_i \in S$, as was provided by the derandomization method. A simple modification of the cost function achieves just that. We follow the same approach as the one used in the hyperbolic cosine algorithm. Only the definition of the cost function, renamed $G(k)$, is different. Given $S_i \in S$, we fix a parameter $\varepsilon_i \in (0,1)$, to be specified later, and define, for $1 \le k \le n$, $G(k) = \sum_{1\le i\le m} G(i,k)$, where

$$G(i,k) = (1+\varepsilon_i)^{p_{i,k}}(1-\varepsilon_i)^{m_{i,k}} + (1+\varepsilon_i)^{m_{i,k}}(1-\varepsilon_i)^{p_{i,k}}.$$

We can verify that the two possible values of $G(k+1)$ average out to $G(k)$ (hence the term *unbiased*). Thus, always picking the assignment of $\chi(v_{k+1})$ that minimizes $G(k+1)$ implies that $G(k+1) \le G(k)$. It is natural to define $G(0) = 2m$. Obviously, $G(i,n) \le G(n) \le 2m$, for any $1 \le i \le m$. Now, observe that

$$G(i,n) = (1-\varepsilon_i^2)^{\frac{|S_i|-|\chi(S_i)|}{2}}\left((1+\varepsilon_i)^{|\chi(S_i)|} + (1-\varepsilon_i)^{|\chi(S_i)|}\right). \qquad (1.3)$$

It follows that

$$(1-\varepsilon_i^2)^{\frac{|S_i|-|\chi(S_i)|}{2}}(1+\varepsilon_i)^{|\chi(S_i)|} \le 2m, \qquad (1.4)$$

and hence

$$|S_i| \ln(1 - \varepsilon_i^2) + |\chi(S_i)| \ln\left(\frac{1+\varepsilon_i}{1-\varepsilon_i}\right) \leq 2\ln(2m).$$

Let $\lambda(x) = \frac{1}{2}\ln((1+x)/(1-x))$ and $f(x) = \lambda(x)^2 + \ln(1-x^2)$. For any $x \in [0,1)$, $f(x) \geq 0$. (This follows trivially from the fact that $f(0) = f'(0) = 0$ and that the derivative of $(1-x^2)f'(x)$ is positive over $(0,1)$.) We derive

$$2|\chi(S_i)|\lambda(\varepsilon_i) - |S_i|\lambda(\varepsilon_i)^2 \leq 2\ln(2m).$$

Since $\lambda(\varepsilon_i)$ varies continuously from 0 to infinity as ε_i goes from 0 to 1, we can choose

$$\lambda(\varepsilon_i) = \sqrt{2\ln(2m)/|S_i|},$$

which gives us $|\chi(S_i)| \leq \sqrt{2|S_i|\ln(2m)}$, as desired.

Theorem 1.2 *In $O(nm)$ time, it is possible to color the elements of a set system (V, S) such that the discrepancy of any $S_i \in S$ is at most $\sqrt{2|S_i|\ln(2m)}$ in absolute value, where $n = |V|$ and $m = |S|$.*

Remark: The unbiased greedy approach entails nothing more than replacing the multiplicative factor $e^{\pm\alpha}$ of the hyperbolic cosine algorithm by the first two terms of its Taylor expansion, that is, $1 \pm \alpha$. This modification implies that the assignment of the next $\chi(v)$ corresponds to a fair game, ie, a random choice multiplies $G(k)$ by 1 on average, as opposed to $\cosh(\alpha)$ in the case of the hyperbolic cosine algorithm. By setting a fixed upper bound on $G(k)$, we thus prevent big sets from overinfluencing the assignment process. Indeed, observe that in the expression for $G(i,n)$ given in (1.3), both the discrepancy and the size of S_i are taken into account.

1.2 The Entropy Method

We consider the particular case of a set system (V, S), where $|V| = |S| = n$ (which is easily generalized to the nonsquare case). The method is based on the use of partial colorings and the pigeonhole principle. The entropy function is used to simplify an otherwise complicated counting argument. The idea is to argue that many colorings have almost the same discrepancy vectors (ie, each $\chi(S_i)$ differs little among the various colorings). Thus, subtracting two such colorings and dividing by two gives a partial coloring, ie, a coloring in $\{-1, 0, 1\}$, with low discrepancy. If we can show that the new coloring uses few zeros, then an appropriate recursion "completes" the coloring without increasing the discrepancy by too much.

Theorem 1.3 *Any set system* (V, S) *such that* $|V| = |S| = n$ *has* $O(\sqrt{n})$ *discrepancy. Some set systems have a matching lower bound.*

The lower bound part of the proof is given in §1.5. Note that the upper bound of $O(\sqrt{n})$ represents a constant number of standard deviations from the random coloring of a given set of size $\Theta(n)$. For this reason, this is often referred to as the *standard deviation bound*. We begin by defining a *partial coloring* of V as a map $\chi : V \mapsto \{-1, 0, 1\}$. As usual, the discrepancy of $S_i \in S$ is defined as

$$\chi(S_i) = \sum_{v \in S_i} \chi(v).$$

Theorem 1.3 is a simple corollary of the following:

Lemma 1.4 *Let* (V, S) *be a set system, where* $|V| = n$ *and* $|S| = m \geq n$. *There exists a partial coloring* χ *of* V, *such that at most* $0.9n$ *elements of* V *are colored zero and, for each* $S_i \in S$,

$$|\chi(S_i)| < c \sqrt{n \ln \frac{2m}{n}},$$

for some constant $c > 0$.

Apply the lemma to (V, S), and let Z be the subset of 0-colored elements in V. Unless Z is empty, apply the lemma to the subsystem $(Z, S|_Z)$, where $S|_Z$ is the collection of subsets of Z of the form $S \cap Z$, for any $S \in S$. Then, iterate in this fashion until every element of V is ± 1-colored. The discrepancy of each $S_i \in S$ will be at most

$$\sum_{k \geq 0} c \sqrt{(0.9)^k n \ln \frac{2m}{(0.9)^k n}} = O(\sqrt{n}),$$

which establishes Theorem 1.3. □

It is easy to generalize the theorem to the case where $m > n$. We find that it is possible to two-color the set system (V, S) so that its discrepancy is in $O(\sqrt{n \ln(2m/n)})$. This represents an improvement over the random coloring provided that $m = n^{1+o(1)}$.

It now remains to prove Lemma 1.4. Let χ_0 be a random two-coloring of V. Given $S_i \in \mathcal{S}$, let

$$\chi_0^*(S_i) = \left\lfloor \left| \frac{\chi_0(S_i)}{c\sqrt{n \ln(2m/n)}} \right| \right\rfloor,$$

for some large enough constant c. By Chernoff's bound,[5] the probability p_k that $\chi_0^*(S_i) = k$ is less than $(2m/n)^{-ck^2}$. Since the function $f(x) = -x \log x$ increases as x goes from 0 to $1/e$, the entropy of $\chi_0^*(S_i)$ satisfies:

$$H(\chi_0^*(S_i)) \stackrel{\text{def}}{=} \sum_{-\infty < k < +\infty} p_k \log \frac{1}{p_k} < p_0 \log \frac{1}{p_0} + \sum_{|k|>0} ck^2 \left(\frac{2m}{n}\right)^{-ck^2} \log \frac{2m}{n}.$$

For c large enough, we easily verify that

$$\sum_{|k|>0} p_k < \frac{n}{cm},$$

and therefore $-p_0 \log p_0 < n/10m$. It follows that

$$H(\chi_0^*(S_i)) < \frac{n}{5m}.$$

Let χ_0^* be the vector

$$(\chi_0^*(S_1), \ldots, \chi_0^*(S_m)).$$

Because of the subadditivity of entropy (Appendix A.3),

$$H(\chi_0^*) < \frac{n}{5}.$$

Thus, by Lemma A.8, there exists a vector $(d_1, \ldots, d_m)$ such that

$$\text{Prob}[\chi_0^* = (d_1, \ldots, d_m)] > 2^{-n/5}.$$

In other words, the set C of two-colorings producing the vector $(d_1, \ldots, d_m)$ is of size greater than $2^{4n/5}$. Pick one coloring χ_1 in C and for each $\chi \in C$ form the partial coloring $\chi' = \frac{1}{2}(\chi - \chi_1)$. Note that

$$|\chi'(S_i)| < \frac{c}{2}\sqrt{n \ln \frac{2m}{n}}, \qquad (1.5)$$

for each $S_i \in \mathcal{S}$. The number of partial colorings with at most $n/10$ nonzeros is equal to

$$\sum_{0 \le k \le n/10} \binom{n}{k} 2^k < 2^{4n/5} < |C|.$$

---

[5]See Lemma A.5.

Therefore, there exists a partial coloring with more than $n/10$ nonzeros that satisfies (1.5), which proves Lemma 1.4. □

1.3 The Beck-Fiala Theorem

We briefly discuss a result commonly known as the Beck-Fiala theorem: It states that efficient colorings are always possible as long as no element of the ground set appears in too many S_i's. Let t be the *degree* of the set system, ie, the maximum number of sets S_i to which any given $v \in V$ belongs. Initialize each $\chi(v_k)$ to 0 and call v_k *undecided*. The algorithm will make the v_k's decided as it goes. A set S_i is said to be *stable* if it contains at most t undecided elements. Because of the degree condition, the number of nonstable sets is strictly less than the number of undecided elements. Thus, if we regard the sequence $(\chi(v_1), \ldots, \chi(v_n))$ as a vector in $\mathbf{R}^n$, by simple linear algebra we can move the vector along (at least) a line while both changing only undecided coordinates and maintaining the discrepancy $\chi(S_i)$ of *each* nonstable set S_i equal to 0.

Let us stop our continuous motion as soon as one (or several) of the $\chi(v_k)$'s becomes equal to ± 1. At that stage, we pull all such v_k's out of the game by declaring them no longer undecided. Note that this might cause new sets S_i to become stable. Obviously, it is still the case that the undecided elements outnumber the nonstable sets, so we can repeat the same process and move the vector of undecided colors accordingly. Iterating in this fashion as long as some v_k remains undecided will eventually make every S_i stable.

Note that the discrepancy of any S_i is 0 until it becomes stable, and from that point on at most t of its elements can have color updates; none of those $\chi(v_k)$ was equal to ± 1 to begin with (else they would not be undecided). So each was in $(-1, 1)$, and therefore the total change on $\chi(S_i)$ amounts to strictly less than $2t$. Since the final value must be integral, $|\chi(S_i)| \leq 2t - 1$ for all $i \leq m$. We have proven the Beck-Fiala theorem.

Theorem 1.5 *The discrepancy of a set system of degree t is less than $2t$.*

1.4 Discrepancy and the VC-Dimension

Range spaces denote particular (finite or infinite) set systems that arise naturally in geometry. Despite their strong geometric connection, range spaces are defined purely in combinatorial terms. In keeping with common usage, we use the notation $\Sigma = (X, \mathcal{R})$, instead of $(V, \mathcal{S})$, to refer to a range space. For example, X might be a set of n points in $\mathbf{R}^2$, while $\mathcal{R}$ is the collection of sets of the form $X \cap D$, where D is a disk. Note that Σ is a subsystem of the infinite geometric set system $(\mathbf{R}^2, \mathcal{R}^o)$, where $\mathcal{R}^o$ denotes the set of all disks.

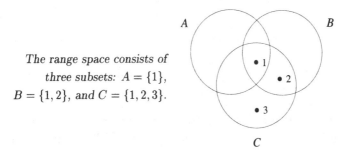

The range space consists of three subsets: $A = \{1\}$, $B = \{1, 2\}$, and $C = \{1, 2, 3\}$.

It is common to consider infinite range spaces of the form $(\mathbf{R}^d, \mathcal{R}^o)$, where $\mathcal{R}^o$ is obtained by letting a group of transformations act on a fixed subset of $\mathbf{R}^d$; the elements of X are called *points* and the subsets in $\mathcal{R}$ are called *ranges*. In practice, the set systems considered in a geometric context are often subsystems of infinite range spaces. A single parameter, called the Vapnik-Chervonenkis dimension (VC-dimension), characterizes the ability of sampling effectively from such range spaces. In any reference to a "range space" it is usually implicit that the VC-dimension is finite. We shall show that the discrepancy of range spaces is smaller than that of more general set systems. If $|X| = n$, then the discrepancy is within $O(n^{1/2-\varepsilon})$, for some small constant $\varepsilon > 0$, which beats the standard deviation bound of $O(\sqrt{n})$.

Primal and Dual Shatter Functions

Let $\Sigma = (X, \mathcal{R})$ be a (finite or infinite) set system. Given $Y \subseteq X$, let $(Y, \mathcal{R}|_Y)$ denote the set system *induced* by Y, ie, $\{Y \cap R \mid R \in \mathcal{R}\}$. Note that although the same set $Y \cap R$ may be obtained for several R's, only one copy appears in $\mathcal{R}|_Y$; in other words, $\mathcal{R}|_Y$ is not a multiset. A subset Y of X is said to be *shattered* (by $\mathcal{R}$) if $\mathcal{R}|_Y = 2^Y$, meaning that every subset of Y (including the empty set) is of the form $Y \cap R$, for some $R \in \mathcal{R}$. The supremum of all sizes of finite shattered subsets of X is called the *Vapnik-Chervonenkis dimension* of Σ, or *VC-dimension* for short. For example, it is easy to see that $d + 1$ is the VC-dimension of the range space formed by points and halfspaces in d-space (Fig. 1.1). One should not be fooled, however; in general, evaluating the VC-dimension is no simple matter.

Fig. 1.1. The VC-dimension of halfplanes and points is three, not four.

If X contains arbitrarily large shattered subsets, then we say that the VC-dimension is infinite. In a Euclidean space of fixed dimension, a rule of thumb is that if Σ is a range space defined by some shape of constant description size (eg, a simplex, an ellipsoid, but not, say, an arbitrary convex set), then the VC-dimension is bounded. We define the *shatter function* $\pi_{\mathcal{R}}$ of a range space Σ as follows: $\pi_{\mathcal{R}}(m)$ is the maximum number of subsets in any subsystem $(Y, \mathcal{R}|_Y)$ induced by some $Y \subseteq X$ of size m.

Lemma 1.6 *If the shatter function of an infinite range space is bounded by a fixed-degree polynomial, then its VC-dimension is bounded by a constant. Conversely, if the VC-dimension is $d \geq 1$ then, for any $m \geq d$,*

$$\pi_{\mathcal{R}}(m) \leq \binom{m}{0} + \binom{m}{1} + \cdots + \binom{m}{d} < \left(\frac{em}{d}\right)^d.$$

Proof: The first part of the lemma is obvious. We prove the inequality on $\pi_{\mathcal{R}}(m)$ by induction on d and m. Let $f(d, m)$ be the maximum value at m of the shatter function of any range space of VC-dimension at most d. Trivially, $f(0, m)$ and $f(d, 0)$ are at most 1, so assume that $d, m > 0$. Fix $Y \subseteq X$ of size m and let $y \in Y$. Let C be the number of distinct sets of the form $Y \cap R$, where $R \in \mathcal{R}$. It might be tempting to say that C is equal to the number A of sets of the form $(Y \setminus \{y\}) \cap R$. But this might fall slightly short of the count, because two distinct R and R' can produce the same set $(Y \setminus \{y\}) \cap R$. For this to happen their restriction to Y must differ by exactly y. Thus, if we define B to be the number of sets $Y \cap R$ that can be expressed as the disjoint union of $Y \cap R'$ ($R' \in \mathcal{R}$) and $\{y\}$, we can then safely write $C = A + B$.

Note that in the definition of B the sets $Y \cap R'$ cannot shatter any subset of $Y \setminus \{y\}$ of size d; otherwise, we could form a subset of X of size $d + 1$ that is shattered by $\mathcal{R}$, which would give a contradiction. Therefore, $B \leq f(d - 1, m - 1)$. Since, obviously $A \leq f(d, m - 1)$, we have the recurrence

$$f(d, m) \leq f(d, m - 1) + f(d - 1, m - 1).$$

Checking that $f(d, m)$ satisfies the inequality of the lemma is routine. We can also solve the recurrence visually by counting the number of paths connecting an integral point on the x-axis between 0 and d to the point (d, m), using only vertical edges and edges oriented at 45 degrees. $\square$

The estimate given by the lemma is optimal. It is not too hard to see that infinite range spaces cannot have sublinear shatter functions. In other words, if $\pi_{\mathcal{R}}(m) = o(m)$, then $\mathcal{R}$ is finite, and hence $\pi_{\mathcal{R}}(m) = O(1)$. Also, by keeping track of the growth of shatter functions, it is quite easy to show that the class of range spaces of bounded VC-dimension is closed under union, intersection, and complementation. More precisely, if $(X, \mathcal{R})$ is of bounded VC-dimension, then so is $(X, \mathcal{S})$, where any $\mathcal{S}$ is a finite combination of unions, intersections, and complementations of subsets of $\mathcal{R}$.

If we represent a (finite) set system by its incidence matrix, transposition has an obvious interpretation: We no longer look at which elements lie in a given set but at which sets contain a given element. In other words, we switch the roles of elements and subsets (or points and ranges). Of course, we can do the same even when the set system is an infinite range space. Given a range space $\Sigma = (X, \mathcal{R})$, this suggests introducing the set $\mathcal{R}^* = \{ R_x \mid x \in X \}$, where $R_x = \{ R \in \mathcal{R} \mid x \in R \}$. The range space $\Sigma^* = (X^*, \mathcal{R}^*)$, where $X^* = \mathcal{R}$, is called the *dual* of Σ.

If Σ is *separable*, meaning that for every $x, y \in X$ there exists $R \in \mathcal{R}$ that contains x but not y (ie, no column appears twice), then duality is involutory; in other words, the dual of $(X^*, \mathcal{R}^*)$ is isomorphic to $(X, \mathcal{R})$. The shatter function of Σ^*, denoted by $\pi_{\mathcal{R}}^*$, is called the *dual shatter function* of $(X, \mathcal{R})$. Although the VC-dimension of a range space might be sometimes quite difficult to evaluate, its dual shatter function is often easier to estimate. For example, in the case of the range space defined by points and balls in d-space, the dual shatter function corresponds to the number of regions into which m balls cut up $\mathbf{R}^d$, which can be shown without difficulty to be $O(m^d)$. To distinguish $(X, \mathcal{R})$ from its dual, we sometimes refer to it as the *primal* range space and we call $\pi_{\mathcal{R}}$ the primal shatter function.

Lemma 1.7 *If a range space has VC-dimension d, then its dual has VC-dimension less than 2^{d+1}.*

Proof: Arguing by contradiction, suppose that the dual range space has VC-dimension at least 2^{d+1}. This implies the existence of 2^{d+1} ranges of $\mathcal{R}$ that are shattered in the dual range space. In other words, there exist $2^{2^{d+1}}$ points of X such that the 2^{d+1}-by-$2^{2^{d+1}}$ incidence matrix A contains all possible column patterns (this is the matrix whose rows are the characteristic vectors of the chosen ranges with respect to the $2^{2^{d+1}}$ points). Let $u_0, \ldots, u_{2^{d+1}-1}$ denote the $(d+1)$-bit binary representations of $0, 1, \ldots, 2^{d+1} - 1$, respectively. Form a 2^{d+1}-by-$(d+1)$ matrix B by making u_0 the first row, u_1 the second row, etc. Each column of B must appear somewhere as a column of A. This shows that the subset of X associated with the columns of A corresponding to those of B is indeed shattered in the primal range space. This subset is of size $d+1$, so we have a contradiction. $\Box$

Beating the Standard Deviation Bound

As usual, let $(X, \mathcal{R})$ be a range space of VC-dimension d, with $|X| = n$. A random two-coloring of X ensures that, with reasonably high probability, no color outnumbers the other by more than $O(\sqrt{n \log n})$. By using structural properties of range spaces, it is possible to reduce this upper bound to $o(\sqrt{n})$. Specifically, we show that the discrepancy of $(X, \mathcal{R})$ is within a polylogarithmic factor of $O(n^{1/2-1/2d})$, for $d > 1$. Unfortunately, the proof is inherently existential (using the pigeonhole principle in a manner similar to the treatment of square matrices in §1.2), and it does not yield an efficient coloring algorithm. Recall that by Lemma 1.6 the primal shatter function of the range space is in $O(m^d)$. We prove the stronger result:

Theorem 1.8 *The discrepancy of a range space whose primal shatter function is bounded by cm^d, for some constants $c > 0, d > 1$, is*

$$O(n)^{1/2-1/2d}(\log n)^{1+1/2d}.$$

Note that the "big-oh" notation hides a constant that depends only on c and d. We begin with a simple technical lemma demonstrating once again the usefulness of partial colorings. Recall that a partial coloring is a map $\chi : X \mapsto \{-1, 0, 1\}$.

Lemma 1.9 *Let $(X, \mathcal{R}_0)$ and $(X, \mathcal{R}_1)$ be two set systems defined on the same ground set X of size n. Assume that*

$$\prod_{R \in \mathcal{R}_0} (|R| + 1) \leq 2^{(n-1)/5},$$

and that $|R| \leq r$, for each $R \in \mathcal{R}_1$. Then there exists a partial coloring $\chi : X \mapsto \{-1, 0, 1\}$ such that

(i) χ *is nonzero over at least one-tenth of X;*
(ii) $\chi(R) = 0$, *for each $R \in \mathcal{R}_0$;*
(iii) $|\chi(R)| \leq \sqrt{2r \ln(4|\mathcal{R}_1|)}$, *for each $R \in \mathcal{R}_1$.*

Proof: Let C be the set of two-colorings of X such that (iii) holds. The argument leading to Theorem 1.1 also shows that $|C| \geq 2^{n-1}$. Order $\mathcal{R}_0$ in arbitrary fashion and, given a two-coloring χ of X, consider the sequence $(\chi(R) : R \in \mathcal{R}_0)$. If $\chi \in C$, then there are at most

$$\prod_{R \in \mathcal{R}_0} (|R| + 1) \leq 2^{(n-1)/5}$$

distinct sequences. (The factor is not $2|R| + 1$ because $|R|$ and $\chi(R)$ always have the same parity.) By the pigeonhole principle, there must be a collection C_1 of at least $2^{n-1}/2^{(n-1)/5}$ colorings of C with the same sequence. Choose some $\chi_0 \in C_1$, and for each $\chi \in C_1$ define the partial coloring

$$\chi'(x) = \frac{\chi(x) - \chi_0(x)}{2}.$$

Notice that each χ' satisfies (ii) and (iii). It remains to show that one of them is nonzero over at least one-tenth of X. The number of partial colorings with at most $n/10$ nonzeros is equal to

$$\sum_{0 \leq k \leq n/10} \binom{n}{k} 2^k < 2^{4(n-1)/5} \leq |C_1|.$$

Therefore, there exists a partial coloring of C_1 satisfying (i).

The reader will appreciate the family resemblance between this proof and the entropy method: two different ways of counting essentially the same things. $\square$

We are now ready to prove Theorem 1.8. As we noticed earlier, the class of range spaces of bounded VC-dimension is closed under union, intersection, and complementation. In particular, the range space $(X, \mathcal{S})$, where $\mathcal{S}$ consists of the sets of the form $R \setminus R'$, for $R, R' \in \mathcal{R}$, has bounded VC-dimension. This is best seen by the fact that its primal shatter function is in $O(m^{2d})$. We need to use a result about range spaces that is proven in Chapter 4. Given some parameter $0 < \varepsilon < 1$ to be determined later, a set $N \subseteq X$ that intersects every $S \in \mathcal{S}$ of size greater than $\varepsilon|X|$ is called an ε-net for $(X, \mathcal{S})$. By Theorem 4.3 (page 172), there exists such a set N of size $O(\varepsilon^{-1} \log n)$. (Better bounds can be found, but they are not needed here.)

We "factor" the range space $(X, \mathcal{R})$ by grouping into the same equivalence class all the sets of $\mathcal{R}$ that have the same restriction to N: Two sets R, R' are in the same class if and only if $N \cap R = N \cap R'$. Let $\mathcal{R}_0$ be the subset of $\mathcal{R}$ obtained by taking one representative from each class. For each $R \in \mathcal{R}$, form the sets $R \setminus R_0$ and $R_0 \setminus R$, where R_0 is the representative in the class of R. Let $\mathcal{R}_1$ denote the collection of all such sets (for each $R \in \mathcal{R}$). Note that no $R_1 \in \mathcal{R}_1$ intersects N. Because N is an ε-net for $(X, \mathcal{S})$, it follows that the size of R_1 cannot exceed εn. We verify that by choosing

$$\varepsilon = \frac{c(\log n)^{1+1/d}}{n^{1/d}},$$

for a large enough constant c and setting $r = \varepsilon n$, the conditions of Lemma 1.9 are satisfied. Given any range $R \in \mathcal{R}$, let R_0 be its representative in $\mathcal{R}_0$. Because $R \cap R_0 = R_0 \setminus (R_0 \setminus R)$, we can express R as the disjoint union:

$$R = (R \setminus R_0) \cup (R_0 \setminus (R_0 \setminus R)).$$

Noting that if $B \subseteq A$,

$$|\chi(A \setminus B)| = |\chi(A) - \chi(B)| \leq |\chi(A)| + |\chi(B)|,$$

it follows that

$$\begin{aligned} |\chi(R)| &\leq |\chi(R \setminus R_0)| + |\chi(R_0)| + |\chi(R_0 \setminus R)| \\ &\leq 2\sqrt{2r \ln(4|\mathcal{R}_1|)} \\ &= O(n^{1/2-1/2d})(\log n)^{1+1/2d}. \end{aligned}$$

Let $Y \subseteq X$ be the set of 0-colored points. If Y is nonempty, we repeat the same argument with respect to $(Y, \mathcal{R}|_Y)$ and iterate in this fashion until all the points of X are colored. In the end, the discrepancy of any subset follows (at worst) a geometric progression summing up to $O(n^{1/2-1/2d})(\log n)^{1+1/2d}$. This completes the proof of Theorem 1.8. $\square$

The bound can be reduced to $O(n^{1/2-1/2d})$ by a more complicated argument. It cannot be improved further. Indeed, by Theorem 3.9 (page 156), the red-blue discrepancy of the range space $(\mathbf{R}^d, \mathcal{R})$, where $\mathcal{R}$ is the set of halfspaces, is $\Omega(n^{1/2-1/2d})$. It is easy to see that the primal shatter function of that range space is in $O(m^d)$.

By using a spanning path argument quite similar to the one given in the next chapter (§2.8), one can prove the theorem below. Again, we must mention that the big-oh notation hides a constant that depends on only c and d. Surprisingly, the upper bound is known to be optimal for all $d > 1$. We omit the proof, which is simply a combinatorial version of the geometric proof of Theorem 2.19 (page 124).

Theorem 1.10 *The discrepancy of a range space whose dual shatter function is bounded by cm^d, for arbitrary constants $c > 0, d > 1$, is $O(n^{1/2-1/2d}\sqrt{\log n})$.*

1.5 Lower Bounds

We discuss spectral techniques for deriving lower bounds on the discrepancy of set systems. This common algebraic thread will persist in our treatment of geometric discrepancy in Chapter 3. There, of course, additional tools, mostly geometric and analytical, will be brought to bear. Although discrepancy questions can be stated purely combinatorially, we are faced here with a situation—unfortunately all too frequent—where counting arguments alone are by and large useless; one notable exception is the matching lower bound of Theorem 1.3 (page 8), which can be derived probabilistically.

All of our techniques rely on asymptotic estimations of the eigenvalues $\lambda_1 \geq \cdots \geq \lambda_n$ of $A^T A$, where A is the incidence matrix of the set system. We show successively how the discrepancy can be bounded from below in terms of (i) the smallest eigenvalue λ_n, (ii) the determinant $\prod \lambda_i$, and (iii) the traces $\sum \lambda_i$ and $\sum \lambda_i^2$. We begin our discussion with a rare case: a set system whose discrepancy can be bounded directly. This warmup exercise nevertheless brings out the spectral flavor that permeates most of this section.

The Hadamard Matrix Bound

It is not hard to exhibit a set system whose discrepancy is $\Omega(\sqrt{n})$. As we just said, this can be established by an elementary, but tedious, counting argument. A more elegant, algebraic proof is given next. Let $H = (h_{ij})$

be a Hadamard matrix[6] of order n. The matrix H is orthogonal, and its elements are all ± 1. Here is a Hadamard matrix of order 8:

$$\begin{pmatrix} 1 & 1 & 1 & 1 & 1 & 1 & 1 & 1 \\ 1 & -1 & 1 & -1 & 1 & -1 & 1 & -1 \\ 1 & 1 & -1 & -1 & 1 & 1 & -1 & -1 \\ 1 & -1 & -1 & 1 & 1 & -1 & -1 & 1 \\ 1 & 1 & 1 & 1 & -1 & -1 & -1 & -1 \\ 1 & -1 & 1 & -1 & -1 & 1 & -1 & 1 \\ 1 & 1 & -1 & -1 & -1 & -1 & 1 & 1 \\ 1 & -1 & -1 & 1 & -1 & 1 & 1 & -1 \end{pmatrix}.$$

The matrix $A = \frac{1}{2}(H + J)$, where J denotes the matrix full of ones, is the incidence matrix of a set system $(V, \mathcal{S})$, ie, each row of A is the n-bit characteristic vector of a distinct set of the system. We show that its discrepancy in the L^∞ norm is at least $\sqrt{n}/2$. Let H_i (resp. J_i) be the i-th column of H (resp. J). Given a coloring $x \in \{-1, 1\}^n$,

$$\|Ax\|_2^2 = (Ax)^T(Ax) = \frac{1}{4} \sum_{i,j} x_i x_j (H_i + J_i)^T (H_j + J_j).$$

Expanding the sum, we find that:

1. By orthogonality, the term $\sum_{i,j} x_i x_j H_i^T H_j$ is equal to $\sum_i x_i^2 H_i^T H_i$, which is $\sum_i x_i^2 n$.

2. Because $J_j = H_1$, we can write $\sum_{i,j} x_i x_j H_i^T J_j$ as

$$\left(\sum_j x_j \right) \sum_i x_i H_i^T H_1.$$

   By orthogonality, this is $(\sum_j x_j) x_1 n$. Obviously, we find the same value for $\sum_{i,j} x_i x_j J_i^T H_j$ as well.

3. The term $\sum_{i,j} x_i x_j J_i^T J_j$ is equal to $(\sum_i x_i)^2 n$.

Putting everything together, we obtain a lower bound on the L^2 norm of Ax.

---

[6]See Appendix B.

$$
\begin{aligned}
4\|Ax\|_2^2 &= n\sum_i x_i^2 + 2n\left(\sum_i x_i\right)x_1 + n\left(\sum_i x_i\right)^2 \\
&= n\left(x_1 + \sum_i x_i\right)^2 + n\sum_{i>1} x_i^2 \\
&\geq n\sum_{i>1} x_i^2 = n(n-1).
\end{aligned}
$$

It follows that at least one coordinate of Ax must exceed $\sqrt{n-1}/2$ in absolute value. This establishes the lower bound of Theorem 1.3 (page 8), for the case where n is a power of 2. The other cases are handled by a standard padding argument.

Although the proof may seem too ad hoc to lend itself to grand statements about lower bounds, it does point the way to the spectral route which we are about to explore now. To minimize $\|Ax\|_2$ for a fixed-length x is a straightforward eigenvalue problem. We formalize this idea below and apply it to the discrepancy of arithmetic progressions.

The Eigenvalue Bound

Let A be the incidence matrix of a set system on n elements; we do not assume that the matrix is square. We consider the mean-square discrepancy $D_2(A)^2$, defined as

$$
\min\left\{\,\|Ax\|_2^2 \,:\, x \in \{-1,1\}^n\,\right\}.
$$

The matrix $A^T A$ is positive semidefinite, and therefore it is diagonalizable and its eigenvalues $\lambda_1 \geq \cdots \geq \lambda_n$ are nonnegative reals. Suppose that $x = x_1 v_1 + \cdots + x_n v_n$, where $\{v_i\}$ is an orthonormal eigenbasis, with v_i associated with λ_i. We have

$$
\begin{aligned}
\|Ax\|_2^2 = x^T A^T A x &= \left(\sum_i x_i v_i\right)^T \left(\sum_i \lambda_i x_i v_i\right) \\
&= \sum_{i=1}^n \lambda_i x_i^2 \geq \lambda_n \|x\|_2^2,
\end{aligned} \tag{1.6}
$$

and thus

$$
D_2(A) \geq \sqrt{n\lambda_n}. \tag{1.7}
$$

The set $\{-1,1\}^n$ of all "colorings" is contained in the Euclidean sphere of radius $\sqrt{n}$ centered at the origin. Geometrically, $A^T A$ transforms the corresponding ball into an ellipsoid. Indeed, expressed over the eigenbasis, the image of a coloring x under the linear transformation $A^T A$ is a vector whose coordinates in the eigenbasis are $(y_1, \ldots, y_n)$, where

$$\left(\frac{y_1}{\lambda_1}\right)^2 + \cdots + \left(\frac{y_n}{\lambda_n}\right)^2 = \|x\|_2^2 = n.$$

Note that by (1.7) the minimum distance from the origin to the ellipsoid's boundary, which is $\lambda_n\sqrt{n}$, is a lower bound (up to a factor of $\sqrt{n}$) on the mean-square discrepancy. To derive a high lower bound on the discrepancy, we therefore must be able to show that the ellipsoid in question is not too "flat."

What we have just done is to relax the constraints $x_i = \pm 1$ into $x^T x = n$. This gives rise to the standard quadratic programming problem:

$$minimize \quad x^T A^T A x,$$

subject to $x^T x = n$. This leads to minimizing the Rayleigh quotient $\|Ax\|_2^2/\|x\|_2^2$, which by the Courant-Fischer characterization of eigenvalues gives precisely the smallest eigenvalue λ_n. This shows that in the relaxation problem the inequality (1.6) cannot be improved.

Roth's $\frac{1}{4}$-Theorem

We use (1.7) to prove a beautiful result on the discrepancy of arithmetic progressions. Van der Waerden's theorem is a classical result in Ramsey theory, which says that any two-coloring of the integers contains an arbitrarily long monochromatic arithmetic progression. Roth established a complementary result by proving that not all arithmetic progressions can be evenly bicolored. This is known as *Roth's $\frac{1}{4}$-theorem*. It is easily derived from the spectral bound of (1.7).

Theorem 1.11 *Any two-coloring of the integers $\{1, \ldots, n\}$ contains an arithmetic progression whose discrepancy is $\Omega(n^{1/4})$.*

Put differently, there is a constant $c > 0$ such that, no matter how we color the first n integers red or blue, there exists an arithmetic progression over which the numbers of red and blue integers differ by at least $cn^{1/4}$ (Fig. 1.2). The bound of $\Omega(n^{1/4})$ is tight.

Note that, to prove any meaningful lower bound, it is crucial to consider arithmetic progressions of different step sizes (ie, distinct differences between consecutive elements). Indeed, any arithmetic progression of step size 10 can be made of low discrepancy by coloring the first 10 elements

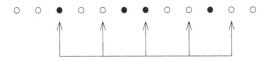

Fig. 1.2. The discrepancy of this arithmetic progression is one.

red, the next 10 elements blue, the following 10 red, etc. The theorem says that neither this coloring scheme nor, for that matter, any other one can be made to handle all step sizes at once.

Why $n^{1/4}$? We know that we need to consider many step sizes. But, of course, we must also deal with long arithmetic progressions, since sparse sets have low discrepancy. We shall occupy the middle ground in trading off step sizes and lengths by choosing progressions of lengths and step sizes roughly $\sqrt{n}$. Notice then that a random coloring guarantees discrepancy in $O(n^{1/4}\sqrt{\log n})$ for those progressions (Theorem 1.1), which is about the bound of the theorem.

We now prove Theorem 1.11. We consider only arithmetic progressions in $\{1,\ldots,n\}$ of length $s = \lfloor\sqrt{n/6}\rfloor$. Such a progression, denoted by $S(p,q)$, is characterized by a starting point p ($1 \le p \le n$) and a step size q ($1 \le q \le 6s$). We construct $S(p,q)$ as follows: Starting at p, we jump by steps of length q and iterate $s-1$ times; so

$$S(p,q) = \{\, p, p+q, p+2q, \ldots, p+(s-1)q \,\}.$$

If we should land past n, we simply wrap around (by performing the jumps modulo n). Note that because $6s(s-1) < n$, there is no risk of reaching p again. This means that $S(p,q)$ is the disjoint union of two "standard" arithmetic progressions.

Given $1 \le q \le 6s$, let A_q be the n-by-n matrix whose p-th row is the characteristic vector of the set $S(p,q)$. The matrix A_q is a circulant matrix[7] obtained by permuting cyclically the characteristic vector of $S(1,q)$. Because it is circulant, the inner product of two column vectors $A_q^{(i)}$ and $A_q^{(j)}$ is equal to the inner product of $A_q^{(i+1)}$ and $A_q^{(j+1)}$ (superscripts are understood to be modulo n). Therefore, $A_q^T A_q$ is also circulant.

---

[7]Recall that an n-by-n matrix $M = (m_{ij})$ is called circulant if each row past the first one derives from the previous row by shifting each element to the right by one position, ie, $m_{i+1,j+1} = m_{i,j}$ (indices modulo n).

We form the incidence matrix A of our set system by stacking up the matrices $A_1, \ldots, A_{6s}$ vertically, one on top of the other:

$$A = \begin{bmatrix} A_1 \\ A_2 \\ \vdots \\ A_{6s} \end{bmatrix}.$$

Note that although A is not circulant, the matrix $M = A^T A$ is equal to $\sum_{q=1}^{6s} A_q^T A_q$ and therefore is itself circulant.

Let $\zeta = e^{2\pi i k/n}$ be an arbitrary n-th root of unity. It is easily verified that $z = (1, \zeta, \ldots, \zeta^{n-1})^T$ is an eigenvector of M. Indeed, observe that the second coordinate of Mz is equal to the first one multiplied by ζ, the third one is the second coordinate multiplied by ζ, and so on. Thus, the vector Mz is obtained by multiplying $(1, \zeta, \ldots, \zeta^{n-1})$ by the first coordinate of Mz. So, clearly z is an eigenvector for M. Let $\lambda(z)$ denote its associated eigenvalue. There are n roots of unity, and the corresponding vectors are orthogonal to each other (see Appendix B); therefore, we have a complete basis of eigenvectors. Also,

$$z^* M z = \lambda(z) z^* z = n\lambda(z),$$

where $*$ denotes the Hermitian transpose. On the other hand,

$$z^* M z = (Az)^* Az = \sum_{q=1}^{6s} \sum_{i=1}^{n} \left| \sum_{j=1}^{n} A_q(i,j)\zeta^{j-1} \right|^2.$$

Note that because $|\zeta| = 1$,

$$\left| \sum_j a_{ij}\zeta^{j-1} \right|^2 = \left| \sum_j a_{ij}\zeta^{j+k} \right|^2,$$

for any k; therefore, all the rows of A_q, for a fixed q, have the same contribution. This yields

$$n\lambda(z) = \sum_{q=1}^{6s} n \left| \sum_{k=0}^{s-1} \zeta^{qk} \right|^2.$$

By the pigeonhole principle, for at least two distinct $1 \leq q_1 < q_2 \leq 6s$, the angles $\arg(\zeta^{q_1})$ and $\arg(\zeta^{q_2})$ fall in an interval of length $2\pi/6s$ (around the unit circle in the complex plane). If this interval contains the angle zero, then for one of the q_i's we have $0 \leq \arg(\zeta^{q_i}) \leq \pi/3s$; otherwise, we have $|\arg(\zeta^{q_0})| \leq \pi/3s$, for $q_0 = q_2 - q_1$. Thus, in general, there exists

$1 \leq q_0 \leq 6s$ such that

$$\left| \arg(\zeta^{q_0 k}) \right| \leq \frac{k\pi}{3s},$$

for each $0 \leq k < s$. This shows that the real part of each $\arg(\zeta^{q_0 k})$ is at least $1/2$. Thus,

$$n\lambda(z) \geq n \left| \sum_{k=0}^{s-1} \zeta^{q_0 k} \right|^2 \geq \frac{ns^2}{4}.$$

Since this is true for any eigenvalue of M, it follows from (1.7) that[8]

$$D_\infty(A)^2 \geq \frac{D_2(A)^2}{6sn} \gg \sqrt{n}.$$

Because of the wrap-around it can be argued that A is not the incidence matrix of a set of arithmetic progressions. As we remarked earlier, however, each set represented by A can be partitioned into two valid arithmetic progressions. If the set has high discrepancy, then so must at least one of its two constituent progressions. Thus, the lower bound on $D_\infty(A)$ completes the proof of Theorem 1.11. □

The View from Harmonic Analysis

We give another proof of Theorem 1.11 (page 20), this time using Fourier transforms as our main tool. As it turns out, the proof is really the same as the previous one, even though it looks quite different on the surface. It is instructive to see why, because it brings together two key tools in discrepancy theory, eigenvalues and Fourier transforms, and their common link, the convolution operator. We will discuss this connection in depth in the next two chapters. Our discussion here is to serve as a kinder, gentler introduction to this material.

Recall that our aim is to show that any two-coloring of the integers $\{1, \ldots, n\}$ contains an arithmetic progression whose discrepancy is $\Omega(n^{1/4})$. Fix a coloring χ:

$$\chi(m) = \begin{cases} 1 & \text{if } m \text{ is red and } 1 \leq m \leq n, \\ -1 & \text{if } m \text{ is blue and } 1 \leq m \leq n, \\ 0 & \text{else.} \end{cases}$$

---

[8] Recall that $x \gg y$ means $x = \Omega(y)$.

As in the previous proof, we are interested in only arithmetic progressions of length within $\sqrt{n}$; put $s = \lfloor\sqrt{n}\rfloor$. Given a step size q, we define the characteristic function c_q of the corresponding arithmetic progression of length $2s + 1$:

$$c_q(m) = \begin{cases} 1 & \text{if } m \text{ is a multiple of } q \text{ and } |m| \leq sq, \\ 0 & \text{else.} \end{cases}$$

Regard $c_q(m)$ as a "comb" centered at 0. Slide it over so that its center coincides with p. The portion of the comb within $[1, n]$ defines an arithmetic progression whose discrepancy we denote by $\Delta_q(p)$. It is immediate that

$$\Delta_q(p) = \sum_{k=1}^{n} \chi(k)c_q(k - p).$$

Since $\chi(k)$ is 0 outside $[1, n]$ and c_q is even, we have

$$\Delta_q(p) = \sum_{k \in \mathbf{Z}} \chi(k)c_q(p - k);$$

in other words, $\Delta_q = \chi \star c_q$. Taking Fourier transforms on the group $\mathbf{Z}$ (see Appendix B), we find that

$$\widehat{c}_q(t) = \sum_{m \in \mathbf{Z}} c_q(m)e^{-2\pi i m t} = \sum_{|k| \leq s} e^{-2\pi i k q t}.$$

By the same pigeonhole argument used in the previous proof, there exists some $1 \leq q(t) \leq bs$ for some fixed b large enough, such that in the sum $\sum_{|k| \leq s} e^{-2\pi i k q(t) t}$ the real part of each summand exceeds some fixed positive constant. Therefore,

$$|\widehat{c}_{q(t)}(t)| \gg s.$$

By the Parseval-Plancherel identity and the convolution theorem,

$$\sum_{q=1}^{bs} \sum_{p \in \mathbf{Z}} \Delta_q(p)^2 = \sum_{q=1}^{bs} \int_0^1 |\widehat{\Delta}_q(t)|^2 \, dt$$

$$= \sum_{q=1}^{bs} \int_0^1 |\widehat{\chi}(t)|^2 \, |\widehat{c}_q(t)|^2 \, dt \geq \int_0^1 |\widehat{\chi}(t)|^2 \, |\widehat{c}_{q(t)}(t)|^2 \, dt$$

$$\gg s^2 \int_0^1 |\widehat{\chi}(t)|^2 \, dt = s^2 \sum_{p \in \mathbf{Z}} |\chi(p)|^2 = ns^2.$$

So, for some step size $q_0 \leq b\sqrt{n}$, we have

$$\sum_{p \in \mathbf{Z}} \Delta_{q_0}(p)^2 \gg ns.$$

Since $\Delta_{q_0}(p)$ is zero for all values of x outside an interval of length $O(n)$, we find that $\Delta_{q_0}(p_0)^2 \gg \sqrt{n}$ for some p_0, which proves Theorem 1.11. $\square$

DISCUSSION

Why are the two proofs of Theorem 1.11 (page 20) really the same in disguise? Recall that the matrix $M = A^T A$ (in the first proof) is circulant. Let F be the Fourier matrix[9] of order n. Because M is circulant, given any coloring $x = (x_1, \ldots, x_n)^T \in \{-1, 1\}^n$, the vector Mx is the convolution of x with a certain vector v. Its Fourier transform FMx is therefore the coordinate-wise product of Fv and Fx. If $Fv = (\lambda_1, \ldots, \lambda_n)^T$, it then follows that $FMx = \Lambda Fx$, where

$$\Lambda = \begin{bmatrix} \lambda_1 & 0 & \ldots & 0 \\ 0 & \lambda_2 & \ldots & 0 \\ \vdots & \vdots & \ddots & \vdots \\ 0 & 0 & \ldots & \lambda_n \end{bmatrix}.$$

By premultiplying by the inverse matrix F^{-1}, we find that $M = F^{-1} \Lambda F$. Taking the Fourier transform diagonalizes our matrix. This is no big surprise since it is a "convolution" matrix.

The moral of the story is this: The eigenvalue method is the most general and direct line of attack on the L^2 norm of the discrepancy. In practice, however, getting a handle on the eigenvalues is no simple matter. But whenever the set system is defined by some form of convolution (a "comb" in the case of arithmetic progressions), the Fourier transform method brings those eigenvalues to the fore (via diagonalization). Geometric discrepancy with respect to boxes or disks is defined by translating (and sometimes rotating or scaling) some fixed object across space and defining one subset of the set system for each position: Translating is just like sliding a comb and acts as a convolution operator in defining the set system. Thus, it is little surprise that Fourier transforms should play a major role.

---

[9]See the discussion of the discrete Fourier transform in Appendix B.

Hereditary Discrepancy and Determinants

The eigenvalue bound is often too weak to be useful because it makes no attempt to exploit the fact that the coloring is a vector with ± 1 coordinates. By introducing the notion of hereditary discrepancy, we are able to use that fact and relate the discrepancy, not only to the minimum eigenvalue, but to the entire spectrum of the incidence matrix.

Let $(V, \mathcal{S})$ be a set system. Given $W \subseteq V$, recall that $\mathcal{S}|_W$ is the collection of subsets of W of the form $S \cap W$, where $S \in \mathcal{S}$. The *hereditary discrepancy* of $(V, \mathcal{S})$, denoted by herdisc $(\mathcal{S})$, is defined to be the maximum value of $D_\infty(\mathcal{S}|_W)$ over all $W \subseteq V$. The motivation behind this notion is that even though some subsystem might have a huge discrepancy, $D_\infty(\mathcal{S})$ itself might be very small by some "fluke." Indeed, by adding only $O(D_\infty(\mathcal{S}))$ new elements to V and the sets of $\mathcal{S}$, we can easily make the discrepancy vanish entirely. Besides its built-in robustness, the hereditary view has an unintended benefit: It allows us to bound the discrepancy in terms of the full spectrum of eigenvalues and not only the smallest one. We consider the product of the eigenvalues (the determinant) in this section and their sum (the trace) later in this chapter.

How much are we giving up by adopting the hereditary viewpoint? In geometry, not much. Indeed, the hereditary discrepancy is particularly well suited for geometric applications, because geometric set systems typically are hereditary themselves: remove points from a set system of points and disks, and you still get a set system of points and disks. And so, in geometry at least, adding the adjective hereditary before the word discrepancy does not narrow the view.

Theorem 1.12 (THE DETERMINANT BOUND) *If A is an n-by-n incidence matrix of a set system, then*

$$\text{herdisc}\,(A) \geq \tfrac{1}{2} |\det A|^{1/n}.$$

Corollary 1.13 herdisc $(A) \geq \frac{1}{2} \max_{k,B} |\det B|^{1/k}$, *where B ranges over all k-by-k submatrices of A.*

To minimize the quadratic form $\|Ax\|_2^2$ is the stuff of linear algebra textbooks. The hereditary discrepancy adds three twists: The vector x has ± 1 coordinates; the norm is not Euclidean; and, if that were not bad enough, all submatrices of A come into play. Our first objective is to find our way back to linear algebra. Once we have done that, we will see that all three twists in fact help produce stronger results than the eigenvalue bound. To prove Theorem 1.12 we introduce a weighted version of the discrepancy.

Recall that

$$D_\infty(A) = \min\Big\{ \|Ax\|_\infty : x \in U \Big\},$$

where $U = \{-1,1\}^n$. A more general definition might allow the range of x to vary, as in

$$D_\infty^c(A) = \min\Big\{ \|Ax\|_\infty : x \in c + U \Big\},$$

where $c \in \mathbf{R}^n$. To establish a lower bound on herdisc (A), a reasonable approach is to bound

$$\text{intdisc}\,(A) \overset{\text{def}}{=} \max\Big\{ D_\infty^c(A) : c \in \{-1,0,1\}^n \Big\},$$

in view of the fact that

$$\text{herdisc}\,(A) \geq \text{intdisc}\,(A). \tag{1.8}$$

This inequality is quite obvious. Think of the hereditary discrepancy in the context of a game: First, your adversary sets any number of x_i's to 0; then you complete the coloring x so as to minimize the (regular) discrepancy. To see the connection with $D_\infty^c(A)$, consider any $c \in \{-1,0,1\}^n$. Form a ± 1-coloring by setting $x_i = -c_i$ for each $c_i \neq 0$ and completing the assignment of x by minimizing the (regular) discrepancy. By definition, the resulting discrepancy Δ is at least $D_\infty^c(A)$. On the other hand, pursuing the game analogy, the assignments $x_i = -c_i$ correspond to the adversary's annulling some of the columns. This shows that $\Delta \leq \text{herdisc}\,(A)$ and establishes (1.8). Note that this reasoning should give no reason to think that the inequality is actually an equality; for example, with the set system $\{a\}$, $\{a,b\}$, $\{b,c\}$, $\{a,c,d\}$, D_∞^c is at most 1 while the hereditary discrepancy is 2.

Relaxing c in the definition of intdisc leads to the *linear discrepancy* of A:

$$\text{lindisc}\,(A) \overset{\text{def}}{=} \max\Big\{ D_\infty^c(A) : c \in [-1,1]^n \Big\}.$$

As one might expect, the benefit of such a relaxation is to make it amenable to linear algebra. Fortunately, relaxing c does not have drastic effects on the discrepancy.

Lemma 1.14

$$\text{lindisc}\,(A) \leq 2\,\text{intdisc}\,(A).$$

We can now finish the proof of the theorem. Given any $c \in [-1,1]^n$, there exists some $x_c \in U$ such that

$$\|A(x_c + c)\|_\infty \leq \operatorname{lindisc}(A).$$

Thus, if Y denotes the set of points $y \in \mathbf{R}^n$ satisfying $\|Ay\|_\infty \leq \operatorname{lindisc}(A)$, the set $Y + U$ covers the entire cube $[-1, 1]^n$ (Fig. 1.3). But the pieces $Y + x$ $(x \in U)$ can be obtained by cutting up a single copy of Y. Formally speaking, Y encloses $[-1, 1]^n$ in $(\mathbf{R}/2\mathbf{Z})^n$. (Topologically, we are identifying the opposite facets of U.) It follows that the volume of Y is at least that of U, ie, 2^n. Obviously, we can assume without loss of generality that A is nonsingular (else, $\det A = 0$ and the theorem is trivial). Then, $Y = A^{-1}[-\operatorname{lindisc}(A), \operatorname{lindisc}(A)]^n$, from which we derive that $\operatorname{vol}(Y) = (2\operatorname{lindisc}(A))^n |\det A^{-1}|$, and hence $\operatorname{lindisc}(A) \geq |\det A|^{1/n}$. In view of (1.8) and Lemma 1.14, the proof of Theorem 1.12 is now complete. $\square$

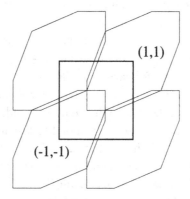

Fig. 1.3. The set $Y + \{-1, 1\}^n$ covers the cube $[-1, 1]^n$.

It remains for us to prove Lemma 1.14. Put

$$t = \max_{c \in \{-1, 0, 1\}^n} D_\infty^c(A).$$

It suffices to show that, given any point $c \in [-1, 1]^n$, there exists $x_0 \in U + c$ such that $\|Ax_0\|_\infty \leq 2\operatorname{intdisc}(A)$. This is obviously true if the coordinates of c are integers (ie, $-1, 0, 1$). Let us say that c is k-*good* if its coordinates are rationals in $[-1, 1]$ whose binary expansions do not extend beyond the k-th bit (ie, all of the lower order bits are 0 past position k). Proceeding by induction, assume that the proposition is true for any k-good point. Now suppose that c is $(k + 1)$-good. It is elementary to see that there always exists a translation vector $a \in U$ such that $b = 2c + a$ falls in the box $[-1,1]^n$. It follows that b is k-good. By induction, there exists $x_1 \in U + b$ such that $\|Ax_1\|_\infty \le 2\operatorname{intdisc}(A)$. Thus, for some $x_2 \in U$, we have (dividing by 2)

$$\|A(c + \gamma)\|_\infty \le \operatorname{intdisc}(A),$$

where $\gamma = (x_2 + a)/2$. Since $\gamma \in \{-1, 0, 1\}^n$, we have $D^{[\gamma]}(A) \le \operatorname{intdisc}(A)$, and therefore $\|A(x_3 + \gamma)\|_\infty \le \operatorname{intdisc}(A)$, for some $x_3 \in U$. Subtracting the last two inequalities yields $\|A(c - x_3)\|_\infty \le 2\operatorname{intdisc}(A)$, which completes the induction. A standard compactness argument finishes the proof of Lemma 1.14. $\square$

Application: Points and Halfplanes

We show how the notion of hereditary discrepancy can be used to prove a tight lower bound for the standard L^∞ (red-blue) discrepancy formed by points and halfplanes. The discrepancy of a square matrix like Hadamard was easily bounded, but in general specific matrices are hopelessly difficult to tackle. Chapter 3 treats the case of several geometric incidence matrices, but these are not square. The determinant bound for the hereditary discrepancy allows us to derive a tight bound on the discrepancy of an important class of square matrices.

Let $A = (a_{ij})$ be the n-by-n incidence matrix of a set system formed by n points in the plane and n closed halfplanes: $a_{ij} = 1$ if the i-th halfplane contains the j-th point, else $a_{ij} = 0$. We prove the following bound, which is optimal:

Theorem 1.15 *There exist n points and n halfplanes in $\mathbf{R}^2$, such that the n-by-n incidence matrix $A = (a_{ij})$ has discrepancy $D_\infty(A) = \Omega(n^{1/4})$.*

The point set $\{p_i\}$ consists of the n integer points in $[1, \sqrt{n}\,]^2$; we assume that n is a large square. The discrepancy vector x is formed by associating its i-th coordinate x_i with the ± 1-color of point p_i. Let us relax the assumption that $x_i = \pm 1$ and instead regard x as any vector in $\mathbf{R}^n$. Given a closed halfplane h bounded above by a nonvertical line, let $f(h)$ denote the sum $\sum_{p_i \in h} x_i$. We define ω to be the unique motion-invariant measure for lines that provides a probability measure for the lines crossing the square $[1, \sqrt{n}\,]^2$; see [265] for details.[10] Alexander [10] has proven that if

---

[10] Intuitively, the probability that a random line hits an object should not depend on

$x_1 + \cdots + x_n = 0$, then

$$\int f(h)^2 \, d\omega(h) \gg \frac{\|x\|_2^2}{\sqrt{n}}. \qquad (1.9)$$

This bound is an easy consequence of the finite differencing method developed in Chapter 3, so we do not reproduce it here.

We subdivide the space of lines crossing $[1, \sqrt{n}\,]^2$ into $N + O(n^2)$ regions, within which $f(h)$ remains invariant. By choosing N large enough, say, $N = 2^n$, we can ensure that the ω-area σ of N of these regions is exactly the same, say around $1/N$, while the other $O(n^2)$ regions have smaller areas. Computing $\int f(h)^2 \, d\omega(h)$ by integrating f^2 over only the equal-area regions produces an absolute error of $O(n^2/N) \sup f^2$. Obviously, $|f(h)|$ cannot exceed

$$|x_1| + \cdots + |x_n| \le \sqrt{n}\, \|x\|_2,$$

by Cauchy-Schwarz, so the error is bounded by $O(n^3 \|x\|_2^2 / N)$. We define B to be the N-by-n matrix whose rows are indexed by the N equal-area regions $\hat{\sigma}$ and are the characteristic vectors of the set of x_i's appearing in (the unique linear form) $f(h)$, given $h \in \hat{\sigma}$. It follows that

$$\left| \|Bx\|_2^2 - \frac{1}{\sigma} \int f(h)^2 \, d\omega(h) \right| = O(n^3) \frac{\|x\|_2^2}{N\sigma}.$$

Because $\sigma = 1/N \pm O(n^2/N^2)$, we have

$$\left| \|Bx\|_2^2 - N \int f(h)^2 \, d\omega(h) \right| = O(n^3 \|x\|_2^2).$$

Lemma 1.16

$$\det B^T B = \Omega \left(N / \sqrt{n} \right)^{n-1}.$$

Proof: Let $\mu_1 \ge \cdots \ge \mu_n \ge 0$ be the eigenvalues of $B^T B$, and let $\{v_i\}$ be an orthonormal eigenbasis, with μ_i corresponding to v_i. Let $(\xi_1, \ldots, \xi_n)$ be the coordinates of x in the basis $\{v_i\}$. The rank of the linear system

$$\begin{cases} x_1 + \cdots + x_n = 0 \\ \xi_j = 0 \quad (j < n - 1) \end{cases}$$

is at most $n - 1$. Feasible solutions lie in the (ξ_{n-1}, ξ_n)-plane, so they intersect the cylinder $\xi_{n-1}^2 + \xi_n^2 = 1$. A solution x at the intersection is of

---

its particular placement but only on its shape. In the plane, the measure for a line $h : ax + by = 1$ has density $d\omega(h) = c(a^2 + b^2)^{-3/2} \, da \, db$, for some normalizing constant $c > 0$ adjusted to make the probabilities sum up to one.

unit length, so

$$\|Bx\|_2^2 = \sum_{i=1}^{n} \mu_i \xi_i^2 = \mu_{n-1}\xi_{n-1}^2 + \mu_n \xi_n^2 \leq \mu_{n-1}$$

and

$$\mu_{n-1} \geq N \int f(h)^2 \, d\omega(h) - O(n^3 \|x\|_2^2) \geq \Omega(N/\sqrt{n}) - O(n^3);$$

hence,

$$\mu_{n-1} \geq \Omega(N/\sqrt{n}). \tag{1.10}$$

Next, we derive a lower bound on the smallest eigenvalue. With N large enough, we can easily assume that, for each point p_i, there exist two lines (adding them on, if necessary, and updating N accordingly), each represented by a distinct row of B, that pass right above and below p_i. The contribution of these two rows to $\|Bx\|_2^2$ is of the form $\Phi^2 + (\Phi + x_i)^2$, which is always at least $x_i^2/2$. We conclude that $\|Bx\|_2^2 \geq \frac{1}{2}\|x\|_2^2$, and so $\mu_n \geq 1/2$. Since $\det B^T B$ is the product of the eigenvalues, the lemma follows from (1.10). $\square$

The Binet-Cauchy formula says that[11]

$$\det B^T B = \sum_{1 \leq j_1 < \cdots < j_n \leq N} \left| \det B \begin{pmatrix} j_1 & j_2 & \cdots & j_n \\ 1 & 2 & \cdots & n \end{pmatrix} \right|^2.$$

This implies the existence of an n-by-n submatrix A of B such that

$$\det A^T A = \left| \det B \begin{pmatrix} j_1 & j_2 & \cdots & j_n \\ 1 & 2 & \cdots & n \end{pmatrix} \right|^2$$

$$\geq \binom{N}{n}^{-1} \det B^T B = \Omega(1)^n \left(\frac{n}{eN}\right)^n \left(\frac{N}{\sqrt{n}}\right)^{n-1};$$

hence,

$$\det A^T A = \Omega(n)^{n/2}. \tag{1.11}$$

Bringing in the hereditary discrepancy $\text{herdisc}(A)$, it follows from Theorem 1.12 (page 26) that

$$\text{herdisc}(A) = \Omega(n^{1/4}).$$

Let A' be the (or any) submatrix of A whose discrepancy is this hereditary

---

[11] The notation following $\det B$ refers to the matrix obtained by picking the rows indexed $j_1, \ldots, j_n$ in B.

discrepancy. The matrix M derived from A by zeroing out the columns not in the submatrix A' is the incidence matrix of a set system of n halfplanes and at most n points. We can make it n-by-n by adding artificial points outside all of the halfplanes (which is possible since they all face down). This completes the proof of Theorem 1.15. $\square$

The Trace Bound

The determinant bound of Theorem 1.12 has two weaknesses: One is that the matrix might have high discrepancy and null determinant (say, one row is duplicated); the other one is that determinants for set systems can be very difficult to estimate asymptotically.[12] We establish a connection between the hereditary discrepancy and the traces of $A^T A$ and its square.

Theorem 1.17 (THE TRACE BOUND)   *If A is an n-by-n incidence matrix and $M = A^T A$, then*

$$\text{herdisc}\,(A) \geq \frac{1}{4}\, c^{n\,\text{tr}\,M^2/\text{tr}^2\,M} \sqrt{\frac{\text{tr}\,M}{n}}\,,$$

for some constant $0 < c < 1$.

How does this compare with the determinant bound? The latter can be rewritten as roughly $\sqrt{(\det M)^{1/n}}$, and so inside the square roots we find the arithmetic mean of the eigenvalues of M being compared against the (never bigger) geometric mean: a sign of progress. There is a correction factor, however. Note that it is inevitable since $\sqrt{\text{tr}\,M/n}$ alone cannot bound the discrepancy: Try the matrix A full of ones to see why. For the trace bound to be of any use, however, it is crucial that the exponent $n\,\text{tr}\,M^2/\text{tr}^2\,M$ in the correction factor be essentially constant. One easily verifies that if θ is the angle between the vectors $(1,\dots,1)$ and $(\lambda_1,\dots,\lambda_n)$ then, by projection,

$$\frac{n\,\text{tr}\,M^2}{\text{tr}^2\,M} = \frac{1}{\cos^2\theta}\,.$$

---

[12] For example, the Riemann hypothesis can be expressed as an upper bound on a very simple determinant. The Redheffer matrix [28] has 1's in the leftmost column and at entry (i,j) if i divides j, and 0's elsewhere. Its determinant is the Mertens function $\sum_{k=1}^{n}\mu(k)$, where $\mu(n)$ is the Möbius function. It is $O(n^{1/2+\varepsilon})$, for any $\varepsilon > 0$, if and only if the Riemann hypothesis is true.

So, to say that the exponent should not be large is another way of requiring that the eigenvalue distribution be fairly uniform. Fortunately, the correction factor is constant in many applications.

Be that as it may, what gives the trace bound its power is that both $\operatorname{tr} M$ and $\operatorname{tr} M^2$ have simple combinatorial meanings. For example, the trace of M is the number of ones in A, ie, the number of incidences in the set system. Similarly, the trace of M^2 is the number of rectangles of ones in A or, equivalently, the number of closed paths (simple and non simple) of length 4 in the bipartite graph corresponding to A. The trace bound follows easily from the lemma below.

Lemma 1.18 *For any* $1 \leq k \leq n$,

$$\operatorname{lindisc}(A) \geq 18^{-n/k}\sqrt{\lambda_k},$$

where $\lambda_1 \geq \cdots \geq \lambda_n \geq 0$ *are the eigenvalues of* $M = A^T A$.

To see why the lemma implies Theorem 1.17, we use a second-moment probabilistic argument. For $x \geq 0$, let $\mathcal{E}_x$ be the event, $\lambda \geq \mathbf{E}\lambda - x$, and let p be its probability. The sequence of derivations,

$$
\begin{aligned}
0 &= \mathbf{E}[\lambda \,|\, \mathcal{E}_x]\, p + \mathbf{E}[\lambda \,|\, \overline{\mathcal{E}}_x]\,(1 - p) - \mathbf{E}\lambda \\
&= (\mathbf{E}[\lambda \,|\, \mathcal{E}_x] - \mathbf{E}\lambda)\, p + (\mathbf{E}[\lambda \,|\, \overline{\mathcal{E}}_x] - \mathbf{E}\lambda)\,(1 - p) \\
&\leq (\mathbf{E}[\lambda \,|\, \mathcal{E}_x] - \mathbf{E}\lambda)\, p - x(1 - p),
\end{aligned}
$$

leads to

$$\mathbf{E}[\lambda \,|\, \mathcal{E}_x] \geq \mathbf{E}\lambda + x(1/p - 1).$$

Consider the random variable $\mathbf{E}[\lambda \,|\, Y]$. Let Y be the event $\mathcal{E}_x$ with probability p and $\overline{\mathcal{E}}_x$ with probability $1 - p$. The conditional variance $\mathbf{var}[\lambda \,|\, Y]$, defined as the variance of the random variable $\mathbf{E}[\lambda \,|\, Y]$, cannot exceed the (unconditional) variance and therefore,

$$
\begin{aligned}
\mathbf{var}\,\lambda &\geq \mathbf{var}\,\mathbf{E}[\lambda \,|\, Y] = \mathbf{E}\left(\mathbf{E}[\lambda \,|\, Y] - \mathbf{E}\,\lambda\right)^2 \\
&\geq (\mathbf{E}[\lambda \,|\, \mathcal{E}_x] - \mathbf{E}\lambda)^2 p + (\mathbf{E}[\lambda \,|\, \overline{\mathcal{E}}_x] - \mathbf{E}\lambda)^2(1 - p) \\
&\geq x^2(1/p - 1)^2 p + x^2(1 - p) \geq x^2(1/p - 1),
\end{aligned}
$$

which shows that

$$p \geq \frac{1}{1 + x^{-2}\,\mathbf{var}\lambda}. \tag{1.12}$$

By setting $x = 3\mathbf{E}\lambda/4$ and k to be about $n/(2\mathbf{var}\lambda/\mathbf{E}^2\lambda + 1)$, we find that $\lambda_k \geq \mathbf{E}\lambda/4$. Since $\mathbf{E}\lambda = \operatorname{tr} M/n$ and $\mathbf{var}\lambda = \operatorname{tr} M^2/n - (\operatorname{tr} M)^2/n^2$, it follows from (1.8) and Lemmas 1.14 and 1.18 that

$$\text{herdisc}\,(A) \geq \frac{1}{2}\,\text{lindisc}\,(A) \geq \frac{1}{2}\,18^{-n/k}\sqrt{\lambda_k} \geq \frac{1}{4}\,18^{-2n\,\text{tr}\,M^2/\text{tr}^2\,M}\sqrt{\frac{\text{tr}\,M}{n}},$$

which proves Theorem 1.17. □

Another interesting expression for the tail of the eigenvalue distribution is obtained by setting $x = \varepsilon\sqrt{\text{tr}\,M^2/n}$ in (1.12). Then,

$$\text{Prob}\Big\{\lambda \geq \text{tr}\,M/n - \varepsilon\sqrt{\text{tr}\,M^2/n}\Big\} \;\geq\; \frac{1}{1 + n\varepsilon^{-2}\,\mathbf{var}\lambda/\text{tr}\,M^2} \geq \frac{1}{1 + \varepsilon^{-2}},$$

which is independent of n. We conclude:

Lemma 1.19 *Let A be an n-by-n $0/1$ matrix, and let $\lambda_1 \geq \cdots \geq \lambda_n$ be the eigenvalues of $M = A^T A$. Then, for any fixed $\varepsilon > 0$,*

$$\lambda_k \geq \text{tr}\,M/n - \varepsilon\sqrt{\text{tr}\,M^2/n},$$

for some $k = \Omega_\varepsilon(n)$.

We now prove Lemma 1.18. A singular-value decomposition of the matrix A allows us to rewrite it as UDV^T, where U (resp. V) is the orthogonal matrix whose columns are the eigenvectors of AA^T (resp. $A^T A$) and D is the n-by-n diagonal matrix whose only nonzero entries are $\sqrt{\lambda_1}, \sqrt{\lambda_2}, \ldots$ (the singular values of A, which are the square roots of the eigenvalues of $A^T A$ or, equivalently, AA^T). Let L be the subspace spanned by the k eigenvectors of $A^T A$ corresponding to $\lambda_1, \ldots, \lambda_k$. The projection of a unit cube in $\mathbf{R}^n$ to a k-dimensional subspace is a convex polytope of volume between c^{-n} and c^n, for some constant $c > 0$. A simple argument that adds the contribution of each of the $\binom{n}{k}2^{n-k}$ k-faces of the cube shows that $c \leq 3$. It follows that

$$\text{vol}\,(A\,\text{proj}_L[-1,1]^n) = \sqrt{\lambda_1 \cdots \lambda_k}\,\text{vol}\,(\text{proj}_L[-1,1]^n) \geq 2^k 3^{-n}\lambda_k^{k/2}. \tag{1.13}$$

Given any $x \in L$ and $y \in L^\perp$, $A^T Ax$ lies in L and so $(Ax)^T(Ay) = (A^T Ax)^T y = 0$. In fact, not only are AL and $A(L^\perp)$ orthogonal, but they span all of $A\mathbf{R}^n$ and therefore $(AL)^\perp = A(L^\perp)$. It easily follows that

$$A\text{proj}_L[-1,1]^n = \text{proj}_{AL}A[-1,1]^n$$

and by (1.13)

$$\text{vol}\,(\text{proj}_{AL}A[-1,1]^n) \geq 2^k 3^{-n}\lambda_k^{k/2}.$$

By definition (page 27), for any $c \in [-1,1]^n$, there exists some $x \in \{-1,1\}^n$ such that $Ac = Ax + y$, where $y \in [-\text{lindisc}\,(A), \text{lindisc}\,(A)]^n$. The image of the cube $[-1,1]^n$ under the transformation given by A is a polyhedron in $\mathbf{R}^n$ whose vertices belong to $A\{-1,1\}^n$. It follows that the polyhedron $A[-1,1]^n$ is covered by the $\leq 2^n$ n-dimensional cubes of side length $2\,\text{lindisc}\,(A)$ centered at the vertices of $A[-1,1]^n$. Projecting onto AL and accounting for the dilation factor of 3^n, we find that

$$\text{vol}\,(\text{proj}_{AL} A[-1,1]^n) \leq 6^n (2\,\text{lindisc}\,(A))^k,$$

which proves Lemma 1.18, and hence the trace bound. □

Application I: Points and Lines

In Chapter 6, we prove the existence of n points and n lines in the plane, all of them distinct, such that each point belongs to $\Theta(n^{1/3})$ lines and each line contains $\Theta(n^{1/3})$ points (Lemma 6.25, page 263). The trace of M is the number of incidences, ie, $\Theta(n^{4/3})$. The trace of M^2 is equal to the number of rectangles of ones in A. Since no proper rectangle can occur (two lines intersect in at most one point), we are left with degenerate rectangles formed by two ones along the same row (or the same column). There are $\Theta(n(n^{1/3})^2) = \Theta(n^{5/3})$ of those. It follows that the trace of M^2 is $O(n^{5/3})$. By the trace bound, the hereditary discrepancy of the set system is at least $\frac{1}{4} c^n \operatorname{tr} M^2 / \operatorname{tr}^2 M \sqrt{\frac{\operatorname{tr} M}{n}}$, which is $\Omega(n^{1/6})$. Notice how the exponent miraculously reduces to a constant!

Theorem 1.20 *There exist n points and n lines in $\mathbf{R}^2$, such that the n-by-n incidence matrix $A = (a_{ij})$ has discrepancy $D_\infty(A) = \Omega(n^{1/6})$.*

One can use the method of partial colorings (page 15) to show that the lower bound is optimal up to within a logarithmic factor. It is interesting to contrast the exponent of $1/6$ for points and lines vs. $1/4$ for points and halfplanes.

Application II: Boxes in Higher Dimension

In fixed dimension, it is a rule of thumb that discrepancies for boxes are $(\log n)^{\Theta(1)}$ if the orientation is fixed and $n^{\Theta(1)}$ if rotations are allowed. If we let the dimension increase, however, we expect this gap to be eventually bridged since in dimension high enough any 0/1 matrix is an incidence matrix for points and boxes. An interesting question thus is: At which dimension do we switch from logarithmic to polynomial? The trace bound indicates that the transition to polynomial discrepancy occurs at dimension as low as $O(\log n)$.

Theorem 1.21 *There exist n points and n axis-parallel boxes in $\mathbf{R}^d$, for any $d = O(\log n)$, such that the n-by-n incidence matrix $A = (a_{ij})$ has discrepancy $D_\infty(A) = 2^{\Omega(d)}$.*

The lower bound is actually more general than stated, as it holds for points and boxes in the Hamming cube $\{0,1\}^d$. Theorem 1.21 follows easily from the lemma below.

Lemma 1.22 *For any n large enough, there exists a set system of n points and n boxes in $\{0,1\}^d$, where $d = \Theta(\log n)$, such that the n-by-n incidence matrix $A = (a_{ij})$ has discrepancy $D_\infty(A) = \Omega(n^{0.0477})$.*

The theorem is essentially a restatement of Lemma 1.22 if $d \geq b \log n$, for some constant $b > 0$. So, assume that $d < b \log n$. Set n_0 to be about $2^{d/b}$ so that we can apply the lemma with respect to n_0 and d. Now, pad the set system to be n-by-n by adding $n - n_0$ points and boxes with no new incidences. The lower bound of $\Omega(n_0^{0.0477})$ is also $\Omega(2^{\Omega(d)})$, which proves Theorem 1.21. □

In view of the trace bound (page 32), Lemma 1.22 follows directly from the existence of m points and n boxes in $\mathbf{R}^{O(\log n)}$ such that: for some constant $c \approx 1.0955$,

   (i) $m = \Theta(n)$ and $\operatorname{tr} M = \Theta(n^c)$ with probability at least $1/2$;
   (ii) $\mathbf{E} \operatorname{tr} M^2 = O(n^{2c-1})$.

For convenience, we introduce a few parameters:

$$\begin{cases} w &= \frac{1-2p+p^9}{1-2p-(1+2p)p^2 \log e}, \text{ where } p = 0.153\,, \\ c &= 2 - (1-p)w\,, \\ g &= n^{c-1}. \end{cases}$$

The dimension d is defined as $w \log n$. The m points are chosen by picking each element of the Hamming cube $\{0,1\}^d$ independently with probability n^{1-w}. (Note that $w \approx 1.067867$, so $n^{1-w} < 1$.) The expected number of points is n and, by Chebyshev's inequality,

Lemma 1.23 *With probability $> 1/2$, the number m of points is $\Theta(n)$.*

A box is specified by a word of length d, over the alphabet $\{0, 1, *\}$, containing exactly pd stars. For example, in dimension 5, the word $0 * 1 * *$ denotes the three-dimensional box $x_1 = 0$, $x_3 = 1$. We construct the n boxes by specifying g groups of parallel boxes.[13] Each group is defined by selecting the location of the stars first and then taking all the corresponding boxes. To select the stars, we pick pd coordinates uniformly at random (without replacement) and make them stars. In our previous example, the group of parallel boxes consists of $0 * 0 * *$, $0 * 1 * *$, $1 * 0 * *$, and $1 * 1 * *$. We have precisely $2^{(1-p)d}g = n$ boxes. Each point in the set system belongs to exactly one box in each of the g groups, so that $\operatorname{tr} M = mg$. By Lemma 1.23, we have the following:

Lemma 1.24 *With probability* $> 1/2$, *the trace of* M *is* $\Theta(n^c)$ *and* (i) *holds.*

To find an upper bound on the trace of M^2, we express it as a sum of four terms:

$$\operatorname{tr} M^2 = O(\sigma_{1,1} + \sigma_{1,2} + \sigma_{2,1} + \sigma_{2,2}),$$

where $\sigma_{i,j}$ is the number of pairs (I, J) such that $I \supseteq J$, where I is the intersection of i distinct boxes and J is a set of j distinct points. To bound these numbers is easy. Any one of the 2^{pd} Hamming cube vertices lying in a given box belongs to the set system with probability n^{1-w}. There are n boxes, so

$$\mathbf{E}\, \sigma_{1,2} = O\left(n(2^{pd}n^{1-w})^2\right) = O(n^{3-2(1-p)w}).$$

Regarding $\sigma_{2,1}$, note that boxes within the same group are disjoint, so only pairs in distinct groups can contribute to $\sigma_{2,1}$. Fix two such groups. Any one of the 2^d points of the Hamming cube belongs to exactly one pair of boxes. Since such a point is picked with probability n^{1-w}, we have $\mathbf{E}\, \sigma_{2,1} = O(g^2 2^d n^{1-w}) = O(n^{2c-1})$. To summarize,

$$\mathbf{E}\, \sigma_{1,1} = \mathbf{E}\operatorname{tr} M = n^c, \quad \mathbf{E}\, \sigma_{1,2} = O(n^{2c-1}), \quad \mathbf{E}\, \sigma_{2,1} = O(n^{2c-1}). \quad (1.14)$$

Finally, we turn to the expectation of $\sigma_{2,2}$: Again, fix two groups of parallel boxes, and let x be the number of stars common to both star patterns. As we just saw, any point of the Hamming cube belongs to exactly one pair of boxes, and this point can be paired with exactly $2^x - 1$

---

[13] By rounding off, if necessary, we can assume that g, d, and pd are all integral.

other points. Each point being picked with probability n^{1-w}, it follows
that

$$\sigma_{2,2} = O(g^2 2^{d+x} n^{2-2w})$$

and, hence,

$$\mathbf{E}\,\sigma_{2,2} = O(n^{2c-w})\,\mathbf{E}\,2^x.$$

What about the expectation of 2^x? Writing

$$N_k \stackrel{\text{def}}{=} N(N-1)\cdots(N-k+1),$$

we use the lower bound $k! > (k/e)^k$ to derive

$$
\begin{aligned}
\mathbf{E}\,2^x &= \sum_{k=0}^{pd} 2^k \binom{pd}{k}\binom{d-pd}{pd-k} \Big/ \binom{d}{pd} = \sum_{k=0}^{pd} \frac{2^k (pd)_k (d-pd)_{pd-k}}{k!(pd-k)!} \Big/ \binom{d}{pd} \\
&\le \sum_{k=0}^{pd} \frac{(2ep^2 d^2)^k (d-pd)_{pd}}{(kd)^k (1-2p)^k (pd)!} \Big/ \binom{d}{pd} \le \sum_{k=0}^{pd} (1-p)^{pd} \left(\frac{2ep^2 d}{(1-2p)k}\right)^k.
\end{aligned}
$$

The function $(A/x)^x$ reaches its maximum value at $x = A/e$, and so

$$\mathbf{E}\,2^x = O(n^{(\log e)p^2 w(1+2p)/(1-2p)} \log n),$$

which implies that

$$\mathbf{E}\,\sigma_{2,2} = O(n^{2c-w+\frac{1+2p}{1-2p}p^2 w \log e} \log n).$$

In view of (1.14),

$$
\begin{aligned}
\mathbf{E}\,\mathrm{tr}\,M^2 &= O\left(n^c + n^{2c-1} + n^{2c-w+\frac{1+2p}{1-2p}p^2 w \log e} \log n\right) \\
&= O\left(n^{2c-1} + n^{2c-1-\frac{p^9}{1-2p}} \log n\right) = O(n^{2c-1}),
\end{aligned}
$$

which satisfies condition (ii). Lemma 1.22 and Theorem 1.21 follow. $\square$

1.6 Bibliographical Notes

Section 1.1: The method of conditional expectations was developed by Raghavan [254] and Spencer [294]. A similar idea is implicit in an earlier work of Erdős and Selfridge [124]. The hyperbolic cosine algorithm is due to Spencer [292, 294]. The unbiased greedy algorithm was first proposed by Beck and Fiala [38] and Beck [32]; it was rediscovered by the author [68]. Not surprisingly, a similar technique can be used to rederive Chernoff-type bounds and prove tail estimates for martingales.

Section 1.2: The fact that n-by-n matrices have discrepancy $O(\sqrt{n})$ (Theorem 1.3, page 8) was established by Spencer [293]. Using the pigeonhole principle on the discrepancy vector is an idea going back to Beck [31]. The use of entropy in the proof follows a suggestion of Boppana (see Alon and Spencer [20]). Spencer's original proof shows that the constant hiding behind the bound $O(\sqrt{n})$ is less than 6.

Section 1.3: Theorem 1.5 (page 10) is due to Beck and Fiala [38]. The bound was (ever so slightly) improved to $2t - 3$ by Bednarchak and Helm [40]. It is conjectured that $O(\sqrt{t})$ is the correct bound. A bound of $O(\sqrt{t} \log n)$ was established by Srinivasan [295], where n is the number of elements in the set system; an earlier bound of $O(\sqrt{t \log t} \log n)$ was obtained by Beck and Spencer; see also [294].

Section 1.4: The notion of VC-dimension was introduced by Vapnik and Chervonenkis [317]. We chose to open the section with it because of its sheer elegance and its historical importance. In most applications, however, bounds on the primal and shatter functions seem more important than the VC-dimension (which, typically, is difficult to compute). The bound on the primal shatter function (Lemma 1.6) was established independently by Sauer [269], Shelah [284], and Vapnik and Chervonenkis [317]—see also [51]. The fact that infinite range spaces cannot have sublinear primal shatter functions appears in Assouad [25].

Dudley [112] observed that any finite number of set-theoretical operations on range spaces keep the VC-dimension bounded. This is useful to prove that certain geometric range spaces are of bounded VC-dimension. The bound on the dual VC-dimension in Lemma 1.7 comes from Assouad [25]. The discrepancy estimates in Theorem 1.8 (page 14) and Theorem 1.10 (page 17) are due to Matoušek, Welzl, and Wernisch [222]. Lemma 1.9 is adapted from Beck [31]. The bound in Theorem 1.8 has been improved to $O(n^{1/2-1/2d})$ by Matoušek [215]. The optimality of the "dual" bound in Theorem 1.10, for $d = 2, 3$, was established by Matoušek [218] and extended to any dimension by Alon, Rónyai, and Szabó [19].

Section 1.5: The lower bound on the discrepancy of the Hadamard matrix comes from Spencer [293]. The case of arithmetic progressions (Theorem 1.11, page 20) was solved by Roth [262], who used the Fourier transform method. The proof based on eigenvalues is due to Lovász and Sós and appears in Beck and Sós' survey [39]. A matching upper bound of $O(n^{1/4})$ was proven by Matoušek and Spencer [221]. An earlier, breakthrough result (weaker by only a polylogarithmic factor) was obtained by Beck [31], who also introduced the partial coloring technique. For an excellent introduction to Ramsey theory, see [149].

The notion of hereditary discrepancy and the determinant bound (Theorem 1.12, page 26) were introduced by Lovász, Spencer, and Vesztergombi [199]. The lower bound for halfplane discrepancy (Theorem 1.15, page 29) was established by Chazelle [75]. A proof of optimality was provided by Matoušek [215]. The lower bound (1.9) was proven by Alexander [10].

The trace bound (Theorem 1.17, page 32) is due to Chazelle and Lvov [79], as are the applications to set systems of lines and boxes in higher dimension [80] (Theorems 1.20 and 1.21).

2

Upper Bound Techniques

 e examine several methods for constructing low-discrepancy point sets. To motivate our discussion, in §2.1 we briefly mention a classical application of discrepancy theory to numerical integration. We introduce the Halton-Hammersley point sets in §2.2 and derive upper bounds on the L^∞ norm of the discrepancy of axis-parallel boxes. We tackle the L^2 norm of the two-dimensional case in §2.3, which leads us to studying the low-discrepancy properties of certain arithmetic progressions modulo 1. This is related to the infinite motion of a billiard ball in a square pool table.

In §2.4 we examine the discrepancy of boxes that are free to be both translated and rotated. We give a simple probabilistic construction, called jittered sampling, which comes close to being optimal. This construction and some close variants of it are popular in computer graphics for solving antialiasing problems.

In §2.5 we consider the problem of placing points on a sphere as "uniformly" as possible. This is a classical problem arising in quadrature, coding theory, tomography, etc. We describe a group-theoretic construction that produces low-discrepancy point sets (especially for integration and, to a lesser extent, for spherical caps). The generation of points is as extraordinarily simple as its analysis is mathematically deep and far-reaching. If you ever harbored any doubt about the unity of mathematics, this is required reading. You will witness all branches of mathematics coming together in spectacular fashion. We also mention in passing the relevance of this material to recent problems in quantum computing.

We include digressions on two important topics related to our construction. We begin with modular forms. These are essential ingredients in our discussion and we felt the need to mention, ever so briefly, their fundamental connection to elliptic curves via their L-functions: This is a pillar of modern mathematics that plays a central role in the recent proof of Fermat's Last Theorem. Another key concept in our discussion is the ubiquitous Laplacian. Combinatorialists often think of it as a function with interesting spectral properties related to graph connectivity. The origins of the concept are a little different. They have to do with least-square optimization via the Dirichlet principle. We discuss how the main concepts in linear and quadratic programming can be rederived from the Laplacian; this is the ideal introduction to Chapter 8.

In §2.8 we discuss the problem of coloring n points in $\mathbf{R}^d$ red and blue, so that within any halfspace no color greatly outnumbers the other. We give quasi-optimal bounds for this case of red-blue discrepancy.

2.1 Numerical Integration and Koksma's Bound

In quasi-Monte Carlo methods, it is common to evaluate an integral of the form

$$I = \int_{[0,1]^d} f(q)\, dq$$

by selecting a sample of points $p_1, \ldots, p_n$ and computing the sum

$$S_n = \sum_{i=1}^{n} f(p_i).$$

Since we naturally estimate the integral I by S_n/n, it is desirable to keep the error

$$\left| \frac{S_n}{n} - \int_{[0,1]^d} f(q)\, dq \right| \tag{2.1}$$

as small as possible. Consider the simpler case where $d = 1$ and $f : [0,1] \mapsto \mathbf{R}$ is a differentiable function of *bounded variation* $V(f)$, where

$$V(f) \overset{\text{def}}{=} \int_0^1 |f'(x)|\, dx.$$

To approximate the integral of f over $[0,1]$ we take a sample of n points,

$$0 \le x_1 \le \cdots \le x_n \le 1,$$

and sum f over the points $x_1, \ldots, x_n$. Interestingly, the error (2.1) is directly related to the discrepancy $\|D\|_\infty$ of the sequence $\{x_i\}$, defined as the supremum of

$$\left| |\{ i \, : \, x_i \le x \}| - nx \right|,$$

over $0 \leq x \leq 1$.

Theorem 2.1

$$\left| \frac{1}{n} \sum_{i=1}^{n} f(x_i) - \int_0^1 f(x) \, dx \right| \leq \frac{V(f)}{n} \|D\|_\infty.$$

Proof: Using summation by parts we have

$$\frac{1}{n} \sum_{i=1}^{n} f(x_i) = f(x_{n+1}) - \frac{1}{n} \sum_{i=1}^{n} i(f(x_{i+1}) - f(x_i)),$$

where $x_{n+1} = 1$. Similarly, integration by parts gives

$$\int_0^1 f(x) \, dx = f(x_{n+1}) - \int_0^1 x \, df(x).$$

Putting $x_0 = 0$, we have

$$\left| \frac{1}{n} \sum_{i=1}^{n} f(x_i) - \int_0^1 f(x) \, dx \right| = \left| \int_0^1 x \, df(x) - \frac{1}{n} \sum_{i=0}^{n} i[f(x_{i+1}) - f(x_i)] \right|$$

$$\leq \sum_{i=0}^{n} \int_{x_i}^{x_{i+1}} \left| x - \frac{i}{n} \right| |df(x)|$$

$$\leq \frac{\|D\|_\infty}{n} \sum_{i=0}^{n} \int_{x_i}^{x_{i+1}} |df(x)|,$$

from which the theorem follows. □

The inequality can be extended to the multidimensional case, but the notion of bounded variation becomes more complex.

2.2 Halton-Hammersley Points

We consider the problem of placing a set P of n points in the unit cube $[0,1]^d$ to minimize the discrepancy with respect to axis-parallel boxes. The dimension d is arbitrary but fixed once and for all, ie, it is independent of n. As usual, the discrepancy of a box $B = \prod_{k=1}^{d} [p_k, q_k)$ is defined as

$$D(B) \stackrel{\text{def}}{=} n \cdot \text{vol}(B) - |P \cap B|.$$

To distinguish it from the combinatorial (ie, red-blue) discrepancy, we call $D(B)$ the *volume discrepancy*.[1]

Theorem 2.2 *There is a set of n points in $[0,1]^d$ such that the discrepancy of any box in $[0,1]^d$ is $O(\log n)^{d-1}$ in absolute value.*

We prove the theorem only for boxes of the form $B_q = \prod_{k=1}^d [0, q_k)$, where $q = (q_1, \ldots, q_d) \in [0,1]^d$. This is clearly sufficient because, on the one hand, $|D(A \setminus B)| \leq |D(A)| + |D(B)|$, while on the other hand any axis-parallel box can be expressed as a constant-size arithmetic expression involving only boxes of the form B_q and the set-difference operation. (The constant depends on d.)

Given a nonnegative integer m, let $\sum_{i \geq 0} b_1(i)\, 2^i$ be its binary decomposition, and let

$$x_1(m) = \sum_{i \geq 0} \frac{b_1(i)}{2^{i+1}} \in [0,1).$$

The numbers $x_1(m)$, for $0 \leq m < n$, form the classical *van der Corput* sequence. We can use it to define the *bit-reversal* point set:

$$\left\{ (x_1(m), m/n) \,\middle|\, 0 \leq m < n \right\}.$$

An example ($n = 8, 16$) is shown in Figure 2.1. Edges have been added to suggest the vertex set of a hypercube in dimension $\log n$ projected onto a plane.

To generalize the construction to d dimensions, we choose $d-1$ relatively prime numbers, say, for simplicity, the first $d-1$ primes: $2 = p_1, p_2, \ldots, p_{d-1}$. The integer m has a unique decomposition in base p_k, $m = \sum_{i \geq 0} b_k(i) p_k^i$, so we can define

$$x_k(m) = \sum_{i \geq 0} \frac{b_k(i)}{p_k^{i+1}}.$$

The point set

$$P = \left\{ \left(x_1(m), \ldots, x_{d-1}(m), \frac{m}{n} \right) : 0 \leq m < n \right\}$$

---

[1]It is sometimes called continuous discrepancy, which is not an entirely felicitous choice of words since the discrepancy is not a continuous function.

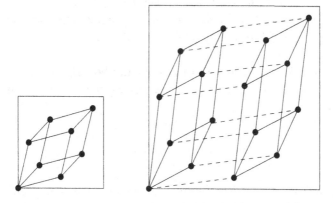

Fig. 2.1. Halton-Hammersley point sets of sizes 8 and 16.

is called *Halton-Hammersley*. There exist a large number of "special" boxes, each of which contains just about the right number of points with respect to volume. Any interval of the form $[A/p_k^j, (A+1)/p_k^j)$, where A and j are nonnegative integers, is called a *special interval of type* (k, j). A box $B = I_1 \times \cdots \times I_d$ is called *special* if

(i) it lies in $[0, 1)^d$,
(ii) for each $k < d$, I_k is a (k, j_k)-interval, for some j_k, and
(iii) $I_d = [0, x)$, for some $x \leq 1$.

Lemma 2.3 *The discrepancy of any special box is less than 1 in absolute value.*

Proof: Given i, k fixed, the interval $[0, 1)$ is naturally partitioned into intervals of the form $[j/p_k^i, (j+1)/p_k^i)$. It is immediate that the sequence $b_k(0), \ldots, b_k(j_k - 1)$ uniquely determines which (k, j_k)-interval contains a given $x_k(m)$. The sequence itself is entirely specified by the residue class of m mod $p_k^{j_k}$. By the Chinese remainder theorem, exactly one integer $0 \leq m < \prod_{k<d} p_k^{j_k}$ is in a given residue class modulo each $p_k^{j_k}$. It follows that any special box B such that $|I_d|n = \prod_{k<d} p_k^{j_k}$ contains exactly one point (note that because $|I_d| \leq 1$, we have $\prod_{k<d} p_k^{j_k} \leq n$ and such a point is thus defined for a valid $m < n$). By cutting the box up if necessary, we now can see that any special box B such that $|I_d|n$ is a multiple of $\prod_{k<d} p_k^{j_k}$, say, by a factor of p, contains exactly p points; its volume is p/n, so its discrepancy is zero. If B is not so lucky, then we clip it by the largest such

box and notice that the only discrepancy comes from the leftover: This is a box with at most one point inside and volume less than $1/n$, so its discrepancy (in absolute value) is less than one. $\square$

We are now in a position to prove that any box $B_q = \prod_{1 \leq k \leq d} [0, q_k)$, where $q = (q_1, \ldots, q_d) \in [0, 1]^d$, has low discrepancy. Any q_k can be expressed in base p_k as

$$q_k = \sum_{i \geq 0} \frac{b_k(i)}{p_k^{i+1}},$$

where the infinite sequence $b_k(i)$, for $i \geq 0$, does not have a suffix of $p_k - 1$'s. Let $h = \lceil \log n \rceil$ and let

$$q_k^* = \sum_{0 \leq i < h} \frac{b_k(i)}{p_k^{i+1}},$$

for $k < d$, and $q_d^* = q_d$. By writing the $b_k(i)$'s in unary it is immediate that, for $k < d$, $[0, q_k^*)$ is the disjoint union of $O(h)$ special intervals. It follows that $\prod_{1 \leq k < d} [0, q_k^*)$ is a disjoint union of $O(h^{d-1})$ special boxes and so, by Lemma 2.3, its discrepancy is $O(h^{d-1})$. The leftover

$$B_q \setminus \prod_{1 \leq k \leq d} [0, q_k^*)$$

is enclosed in the union $\bigcup_{1 \leq k < d} C_k$, where

$$C_k = [0, 1) \times \cdots \times [0, 1) \times [q_k^*, q_k) \times [0, 1) \times \cdots \times [0, 1).$$

Each C_k has volume at most $1/p_k^h \leq 1/n$. Since each interval $[q_k^*, q_k)$ is enclosed in a special interval of type (k, h), the box C_k is itself enclosed in a special box, and hence its discrepancy is bounded by a constant, and so is that of any subset of it. The leftover can be partitioned into $O(1)$ boxes each within some C_k, and hence of discrepancy $O(1)$. This shows that $|D(B_q)| = O(\log n)^{d-1}$, which concludes the proof of Theorem 2.2 (page 44). $\square$

2.3 Arithmetic Progressions in R/Z

Consider a set P of n points in the unit square. Given a box B_q of the form $[0, q_1) \times [0, q_2)$, where $q = (q_1, q_2)$, the discrepancy of B_q is

$$D(B_q) = n \cdot \text{area}(B_q) - |P \cap B_q|.$$

We define the L^2-norm discrepancy of P as

$$D_2(P) \stackrel{\text{def}}{=} \sqrt{\int_{[0,1]^2} D(B_q)^2 \, dq}.$$

Halton-Hammersley construction yields an $O(\log n)$ bound on the L^2-norm discrepancy. By using tools from the theory of diophantine approximation, it is possible to reduce the bound to $O(\sqrt{\log n})$. The optimality of the bound is established in Chapter 3.

Theorem 2.4 *It is possible to find a set P of n points in $[0,1]^2$ such that $D_2(P) = O(\sqrt{\log n})$.*

The set P is particularly simple. For example, one can choose the set of $n = 2k - 1$ points of the form

$$\left(\{ j\varphi \}, \frac{|j|}{n} \right),$$

for all j ($|j| < k$), where $\{x\} \stackrel{\text{def}}{=} x \pmod 1$ is the fractional part of x and $\varphi = \frac{1}{2}(\sqrt{5} + 1)$ is the golden ratio. Before proving this result, we make a short digression to illustrate the key concept of ergodicity and its relation to Fourier series.

Weyl's Ergodicity Criterion

Statistical mechanics operates under the fundamental assumption that the distribution in phase space mimics the distribution in time. Thus, a single particle evolving in time is modeled by a cloud of points in space, where at any given time each point is weighted by the probability that the particle should be there. The equivalence of phase and time averages is called the *ergodic principle*; it underlies much of our discussion of rapidly mixing Markov chains in Chapter 9.

To take a concrete example, consider a particle moving on the unit interval at discrete intervals (Fig. 2.2). At time n, the particle is at position

$$\{ n\gamma \} \stackrel{\text{def}}{=} n\gamma \pmod 1,$$

where γ is a fixed real. (A similar effect is obtained by kicking a billiard ball with a slope of γ and monitoring its hits against the vertical walls.) Let f be a periodic, continuously differentiable function from $\mathbf{R}$ to $\mathbf{C}$ of period 1. The ergodic principle states that the time average

$$\frac{1}{n}\sum_{k=0}^{n-1} f(\{k\gamma\})$$

should converge to the phase average $\int_0^1 f(x)\,dx$.

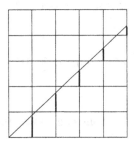 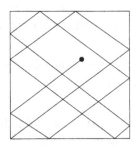

Fig. 2.2. The values of $n\gamma$ (mod 1) are represented by the vertical bars in the integral lattice. They can also be viewed as the hits of a billiard ball along the left and right sides of the table.

Theorem 2.5 (Weyl)   *If γ is irrational, then the sequence $\{n\gamma\}$ obeys the ergodic principle.*

Proof: The proof is based on *Weyl's criterion*, which characterizes the uniformity of infinite sequences (u_n) in $[0,1)$ in terms of the vanishing of the exponential sums

$$\frac{1}{N}\sum_{n=1}^{N} e^{2\pi i z u_n}\,,$$

for all integers $z \neq 0$, as N grows to infinity. Let $S_\ell = \sum_{|k|\leq\ell}\widehat{f}(k)e^{2\pi ikx}$ be the partial sum of the Fourier series expansion[2] of f, where

$$\widehat{f}(k) = \int_0^1 f(x)e^{-2\pi ikx}\,dx.$$

Let $f_m(x) = \frac{1}{m}\sum_{\ell=0}^{m-1} S_\ell$. By Fejér's theorem, we know that f_m converges uniformly towards f, ie,[3]

$$\|f - f_m\|_\infty \leq \varepsilon(m),$$

---

[2]See Appendix B.
[3]$\|g\|_\infty = \sup_x |g(x)|.$

for some $\varepsilon(m)$ going to 0 as $m \to \infty$. This implies that

$$\left| \frac{1}{n} \sum_{k=0}^{n-1} f(\{k\gamma\}) - \int_0^1 f(x)\, dx \right| \leq \left| \frac{1}{n} \sum_{k=0}^{n-1} f_m(\{k\gamma\}) - \int_0^1 f(x)\, dx \right| + \varepsilon(m).$$

Note that

$$\frac{1}{n} \sum_{k=0}^{n-1} f_m(\{k\gamma\}) = \sum_{\ell=0}^{m-1} \sum_{|j| \leq \ell} \widehat{f}(j) A_j,$$

where

$$A_j = \frac{1}{nm} \sum_{k=0}^{n-1} e^{2\pi i j k \gamma}.$$

If $j = 0$, then $A_j = 1/m$; otherwise,

$$|A_j| = \frac{1}{nm} \left| \frac{e^{2\pi i j n \gamma} - 1}{e^{2\pi i j \gamma} - 1} \right|.$$

Because γ is irrational, the denominator never vanishes; so for any given $j > 0$, $|A_j|$ tends to 0 as n goes to infinity. It follows that, for fixed m,

$$\frac{1}{n} \sum_{k=0}^{n-1} f_m(\{k\gamma\})$$

tends to $\widehat{f}(0) m A_0 = \widehat{f}(0) = \int_0^1 f(x)\, dx$, as n goes to infinity, from which the theorem follows. $\square$

Continued Fractions

Let $[a_0, \ldots, a_n]$ denote the expression

$$a_0 + \cfrac{1}{a_1 + \cfrac{1}{a_2 + \cfrac{1}{\ddots + \cfrac{1}{a_{n-1} + \cfrac{1}{a_n}}}}}$$

Given any real $\gamma > 0$, we set $a_0 = \lfloor \gamma \rfloor$ and, if γ is not an integer, we write $\alpha_0 = 1/(\gamma - \lfloor \gamma \rfloor)$. Noticing that $\gamma = [a_0, \alpha_0]$, we extend this idea and consider the general recurrence relations $a_n = \lfloor \alpha_{n-1} \rfloor$ and

$$\alpha_n = \frac{1}{\alpha_{n-1} - \lfloor \alpha_{n-1} \rfloor} \, .$$

Again, we stop the recurrence at n if α_n is an integer. We easily verify that

$$\gamma = [a_0, a_1, \ldots, a_n, \alpha_n].$$

Letting n run to infinity, we call the infinite sequence $[a_0, a_1, \ldots]$ the *continued fraction expansion* of γ. The integers a_i are called the *partial quotients* of the expansion; note that $a_i \geq 1$ for $i \geq 1$. The sequence of *convergents*, $r_n = [a_0, a_1, \ldots, a_n]$, for $n = 0, 1$, etc, provides an (excellent) approximation of γ by rationals. Each r_n is a rational p_n/q_n, where the monotonically increasing sequences of p_n and q_n are defined by the recurrence $p_{-2} = q_{-1} = 0$, $q_{-2} = p_{-1} = 1$, and for $n \geq 0$

$$\begin{cases} p_n &=& a_n p_{n-1} + p_{n-2}, \\ q_n &=& a_n q_{n-1} + q_{n-2}. \end{cases}$$

A simple observation is that if γ is a rational u/v, then the partial quotients are obtained by applying Euclid's GCD algorithm to u and v. Two interesting facts follow immediately: (i) The continued fraction expansion of a rational is finite (though not unique); (ii) the fractions p_n/q_n are irreducible (whether γ is rational or not). A theorem of Lagrange says that the expansion is periodic if and only if γ is irrational and algebraic of degree 2. For example, the golden ratio $\frac{1}{2}(\sqrt{5}+1)$ has the expansion $[1, 1, 1, \ldots]$, and its convergents are the ratios of consecutive Fibonacci numbers. The convergence rate for continued fractions is provided by the following bounds: For any $n \geq 0$,

$$\frac{1}{2q_{n+1}^2} \leq \left| \gamma - \frac{p_n}{q_n} \right| \leq \frac{1}{q_n^2} \, . \tag{2.2}$$

Continued fractions provide the best possible approximation by rationals in the following sense: For any integers $1 \leq a < q_{n+1}$ and $b > 0$,

$$|a\gamma - b| \geq |q_n \gamma - p_n| \, . \tag{2.3}$$

As an aside, let us mention an important development in the theory of diophantine approximation. By tightening the Thue-Siegel inequality, Roth proved the remarkable fact that, for algebraic numbers, continued fractions are as good approximants as we can hope for. Let γ be an irrational algebraic number. By (2.2) it is obvious that the inequality

$$\left|\gamma - \frac{p}{q}\right| < \frac{c(\gamma)}{q^2},$$

where $c(\gamma)$ is a constant independent of n, has an infinite number of integer solutions in p, q. Roth's result says that if we replace the exponent 2 by any larger number, the number of solutions becomes finite. This result has very powerful consequences; for example, it immediately shows that the number $\sum_{i \geq 0} 10^{-3^i}$ is transcendental. Indeed, it obviously admits an unbounded number of approximations of the type $|\gamma - p/q| < O(1/q^3)$.

We close this brief excursion into the world of continued fractions by stating the main result of use to us later. Let

$$\langle\gamma\rangle \overset{\text{def}}{=} \min\left\{\gamma - \lfloor\gamma\rfloor, \lceil\gamma\rceil - \gamma\right\}$$

denote the distance from γ to the nearest integer. We now show that if γ is a "nice" irrational number, then $n\gamma$ stays reasonably far from any integer.

Lemma 2.6 *Let γ be a positive irrational whose continued fraction expansion has bounded partial quotients. Then, $\langle n\gamma\rangle > c/n$, for some constant $c = c(\gamma) > 0$.*

Proof: We can assume that n is large enough. Since the q_i's are monotonically increasing, $q_k \leq n < q_{k+1}$, for some $k > 0$. If b is the nearest integer to $n\gamma$, then by (2.3) we have $\langle n\gamma\rangle = |n\gamma - b| \geq |q_k\gamma - p_k|$, and hence, by (2.2),[4]

$$\langle n\gamma\rangle \geq \frac{q_k}{2q_{k+1}^2} \geq \frac{q_k}{2(a_{k+1}q_k + q_{k-1})^2} \geq \frac{1}{2q_k(a_{k+1} + 1)^2} \gg \frac{a_{k+1}^{-2}}{n}.$$

The lemma follows from the fact that, for any k, the partial quotient a_{k+1} is bounded above by a constant. $\square$

Note that if we are free to choose γ in the application of the lemma, the golden ratio $\frac{1}{2}(\sqrt{5}+1)$ is as good a choice as any, since its partial quotients are all 1. Also, by the pigeonhole principle, it is obvious that the lemma is the best possible asymptotically.

---

[4] Recall that the notation $x \ll y$ means $x = O(y)$, while $x \gg y$ denotes $x = \Omega(y)$.

Irrational Lattices

Let $\gamma > 0$ be an irrational, chosen once and for all, whose continued fraction expansion has bounded partial quotients. We construct a $(2n-1)$-point set

$$P = \left\{ \left(\{j\gamma\}, \frac{|j|}{n} \right) : |j| < n \right\}.$$

For technical convenience, we duplicate the point of P at the origin, which brings the number of points to $2n$. Discrepancy with respect to this multiset differs from the original P by at most one, which is negligible. Fix some integer $0 < r < n$. We are interested in the discrepancy $D(B_x)$ of the semiopen box

$$B_x = [0, x) \times [0, r/n),$$

where $x > 0$, ie,

$$D(B_x) = 2n \cdot \text{area}\,(B_x) - |P \cap B_x|.$$

We could easily show that $|D(B_x)| = O(\log n)$, but our purpose here is to prove something much stronger, ie,

$$\int_0^1 D(B_x)^2 \, dx \ll \log r.$$

We have

$$D(B_x) = 2rx - \sum_{j=0}^{r-1} \Big(\chi_x(\{j\gamma\}) + \chi_x(\{-j\gamma\}) \Big),$$

where $\chi_x(z)$ denotes the characteristic function of $[0, x)$, ie,

$$\chi_x(z) = \begin{cases} 1 & \text{if } z < x, \\ 0 & \text{otherwise.} \end{cases}$$

We easily verify that for all $0 \le x, z \le 1$,

$$\chi_x(z) = x + \{z - x\} - \{z\},$$

and therefore

$$D(B_x) = \sum_{j=0}^{r-1} \Big(\{j\gamma\} - \{j\gamma - x\} + \{-j\gamma\} - \{-j\gamma - x\} \Big).$$

By (B.1) in Appendix B.2, at nonintegral z,

$$\{z\} = \frac{1}{2} - \sum_{m \ne 0} \frac{e^{2\pi i m z}}{2\pi i m}.$$

Since the sum on the right-hand side is an odd function, we derive

$$D(B_x) = \sum_{m \neq 0} \frac{e^{-2\pi i m x} - e^{2\pi i m x}}{2\pi i m} \sum_{j=0}^{r-1} e^{2\pi i m j \gamma}.$$

To have a sine function appear in the first sum is the reason we included both positive and negative indices j. Indeed, we can easily compute the m-th Fourier coefficient of $D(B_x)$. By Parseval-Plancherel, it now easily follows that

$$\int_0^1 D(B_x)^2 \, dx \ll \sum_{m \neq 0} \frac{1}{m^2} \left| \sum_{j=0}^{r-1} e^{2\pi i m j \gamma} \right|^2. \tag{2.4}$$

We recognize a variant of the Dirichlet kernel,[5]

$$\left| \sum_{j=0}^{r-1} e^{2\pi i m j \gamma} \right| = \left| \frac{e^{2\pi i m r \gamma} - 1}{e^{2\pi i m \gamma} - 1} \right| = \left| \frac{\sin \pi m r \gamma}{\sin \pi m \gamma} \right| \leq \frac{1}{|\sin \pi \langle m \gamma \rangle|},$$

which thus remains bounded unless $\langle m \gamma \rangle$ becomes very small. In that case, a Taylor expansion around 0 shows that it is bounded by $1/\langle m \gamma \rangle$. It follows that

$$\left| \sum_{j=0}^{r-1} e^{2\pi i m j \gamma} \right| \ll \min \left\{ r, \frac{1}{\langle m \gamma \rangle} \right\},$$

and by (2.4),

$$\int_0^1 D(B_x)^2 \, dx \ll \sum_{m \neq 0} \frac{1}{m^2} \min \left\{ r^2, \frac{1}{\langle m \gamma \rangle^2} \right\},$$

which we can also write as

$$\int_0^1 D(B_x)^2 \, dx \ll \sum_{k=1}^{\infty} 2^{-2k} \sum_{2^{k-1} \leq m < 2^k} \min \left\{ r^2, \frac{1}{\langle m \gamma \rangle^2} \right\}.$$

Divide the interval $[0, 1]$ into subintervals, called *bins*, of width $b/2^k$, for some constant $b > 0$. If we choose b small enough, then no more than two values of $m \in [2^{k-1}, 2^k)$ can be such that each $\langle m \gamma \rangle$ falls into the same bin. Indeed, suppose that there are three of them. First, note that for these three values of m, the number $\{m \gamma\}$ can lie in only one of two bins. So, by the pigeonhole principle, at least two values of m have $\{m \gamma\}$ lie in the

---

[5] See Appendix B.2.

same bin. If m_1 and m_2 are these integers ($m_1 < m_2$), then

$$\left\langle (m_1 - m_2)\gamma \right\rangle \leq \left\{ (m_1 - m_2)\gamma \right\} \leq \frac{b}{2^k}.$$

But since $0 < m_1 - m_2 < m$, choosing b small enough contradicts Lemma 2.6, which proves our claim. It immediately follows that

$$\int_0^1 D(B_x)^2\, dx \ll \sum_{k=1}^{\infty} \sum_{p=1}^{\infty} \min\left\{ \left(\frac{r}{2^k}\right)^2, \frac{1}{p^2} \right\}.$$

Breaking up the sum around $k = \log r$, we find that the first part amounts to $O(\log r)$. To handle the second part, we break the inner sum around $p = 2^k/r$ and find a sum of the form $\sum_{k > \log r} r 2^{-k} = O(1)$. This shows that

$$\int_0^1 D(B_x)^2\, dx \ll \log r.$$

If we now allow r to be an arbitrary real between 0 and n, we can always round it to an integer and add only a constant additive error to the discrepancy bound above. This completes the proof of Theorem 2.4 (page 47). □

2.4 Jittered Sampling

We now give a simple probabilistic construction for keeping the volume discrepancy of boxes as low as possible. We allow the boxes to be rotated. All of the main ideas can be explained in two dimensions, so this is where we confine our discussion. Recall that, given a set P of n points, the discrepancy of a box B is

$$D(B) = n \cdot \text{area}\,(B) - |P \cap B|.$$

Theorem 2.7 *It is possible to place n points in the unit square $[0,1]^2$, so that any (rotated) box has discrepancy $O(n^{1/4}\sqrt{\log n})$ in absolute value.*

Given n points in $[0,1]^2$ and three regions $A \subseteq R \subseteq B \subseteq [0,1]^2$, trivially,

$$\begin{cases} |P \cap R| - n \cdot \text{area}\,(R) \leq |P \cap B| - n \cdot \text{area}\,(B) + n \cdot \text{area}\,(B \setminus A), \\ |P \cap R| - n \cdot \text{area}\,(R) \geq |P \cap A| - n \cdot \text{area}\,(A) - n \cdot \text{area}\,(B \setminus A); \end{cases}$$

hence,

$$|D(R)| \leq \max\left\{ |D(A)|, |D(B)| \right\} + n \cdot \text{area}\,(B \setminus A). \qquad (2.5)$$

Lemma 2.8 *There exists a collection Q of n^c convex quadrilaterals,[6] for some constant $c > 0$, such that the following holds: Given any rectangle $R \subseteq [0,1]^2$, there exist $A, B \in Q$ such that $A \subseteq R \subseteq B$ and area $(B \setminus A) = O(1/n)$.*

Proof: We can assume that n is large enough. Subdivide the square $[-1,2]^2$ into an $n^3 \times n^3$ square grid, and let Q consist of the empty set, together with the convex hulls of every 4-tuple of grid points. To see why this construction works, consider a rectangle $R \subseteq [0,1]^2$. If both of its sides are of length at least $1/n$, then place eight little squares of side length $1/n^2$ around its four corners as indicated in Figure 2.3. Note that the inner squares are mutually disjoint and that each of them contains at least one grid point. Pick one in each inner square and form their convex hull. Repeat the operation with the outer squares (Fig. 2.3).

- The two convex hulls belong to Q.
- One lies inside R while the other is enclosed in R.
- The difference between the two squares has an area $O(1/n^2)$.

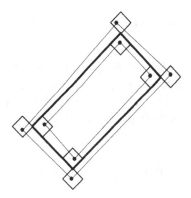

Fig. 2.3. Sandwiching a rectangle between two quadrilaterals.

The two convex hulls play the role of A and B in the lemma. Suppose now that one or both sides of R is shorter than $1/n$. Then, we place only the outer squares and set A to be the empty set. The area of $B \setminus A$ is $O(1/n)$, and the proof is complete. $\square$

---

[6]By abuse of terminology the set of quadrilaterals might include the empty set.

We are now ready to establish the desired upper bound. The first step is to subdivide $[0,1]^2$ into n equal-area rectangular cells, in such a way that no line intersects more than roughly $\sqrt{n}$ cells. If $n = k^2$, a simple square grid with $(1/n)$-area cells does the trick; otherwise, we apply the construction for $k = \lfloor \sqrt{n} \rfloor$. Finally, we scale down the square into an $x \times 1$ rectangle, where $x = k^2/n$, and we subdivide the $(1-x) \times 1$ leftover strip into $n - k^2$ equal-area rectangles. It is obvious that all n rectangles have area $1/n$, and that a line can cut only $O(k)$ of them.

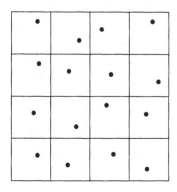

Fig. 2.4. Placing a point at random in each grid cell.

Next, we define the point set P by placing one point at random (uniformly and independently) in each cell. Figure 2.4 illustrates the case where n is a perfect square. Let $\mathcal{Q}$ be as in Lemma 2.8. Given $S \in \mathcal{Q}$, let $C_1, \ldots, C_m$ be the cells that intersect the boundary of S. Note that $m = O(\sqrt{n})$. Obviously,

$$D(S) = \sum_{i=1}^{m} D(S \cap C_i).$$

Introducing the random variable $y_i = |P \cap S \cap C_i|$, we have $D(S \cap C_i) = p_i - y_i$, where $p_i = \mathbf{E}y_i$. The probability that $D(S \cap C_i)$ is equal to p_i (resp. $p_i - 1$) is $1 - p_i$ (resp. p_i). By Lemma A.4, for any $\Delta > 0$,

$$\mathrm{Prob}[\,|D(S)| > \Delta\,] < 2e^{-2\Delta^2/m};$$

so, for any constant $c > 0$, there is a constant $b = b(c)$ such that the probability that S has *low* discrepancy, meaning here that $|D(S)| < bn^{1/4}\sqrt{\log n}$, is at least $1 - 1/2n^c$. Since $|\mathcal{Q}| \leq n^c$, with probability at least $1/2$ each quadrilateral of $\mathcal{Q}$ has low discrepancy.

Now, let R be an arbitrary box in $[0,1]^2$. By Lemma 2.8, it is sandwiched between two low-discrepancy quadrilaterals (we view the empty set as a quadrilateral of zero discrepancy) and the area in between is $O(1/n)$. It follows from (2.5) that $|D(R)| = O(n^{1/4}\sqrt{\log n})$, which proves Theorem 2.7. □

2.5 An Orbital Construction for Points on a Sphere

We examine the problem of placing points on S^2 (the unit-radius sphere in three dimensions) to minimize the volume discrepancy[7] of spherical caps. We define a spherical cap C as the intersection of S^2 with a closed halfspace. Given a set P of n points on S^2, the discrepancy of C is defined as

$$D_P(C) = \frac{n|C|}{4\pi} - |P \cap C|,$$

where $|C|$ denotes the area of C and the normalizing factor 4π is the area of S^2. The discrepancy for P is $D(P) = \sup_C |D_P(C)|$.

A probabilistic construction similar to *jittered sampling* (see §2.4) achieves a discrepancy within $O(n^{1/4}\sqrt{\log n})$, which is optimal within a polylogarithmic factor. The basic idea is to partition the sphere into pieces of area $4\pi/n$ and diameter $O(1/\sqrt{n})$, and throw one point in each piece randomly. The purpose of this section is to explore a completely different route to achieve low discrepancy. We place points on the sphere by iterating on a very straightforward process, somewhat in the spirit of the construction of §2.3 based on arithmetic progressions modulo 1. It is a rather fundamental technique, which we will encounter again in Chapter 9 in our discussion of explicit constructions of expanders. The method is simple but not optimal in all applications. One reason for discussing it in some detail is that it is a vehicle for exploring some of the most powerful machinery used in discrepancy theory (and, in fact, in all of mathematics). Our discussion covers a wide front. As much as we could, we have tried to make the presentation self-contained by providing sufficient background material.

Let R_1, R_2, R_3 be rotations of angle $\arccos(-3/5)$ around the x-, y-, z-axes, respectively. Form all nontrivial reduced words of length at most s over the alphabet,

$$\{\, R_1, R_1^{-1}, R_2, R_2^{-1}, R_3, R_3^{-1} \,\}.$$

---

[7]In this case, "spherical-area discrepancy" would be more descriptive.

(A word is said to be reduced if it does not contain any consecutive pair of the form XX^{-1}.) Given a starting point p_0 (not on any of the axes x, y, z), we form the set $P_0 = \{p_1, \ldots, p_n\}$, which consists of all points of the form wp_0, where w is such a word. Straightforward counting shows that P_0 consists of $n = \frac{3}{2}(5^s - 1)$ points.[8]

A note of timely relevance. As we discuss below, there is a close relation between the rotations of the sphere and the 2-by-2 unitary matrices with unit determinant. This implies a connection between this material and the generation of universal bases, a problem which arises in quantum computing.

Theorem 2.9 *The maximum discrepancy $D(P_0)$ for spherical caps of the set P_0 is $O(n \log n)^{2/3}$.*

Even though this upper bound is far from optimal, the method behind it is interesting because it is constructive and completely straightforward. Its analysis is not, however, but it uses several techniques that fit beautifully into the theory of pseudorandomness and the discrepancy method. The construction is almost optimal for quadrature (ie, within a logarithmic factor). Here is what we mean by that: Given any real-valued function $f \in L^2(S^2)$, we define its *operator discrepancy* δ by[9]

$$\delta f(p_0) \stackrel{\text{def}}{=} \frac{1}{n} \sum_{i=1}^{n} f(p_i) - \frac{1}{4\pi} \int_{S^2} f(p) \, d\omega(p),$$

where ω is the Lebesgue measure on S^2.

Theorem 2.10 *The operator discrepancy of any function $f \in L^2(S^2)$ satisfies*

$$\frac{\|\delta f\|_2}{\|f\|_2} \ll \frac{\log n}{\sqrt{n}}.$$

This upper bound is optimal within a factor of $\log n$. Both proofs rely

---

[8]This is trivially an upper bound. It is also the exact bound because the subgroup generated is free, ie, it has no nontrivial relations among its generators, but there is no need to be concerned about this here.

[9]It is useful to clarify a point of notation: p_0 might look like a point fixed once and for all, but it is really a variable. The subscript 0 is only there to distinguish the point from its "descendants," $p_1, \ldots, p_n$. In the definition of $\delta f(p_0)$, the point p_0 is the variable and $p_1, \ldots, p_n$ depend on p_0.

on a detailed analysis of the spectrum of the operator

$$Tf(p) \overset{\text{def}}{=} \sum_{i=1}^{3} \Big(f(R_i p) + f(R_i^{-1} p) \Big)$$

and, more specifically, on the fact that the second largest eigenvalue of T in absolute value is at most $2\sqrt{5}$. We will show how this fact can be derived from the classical Ramanujan Conjectures proven in the mid-seventies by Deligne. Short of proving the conjectures themselves, we will present the complete proofs of both theorems in a manner that is mostly self-contained. We begin our discussion with an introduction to quaternions and spherical harmonics.

Quaternions and SO(3)

An attractive feature of the set P_0 is that it can be easily computed (using real arithmetic). Linear fractional transformations in the complex plane are maps

$$z \in \mathbf{C} \;\mapsto\; \frac{az + b}{cz + d},$$

where $a, b, c, d \in \mathbf{C}$ and $ad - bc = 1$. If, in addition, the matrix

$$\begin{pmatrix} a & b \\ c & d \end{pmatrix}$$

is unitary, then, by stereographic projection, the linear fractional transformation corresponds to a rotation of the sphere, ie, a member of the special orthogonal group SO(3). In particular, this implies that the points of P_0 can be computed by means of the three unitary transformations:

$$M_1 = \frac{1}{\sqrt{5}} \begin{pmatrix} 1 + 2i & 0 \\ 0 & 1 - 2i \end{pmatrix}, \qquad M_2 = \frac{1}{\sqrt{5}} \begin{pmatrix} 1 & 2 \\ -2 & 1 \end{pmatrix},$$

and

$$M_3 = \frac{1}{\sqrt{5}} \begin{pmatrix} 1 & 2i \\ 2i & 1 \end{pmatrix}.$$

Let us give some background and more details for the sake of the reader who is not so familiar with these concepts. Quaternions are the set H of "numbers" of the form

$$\alpha = a_0 + a_1 \mathbf{i} + a_2 \mathbf{j} + a_3 \mathbf{k},$$

where each $a_i \in \mathbf{R}$, and $\mathbf{i}^2 = \mathbf{j}^2 = \mathbf{k}^2 = -1$, $\mathbf{ij} = -\mathbf{ji} = \mathbf{k}$, $\mathbf{jk} = -\mathbf{kj} = \mathbf{i}$, and $\mathbf{ki} = -\mathbf{ik} = \mathbf{j}$. The norm $N(\alpha)$ is the sum $a_0^2 + a_1^2 + a_2^2 + a_3^2$. Unit quaternions are those of norm 1. A nonzero quaternion has an inverse, which is $\bar{\alpha}/N(\alpha)$, where $\bar{\alpha}$ is the conjugate $a_0 - a_1\mathbf{i} - a_2\mathbf{j} - a_3\mathbf{k}$. The set H forms a division ring for addition and multiplication. It is not a field (in the regular sense of the term) because it is not commutative.

There is a natural correspondence between unit quaternions and rotations of the sphere S^2. The quaternion α corresponds to the rotation of angle $2 \arccos a_0$ around the (oriented) axis (a_1, a_2, a_3). If $a_1 = a_2 = a_3 = 0$, then the rotation becomes the identity. Note that unfortunately the correspondence is not bijective, since α and $-\alpha$ correspond to the same rotation. This is a minor inconvenience, however, especially in view of the attractive feature that multiplying two unit quaternions is the same as composing the corresponding rotations. Nonunit quaternions also correspond to rotations: divide the quaternion by the square root of its norm. To summarize, the multiplicative group of nonzero quaternions maps homomorphically to $SO(3)$. If α is a unit quaternion, it is convenient to represent it as the matrix $M(\alpha)$ with complex entries,

$$M(\alpha) = \begin{pmatrix} a_0 + a_1 i & a_2 + a_3 i \\ -a_2 + a_3 i & a_0 - a_1 i \end{pmatrix}.$$

We easily check that this correspondence commutes with multiplication, ie, $M(\alpha\beta) = M(\alpha)M(\beta)$. Any matrix in the *special unitary group* $SU(2)$, ie, any 2-by-2 unitary matrix with determinant 1, can be represented as $M(\alpha)$, so the multiplicative subgroup of unit quaternions is isomorphic to $SU(2)$. It follows from our discussion that multiplying two elements of $SU(2)$ can be done by multiplying the corresponding quaternions. Equivalently, this can be done by composing their associated linear fractional transformations. Indeed, we easily verify that the map from the matrix of $SL(2, \mathbf{C})$ (ie, complex matrix with determinant 1),

$$\begin{pmatrix} a & b \\ c & d \end{pmatrix},$$

to the linear fractional transformation,

$$z \in \mathbf{C} \mapsto \frac{az + b}{cz + d},$$

where $ad - bc = 1$, is an isomorphism. (Actually, switching to $-a$, $-b$, $-c$, $-d$ gives the same transformation, so the isomorphism is only with the subgroup $PSL(2, \mathbf{C}) = SL(2, \mathbf{C})/\{\pm I\}$.) To summarize, we have exhibited a correspondence between the rotations of the sphere and (a subset of the) linear fractional transformations. For the time being, what we should retain from this discussion is that SO(3) is isomorphic to SU(2)/{±I}.

Geometrically, transforming a point of S^2 by rotating S^2 is equivalent to moving its stereographic projection (centered at the north pole and aimed toward the tangent plane passing through the south pole) to its image under the corresponding linear fractional transformation.

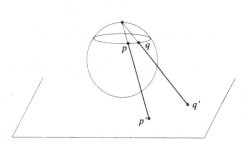

Spherical Harmonics and the Laplacian

Let T be a linear operator from $L^2(S^2)$ to itself.[10] Think of T as the infinite-dimensional analogue of a linear transformation in Euclidean space. The concept of a symmetric (or Hermitian) matrix has its counterpart in the world of operators. We say that T is *self-adjoint* if

$$\langle Tf, g \rangle = \langle f, Tg \rangle,$$

where

$$\langle f, g \rangle = \int_{S^2} \overline{f(p)} g(p) \, d\omega(p),$$

where, we recall, ω is the Lebesgue measure on the sphere. (Conjugacy is needed for only complex-valued functions, of course.) An important property for us is that, because it is self-adjoint, T can be diagonalized and all of its eigenvalues are real. Also, the eigenvectors corresponding to distinct eigenvalues are orthogonal. (Note that this is the precise counterpart to real-symmetric or Hermitian matrices.)

In spherical coordinates, a point is represented by

$$(\sin\theta\cos\varphi, \sin\theta\sin\varphi, \cos\theta),$$

where $\theta \in [0, \pi]$ is the angle between the point and the north pole (latitude)

---

[10]This means that T maps a complex-valued function that is square-integrable on the unit sphere to another function with the same property. Furthermore, T obeys the usual linearity conditions: $T\lambda f = \lambda T f$ and $T(f + g) = Tf + Tg$.

and $\varphi \in [0, 2\pi)$ is the polar angle of the projection of the point on the equator plane (longitude). From the simple form of the Laplacian in two-dimensional Cartesian coordinates,

$$\frac{\partial^2 f}{\partial x^2} + \frac{\partial^2 f}{\partial y^2},$$

we must now deal with the forbidding-looking expression

$$\Delta f = \frac{1}{\sin^2 \theta} \frac{\partial^2 f}{\partial \varphi^2} + \frac{1}{\sin \theta} \frac{\partial}{\partial \theta} \left(\sin \theta \frac{\partial f}{\partial \theta} \right).$$

Simple integration by parts shows that the Laplacian is self-adjoint. In the plane, the standard harmonic $e^{i(k_1 x + k_2 y)}$ is an eigenfunction of the Laplacian: Its eigenvalue is $-(k_1^2 + k_2^2)$. Of course, things are not nearly as simple on the sphere. To examine the spectrum of Δ, we begin with a particularly nice family of *surface spherical harmonics*, as the eigenfunctions of the Laplacian on the sphere are called. These are the eigenfunctions of Δ that are independent of the longitude φ; they are called the *zonal harmonics* and are defined by means of the *Legendre polynomial* (of the first kind) of degree n:

$$P_n(x) = \frac{1}{2^n n!} \frac{d^n}{dx^n} (x^2 - 1)^n.$$

It is immediate from the definition that the Legendre polynomials obey the recurrence relation:

$$(2n + 1)P_n(x) = \frac{d}{dx}P_{n+1}(x) - \frac{d}{dx}P_{n-1}(x). \tag{2.6}$$

Historically, these polynomials were obtained by Legendre while expanding the potential function between two points. Specifically, given two points p and q in the complex plane forming an angle θ between them, consider the potential

$$\frac{1}{|p - q|} = \frac{1}{\sqrt{|p|^2 + |q|^2 - 2|p||q| \cos \theta}}$$

and expand it as a power series in $|p/q|$,

$$\frac{1}{|q|} \sum_{n=0}^{\infty} \left| \frac{p}{q} \right|^n P_n(\cos \theta).$$

The coefficient of $|p/q|^n$ is precisely the Legendre polynomial of degree n

evaluated at $\cos\theta$. We have $P_n(1) = 1$ and, in general,

$$P_n(x) = \sum_{j=0}^{\lfloor n/2 \rfloor} (-1)^j \frac{(2n-2j)!}{2^n j!(n-j)!(n-2j)!} x^{n-2j}. \tag{2.7}$$

The orthogonality of the family is expressed by

$$\int_{-1}^1 P_l(x)P_m(x)\,dx = \begin{cases} \frac{2}{2l+1} & \text{if } l = m, \\ 0 & \text{else.} \end{cases}$$

It is not difficult to verify that the function

$$Y_n^0(\theta, \varphi) \overset{\text{def}}{=} \sqrt{\frac{2n+1}{4\pi}}\, P_n(\cos\theta)$$

is a surface spherical harmonic associated with the eigenvalue $n(n+1)$. It is independent of the longitude φ and, for this reason, is often called a *zonal harmonic*. The factor $\sqrt{(2n+1)/4\pi}$ is needed to ensure that the function has L^2 norm 1. The space H_n of surface spherical harmonics associated with the eigenvalue $n(n+1)$ has dimension $2n+1$: We have the zonal harmonic, which makes one, so we need another $2n$ eigenfunctions to be able to form a basis for H_n. In particular, it is intuitively clear that we need harmonics that are not axially symmetric around the polar axis. For this we define the *associated Legendre function* (of degree n):

$$P_n^k(x) = (1-x^2)^{k/2} \frac{d^k}{dx^k} P_n(x),$$

for $0 \leq k \leq n$. Note that $P_n = P_n^0$. We have the slightly more complicated orthogonality relation:

$$\int_{-1}^1 P_l^k(x)P_m^k(x)\,dx = \begin{cases} \frac{2}{2l+1}\frac{(l+k)!}{(l-k)!} & \text{if } l = m, \\ 0 & \text{else.} \end{cases} \tag{2.8}$$

We are now ready to define a complete basis of eigenfunctions for H_n. In addition to $Y_n^0(\theta, \varphi)$, for any $1 \leq k \leq n$, we have

$$\begin{cases} Y_n^k(\theta, \varphi) &= N_{n,k}P_n^k(\cos\theta)\cos(k\varphi), \\ Y_n^{-k}(\theta, \varphi) &= N_{n,k}P_n^k(\cos\theta)\sin(k\varphi). \end{cases} \tag{2.9}$$

To ensure that the basis is orthonormal, we must adjust the normalizing factors $N_{n,k}$ in accordance with (2.8): For $k \geq 0$, we have $N_{n,k} = N_{n,-k}$ and

$$N_{n,k} = \sqrt{\frac{(2n+1)(n-k)!}{4\pi(n+k)!}}.$$

Notice that

$$Y_n^k(0, \varphi) = \begin{cases} N_{n,0} & \text{if } k = 0, \\ 0 & \text{else,} \end{cases} \qquad (2.10)$$

and

$$\begin{aligned} \int_{S^2} Y_i^k Y_j^l &= \int_0^\pi \int_0^{2\pi} \overline{Y_i^k(\theta, \varphi)} \, Y_j^l(\theta, \varphi) \sin\theta \, d\varphi \, d\theta \\ &= \begin{cases} 1 & \text{if } i = j \text{ and } k = l, \\ 0 & \text{else.} \end{cases} \end{aligned}$$

For $k = 0$, the function Y_n^k is the zonal harmonic, which is symmetric around the polar axis and vanishes on n latitude circles. At the opposite end, the two functions $Y_n^{\pm n}(\theta, \varphi)$ are, up to a constant factor, equal to $\cos(n\varphi)$ or $\sin(n\varphi)$: They are called the *sectorial harmonics*. The other functions $Y_n^k(\theta, \varphi)$ are the *tesseral harmonics*: They vanish along $n - |k|$ latitude circles and $|k|$ longitude circles, and form a tessellated pattern. As a whole the functions Y_n^k are called the *ultraspherical harmonics*.

By Weierstrass' theorem, continuous functions in $L^2(S^2)$ can be uniformly approximated by polynomials of the form $\sum a_{k,l,m} x^k y^l z^m$, with $x^2 + y^2 + z^2 = 1$. It is not very difficult to derive from this fact that not only does $\{ Y_n^k : |k| \leq n \}$ form an orthonormal basis for H_n, but the infinite family $\{ Y_n^k : n \geq 0, |k| \leq n \}$ is a basis for all $L^2(S^2)$:

$$L^2(S^2) = \bigoplus_{n=0}^{\infty} \{ Y_n^k : |k| \leq n \}.$$

Note that closure is implied and so equality is in the L^2 sense, meaning that two functions are said to be equal if and only if the L^2 norm of their difference is zero. Thus, the two functions can differ at some places, in particular at discontinuities.

Spectrum of Self-Adjoint Operators

Suppose that T is a self-adjoint operator that commutes with the Laplacian, meaning that the two operators $T\Delta$ and ΔT are equal. Then, T leaves each H_n invariant. To see why is quite easy. By definition, any $f \in H_n$ satisfies $\Delta f = n(n+1)f$; therefore, $(T\Delta)f = n(n+1)Tf$. Since T and Δ commute, Tf is thus an eigenfunction of Δ with eigenvalue $n(n+1)$, and thus belongs to H_n. We can use the same line of reasoning to derive a more general result. This is classical stuff and math-inclined readers can skip the following proof, which we include only for completeness.

Lemma 2.11 *If K and T are two self-adjoint operators that commute with Δ and with each other, then there exists an orthonormal basis for H_n that diagonalizes both K and T.*

Proof: Because K and T are both self-adjoint, they can be diagonalized and their eigenvalues are reals. But why should they have a common eigenbasis? First, observe that if the eigenvalues of, say, T, were all distinct, then any eigenbasis for T would also be an eigenbasis of K. Indeed, if ϕ were an eigenfunction for T with eigenvalue λ, then, by the previous argument,

$$(KT)\phi = \lambda K\phi = TK\phi,$$

so $K\varphi$ would be an eigenfunction for T with eigenvalue λ. By the distinctness of eigenvalues, therefore, $K\phi$ would be equal to a multiple of ϕ. So, every eigenfunction of T would also be an eigenfunction of K, and our claim would follow readily.

Suppose now that the eigenvalues are not distinct. Then, given an eigenvalue λ of T, let $\mathcal{T}_\lambda$ be the maximal subspace of H_n consisting of eigenfunctions of T with eigenvalue λ. By the previous argument (restricted to H_n, which is invariant under both T and K), K leaves $\mathcal{T}_\lambda$ invariant, so we can define an eigenbasis B_λ for K within $\mathcal{T}_\lambda$. But, within $\mathcal{T}_\lambda$, T is the identity (up to a scaling factor); therefore, B_λ is also an eigenbasis for T. The union of all $\{B_\lambda\}$ is therefore a common eigenbasis for T and K that spans H_n. □

The proofs of Theorems 2.9 and 2.10 (page 58) rely on a careful examination of the spectrum of the operator T defined by

$$Tf(p) = \sum_{i=1}^{3}\Big(f(R_i p) + f(R_i^{-1}p)\Big);$$

T is called a *Hecke operator*. (Why Hecke felt the need to invent such things is explained in §2.6.) It is obviously self-adjoint, and constant functions are eigenfunctions with eigenvalue 6. A nontrivial result says that, besides 6, all of the other eigenvalues of T, which are all real, lie in a small interval.

Lemma 2.12 *Let $d = 6$ be the number of rotations involved in the definition of T. Besides d, all of the other eigenvalues of T are at most $2\sqrt{d-1}$ in absolute value.*

We use the parameter d because the lemma applies to an infinite number of values of d, and not just $d = 6$. The upper bound of $2\sqrt{d-1}$ is known as the *Ramanujan bound*, as its proof is directly connected to the Ramanujan Conjectures.

Harmonic Analysis on a Tree

Iterating the operator T can be best understood by considering an infinite undirected tree $\mathcal{T}$ of degree $d = 6$. Think of p_0 as its root; the neighbors of p_0 are the points

$$R_1 p_0, R_1^{-1} p_0, R_2 p_0, R_2^{-1} p_0, R_3 p_0, R_3^{-1} p_0.$$

From each of these d new points, we obtain $d - 1$ more neighbors by repeating the construction. Iterating on this process produces the infinite tree $\mathcal{T}$. Let A be its (infinite) adjacency matrix.[11] Our first observation is that this is the matrix of the Hecke operator T restricted to the nodes of $\mathcal{T}$. Indeed, the action of T on a function f is to "replace" the value of f at any node of the tree by the sum of the value at its neighbors. In other words, if $p_0, p_1, \ldots$, are the nodes of the tree, the operator T has the effect of mapping the vector $u = [f(p_0), f(p_1), \ldots]^T$ to Au, which is to say,[12]

$$Au = [Tf(p_0), Tf(p_1), \ldots]^T.$$

Note that even though the p_i's depend on the particular placement of p_0 (as does the whole tree $\mathcal{T}$), the value of $Tf(p_i)$ depends only on f and p_i.

Fix some integer $s \geq 0$; let g_i be the sum $\sum f(p_j)$ over all nodes p_j at a distance from p_i (in the tree) positive but at most equal to s, and let A_s denote the operator (on $\mathcal{T}$) that maps u to the vector $[g_0, g_1, \ldots]^T$. Because understanding A_s is the key to proving the bound on the operator discrepancy of Theorem 2.10 (page 58), we investigate its spectrum. Unfortunately, this poses some slight technical difficulties, which we can overcome by defining the intermediate operator B_s. This operator is defined on $\mathcal{T}$ just like A_s: The only difference is that for a given p_i the sum is taken *only* over those nodes p_j whose distance to p_i is at most s and has the same parity as s. The advantage of B_s over A_s is that it lends itself naturally to a recurrence relation. Furthermore, we have the obvious relations: $B_0 = I$, $B_1 = A$ and, for $s > 0$,

$$A_s = B_s + B_{s-1} - I. \tag{2.11}$$

---

[11] The entry (i, j) is 1 or 0, depending on whether nodes i and j are adjacent in the tree.

[12] The reader should be careful not to confuse the Hecke operator with the transpose symbol.

To obtain B_{s+1}, make p_i the root of $\mathcal{T}$, and rank the levels of the tree in ascending order, beginning at level 0 (the root). For concreteness, suppose without loss of generality that s is even. Apply A to u first. All of the even-ranked nodes acquire the sum of the values at their neighbors, all of which are odd-ranked. This effectively "pushes" the values of odd-ranked nodes into the even-ranked nodes. So, it would seem that summing up all of the new values at the nodes of even rank $\leq s$, ie, computing $B_s A$, would give us B_{s+1}. The only problem is that every node of odd rank less than s is actually pushed d times, which is $d-1$ times too many. Note that this does not affect the nodes on level $s+1$. Indeed, with respect to the nodes of interest to us, ie, those of rank $\leq s+1$, the nodes of rank $s+1$ are pushed only once. Thus, we now must subtract $(d-1)B_{s-1}$ to derive the proper value of B_{s+1}, ie,

$$B_{s+1} = B_s A - (d-1)B_{s-1}.$$

To solve this equation, let us "pretend" for the time being that A and B_s are real numbers and that

$$|A| \leq 2\sqrt{d-1}. \qquad (2.12)$$

We will justify all of this later. Meanwhile, we solve the equation by looking for solutions of the form $B_s = x^s$, which yields the quadratic equation

$$x^2 - Ax + d - 1 = 0,$$

whose solutions are

$$\frac{A}{2} \pm \frac{i}{2}\sqrt{4(d-1) - A^2}.$$

These two numbers are conjugate with modulus $\sqrt{d-1}$, so we can write them as $\sqrt{d-1}\,e^{\pm i\alpha}$. By equating the real parts of both expressions, we find

$$A = 2\sqrt{d-1}\cos\alpha. \qquad (2.13)$$

All solutions of the equation are of the form

$$B_s = C_1(d-1)^{s/2}e^{i\alpha s} + C_2(d-1)^{s/2}e^{-i\alpha s}.$$

Plugging in the values of $B_0 = 1$ and $B_1 = A$, we find that $C_1 + C_2 = 1$ and

$$C_1\sqrt{d-1}\,e^{i\alpha} + C_2\sqrt{d-1}\,e^{-i\alpha} = A.$$

By (2.13), this implies that $C_1 = e^{i\alpha}/(2i\sin\alpha)$ and $C_2 = -e^{-i\alpha}/(2i\sin\alpha)$, and hence

$$B_s = (d-1)^{s/2} \frac{\sin(s+1)\alpha}{\sin \alpha}. \qquad (2.14)$$

From (2.11) we now conclude that

$$A_s = (d-1)^{s/2} \left(\frac{\sin(s+1)\alpha}{\sin \alpha} + \frac{\sin s\alpha}{\sqrt{d-1}\sin \alpha} \right) - 1. \qquad (2.15)$$

Note that the identity I in (2.11) becomes 1 in the relation above. We now must justify our assumption that A and A_s are real numbers. Informally, the answer is this: If we restrict ourselves to eigenfunctions of T, then the operator acts as a multiplicative scalar, and, in some sense, A can be thought of as a real number. We flesh out this intuition in the following.

Operator Discrepancy

It is clear that T commutes with the Laplacian Δ. So, by the results of the previous section, there exists an orthonormal basis for H_m, denoted by $\{ \Phi_{m,j} : |j| \leq m \}$, that diagonalizes T. Let $\rho_{m,j}$ be the eigenvalues of T associated with $\Phi_{m,j}$. Note that $\Phi_{0,0}$ is the constant function equal to $1/\sqrt{4\pi}$ everywhere (so that its L^2 norm is 1) and H_0 is the space of constant functions. Because $L^2(S^2) = \bigoplus_{m=0}^{\infty} H_m$, it follows that

$$L^2(S^2) = \bigoplus_{m=0}^{\infty} \{ \Phi_{m,j} : |j| \leq m \}, \qquad (2.16)$$

which means that any function $f \in L^2(S^2)$ can be expanded over the basis $\{\Phi_{m,j}\}$:

$$f(p) = \sum_{m=0}^{\infty} \sum_{|j|\leq m} \widehat{f}(m,j)\Phi_{m,j}(p). \qquad (2.17)$$

To find the Fourier coefficients $\widehat{f}(m,j)$, we form the inner product with the function $\Phi_{m,j}$. (This is the same as projecting a vector along an axis to find its coordinate.) By orthonormality of the basis,

$$\widehat{f}(m,j) = \int_{S^2} f(q)\Phi_{m,j}(q)\,d\omega(q). \qquad (2.18)$$

Now, recall the meaning of A_s: Given a vector u whose coordinates are the values at the nodes of $\mathcal{T}$, the first coordinate of $A_s u$ is equal to the sum of the values at the nodes (distinct from p_0) lying within distance s of p_0. Let us now set s to the earlier value given by $n = \frac{3}{2}(5^s - 1)$. Then, the set of nodes corresponds precisely to $P_0 = \{p_1, \ldots, p_n\}$.

Suppose that $m > 0$, ie, $\Phi_{m,j}$ is nonconstant. By Lemma 2.12 (page 65), $|\rho_{m,j}| \leq 2\sqrt{d-1}$; therefore $\rho_{m,j}$ can be written as $2\sqrt{d-1}\cos\alpha$, for some angle α. Since T acts as a multiplicative scalar with respect to $\Phi_{m,j}$, if we form the vector $u = [\Phi_{m,j}(p_0), \Phi_{m,j}(p_1), \ldots]^T$, then by definition $Au = \rho_{m,j}u$. In other words, A acts like a real number that, furthermore, satisfies the bound (2.13). So, as long as we stick to $\Phi_{m,j}$, we can then apply (2.15). The relation is independent of p_0, so it appears that $\Phi_{m,j}$ is an eigenfunction for the operator[13]

$$T^s : f(p_0) \mapsto \frac{1}{n}\sum_{i=1}^{n} f(p_i).$$

Its associated eigenvalue $\rho_{m,j}^{(s)}$ satisfies

$$n\rho_{m,j}^{(s)} = (d-1)^{s/2}\left(\frac{\sin(s+1)\alpha}{\sin\alpha} + \frac{\sin s\alpha}{\sqrt{d-1}\sin\alpha}\right) - 1.$$

Since $|\sin n\alpha/\sin\alpha| = O(n)$, we find that $|\rho_{m,j}^{(s)}| \ll s(d-1)^{s/2}/n$, for $m > 0$. Using the fact that $d = 6$ and $n = \frac{3}{2}(5^s - 1)$, it follows that, for $m > 0$,

$$|\rho_{m,j}^{(s)}| \ll \frac{\log n}{\sqrt{n}}. \tag{2.19}$$

Given $f \in L^2(S^2)$, we know from (2.17) that

$$f(p_0) = \sum_{m=0}^{\infty}\sum_{|j|\leq m} \widehat{f}(m,j)\Phi_{m,j}(p_0).$$

It follows from (2.18) that the first term in the sum (which intuitively is the constant-function "component" of f) satisfies

$$\widehat{f}(0,0)\Phi_{0,0}(p_0) = \frac{1}{4\pi}\int_{S^2} f(p)\,d\omega(p).$$

Therefore, going back to the Fourier series expansion, we find that

$$f(p_0) - \frac{1}{4\pi}\int_{S^2} f(p)\,d\omega(p) = \sum_{m=1}^{\infty}\sum_{|j|\leq m} \widehat{f}(m,j)\Phi_{m,j}(p_0).$$

Applying the operator T^s to the function f, we find that at p_0

---

[13]We divide by n to make it an averaging operator.

$$\delta f(p_0) = T^s f(p_0) - \frac{1}{4\pi} \int_{S^2} f(p) \, d\omega(p)$$

$$= \sum_{m=1}^{\infty} \sum_{|j| \leq m} \widehat{f}(m, j) \rho_{m,j}^{(s)} \Phi_{m,j}(p_0).$$

Applying Parseval-Plancherel, ie, taking the L^2 norm, it follows from (2.19) that

$$\|\delta f\|_2^2 = \sum_{m=1}^{\infty} \sum_{|j| \leq m} |\widehat{f}(m, j)|^2 \, |\rho_{m,j}^{(s)}|^2$$

$$\ll \frac{(\log n)^2}{n} \sum_{m=1}^{\infty} \sum_{|j| \leq m} |\widehat{f}(m, j)|^2$$

$$\ll \frac{(\log n)^2}{n} \|f\|_2^2,$$

which establishes Theorem 2.10 (page 58). $\square$

Spherical Cap Discrepancy

We are now equipped with all of the tools needed to prove Theorem 2.9 (page 58). Again, we find that the operators of interest to us commute with the Laplacian, so we can decompose any function of $L^2(S^2)$ into common eigenfunctions. The advantage of doing this is that we have a complete understanding of the eigenspaces of the Laplacian. Commuting with the Laplacian is thus the key to being able to exploit harmonic analysis in S^2.

Given two points $p, q \in S^2$, the spherical distance from p to q (measured along the great circle passing through the two points) is denoted by $d(p, q)$. It is equal to the angle $\angle(Op, Oq)$; see Figure 2.5. Fix some small positive parameter ε once and for all. We use the notation $C(p, \varepsilon)$ to denote the spherical cap centered at p of (spherical) radius ε. Let $k(p, q) : S^2 \mapsto \mathbf{R}$ be the function defined as follows:

$$k(p, q) = \begin{cases} \frac{1}{2\pi(1 - \cos \varepsilon)} & \text{if } d(p, q) < \varepsilon, \\ 0 & \text{else.} \end{cases}$$

The denominator $2\pi(1 - \cos \varepsilon)$ is equal to

$$\int_0^{\varepsilon} 2\pi \sin \theta \, d\theta,$$

which is the area of $C(p, \varepsilon)$. The function is used as the *kernel* of the

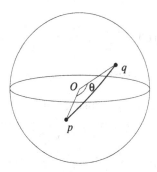

Fig. 2.5. The distance between two points.

integral operator

$$Kf(p) = \int k(p,q) f(q) \, d\omega(q).$$

Note that $k(p,q)$ is only a function of $p - q$, so the operator K acts as a convolution. Fix a spherical cap C of center c and radius ρ once and for all, and let χ_C denote its characteristic function. We are interested in the convolution $K\chi_C(p)$, which indicates what fraction of $C(p, \varepsilon)$ is occupied by the spherical cap C. Let $C_1 \subseteq C \subseteq C_2$ be two spherical caps sandwiching C, of radius $\max\{0, \rho - 2\varepsilon\}$ and $\rho + 2\varepsilon$, respectively, and centered at c. Theorem 2.9 follows directly from

Lemma 2.13 *For any* $i = 1, 2$,

$$\left| \sum_{j=1}^{n} K\chi_{C_i}(p_j) - \frac{n|C|}{4\pi} \right| \ll \varepsilon n + \sqrt{\frac{n}{\varepsilon}} \log n,$$

where, as we recall, $|C|$ denotes the area of C. Indeed, observe that $\chi_C(p) = 0$ implies that p is at least 2ε away from C_1. Thus, $C(p, \varepsilon)$ does not intersect C_1 and $K\chi_{C_1}(p) = 0$. Similarly, $\chi_C(p) = 1$ implies that p lies within C_2 at least 2ε away from the boundary, and therefore $C(p, \varepsilon) \subset C_2$ and $K\chi_{C_2}(p) = 1$. It follows that

$$K\chi_{C_1}(p) \leq \chi_C(p) \leq K\chi_{C_2}(p),$$

and therefore

$$\sum_{j=1}^{n} K\chi_{C_1}(p_j) \leq |P \cap C| \leq \sum_{j=1}^{n} K\chi_{C_2}(p_j).$$

We derive

$$\left| |P \cap C| - \frac{n|C|}{4\pi} \right| \ll \varepsilon n + \sqrt{\frac{n}{\varepsilon}} \log n.$$

Setting $\varepsilon = (\log n)^{2/3}/n^{1/3}$ proves Theorem 2.9. $\square$

Again we prove Lemma 2.13 by using the bound on the spectrum of T given in Lemma 2.12 (page 65). Recall that T is the Hecke operator

$$Tf(p) = \sum_{i=1}^{3} \Big(f(R_i p) + f(R_i^{-1} p) \Big).$$

Because T commutes with Δ, it leaves each H_m invariant; and so, as before, it suffices to study the restriction of T to each H_m. It is easy to see that the convolution operator K is self-adjoint and commutes with both Δ and T. Therefore, by Lemma 2.11, there exists an orthonormal basis for H_m, denoted by $\{\phi_{m,j} : |j| \leq m\}$, that diagonalizes both T and K. Let $\rho_{m,j}$ and $\lambda_{m,j}$ be the eigenvalues of T and K, respectively, associated with $\phi_{m,j}$. Recall that H_0 is the space of constant functions; therefore, $\phi_{0,0} = 1/\sqrt{4\pi}$. (Note that, in general, $\Phi_{m,j}$ and $\phi_{m,j}$ need not be the same functions.) We have

$$L^2(S^2) = \bigoplus_{m=0}^{\infty} \{\phi_{m,j} : |j| \leq m\}, \tag{2.20}$$

which means that any function $f \in L^2(S^2)$ can be expanded over the basis $\{\phi_{m,j}\}$:

$$f(q) = \sum_{m=0}^{\infty} \sum_{|j| \leq m} \widehat{f}(m,j) \phi_{m,j}(q).$$

As usual, the Fourier coefficients are given by

$$\widehat{f}(m,j) = \int_{S^2} f(q) \phi_{m,j}(q) \, d\omega(q). \tag{2.21}$$

Applying this to $f(q) = k(p,q)$ yields

$$k(p,q) = \sum_{m=0}^{\infty} \sum_{|j| \leq m} \widehat{k}_p(m,j) \phi_{m,j}(q),$$

where

$$\widehat{k}_p(m,j) = \int_{S^2} k(p,z) \phi_{m,j}(z) \, d\omega(z). \tag{2.22}$$

Lemma 2.14 *If D is the spherical cap $C(p,\varepsilon)$ centered at p of angular* *radius ε, then*

$$\int_D \phi_{m,j}(z)\, d\omega(z) = 2\pi \phi_{m,j}(p) \int_{\cos \varepsilon}^1 P_m(x)\, dx.$$

Proof: Let $\{Y_m^k\}$ be the family of ultraspherical harmonics defined with p as the north pole. Since $\phi_{m,j} \in H_m$, we can express the function in that basis as

$$\phi_{m,j}(z) = \sum_{|l| \leq m} \alpha_l Y_m^l(z).$$

Suppose that $l > 0$. From (2.9) it then follows that

$$\begin{aligned}
\int_D Y_m^l(z)\, d\omega(z) &= \int_D Y_m^l(\theta, \varphi) \sin \theta\, d\theta\, d\varphi \\
&= N_{m,l} \int_D P_m^l(\cos \theta) \cos(l\varphi) \sin \theta\, d\theta\, d\varphi.
\end{aligned}$$

Because of the rotational symmetry around the (new) polar axis, the expression above vanishes. The reasoning applied to the case $l < 0$ leads to the same conclusion. On the other hand, the case $l = 0$ gives

$$\begin{aligned}
\int_D Y_m^0(z)\, d\omega(z) &= N_{m,0} \int_D P_m(\cos \theta) \sin \theta\, d\theta\, d\varphi \\
&= -2\pi N_{m,0} \int_1^{\cos \varepsilon} P_m(x)\, dx,
\end{aligned}$$

so we derive

$$\int_D \phi_{m,j}(z)\, d\omega(z) = 2\pi \alpha_0 N_{m,0} \int_{\cos \varepsilon}^1 P_m(x)\, dx.$$

Because p is the north pole, $\theta = 0$, and so it follows from (2.10) that $\phi_{m,j}(p) = \alpha_0 N_{m,0}$. This proves the lemma. $\square$

The lemma indicates that the right-hand side of (2.22) can be expressed as

$$\frac{1}{2\pi(1 - \cos \varepsilon)} \int_{C(p,\varepsilon)} \phi_{m,j}(z)\, d\omega(z) = \frac{\phi_{m,j}(p)}{1 - \cos \varepsilon} \int_{\cos \varepsilon}^1 P_m(x)\, dx.$$

Because $\phi_{m,j}$ is an eigenfunction of the operator K, the right-hand side of (2.22) is also equal to $\lambda_{m,j}\, \phi_{m,j}(p)$, and therefore

$$\lambda_{m,j} = \lambda_m = \frac{1}{1 - \cos \varepsilon} \int_{\cos \varepsilon}^1 P_m(x)\, dx \qquad (2.23)$$

does not depend on j at all. Thus, (2.22) can be rewritten as

$$\widehat{k}_p(m,j) = \lambda_m \phi_{m,j}(p).$$

To summarize, the expansion of k becomes

$$k(p,q) = \sum_{m=0}^{\infty} \lambda_m \sum_{|j|\le m} \phi_{m,j}(p)\phi_{m,j}(q), \qquad (2.24)$$

and we have the important "eigenvalue" relation,

$$\int_{S^2} k(p,q)\phi_{m,j}(q)\,d\omega(q) = \lambda_m\,\phi_{m,j}(p). \qquad (2.25)$$

We are now in a position to prove Lemma 2.13. From (2.24) we know that

$$K\chi_C(p) = \int_{S^2} \sum_{m=0}^{\infty} \lambda_m \sum_{|j|\le m} \phi_{m,j}(p)\phi_{m,j}(q)\chi_C(q)\,d\omega(q),$$

which we can rewrite as

$$\sum_{m=0}^{\infty} \lambda_m \sum_{|j|\le m} \phi_{m,j}(p) \int_{S^2} k'(c,q)\phi_{m,j}(q)\,d\omega(q),$$

where c is the center of C and $k'(p,q)$ is a function identical to $k(p,q)$, except for constant factors and a different cap radius (ρ for k' and ε for k). Thus, the same analysis leads to a relation similar to (2.25): $\phi_{m,j}$ is an eigenfunction of the integral operator

$$K'f(p) \overset{\text{def}}{=} \int_{S^2} k'(p,q)f(q)\,d\omega(q),$$

whose corresponding eigenvalue $\lambda_m(C)$ does not depend on j. It follows that

$$\int_{S^2} k'(c,q)\phi_{m,j}(q)\,d\omega(q) = \lambda_m(C)\phi_{m,j}(c),$$

and hence

$$K\chi_C(p) = \sum_{m=0}^{\infty} \lambda_m(C)\lambda_m \sum_{|j|\le m} \phi_{m,j}(c)\phi_{m,j}(p).$$

The first term in the sum is

$$\lambda_0\,\phi_{0,0}(p)\,\lambda_0(C)\,\phi_{0,0}(c).$$

Recall from (2.25) that

$$\lambda_0 \, \phi_{0,0}(p) = \int_{S^2} k(p,q)\phi_{0,0}(q) \, d\omega(q) = \frac{1}{2\sqrt{\pi}} \int_{S^2} k(p,q) \, d\omega(q) = \frac{1}{2\sqrt{\pi}} \, .$$

Similarly,

$$\lambda_0(C) \, \phi_{0,0}(c) = \frac{|C|}{2\sqrt{\pi}} \, .$$

It then follows that

$$K\chi_C(p) - \frac{|C|}{4\pi} = \sum_{m=1}^{\infty} \lambda_m(C)\lambda_m \sum_{|j|\le m} \phi_{m,j}(c)\phi_{m,j}(p).$$

Of course, we have a similar identity if we replace C by C_1 or C_2. Since for any $i = 1, 2$, the area $|C_i|$ differs from $|C|$ by $O(\varepsilon)$, Lemma 2.13 will be established if we can prove the following.

Lemma 2.15 *For any $i = 1, 2$,*

$$\left| \sum_{m=1}^{\infty} \lambda_m(C_i)\lambda_m \sum_{|j|\le m} \phi_{m,j}(c) \sum_{l=1}^{n} \phi_{m,j}(p_l) \right| \ll \sqrt{\frac{n}{\varepsilon}} \log n,$$

where c is the common center of C, C_1, C_2.

Proof: Without loss of generality, we assume that $i = 1$. Let S denote the left-hand side of the inequality. We know from (2.19) that any eigenvalue $\rho_{m,j}^{(s)}$ $(m > 0)$ of the operator

$$T^s : f(p_0) \mapsto \frac{1}{n} \sum_{l=1}^{n} f(p_l)$$

satisfies $|\rho_{m,j}^{(s)}| \ll (\log n)/\sqrt{n}$. Since $\phi_{m,j}$ is an eigenfunction for T within H_m, it is also an eigenfunction for T^s. Therefore, by Cauchy-Schwarz,

$$S \ll \sqrt{n} \log n \sum_{m=1}^{\infty} |\lambda_m(C_i)\lambda_m| \sum_{|j|\le m} |\phi_{m,j}(c)\phi_{m,j}(p_0)|$$

$$\ll \sqrt{n} \log n \sum_{m=1}^{\infty} |\lambda_m(C_i)\lambda_m| \sqrt{\sum_{|j|\le m} |\phi_{m,j}(c)|^2} \sqrt{\sum_{|j|\le m} |\phi_{m,j}(p_0)|^2} \, .$$

It is not difficult to show that

$$\sum_{|j|\le m} |\phi_{j,m}(p)|^2 = \frac{2m+1}{4\pi},$$

for any $p \in S^2$. Indeed, since both $\{\phi_{j,m}\}$ and the ultraspherical functions $\{Y_m^j\}$ form orthonormal bases for H_m, there exists an orthonormal $(2m+1)$-by-$(2m+1)$ matrix to go from one basis to the other. It follows that, for any point $p \in S^2$, the vectors $(\phi_{j,m}(p) : |j| \leq m)$ and $(Y_m^j(p) : |j| \leq m)$ have the same L^2 norm. The claim then follows from the classical bound

$$\sum_{|j| \leq m} |Y_m^j(\theta, \varphi)|^2 = \frac{2m+1}{4\pi},$$

which can be derived from the addition law for ultraspherical polynomials [200]. This allows us to simplify the upper bound for S into

$$S \ll \sqrt{n} \log n \sum_{m=1}^{\infty} m |\lambda_m(C_i) \lambda_m|. \tag{2.26}$$

From (2.23) we find that

$$\lambda_m = \frac{1}{1 - \cos \varepsilon} \int_{\cos \varepsilon}^{1} P_m(x)\, dx$$

and

$$\lambda_m(C_i) = 2\pi \int_{\cos \rho}^{1} P_m(x)\, dx.$$

In particular, observe that because $|P_m(x)| \leq 1$ over $[-1, 1]$, $|\lambda_m| \leq 1$. With the identity (2.6),

$$(2m+1)P_m(x) = \frac{d}{dx}P_{m+1}(x) - \frac{d}{dx}P_{m-1}(x),$$

we easily evaluate the desired integrals, which yields

$$\lambda_m = \frac{1}{(2m+1)(1 - \cos \varepsilon)} \Big(P_{m-1}(\cos \varepsilon) - P_{m+1}(\cos \varepsilon) \Big)$$

and

$$\lambda_m(C_i) = \frac{2\pi}{2m+1} \Big(P_{m-1}(\cos \rho) - P_{m+1}(\cos \rho) \Big).$$

From the inequality (see page 167 in [301])

$$\left| P_{m+1}(\cos \theta) - P_{m-1}(\cos \theta) \right| \ll \sqrt{\frac{\sin \theta}{m}}$$

and the above-noted fact that $|\lambda_m| \leq 1$, we derive

$$|\lambda_m| \ll \min \left\{ 1, \frac{1}{1 - \cos \varepsilon} \sqrt{\frac{\sin \varepsilon}{m^3}} \right\}.$$

Taking Taylor expansions gives

$$|\lambda_m| \ll \min\left\{1, \sqrt{\frac{1}{(m\varepsilon)^3}}\right\}.$$

Similarly, we have

$$|\lambda_m(C_i)| \ll \sqrt{\frac{\sin\rho}{m^3}}.$$

Returning to (2.26), we derive that

$$\begin{aligned}
S &\ll \sqrt{n}\log n \sum_{m=1}^{\infty} \min\left\{\sqrt{\frac{\sin\rho}{m}}, \sqrt{\frac{\sin\rho}{m^4\varepsilon^3}}\right\} \\
&\ll \sqrt{n}\log n \sum_{1\le m\le 1/\varepsilon} \frac{1}{\sqrt{m}} + \sqrt{n}\,\frac{\log n}{\varepsilon^{3/2}} \sum_{m>1/\varepsilon} \frac{1}{m^2} \\
&\ll \sqrt{\frac{n}{\varepsilon}}\log n,
\end{aligned}$$

which completes the proofs of Lemma 2.15 and Theorem 2.9 (page 58). $\square$

Hecke Operators and the Ramanujan Bound

Let $H(\mathbf{Z})$ be the ring of quaternions of the form $\alpha = a_0 + a_1\mathbf{i} + a_2\mathbf{j} + a_3\mathbf{k}$, for $a_0, a_1, a_2, a_3 \in \mathbf{Z}$. Recall that the norm of α is $N(\alpha) = a_0^2 + a_1^2 + a_2^2 + a_3^2$. Given an integer n, let S_n be the set of quaternions in $H(\mathbf{Z})$ whose norm is equal to n. The set S_n has great number-theoretic relevance, since its size is obviously the number of representations of the number n as a sum of four squares. A result of Jacobi says this number is precisely

$$|S_n| = 8 \sum_{\substack{d\,|\,n \\ 4\,\nmid\,d}} d. \tag{2.27}$$

For example, if $n = 4$, then we have all permutations of $(\pm 2, 0, 0, 0)$ and $(\pm 1, \pm 1, \pm 1, \pm 1)$, which gives $8 + 16 = 24 = 8(1+2)$ combinations. For our purposes, one interesting consequence of this result is that if n is prime, then the size of S_n is $8(n+1)$.

Fix a prime p congruent to 1 modulo 4. It is easy to enumerate the elements of S_p. Let S_p^0 be the subset of those α such that a_0 is positive and odd. Note that any element of S_p must have exactly one odd coefficient a_i with all of the others even. (To see this, observe that, taken modulo 4, the norm of α is the sum of the parity of each a_i.) Any α of S_p can be brought into S_p^0 by multiplying it by the appropriate unit. For example, the number $2 + 2i + j + 2k \in S_{13}$ is brought over to S_{13}^0 by premultiplying it by $-j$. Since $|S_p| = 8(p+1)$ and there are eight units of the form $\pm i$, $\pm j$, $\pm k$, the set S_p^0 consists of $p+1$ elements, more precisely, $(p+1)/2$ elements together with their conjugates. Let $\alpha_0, \ldots, \alpha_p$ denote the elements of S_p^0. We define the Hecke operator,

$$T_p : f(z) \mapsto \sum_{\alpha \in S_p^0} f(\alpha z),$$

where $f \in L^2(S^2)$ and αz means the point $z \in S^2$ "rotated" by the quaternion α, as discussed earlier. In the case $p = 5$, we find that

$$S_5^0 = \{\, 1 \pm 2i, 1 \pm 2j, 1 \pm 2k \,\}.$$

Since $2\arccos(1/\sqrt{5}) = \arccos(-3/5)$, we recognize the set of rotations R_i, R_i^{-1} defined earlier, and the operator coincides with

$$Tf(z) = \sum_{i=1}^{3} \Big(f(R_i z) + f(R_i^{-1} z) \Big).$$

Therefore, to prove Lemma 2.12 (page 65), it suffices to show the following.

Lemma 2.16 *The second largest eigenvalue (in absolute value) of T_p is at most $2\sqrt{p}$.*

To prove the lemma we examine the spectrum of the operator T_n (generalized to any n) as n goes to infinity. This might seem rather paradoxical, since we did just about the opposite when studying the discrepancy of $P_0 = \{p_1, \ldots, p_n\}$. From our previous discussion, it is clear that we can rewrite $T_p f(z)$ as

$$\frac{1}{2} \sum_{\substack{N(\alpha) = p \\ \alpha \equiv 1 \,(\mathrm{mod}\ 2)}} f(\alpha z),$$

where the residue class of a quaternion is defined as the quaternion whose coefficients are the relevant residue classes. The factor $1/2$ is due to the fact that we now allow a_0 to be negative. Fortunately, the two quaternions $a_0 + a_1 i + a_2 j + a_3 k$ and $-a_0 + a_1 i + a_2 j + a_3 k$ produce the same rotation in SO(3). We generalize the operator to nonprimes n:

$$T_n : f(z) \mapsto \frac{1}{2} \sum_{\substack{N(\alpha) = n \\ \alpha \equiv 1 \,(\mathrm{mod}\ 2)}} f(\alpha z).$$

Lemma 2.17 *Any integral quaternion $\alpha \equiv 1 \pmod 2$ such that $N(\alpha) = p^s$ can be represented uniquely as*

$$\alpha = \pm p^j w,$$

where w is a reduced word over the alphabet $\alpha_0, \ldots, \alpha_p$ and $s = 2j + m$, with m the length of w.

Proof: Recall that a word is said to be reduced if it has no consecutive conjugate pairs. To prove the lemma we begin by recalling some basic facts about $H(\mathbf{Z})$ [109]. The ring $H(\mathbf{Z})$ is left and right Euclidean.[14] If $s > 1$, then $N(\alpha)$ is odd and not prime. It is shown in [109] that α cannot then be prime. Thus it can be written as $\beta\gamma$, where $N(\beta), N(\gamma) > 1$. Since $N(\beta)N(\gamma) = N(\alpha)$, it follows that $N(\beta) = p^{s-l}$ and $N(\gamma) = p^l$ ($l > 0$). Trivially, this shows by induction on s that α is a product of two quaternions, one of which has norm exactly p. Thus we can assume that $l = 1$. Since $N(\gamma) = p$, as we saw earlier, there exists a unit ε_0 such that

$$\alpha = \beta\varepsilon_0\delta,$$

for some $\delta \in S_p^0$. Repeating this process with respect to $\beta\varepsilon_0$ and iterating leads to the factorization

$$\alpha = \varepsilon w_0,$$

where ε is a unit and w_0 is a word of length s over the alphabet $S_p^0 = \{\alpha_0, \ldots, \alpha_p\}$. If we now reduce the word by replacing every consecutive conjugate pair by p, we find

$$\alpha = p^j \varepsilon w,$$

where w is of length $s - 2j$ (the factor 2 comes from the fact that every reduction produces one p but deletes two letters from the word).

We prove the uniqueness of the factorization by a counting argument. The number of reduced words of length $l \geq 1$ is $(p+1)p^{l-1}$, so the total number of factorizations of elements of norm p^s (not just those congruent

---

[14] In the commutative case, a *Euclidean ring* is simply a ring where Euclid's division algorithm is possible. Specifically, there is a nonnegative integer function W, called a *valuation*, such that, given any two $\alpha, \beta \neq 0$ in the ring, there exist a quotient q and a remainder r, such that $r = 0$ or $W(r) < W(\beta)$ and

$$\alpha = q\beta + r.$$

For this definition to make sense in the noncommutative case, such as $H(\mathbf{Z})$, we distinguish between left and right, depending on whether $\alpha = q\beta + r$ or $\alpha = \beta q + r$; hence the terminology left and right Euclidean.

to 1 mod 2) is

$$8 \sum_{0 \leq j < s/2} (p+1)p^{s-2j-1} + 8\delta(s),$$

where $\delta(s)$ is 1 (resp. 0) if s is even (resp. odd). This gives a total of $8(p^{s+1} - 1)/(p-1)$ factorizations, which is precisely the number given by the Jacobi bound (2.27), for $n = p^s$, ie, the number of integral quaternions of norm n. The factorization that we found is therefore unique.

Note that the assumption that $\alpha \equiv 1 \pmod 2$ was not used at all. We use it now to show that $\varepsilon = \pm 1$. Because each $\alpha_i \in S_p^0$ is such that $\alpha_i \equiv 1 \pmod 2$, we have $w_0 \equiv 1 \pmod 2$. It follows that $\alpha = \varepsilon w_0 \equiv \varepsilon \pmod 2$. By our assumption that $\alpha \equiv 1 \pmod 2$, we find that $\varepsilon = \pm 1$. $\square$

The lemma gives us a natural interpretation of the operator T_{p^s}: Given a function f whose values are stored at the nodes of an infinite $(p+1)$-regular tree, the value of $T_{p^s} f$ at any given node is equal to the sum of the values of f at the nodes whose distance to the given node is at most s and has the same parity as s. But this is precisely the definition of the operator B_s, which we used to study the operator discrepancy. We saw that if $\{ \Phi_{m,j} : |j| \leq m \}$ is an orthonormal basis for H_m that diagonalizes T_p (which was then called T), then, with respect to any given $\Phi_{m,j}$, the operator T_{p^s} acts as a multiplicative scalar B_s. We can always write the eigenvalue associated with $\Phi_{m,j}$ as $\rho_{m,j} = 2\sqrt{p} \cos \theta$. Assume from now on that $m > 0$, ie, $\Phi_{m,j}$ is nonconstant. To prove Lemma 2.16 (page 78), it suffices to show that θ is a real number. From (2.14) we know that

$$B_s = p^{s/2} \frac{\sin(s+1)\theta}{\sin \theta}.$$

Fix $p_0 \in S^2$ (this is not to be confused with the prime p) such that $\Phi_{m,j}(p_0) \neq 0$. We will have reached our goal once we can prove the following.

Lemma 2.18 *For any fixed $\varepsilon > 0$,*

$$|T_{p^s} \Phi_{m,j}(p_0)| \ll p^{s(1/2+\varepsilon)}.$$

Note that the constant factor hiding behind $\ll$ depends on ε, p_0, m, j. The reason why our goal is reached is that the multiplicative scalar B_s cannot exceed (in absolute value) the second-largest eigenvalue of T_{p^s} (up to a multiplicative factor independent of s). It follows that

$$p^{s/2} \left| \frac{\sin(s+1)\theta}{\sin \theta} \right| \ll p^{s(1/2+\varepsilon)},$$

and so

$$\left| \frac{\sin(s+1)\theta}{\sin \theta} \right| \ll p^{s\varepsilon}.$$

But if θ is not real, the left-hand side of the inequality grows as $e^{sg(\theta)}$, where $g(\theta)$ is positive and does not depend on ε. By setting ε small enough and letting s tend to infinity, we derive a contradiction. This shows that θ is real, and therefore $|\rho_{m,j}| = 2\sqrt{p} \, |\cos \theta| \leq 2\sqrt{p}$, which proves Lemma 2.16, and hence Lemma 2.12 (page 65). $\square$

The Modular Group

We now prove Lemma 2.18 by appealing to the Ramanujan conjectures. To give a complete, self-contained proof would take us far beyond the intended scope of this book. The reader will no doubt bemoan the brevity of our treatment, or be thankful for it, or both.[15] This section introduces the cast of characters. It includes enough definitions and basic facts to understand how Lemma 2.18 on page 80 follows from Deligne's work on the Fourier coefficients of modular forms. The next section goes deeper into arithmetic algebraic geometry and attempts to tie together the major threads of mathematics that make the whole story so compelling. Our emphasis is on conveying the intuition behind the results rather than providing a technically complete exposition.

We begin with a quick review of linear fractional maps (also known as Möbius transformations). As we saw earlier in this chapter, these are maps from the complex plane to itself of the form

$$f(z) = \frac{az + b}{cz + d},$$

where $a, b, c, d \in \mathbf{C}$ and $ad - bc = 1$. It is convenient to compactify $\mathbf{C}$ by adding a point at infinity and setting $f(\infty) = a/c$, which is equivalent to working on the Riemann sphere.[16] A linear fractional function carries any

---

[15]To keep the flow of our presentation, we have demoted many points in the discussion—like this one—to the status of a footnote.

[16]The concept of a Riemann surface is central to the entire subject. The notion was introduced to disambiguate functions like $\sqrt{z}$ (which root are we talking about?). We

circle in $\mathbf{C}$ into a circle (where lines are considered special circles). Furthermore, it respects circle inversion: Given a circle C with center c and radius r, the inversion of p with respect to C is the point q on the ray (c, p) such that $|cp| \cdot |cq| = r^2$; it can be shown that $f(q)$ is the inversion of $f(p)$ with respect to the circle $f(C)$. In fact this property characterizes linear fractional transformations up to conjugacy ($z \mapsto \bar{z}$). A more common characterization is to identify them with the one-to-one conformal mappings of the complex plane (or rather, to be accurate, of the Riemann sphere).[17] This means that the angle between two crossing curves is preserved under any linear fractional transformation (Fig. 2.6).

Another property of these transformations has to do with the isometries of the hyperbolic plane (in the upper halfplane model).[18] Here is what we mean. To begin with, we can check that the transformations preserve the

---

create two branches, one for each root, carefully ensuring that their analytic properties are still preserved. So, think of a (compact) Riemann surface as a sphere, a torus, or even a surface with more handles, in which we can perform complex analysis. This means that we have a connection between the surface and $\mathbf{C}$. The surface is decomposed into (overlapping) open patches U_i, each one of which is supplied with a homeomorphism $z_i : U_i \mapsto \mathbf{C}$. The simplest example is the Riemann sphere (topologically equivalent to the complex projective line), which can be parameterized by two patches: the complements of the north and south poles, with the homeomorphisms being provided by the corresponding stereographic projections.

Riemann surfaces are used as domains of definition of functions to make them single-valued, eg, $f : z \mapsto \sqrt{z}$. To say that a function f is analytic on the surface means that, at any point p on any patch U_i, the function $f \circ z_i^{-1}$ is analytic at $z_i(p)$; of course, the z_i's must be consistent within the overlaps. In our example, say, the number $4 \in \mathbf{C}$ must map back to at least two points p, q on the surface, ie, $z_i(p) = z_j(q) = 4$, where $f(p) = 2$ and $f(q) = -2$.

Obviously, we can use the same Riemann surface to characterize the points on the curve $x - y^2 = 0$ in the complex plane. Viewing the complex points of an algebraic curve as a Riemann surface is a crucial idea. A point p on the surface corresponds to a point $p^* = (x, y) \in \mathbf{C}^2$ on the curve: Given a patch U_i enclosing p, it is typical to choose z_i as the map $p \mapsto x$. So, moving locally around x in $\mathbf{C}$ corresponds, through the inverse of some z_i, to a local motion around p on the surface. Doing the same thing at the same x but with a different (U_j, z_j) gives a smooth motion elsewhere on the surface.

In our discussion, Riemann surfaces will arise almost exclusively as quotient spaces. In many ways, this simplifies matters. For example, think of the group G generated by $z \mapsto z + w$, for a fixed $w \in \mathbf{C}$. The quotient space $\mathbf{C}/w\mathbf{Z}$ is defined by identifying any two complex numbers whose difference is a multiple of w. This Riemann surface (which is not compact) has the topology of a cylinder, and its analytic functions coincide with the analytic functions on $\mathbf{C}$ of period w, eg, $z \mapsto e^{2\pi i z/w}$. This means that complex analysis on the cylinder is nothing more than the analysis of singly-periodic functions in $\mathbf{C}$: no parameterization into $\mathbf{R}^3$, no differential geometry needed.

[17] Note that bijectivity is important because a function like $z \mapsto e^z$ certainly is conformal.

[18] See Chapter 4 for a primer in hyperbolic geometry.

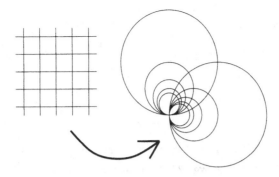

Fig. 2.6. A simple conformal mapping: $z \mapsto 1/z$.

cross-ratios of four points, z_1, z_2, z_3, z_4, ie,

$$\frac{(z_1 - z_3)/(z_1 - z_4)}{(z_2 - z_3)/(z_2 - z_4)}.$$

Next, let us pick these four points on a circle in clockwise order and observe that the cross-ratio is a positive number.[19] Finally, take two distinct points z_1, z_2 and draw the hyperbolic line connecting them (which is provided by the circle passing through them and orthogonal to the real line). Now, consider the cross-ratio formed by z_1, z_2 and the two intersections a_1, a_2 with the real line.

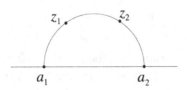

Fig. 2.7. The hyperbolic metric.

With the proper orientation, the cross-ratio is positive and its logarithm provides the measure of the hyperbolic distance; we need the log to ensure the natural additive property of distances.

---

[19]That we get a real number is obvious: Think of the linear fractional transformation that sends three of these points to the real line; note that it takes three points to characterize such a map. Since circles remain circles, the map must send the fourth point to the real line as well. □

The correspondence from

$$z \in \mathbf{C} \;\mapsto\; \frac{az+b}{cz+d}$$

to the matrix

$$\begin{pmatrix} a & b \\ c & d \end{pmatrix}$$

is homomorphic: Composing two transformations is like multiplying their matrices. For our purposes, we may limit ourselves to the case where a, b, c, d are reals. So, the linear fractional transformations of interest can be identified with $\mathrm{PSL}(2, \mathbf{R})$; see page 60 for definitions. This provides the orientation-preserving isometries of the upper halfplane (viewed in the hyperbolic sense). This short excursion into hyperbolic geometry is not just cultural; it will soon help us to understand better why modular forms are so remarkable.

With this identification, it becomes quite easy to classify the linear fractional transformations with real coefficients. By the theory of Jordan canonical forms we know that, outside of the scalar case ($a = d \neq 0$ and $b = c = 0$), all matrices are conjugate to

$$\begin{pmatrix} a & 1 \\ 0 & a \end{pmatrix} \qquad \text{or} \qquad \begin{pmatrix} a & 0 \\ 0 & b \end{pmatrix}, \text{ with } a \neq b.$$

The first case is called *parabolic*; because the determinant is 1, the absolute value of the trace is 2. On the Riemann sphere, f has only one fixed point p, and we have

$$\frac{1}{f(z) - p} = \frac{1}{z - p} + \frac{c}{a - cp}.$$

Besides parabolic transformations, the standard classification includes *elliptic* maps ($|\operatorname{trace}| < 2$) and *hyperbolic* ones (all of the others).

We are now ready to move to the arithmetic part of the story. The subgroup $\Gamma = \mathrm{SL}(2, \mathbf{Z})$ of 2-by-2 matrices with coefficients in $\mathbf{Z}$ and determinant 1 is called the *full modular group*.[20] Of course, we are more interested in $\mathrm{PSL}(2, \mathbf{Z}) = \mathrm{SL}(2, \mathbf{Z})/\{\pm I\}$, which we refer to as the *modular group* $\overline{\Gamma}$. (We've been through this distinction before, so there is no need justifying it again.)

Any transformation in Γ maps the upper halfplane

$$\mathcal{H} = \{\, z \in \mathbf{C} \mid \Im(z) > 0 \,\}$$

---

[20]Can you see why it is a group? Think unit determinant, Cramer....

conformally onto itself. The modular group is a discontinuous subgroup. The adjective is meant to suggest that the group does not act continuously on $\mathcal{H}$, ie, applying to a given point an infinite sequence of transformations from the group can produce only a finite number of points within a finite region. The modular group introduces all sorts of beautiful symmetries into the upper halfplane. To see this, take the quotient space[21] $\Gamma\backslash\mathcal{H}$. This consists of identifying any pair of points in $\mathcal{H}$ that can be mapped to each other by some transformation in Γ.

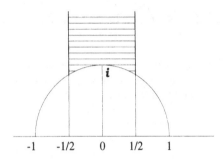

Fig. 2.8. The dashed area is a fundamental domain of the modular group.

Equivalence classes form orbits, and $\Gamma\backslash\mathcal{H}$ is the space of all orbits. Visually and topologically it corresponds to a fundamental domain (defined to include exactly one point per orbit), the most famous of which is shown in Figure 2.8. It is easy to see that the modular group $\overline{\Gamma}$ is generated by

$$\begin{cases} \sigma: & z \mapsto -1/z\,, \\ \tau: & z \mapsto z+1\,, \end{cases}$$

with the relations $\sigma^2 = (\sigma\tau)^3 = 1$. As is suggested by Figure 2.9, the images of the fundamental domain by the linear fractional transformations of $\overline{\Gamma}$ form a triangular tiling of the upper halfplane; note that in the hyperbolic plane these are bona fide triangles bounded by actual lines.

Each triangle has exactly one vertex (call it v) that lies either on the real line (in fact in $\mathbf{Q}$) or at ∞; for example, v is at ∞ in Figure 2.8. (To be rigorous, we should really replace $\mathcal{H}$ by $\mathcal{H} \cup \mathbf{Q} \cup \{\infty\}$ in our discussion of the modular group, but in view of what is coming ahead these are truly cosmetic issues....) The stabilizer of v, ie, the subgroup that leaves it fixed,

---

[21] We write $\Gamma\backslash\mathcal{H}$ and not $\mathcal{H}/\Gamma$ because the action comes from the left. But do keep in mind that the set being modded out is $\mathcal{H}$ and *not* Γ.

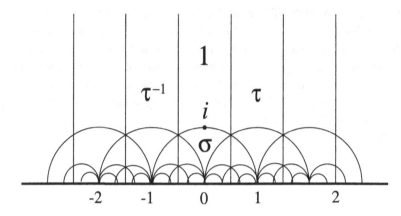

Fig. 2.9. The quotient space of $\mathcal{H}$ by the modular group produces this tiling. With the fundamental domain labeled 1, one can see how neighboring copies are derived from it via the generators σ and τ.

is an infinite cyclic group. The two edges incident to v share the same tangent at that point, which thus is given the geometrically suggestive name of *cusp*. The set of all cusps, ie, vertices on the boundary $\mathbf{R} \cup \{\infty\}$ of the tiling in Figure 2.9, is exactly $\mathbf{Q} \cup \{\infty\}$.

The two other vertices are of the *elliptic* kind, meaning that they are fixed points of an elliptic element of Γ. They are in the $\overline{\Gamma}$-orbit[22] of $e^{\pi i/3}$.

The modular group encapsulates the symmetries of the upper halfplane relative to isometries in hyperbolic geometry. To understand the group, and hence the symmetries in question, we must study its subgroups $\Gamma' \subset \Gamma$ of finite index (ie, with cosets of finite size). A point x_0 on the real axis is called a *cusp* of Γ' if the stabilizer of x_0 is a free cyclic group generated by a parabolic transformation. Again, the cusps are points of the fundamental domain that lie on the real line or at infinity. They are incident upon an infinite number of edges bounding reflections of the fundamental domain.

Of particular interest are the *principal congruence subgroups*. For any

---

[22]It is easy to show that the elements of Γ that fix some point of $\mathcal{H}$ form a finite cyclic group, all nontrivial elements of which are elliptic. If γ is an elliptic element in $\mathrm{SL}(2,\mathbf{R})$, we have $|\,\mathrm{trace}\,| < 2$, and so its characteristic polynomial has complex conjugate roots of modulus 1. If γ has finite order and belongs to Γ, these must be roots of unity that lie in a quadratic field; therefore their order must be 2, 3, 4, or 6. It follows that elliptic points, ie, points of $\mathcal{H}$ with a nontrivial stabilizer in $\overline{\Gamma}$, are points in the $\overline{\Gamma}$-orbits of i and $e^{\pi i/3}$.

integer $N > 0$, the subgroup $\Gamma(N)$ consists of the set of matrices of Γ such that

$$\begin{pmatrix} a & b \\ c & d \end{pmatrix} \equiv \begin{pmatrix} 1 & 0 \\ 0 & 1 \end{pmatrix} \pmod{N}.$$

Note that $\Gamma(N)$, as the kernel of the homomorphism reducing the matrices mod N, is actually a normal subgroup.[23] It is infinite, but its index is finite, which is a short way of saying that the quotient group $\Gamma/\Gamma(N)$ is of finite order.[24] To get the proper geometric picture, you must keep in mind that subgroups miss transformations, and so the "smaller" the subgroup the "bigger" the fundamental domain (Fig. 2.10).

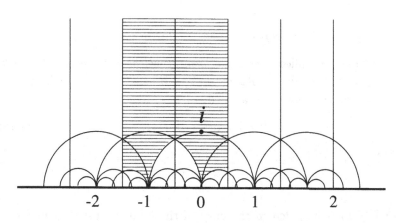

Fig. 2.10. A fundamental domain for $\Gamma(2)\backslash\mathcal{H}$. The index of the subgroup is 6, and that is how many copies of the modular group's fundamental domain we need. The cusps are $-1, 0, \infty$.

Any subgroup Γ' of Γ that contains $\Gamma(N)$ is called a *congruence subgroup*. The subgroup $\Gamma_0(N) \subset \Gamma$, albeit not normal, is crucial to the theory. It is defined by the condition $c \equiv 0 \pmod{N}$. In other words, its elements are

---

[23] A normal subgroup F is one that coincides with all of its conjugacy classes, ie, $F = \gamma F \gamma^{-1}$. Just as ideals are useful for creating new rings, normal subgroups are useful for creating new subgroups: this is because they form the kernel of the group homomorphism formed by "modding out." The abelian case is trivial, of course, because all subgroups are normal.

[24] There is a formula for computing it: The index of $\Gamma(N)$ is $N^3 \prod_{p|N} (1 - p^{-2})$.

of the form

$$\begin{pmatrix} \star & \star \\ 0 & \star \end{pmatrix} \quad (\bmod\ N).$$

Modular Forms

Having defined the basic symmetries of the upper halfplane, we now consider functions that are invariant under such symmetries. This leads us to modular functions and then to (holomorphic) modular forms. Strictly speaking, we need only the latter here, so it makes sense for us to begin with them. Recall that a function $f(x+iy) = u(x+iy) + iv(x+iy)$ is said to be *holomorphic* (or analytic) in an open set $D \subseteq \mathbf{C}$ if it is differentiable in the complex sense over D so, in particular, it obeys the Cauchy-Riemann equations:

$$\partial u / \partial x = \partial v / \partial y \quad \text{and} \quad \partial u / \partial y = -\partial v / \partial x.$$

The function f is called a *modular form of weight k* (for even integral k, with respect to the full modular group Γ) if it is holomorphic in $\mathcal{H}$ and the following two conditions are satisfied: First,

$$f(\gamma z) = (cz + d)^k f(z), \tag{2.28}$$

for any

$$\gamma = \begin{pmatrix} a & b \\ c & d \end{pmatrix} \in \Gamma.$$

Second, $f(z)$ is bounded in the cusp of the fundamental domain for Γ. Since $f(z+1) = f(z)$ by (2.28), f can be expanded as a Fourier series,

$$f(q) = \sum_{n \in \mathbf{Z}} a_n q^n,$$

where $q = e^{2\pi i z}$. The second condition means that f is holomorphic at the cusp at infinity. For any point z at infinity in the upper halfplane (take $z = \infty \times i$), we have $q = 0$, and so, for convergence, a_n must be zero for all $n < 0$, which gives us $f(\infty) = a_0$. Finally, we say that a modular form f for Γ is a *cusp form* if the Fourier coefficient a_0 is itself zero. Note that in the definition it is essential that k be even. This is because by setting $a = d = -1$ and $b = c = 0$ we find that $f(z) = (-1)^k f(z)$; so f would be trivial if k were odd.

Perhaps we should explain how these seemingly arbitrary definitions come about. But before we do so, we should reassure the reader that such functions do exist. Probably the single most important example of a modular form of weight $k > 2$ is the *Eisenstein series*

$$G_k(z) = \sum_{\substack{m,n \in \mathbf{Z} \\ (m,n) \neq (0,0)}} \frac{1}{(m+nz)^k},$$

to which we shall come back shortly. The reader is encouraged to check that the modularity condition (2.28) holds, which is both straightforward and instructive.

To build some intuition about modular forms, three observations might help: First, note that condition (2.28) is not as scary as it looks. After all, recall that σ and τ generate $\overline{\Gamma}$, and so the two identities

$$\begin{cases} f(-1/z) = z^k f(z), \\ f(z+1) = f(z) \end{cases}$$

alone imply (2.28).

Our second observation explains the use of the word "form." Given $\gamma \in \Gamma$, differentiating γz yields (forgive the overloaded use of the symbol d)

$$d(\gamma z) = d \frac{az+b}{cz+d} = \frac{a(cz+d) - c(az+b)}{(cz+d)^2} \, dz = (cz+d)^{-2} \, dz.$$

So, another way of stating the functional equation (2.28) is to say that the $\frac{k}{2}$-fold differential form $f(z) \, (dz)^{k/2}$ is invariant under the action of the modular group.

Finally, our third observation pertains to *modular functions*. These are just like modular forms, except that $k = 0$ and the analyticity condition is dropped. Setting the weight k to 0 makes eminent sense, since we then have true invariance under the action of the modular group and not the weaker version expressed by (2.28). In fact, who needs modular forms when you can have modular functions? There is a catch, however. The price that we pay for perfect invariance is analyticity: It must go! Indeed, only trivial modular functions are analytic. So, the Fourier series expansion of a modular function looks like

$$\sum_{n > -m} a_n e^{2\pi i n z},$$

for some finite m. The function is meromorphic but typically not analytic.[25]

Modular functions are nicer but harder to construct than modular forms. On the other hand, by dividing two modular forms of the same weight we can hope to get a modular function. On a much more pedestrian level, this is what we do to define functions in projective space (ie, invariant under scaling of variables). We take two homogeneous polynomials of degree k (the analogue of modular forms) and divide them to produce a rational function (the analogue of a modular function).

Modular functions are meromorphic functions invariant under a group of symmetries of the manifold where they are defined. They are part of a larger group of functions, defined on more general manifolds, that are called *automorphic* functions. The beauty of the concept of automorphic functions is to blend several ingredients together: the geometry and topology of Riemann surfaces; the algebra of their groups of symmetries; and the analysis of meromorphic functions.

By way of analogy, consider *elliptic functions*. These are defined as doubly periodic functions, ie, functions on a torus as a Riemann surface. This means that they are invariant under certain translations; ie, the fundamental domains associated with their invariance are parallelograms. So, if you will, elliptic functions are to a certain discrete subgroup of isometries in the Euclidean plane (translations) what modular functions are to a certain discrete subgroup of isometries in the hyperbolic plane (the modular group). The hyperbolic plane has a richer structure than its Euclidean counterpart. For the same reason, modular functions have more to say than elliptic ones.

Certainly the modular group, being noncommutative, is bound to be more interesting than the abelian group that tiles the Euclidean plane with parallelograms. What we cannot explain at this point (but we will later) is why the hyperbolic plane allows an arithmetic perspective on algebraic curves. What makes hyperbolic geometry more conducive to number theory

---

[25] By definition, a meromorphic function may have a discrete set of poles $\{a\}$ at which the function $f(z)$ is not analytic but $(z - a)^m f(z)$ is, for some integer $m > 0$. So, to express the function locally as a power series requires the use of a finite number of negative exponents: the so-called Laurent series expansion. Here is some background on the Fourier series expansion for modular functions. In general, these have the form

$$\sum_{n \in \mathbf{Z}} a_n(y) e^{2\pi i n z},$$

where $y = \Im(z)$. The function is holomorphic over $\mathcal{H}$ if the a_n's are constants independent of y. To be meromorphic at the cusps, however, one needs the additional condition that only a finite number of a_n $(n < 0)$ can be nonzero. Plug $z = i\infty$ and you will immediately see why!

than Euclidean geometry is, for example, the magic at the heart of the Shimura-Taniyama conjecture and, hence, Fermat's Last Theorem. What we can explain right away, however, is what makes the hyperbolic plane geometrically "superior."

The "problem" with the Euclidean plane is that its symmetries are dreadfully dull. Remember those platonic solids from high school? Perhaps the most remarkable thing about them is how few there are: only five! The hyperbolic plane, by contrast, has "so much more space" than its Euclidean counterpart that it can be tiled by using just about any regular convex shape. Let's face it, those outer-space aliens with long antennas who live in hyperbolic geometry have much more fun tiling their kitchen floors than we do. The intrinsic symmetries of the floors (given by the modular group and all of its subgroups) are enormously rich and complex. Modular functions give us a vehicle for carrying over all of those symmetries into the analytic arena.

Our discussion has been limited to Γ, but it generalizes naturally to congruence subgroups of Γ. A modular form of weight k and level N is defined by relaxing (2.28) to $\gamma \in \Gamma'$, for some congruence subgroup $\Gamma' \supseteq \Gamma(N)$; the level is defined as the minimum such N. A cusp form f for Γ' is defined similarly. Of course, we now require that the modular form (resp. cusp form) should be holomorphic (resp. vanish) at *each* of the cusps.

Here is a classical example of a modular form for a congruence subgroup:

$$\left(\sum_{n \in \mathbf{Z}} q^{n^2} \right)^4 , \tag{2.29}$$

where $q = e^{2\pi i z}$. This is a modular form of weight 2 for $\Gamma_0(4)$. (Unlike the case of the Eisenstein series, the modularity condition holds for nontrivial reasons.) Note that its n-th coefficient in the expansion in q is exactly the number of ways that n can be written as a sum of four squares. This puts our discussion of Jacobi's result (page 77) under the bright spotlight of modular forms. Incidentally, we can replace 4 by k and ask about odd k. So we do need a theory of modular forms of odd weight, after all. Well, this book does not.... A general philosophy behind the use of modular forms is that nontrivial identities hide behind the modularity condition. When expressed in terms of the Fourier coefficients, these identities in turn reveal deep arithmetical significance.

We cannot close this discussion without mentioning a crucial property of modular forms: for given weight and level, they form a finite-dimensional vector space over $\mathbf{C}$, so we can use linear algebra to explore them. The dimension can be computed by using a fundamental result from the theory of algebraic curves: the *Riemann-Roch theorem*. First, some background.

Recall from basic complex analysis that to be analytic over all $\mathbf{C}$ is a rather strong condition. Cauchy's theorem implies that specifying an analytic function along a single contour entirely specifies the function inside. By Liouville's corollary, to be analytic and bounded implies to be constant. Adding compactness can only make things worse. Predictably, the only functions that are holomorphic over a compact Riemann surface are constant; not much of a theory to be built on these grounds. To be meromorphic gives more freedom, but not a great deal more. An immediate consequence of our previous remark is that the number of poles equals the number of zeros (counting multiplicities). On the simplest compact Riemann surface, the Riemann sphere, the only meromorphic functions are the rational functions, ie, fractions of polynomials.

Surfaces of higher genus g are somewhat more interesting.[26] The Riemann-Roch theorem counts how many meromorphic functions have a specified number of poles and zeros. Rather than stating it in full generality, we illustrate it by an example. Pick two points p, q on a compact Riemann surface of genus g, and consider the space $S(31, 62)$ of meromorphic functions with at worst 31 poles at p and 62 poles at q (and, hence, 93 zeros elsewhere). The Riemann-Roch theorem says that if, say, $g = 40$, then $S(31, 62)$ is a vector space of dimension $31 + 62 + 1 - g = 54$.

---

[26] Recall that the genus is the number of "holes" in the surface. There are many ways to define this notion. For example, we can triangulate the surface to form a simplicial cell complex with V vertices, E edges, and F faces. The genus g is defined by the formula $V - E + F = 2 - 2g$. In the language of homology theory, g is half the first Betti number, ie, half the number of independent nontrivial cycles. Formally, the i-th Betti number is the rank of the homology group $H_i = \ker(\partial_i)/\operatorname{im}(\partial_{i+1})$, where ∂_i is the standard boundary map with $\partial_{i-1} \circ \partial_i = 0$. In the case of interest, $i = 1$, here is what it means. Consider any subset of oriented edges and associate a linear form with it (with integer coefficients). The boundary map ensures that the image is null when the edges form a closed chain. Geometrically, what is happening is obvious: H_1 is the space of closed chains, with the proviso that two chains that differ only by the boundary of a bunch of faces are considered equal. In the case of an orientable surface, these groups have no torsion parts and the Betti numbers (ie, the number of copies of $\mathbf{Z}$ that they consist of) tell the whole story—see Appendix B for background on the classification of finitely generated abelian groups. For completeness, let us also add that, in the language of homotopy theory, H_1 is simply the abelianization of the fundamental group.

Deligne's Spectral Bounds for Cusp Forms

We are ready to prove Lemma 2.18. Deligne proved that if f is a cusp form of weight k for a congruence subgroup Γ', then[27]

$$|a_n| \ll_\varepsilon n^{(k-1)/2+\varepsilon}, \qquad (2.30)$$

for any fixed $\varepsilon > 0$, where the a_n's are the Fourier coefficients of f at any cusp. (The constant factor in the inequality depends on the function f.) Here is how we connect this result to Lemma 2.18 (page 80). The crux is the fact that, given a fixed $p_0 \in S^2$ and $f \in H_m$, the function

$$F(z) \stackrel{\text{def}}{=} \sum_{\substack{\alpha \in H(\mathbf{Z}) \\ \alpha \equiv 2 \ (\text{mod } 4)}} N(\alpha)^m f(\alpha p_0) e^{2\pi i N(\alpha) z/16}$$

is a cusp form[28] of weight $2m + 2$ for the congruence group $\Gamma(4)$. So, we can rewrite $F(z)$ as

$$F(z) = \sum_n a_n e^{2\pi i n z/16},$$

where

$$a_n = \sum_{\substack{N(\alpha)=n \\ \alpha \equiv 2 \ (\text{mod } 4)}} n^m f(\alpha p_0).$$

Obviously, a_n is the n-th Fourier coefficient in the expansion of $F(16z)$ so, by (2.30),

$$|a_n| \ll n^{m+1/2+\varepsilon},$$

for any fixed $\varepsilon > 0$. Writing $n = 4p^s$, we derive

$$\left| \sum_{\substack{N(\alpha) = 4p^s \\ \alpha \equiv 2 \ (\text{mod } 4)}} f(\alpha p_0) \right| \ll p^{s(1/2+\varepsilon)}.$$

The condition $\alpha \equiv 2 \ (\text{mod } 4)$ implies that α is congruent to 0 mod 2 (all its coefficients are even), so we can write $\alpha = 2\beta$. Since β and 2β correspond to the same rotation, we have $f(\alpha p_0) = f(\beta p_0)$ and, therefore,

$$\left| \sum_{\substack{N(\beta) = p^s \\ \beta \equiv 1 \ (\text{mod } 2)}} f(\beta p_0) \right| \ll p^{s(1/2+\varepsilon)}.$$

---

[27] The proof makes full use of SGA (Grothendieck's work in algebraic geometry), and according to Zagier, would run over 2000 pages if written out in full.

[28] We omit the proof, but we should note that this can be seen as a grand generalization of the fact that the function in (2.29) is modular; see the similarity in the case $m = 0$.

The constant factors in the inequality depend on f, p_0, and ε. The left-hand side is precisely $2|T_{p^s}f(p_0)|$, so the inequality establishes Lemma 2.18 (page 80). □

2.6 A Review of Arithmetic Algebraic Geometry *

Modular forms can be used to encode the arithmetic aspects of a wide variety of problems. Why this is useful is that modular forms constitute vector spaces of finite dimension (thank our good old Hecke operators for that!). So, as soon as we are faced with a number of modular forms in excess of the dimension, we know that there must exist linear relations among them. This yields extra information that we can exploit, typically by looking at the Fourier coefficients of the modular forms in question. These coefficients can encode information as diverse as energy levels in physics, sums over the divisors of integers, the number of solutions of Diophantine equations, special values of zeta functions, etc.

This section is meant as a grand tour of modular forms, using as our main vehicle Wiles' recent proof of Fermat's Last Theorem (FLT) or, rather, his proof of the *Shimura-Taniyama conjecture*.[29] In truth, this short introduction, which requires only elementary knowledge of mathematics, barely scratches the surface of this deep subject.[30] Our aim is merely to place Deligne's bound in its broader mathematical context, and to shed light on the role of modular forms in the orbital construction of points on a sphere. Our discussion covers a large amount of territory. For this reason, all proofs have been omitted, but references to where they can be found have been included.[31]

The Shimura-Taniyama conjecture says that that all rational elliptic curves are modular, ie, are modular forms in disguise. The so-called semistable restriction of the conjecture has been established by Wiles [323],

---

[29]There is a debate swirling around the name of the conjecture. Some prefer the neutral "Modularity conjecture," others the witty "*** conjecture," where *** hides your favorite subsequence of the names Shimura, Taniyama, and Weil. Big conjectures often raise big issues, not all of them necessarily mathematical.

[30]This author is hardly an expert on the said subject. But, with apology to Oscar Wilde, he is always ready to give to those who are more experienced than himself the full benefits of his inexperience.

[31]While writing these pages I remained aware of the "curse of the encyclopedia," which is to be trivial to the expert and incomprehensible to the uninitiated. I hope readers will wish to give it a try, and see for themselves whether I managed to avoid the "curse." I can promise them only one thing: The subject matter might be demanding but, as mathematics goes, it is as beautiful as it gets.

with the proof completed by Taylor and Wiles [306]. Semistability is a minor technical restriction that has been completely removed.[32] The conjecture thus appears now to be a theorem. It is truly an amazing theorem, defying common sense. To convey this sense of awe, we begin with an example due to Eichler and Shimura. Consider the product

$$\Phi(z) = q \prod_{n \geq 1} (1 - q^n)^2 (1 - q^{11n})^2,$$

where $q = e^{2\pi i z}$. We can check that $\Phi(z)$ is a cusp form of weight 2 for the congruence subgroup $\Gamma_0(11)$, ie, defined by the condition $c \equiv 0 \pmod{11}$. If a_n is the n-th Fourier coefficient of Φ, ie, $\Phi(z) = \sum_{n \geq 1} a_n q^n$, then for any prime $p \neq 11$,

$$a_p \Phi(z) = \Phi(pz) + \sum_{0 \leq j \leq p} \Phi((z+j)/p),$$

for any $z \in \mathcal{H}$. Remarkably, these Fourier coefficients lead trivially to the number of points with coordinates in any Galois field[33] $\mathbf{F}_p$ on the elliptic curve

$$y^2 = x^3 - 4x^2 + 16.$$

In particular, for any odd prime $p \neq 11$, that number is precisely $p - a_p$. We have many of the ingredients of the stew in this example: a modular form, an elliptic curve, and a Fourier series counting its zeros in finite fields. What is missing is just any clue to this nagging question: What in the world is going on?

The key to the answer is the link between modular forms and elliptic curves, which is their L-functions. In fact, what the Shimura-Taniyama conjecture really says is that, viewed through the prism of their L-functions, rational elliptic curves and (certain) modular forms are one and the same thing. The two notions are so completely different it is truly remarkable that they should actually coincide. But this is just one of many wonders. Since modular forms are still fresh in the minds of the readers (those who are still with us, that is), our story begins with them.

L-Functions of Modular Forms

Let $f(z)$ be a modular form of weight k for Γ. Recall that the modular group $\overline{\Gamma} = \mathrm{PSL}(2, \mathbf{Z})$ is generated by $\sigma(z) = z + 1$ and $\tau(z) = -1/z$.

---

[32] Work by Breuil, Conrad, Diamond, and Taylor, building on Wiles' results.

[33] A brief review of finite fields is given in §9.2 on page 319.

As we already remarked, the invariance condition (2.28) applied to σ, ie, $f(z+1) = f(z)$, implies a Fourier series expansion,

$$f(z) = \sum_{n \geq 0} a_n e^{2\pi i n z},$$

for $z \in \mathcal{H}$ and $a_n \in \mathbf{C}$. It is known that $|a_n| = O(n^{k-1})$ in general, and $|a_n| = O(n^{k/2})$ for cusp forms.[34] So, the Dirichlet series

$$L(f, s) = \sum_{n \geq 1} \frac{a_n}{n^s}$$

converges absolutely for $\Re(s) > k$, in general, and for $\Re(s) > k/2 + 1$ in the case of cusp forms. This defines the *L-function* of f. Applying (2.28) to τ gives $f(-1/z) = z^k f(z)$. From this, Hecke derived a functional equation for L. Conversely, he also showed that Dirichlet series that satisfy a certain functional equation (together with some analytic conditions) correspond to modular forms. Weil extended this result to the subgroup $\Gamma_0(N)$. Here are some of the details.

The bridge between Fourier series expansions and Dirichlet series is the *Mellin transform*. We define the Mellin transform of a function φ on the positive real axis as

$$M(\varphi, s) = \int_0^\infty \varphi(t) t^{s-1} \, dt.$$

A simple calculation shows that

$$L(f, s) = \frac{(2\pi)^s}{\Gamma(s)} M(f_1, s),$$

where $f_1(t)$ denotes $f(it) - a_0$ (obviously a_0 has to go since, for the sake of convergence, it cannot appear in the Dirichlet series) and

$$\Gamma(s) = \int_0^\infty e^{-t} t^{s-1} \, dt$$

is the gamma-function (ie, the Mellin transform of e^{-t}). The function $L(f, s)$ is defined a priori only for $\Re(s)$ large enough, but it can be extended to the whole complex plane by meromorphic continuation. Suppose for simplicity that $a_0 = 0$ and $k > 2$, and define

$$\Lambda(s) = (2\pi)^{-s} \Gamma(s) L(f, s).$$

From the relation $f(-1/z) = z^k f(z)$, it follows from elementary complex

---

[34]Of course, as we just saw, tighter bounds follow from Deligne's work.

integration that

$$\Lambda(s) = \Lambda(k - s). \tag{2.31}$$

Conversely, one may wonder whether any Dirichlet series with meromorphic continuation and a functional equation like (2.31) arises from some modular function. With the right convergence assumptions, we can essentially reverse our steps and answer yes. This is true if f is modular for the full modular group Γ. In the case where f is modular for only, say, $\Gamma_0(N)$, things are more complicated. Unlike the full modular group, $\Gamma_0(N)$ cannot be defined by only two generators, which leads to more than one functional equation. Roughly speaking, Weil showed that if the functional equation holds for enough "twists"

$$\sum_{n \geq 1} \frac{\chi_k(n) a_n}{n^s}$$

of the standard Dirichlet series (together with some analytic properties), then the corresponding Fourier series expansion $\sum a_n q^n$, where $q = e^{2\pi i z}$, is modular for $\Gamma_0(N)$. These twisted series are defined by using multiplicative characters $\chi_k(n)$. Such characters play an important role in the story, so we must discuss them at least a little.[35] They are the multiplicative versions of the additive characters used in Fourier analysis ($n \mapsto e^{2\pi i n/k}$, remember?). They are periodic functions that map the integers to the complex units. Their period is k, so it suffices to define them over the integers mod k. We require that χ_k respect multiplication,[36] and that $\chi_k(n) = 0$ if n and k are not relatively prime. It is interesting to see both additive and multiplicative characters at work simultaneously. In some way, L-functions express the subtle interplay between additive and multiplicative Fourier analysis.

Hecke Operators and Euler Products

It is a standard (easy) fact of analytic number theory that, given any multiplicative function $n \mapsto a_n$, if $\sum_n a_n n^{-s}$ converges absolutely for $\Re(s) > \sigma$, then

$$\sum_{n \geq 1} \frac{a_n}{n^s} = \prod_p \left(1 + a_p p^{-s} + a_{p^2} p^{-2s} + \cdots \right),$$

---

[35] See also our discussion of characters in §9.1.

[36] Multiplicative functions, ie, such that $f(mn) = f(m)f(n)$ whenever m and n are relatively prime, abound in number theory: Two famous ones are the Euler totient function and the Möbius function.

for $\Re(s) > \sigma$, where the sum runs over all primes p. Note that if a_n is strongly multiplicative, ie, $a_{mn} = a_m a_n$ for all m, n, then clearly the Euler product takes on the standard form:

$$\sum_{n \geq 1} \frac{a_n}{n^s} = \prod_p \frac{1}{1 - a_p p^{-s}}.$$

In the context of modular forms, Euler products are to be sought in intermediate form. Specifically, let $\sum_n a_n n^{-s}$ be the L-function of a modular form for $\Gamma_0(N)$ with Fourier coefficients a_n. We investigate which conditions ensure that

$$\sum_{n \geq 1} \frac{a_n}{n^s} = \prod_p \frac{1}{1 - P_p(p^{-s})},$$

where each P_p is a polynomial of degree 2 vanishing at 0. The answer is, roughly, those forms whose Fourier coefficients satisfy

(i) the standard (weak) multiplicative property, $a_{mn} = a_m a_n$, for any n, m relatively prime, and

(ii) some suitable recurrence relation involving the a_n's indexed by prime powers n.

To understand why, we need to come back to our old friends, the Hecke operators; the reader will happily remember them from our points-on-the-sphere days. Recall how Eisenstein series—our basic example of a modular form—were defined by summing $1/(m + nz)^k$ over the integer lattice. This basic relation between modular forms and functions defined on lattices means that operators on lattices can be used to build operators on modular forms. In particular, given a lattice Λ, one can define an operator from the abelian group of linear sums in Λ into itself simply by summing over all sublattices of a certain index. If m and n are relatively prime, it is easy to see that composing the operators of index n and m gives the operator of index nm. There we have the desired multiplicative property.

Hecke defined operators that act linearly on the space of cusp forms for $\Gamma_0(N)$ of weight 2. These operators satisfy relations remarkably similar to (i, ii). Furthermore, they commute, so we can find a common eigenbasis to all of them and decompose cusp forms accordingly. That is essentially what we did for the points on the sphere. Unsurprisingly, these functions are called *eigenforms*. For normalization purposes, set their first Fourier coefficient to be 1. Then everything falls into place. If we fix an eigenform and look at the eigenvalues corresponding to all the Hecke operators, what we see are precisely the Fourier coefficients of the eigenform. In other words, an eigenform is completely specified by the "scaling" action of the Hecke operators on it. Collecting all of those fun facts, we find that:

1. From the viewpoint of an eigenform, Hecke operators act like scalars.
2. These scalars are nothing but the Fourier coefficients of the form, and hence the coefficients of its Dirichlet series.
3. Hecke operators satisfy relations that are similar to the conditions required on the coefficients of a Dirichlet series to admit an Euler product expansion.

The punchline: the Dirichlet series of a normalized eigenform has an Euler product expansion. So now the formidable problem of finding cusp forms with Euler product expansions has been reduced to the much tamer task of finding simultaneous eigenforms for linear operators. Standard linear algebra[37] suggests defining some suitable positive-definite Hermitian form relative to which the Hecke operators are self-adjoint. (Again, the alert reader will fondly remember that it is essentially what we did when placing points on a sphere, with help from the Laplacian.) Here, the Hermitian form comes from something called the *Petersson inner product*. If f and g are modular of weight k for some congruence subgroup Γ' and at least one of them is a cusp form, then their Petersson inner product is defined as

$$\int_D f(z)\overline{g(z)}\, y^{k-2}\, dx\, dy,$$

where $z = x + iy$ and D is a fundamental domain (any one for that matter).

The importance of Hecke operators to the theory of modular forms cannot be overemphasized. To have at our disposal a commutative algebra of operators acting on the space of modular forms of weight k is a godsend. It allows us to find a canonical basis of *simultaneous* eigenvectors. The Fourier coefficients of these eigenforms are algebraic integers with multiplicative properties. This leads to Euler product expansions for their associated Dirichlet series. We also obtain analytic continuations to the whole complex plane with functional equations reminiscent of the one for the Riemann zeta function.

Enough said about modular forms. We now turn to the most important player in the whole story, the elliptic curve, and show how—rather improbably—it connects to modularity.

---

[37] The funny thing is, according to J. S. Milne (Lecture Notes on Elliptic Curves, 1996, page 137), Hecke had trouble doing that because the needed linear algebra, ie, a certain inner product to be defined next, was not available to him....

Elliptic Curves

An elliptic curve $\mathcal{C}$ is defined by the cubic equation in x and y,

$$y^2 = 4x^3 - ax - b.$$

To be nonsingular the curve must have a nonvanishing discriminant, ie, $a^3 - 27b^2 \neq 0$. This keeps the right-hand side from having multiple roots in x. As cubic curves go, this definition hardly seems general, but actually it is as long as we work over a field of characteristic $\neq 2, 3$. (We can prove this algebraically by changing coordinates.) Of course, for a projective curve, the equation must be made homogeneous: $Y^2 Z = 4X^3 - aXZ^2 - bZ^3$. An historical aside: The term "elliptic" is a bit of a misnomer; it is derived from the integral we get by computing the arc length of an ellipse.

We can study the equation over any field, but it makes the most sense to start with an algebraically closed one like $\mathbf{C}$, especially since this puts the full power of complex analysis in our hands. Viewed as a function $y = y(x)$ over $\mathbf{C}$, it is necessary to disambiguate the two different values of y corresponding to the same x.

The time-honored way to do that, of course, is to transform the domain of definition into a Riemann surface where the function becomes single-valued but keeps its fundamental analytic features. Not only does this lift ambiguities about y, but it also gives the function a topological structure. In this case, the surface is compact and its genus is 1, so it looks like a torus. Remarkably, the converse is true. Compact Riemann surfaces of genus 1 correspond exactly to elliptic curves. This correspondence provides a parameterization of an elliptic curve as a function from $\mathbf{C}$ (where the parameter z lives) to $\mathbf{C} \times \mathbf{C}$ (where the solutions $(x(z), y(z))$ live).[38] As we come to expect of a torus, the map is doubly periodic. We explain all of this in the following.

Take two complex numbers ω_1, ω_2, which are not collinear in the complex plane. They define a lattice $\Lambda = \omega_1 \mathbf{Z} + \omega_2 \mathbf{Z}$. The fundamental domain of $\mathbf{C}/\Lambda$ is a parallelogram with opposite sides identified. Topologically it is a torus (Fig. 2.11). Recall that elliptic functions are, by definition, meromorphic and doubly periodic. The simplest ones relative to Λ are of the form

$$\sum_{\omega \in \Lambda} \varphi(z + \omega),$$

---

[38] Recall that we have compactified $\mathbf{C}$, so to be fully rigorous we should talk about the Riemann sphere instead of $\mathbf{C}$. For simplicity we shall be frequently sloppy and hang on to the notation $\mathbf{C}$ when we should not.

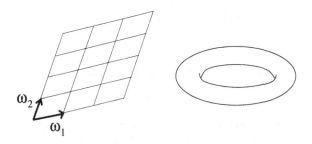

Fig. 2.11. Tiling the plane with parallelograms provides a covering of the torus.

for some meromorphic function φ. By straightforward application of the residue theorem of complex analysis, we see that the sum of the residues over a parallelogram is zero and so, by Liouville' theorem, to be nonconstant the function needs to have a multiple pole or several simple poles. Toying around with this type of requirement, we are quickly led to one of the big hitters in this ball game: the *Weierstrass function*,

$$\wp(z) = \frac{1}{z^2} + \sum_{\omega \in \Lambda \setminus \{0\}} \left(\frac{1}{(z - \omega)^2} - \frac{1}{\omega^2} \right),$$

which can be easily shown to converge absolutely and uniformly over any compact subset of $\mathbf{C} - \Lambda$. Its derivative is

$$\wp'(z) = -2 \sum_{\omega \in \Lambda} \frac{1}{(z - \omega)^3}.$$

Obviously, both $\wp(z)$ and $\wp'(z)$ are elliptic relative to Λ. Their relation to elliptic curves comes from the functional equation that they satisfy:

$$\wp'(z)^2 = 4\wp(z)^3 - g_2\wp(z) - g_3,$$

for constants g_2, g_3 that depend only on the lattice Λ. By simple manipulation we find that

$$\wp(z) = \frac{1}{z^2} + 3G_4 z^2 + 5G_6 z^4 + 7G_8 z^6 + \cdots,$$

where, for even $k > 2$,

$$G_k = G_k(\Lambda) = \sum_{\substack{m,n \in \mathbf{Z} \\ (m,n) \neq (0,0)}} \frac{1}{(m\omega_1 + n\omega_2)^k}.$$

This seems to generalize our earlier definition of Eisenstein series. Could modular forms be far behind? The constants g_i are defined by

$$g_2 = 60G_4 \qquad \text{and} \qquad g_3 = 140G_6. \qquad (2.32)$$

Not only does this allow us to parameterize the elliptic curve

$$z \mapsto (\wp(z), \wp'(z)),$$

but in fact we can characterize the set of all elliptic functions as the rational functions of $\wp(z)$ and $\wp'(z)$, ie, fractions of polynomials in $\wp(z)$ and $\wp'(z)$. We omit the proof of this important theorem (which is actually quite easy). The reason we even mention it is to draw the reader's attention to how much simpler periodic meromorphic functions are than their real counterparts. Over the reals, (nice) periodic functions admit Fourier series expansions in terms of $\sin x$ and $\cos x$, but these require an infinite number of terms. For elliptic functions, on the other hand, $\wp(z)$ and $\wp'(z)$ need only appear with a finite number of coefficients attached to them!

With the parameterization discussed above, the Weierstrass function establishes a map from lattices in $\mathbf{C}$ to elliptic curves. So, we know how to go from a Riemann surface of genus one of the form $\mathbf{C}/\Lambda$ to an elliptic curve. Can we go the other way around? Our first observation is that any nonzero, complex-valued scaling of the basis elements of the lattice yields the same elliptic curve, so we can normalize the lattice into the form $\Lambda(z_0) = m + nz_0$, for $z_0 \in \mathbf{C}$. Furthermore, we can easily check that $\Lambda(z_0) = \alpha\Lambda(z_0')$, for some nonzero $\alpha \in \mathbf{C}$, if and only if z_0 and z_0' are equivalent for the modular group (ie, are in the same orbit). Now, a simple but important observation is that the Riemann surfaces $\mathbf{C}/\Lambda$ and $\mathbf{C}/\Lambda'$ are isomorphic if and only if $\Lambda' = \alpha\Lambda$, for some $\alpha \neq 0$.[39] Therefore, the $\overline{\Gamma}$-orbit of a given z_0 leads to isomorphic elliptic curves via $\Lambda(z_0)$.

Given an elliptic curve $\mathcal{C}$: $y^2 = 4x^3 - ax - b$, there exists $z_0 \in \mathcal{H}$ such that $a = \alpha^2 g_2(z_0)$ and $b = \alpha^3 g_3(z_0)$, where the g_i are defined as in (2.32) for the lattice $\Lambda(z_0)$. The curve $\mathcal{C}$ is isomorphic[40] to $y^2 = 4x^3 - g_2(z_0)x - g_3(z_0)$. Therefore, as a Riemann surface, $\mathcal{C}$ is itself isomorphic to $\mathbf{C}/\Lambda(z_0)$, and the class of z_0 in $\Gamma \backslash \mathcal{H}$ is uniquely determined by $\mathcal{C}$. In other words, we have a parameterization of the space of all elliptic curves over $\mathbf{C}$ by points in a fundamental domain of the modular group.

---

[39]This is a reminder that Riemann surfaces are more than just topological objects. These particular Riemann surfaces, being parallelograms, are homeomorphic yet not necessarily isomorphic.

[40]This point might dispel the suspicion the reader might have had about one number z_0 parameterizing a family of curves that is defined by two parameters a, b. All the curves specified by $\alpha^2 a$ and $\alpha^3 b$, for $\alpha \neq 0$, are actually isomorphic.

Now a closing note to reconnect our discussion to modular functions. If we define the discriminant function

$$\Delta(z) \overset{\text{def}}{=} g_2(z)^3 - 27g_3(z)^2, \qquad (2.33)$$

then it can be shown that the value at z_0 of the function

$$j(z) = \frac{(12g_2(z))^3}{\Delta(z)} \qquad (2.34)$$

characterizes the elliptic curve: It is called its *j-invariant*. It is a modular function for Γ with a simple pole at the cusp. In fact, it has an expansion in $q = e^{2\pi i z}$ of the form

$$j(q) = \frac{1}{q} + 744 + \sum_{n \geq 1} c_n q^n,$$

for integral c_n. Moreover, the modular functions for Γ are precisely the rational functions of j. From an arithmetic point of view, the j-function is particularly powerful. For example, consider a complex elliptic curve $\mathcal{C} = \mathbf{C}/\Lambda(z)$ defined over some number field[41] $\mathbf{F}$. It is isomorphic to the curve

$$y^2 = 4x^3 - ax - b,$$

where $a, b \in \mathbf{F}$. Obviously, the function j takes on an algebraic value at z, since

$$j(z) = \frac{(12a)^3}{a^3 - 27b^2} \, .$$

Most often, however, $j(z)$ is transcendental. Indeed, by a result of Schneider, it is known that, for any $z \in \mathcal{H}$, the numbers z and $j(z)$ are simultaneously algebraic if and only if z is a quadratic number.

The Riemann Surface of the Curve $X_0(N)$

In the last section we took a torus and showed that by looking at it the right way we could view it as an elliptic curve.[42] The fundamental domain of

---

[41] A number field is an extension field derived from $\mathbf{Q}$ by adjoining to it roots of polynomials with rational coefficients. It forms a vector space whose dimension is called its *degree*. For example, adjoining $i = \sqrt{-1}$ gives the Gaussian numbers $\mathbf{Q} + i\mathbf{Q}$, whose degree is 2.

[42] The reader might be forgiven for feeling that the terminology has gone awry. The problem is that geometers are uneasy about curves being surfaces, while analysts find nothing more natural. Blame it on the fact that a complex number is a point in the plane.

$\Gamma' \backslash \mathcal{H}$, where Γ' is a congruence subgroup, can have an arbitrarily complex topology (as a Riemann surface). Can we somehow view it, too, as an elliptic curve? Obviously not. When the genus is higher than 1 we cannot hope to get an elliptic curve. But what about an algebraic curve? The answer is yes. Not only that, but if Γ' is chosen as $\Gamma_0(N)$, then the quotient Riemann surface $\Gamma_0(N) \backslash \mathcal{H}$, with cusps added, which is denoted by $X_0(N)$, is an algebraic curve that can be given with *rational* coefficients.[43] This is hard to prove but easy to explain.

We begin with the boring case: $X_0(1)$. This is simply the fundamental domain of the modular group shown in Figure 2.8. You can triangulate it by hand and check for yourself, by using the formula $V - E + F = 2 - 2g$, that its genus is 0. It is just the Riemann sphere. The function $j(z)$ defined in (2.34), being modular for Γ, provides a bijection from $X_0(1)$ to the Riemann sphere. As we saw earlier, the modular functions for Γ are precisely the rational functions of j. We are done.

Before we move on to the more difficult case, $N > 1$, we should mention that the discriminant function $\Delta(z)$ of (2.33) is a modular form of weight 12 for Γ. It is actually a cusp form and, as it turns out, the lowest possible one of that weight. A remarkable formula of Jacobi says that

$$(2\pi)^{-12} \Delta(z) = q \prod_{n \geq 1} (1 - q^n)^{24}.$$

If we expand the right-hand side in q, we obtain

$$q \prod_{n \geq 1} (1 - q^n)^{24} = \sum_{n \geq 1} \tau(n) q^n,$$

where $\tau(n)$ is called the *Ramanujan function*. It satisfies all sorts of properties, in particular, multiplicativity, ie, $\tau(nm) = \tau(m)\tau(n)$ whenever n and m are relatively prime (those multiplicative functions cropping up in a Dirichlet series again). This was conjectured by Ramanujan and proven by Mordell.

What about $X_0(N)$, for $N > 1$? It can be interpreted as an algebraic curve with rational coefficients. This is derived from the observation that the functions $j(z)$ and $j(Nz)$ satisfy an algebraic equation with coefficients in **Q**. The whole field of modular functions over $X_0(N)$, ie, relative to $\Gamma_0(N)$, is in fact generated by $j(z)$ and $j(Nz)$.

We are talking here about isomorphisms between Riemann surfaces $X_0(N)$ and algebraic curves. Anticipating our discussion ahead, we might change

---

[43] Its genus is 1 for $N = 11, 14, 15, 17, 19, 20, 21, 24, 27, 32, 36, 49$.

our viewpoint slightly and consider $X_0(N)$ as the domain of parameterization of an algebraic curve. One big difference is that the correspondence remains onto but not necessarily one-to-one. This means, in particular, that in the case of an elliptic curve the candidate $X_0(N)$ can be of genus higher than one. The question we now ask is: Given $a, b \in \mathbf{Q}$, does there exist N and two modular functions f_1, f_2 for $\Gamma_0(N)$ parameterizing the elliptic curve $y^2 = 4x^3 - ax - b$ through the identity

$$f_2(z)^2 = 4f_1(z)^3 - af_1(z) - b?$$

The Shimura-Taniyama conjecture says yes. To put this in perspective, recall that, via Weierstrass, parallelograms provide parameterizations for *all* complex elliptic curves. Via modular functions, the Shimura-Taniyama conjecture says that any rational elliptic curve can be parameterized from some suitable $X_0(N)$.[44] So, it appears that the "arithmetic" behind the words, arithmetic geometry, is revealed through the study of hyperbolic space.

We could end this appendix here but we would be missing an important chunk of the story, ie, the role played by L-functions. The two modular functions f_1, f_2 discussed above can be combined to define a modular form of weight 2 whose L-function is the same as that of the elliptic curve. But, what exactly is the L-function of an elliptic curve? The definition is quite different from the one we gave for modular forms, and that the two notions should be closely related is quite amazing. Our first order of business is to discuss one of the most beautiful features of an elliptic curve: its group structure.

The Addition Law of Elliptic Curves

To look at an elliptic curve over a Riemann surface brings analysis and topology into the picture. But perhaps the most amazing property of an elliptic curve is algebraic: It has an abelian group embedded in it. The algebraic structure of an elliptic curve is how we get L-functions into play.

We explain. To define an additive group over the points of an elliptic curve $\mathcal{C}$, we could simply use the parameterization from $\mathbf{C}/\Lambda$ to $\mathcal{C}$ and carry over the addition law in the obvious way, ie, to add the points $(\wp(z_1), \wp'(z_1))$

---

[44]The reader might be wondering why we seem to have a fixation on $X_0(N)$ and not, say, $X(N)$. The reason is that the covering of an elliptic curve by $X_0(N)$ is, in some sense, as tight as possible. Remember that $\Gamma(N) \subset \Gamma_0(N)$ and there are coverings $X(N) \mapsto X_0(N)$, and hence $X(N) \mapsto \mathcal{C}$. Moreover, $X(N)$ might not in general be defined over $\mathbf{Q}$ but only over a number field containing N-th roots of unity.

and $(\wp(z_2), \wp'(z_2))$ we take the point $(\wp(z_1 + z_2), \wp'(z_1 + z_2))$. This works, but it misses the whole point, ie, the beauty of the geometry behind it.

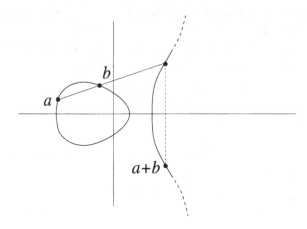

Fig. 2.12. The group structure of an elliptic curve.

Adding points a and b is easily visualized over the real projective plane (Fig. 2.12). If $a \neq b$, we draw the line connecting them and look at the third intersection point. As its name does not indicate, the curve is a cubic; so if we have two intersections with a line, we expect a third one (x, y) by Bézout's theorem. We define $a + b$ to be the point $(x, -y)$. If $a = b$, we do the same with the tangent line at a. The identity is the point at infinity. These definitions are trivial. What is less so is that they do indeed give us an abelian group $(\mathcal{C}(\mathbf{C}), +)$.[45] Commutativity is obvious, but what about associativity (Fig. 2.13)?

Of course, we can define a similar group structure over all kinds of fields (not just the reals). The group $\mathcal{C}(\mathbf{Q})$ of rational points on the curve $\mathcal{C}$ is striking because, as Mordell showed, it is finitely generated (in fact, Mordell-Weil's theorem says that this remains true over any number field). So, following the standard classification of finitely generated abelian groups, we know that it must consist of (at most) a torsion subgroup (ie, a direct sum of finite cyclic groups of the form $\mathbf{Z}/n\mathbf{Z}$) and an infinite free group $\mathbf{Z}^r$. How to compute the number r, which is called the rank of the group $\mathcal{C}(\mathbf{Q})$, remains a major open problem to this day. Computing the torsion part is easier, but classifying all possibilities is very difficult. The problem was solved only in the late 1970's by Mazur, who showed (among other

---

[45]The notation $\mathcal{C}(S)$ refers to the set of points of $\mathcal{C}$ with coordinates in S.

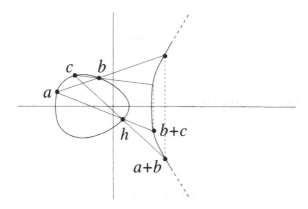

Fig. 2.13. The beautiful geometry behind associativity: The intersection h of the segments connecting a to $b + c$ and c to $a + b$ lies on the curve!

things) that the torsion subgroup of an elliptic curve over $\mathbf{Q}$ is one of only 15 possible groups; each one consists of one or two cyclic groups and their maximum cardinality is 16.

To understand the group of rational points we follow a local approach, ie, we study its subgroup obtained by reducing modulo primes p. (This may sound like wishful thinking, but the theory of p-adic numbers is there to help with this sort of extension from local to global.) Obviously, we must assume that the denominators in the coefficients of the curve are not multiples of p. This implies that they have inverses mod p, so we might as well assume that all of the coefficients are integral, which we take mod p when reducing at prime p. By requiring a nonvanishing discriminant we ensured that in the elliptic equation the polynomial in x has no multiple roots over the complex numbers. But this may no longer be true once we reduce mod p. We call this a *bad reduction at p*. Fortunately, such "bad" primes divide the discriminant, so there are only a finite number of them.

Zeta Functions of Number Fields

The L-function of an elliptic curve is a global object which is a collection of local pieces: the zeta functions of the curve over finite fields. Instead of rushing into definitions (which are not so simple) and struggling to explain their purposes, it is better to follow a more historical path, if only to appreciate how new concepts often emerge from older ones by mere analogy. So, we begin our story with the Riemann zeta function. As is well known, for $\Re(s) > 1$,

$$\zeta(s) \stackrel{\text{def}}{=} \sum_{n \geq 1} \frac{1}{n^s} = \prod_p \frac{1}{1 - p^{-s}}, \tag{2.35}$$

where the product is over all primes p. This follows trivially from the unique factorization of the integers and the expansion,

$$\frac{1}{1 - x} = \sum_{i \geq 0} x^i.$$

The Riemann zeta function can be extended to a meromorphic function over the whole complex plane with a simple pole at $s = 1$. It satisfies the functional equation

$$\Lambda(s) = \Lambda(1 - s),$$

where

$$\Lambda(s) \stackrel{\text{def}}{=} \pi^{-s/2} \Gamma(s/2) \zeta(s).$$

It has "obvious" zeros at $s = -2n$ for $n > 0$, and the famous *Riemann hypothesis* says that all the others, of which there are infinitely many, lie on the line $\Re(s) = 1/2$. The importance of this hypothesis—and the difficulty in settling it—is hard to overestimate. Scores of fundamental mathematical results would follow directly from it.

Dedekind generalized this idea to any number field by re-interpreting the term p in the denominator of the Riemann zeta function (2.35) as the order of the quotient ring formed by the integers of that field modulo some nonzero prime ideal. We still have equality with the corresponding Dirichlet series because of the unique factorization of nonzero ideals over the integers of number fields.

We open a brief parenthesis to explain what this all means. Throughout our discussion, rings are assumed to be commutative. An ideal of a ring is a subgroup (for addition) closed under multiplication by any ring element. Just as normal subgroups are useful for manufacturing new groups, ideals are useful for creating ring homomorphisms and, hence, subrings. Indeed, modding out the ring by the ideal (ie, identifying ring elements that differ only by an ideal element) creates another ring.

Recall that a number field K is a finite-degree extension of the rationals. It can be viewed either as $\mathbf{Q}[X]$ modulo some polynomial or as the adjunction to $\mathbf{Q}$ of the roots of a polynomial in some algebraic closure. Taking the latter view, consider the numbers that are roots of monic polynomials (ie, highest degree coefficient $= 1$) with integral coefficients: they form a ring R, called the *integers*[46] of K. If $\mathcal{I}$ is a nonzero ideal of R, then the quotient ring $R/\mathcal{I}$ is actually finite. Its number of elements defines the *norm* of the ideal $\mathcal{I}$ and is denoted by $N(\mathcal{I})$. By direct analogy with $\mathbf{Z}/p\mathbf{Z}$, if the ideal is *prime* (ie, $fg \in \mathcal{I}$ implies that f or g is in $\mathcal{I}$) then $R/\mathcal{I}$ is a finite field, and therefore $N(\mathcal{I}) = p^m$, for some prime p (see page 319 for a review of finite fields). A wonderful property of the ring of integers is that any nonzero ideal of R can be written uniquely as a product of prime ideals. This is the generalization in the language of ideals of the unique factorization[47] of integers in $\mathbf{Z}$. So, by analogy with the Riemann zeta function, the *Dedekind zeta function* defined below should be expressible as both an Euler product and an infinite sum:

$$\zeta_K(s) \overset{\text{def}}{=} \sum_{\mathcal{I} \neq 0} \frac{1}{N(\mathcal{I})^s} = \prod_{\mathcal{P}} \frac{1}{1 - N(\mathcal{P})^{-s}},$$

where the sum (resp. product) is over all nonzero ideals (resp. prime ideals). The function $\zeta_K(s)$ converges absolutely for $\Re(s) > 1$ and has a meromorphic continuation over the whole complex plane. Obviously, the case $K = \mathbf{Q}$ gives us back the Riemann zeta function. One can formulate a Riemann hypothesis for number fields by direct analogy. No one has a clue on how to go about proving it, however, so we turn to function fields, instead, which are far less mysterious.

Zeta Functions of Curves

Function fields provide the setting for the (proven) Riemann hypothesis for algebraic curves, which can then be generalized to arbitrary varieties. Consider the case of an affine elliptic curve $\mathcal{C}$ over a finite field of the type introduced in the previous section. Let $f(X, Y) = Y^2 - g(X)$ denote its defining polynomial, and let

$$F(\mathcal{C}) = \mathbf{F}_p[X, Y]/(f(X, Y))$$

---

[46] Unfortunately, to determine which elements of K are integers is usually not so easy. There are exceptions. For example, consider the cyclotomic field obtained by adjoining to $\mathbf{Q}$ the root of unity $\zeta = e^{2\pi i/p}$, for some odd prime p. Its integers are the integral combinations $n_0 + n_1\zeta + \cdots + n_{p-2}\zeta^{p-2}$. Note that we stop at $p - 2$ because, trivially, ζ_{p-1} is a linear combination of the other roots.

[47] Unique factorization should not be taken for granted. Of course, some rings enjoy this property, eg, $\mathbf{Q}[x_1, \ldots, x_n]$ and $\mathbf{Z} + \mathbf{Z}\sqrt{-1}$. The latter case is an accident. Most rings of integers do not have unique factorization. For example, in the ring $\mathbf{Z} + \mathbf{Z}\sqrt{-5}$, we have $6 = 2 \times 3 = (1 + \sqrt{-5})(1 - \sqrt{-5})$.

be the quotient ring of polynomials in X, Y with coefficients in $\mathbf{F}_p$ modulo the ideal generated by $f(X,Y)$. This is like adjoining $\sqrt{g}$ to the base ring, so it is a quadratic extension of $\mathbf{F}_p[X]$. Now, consider any nonzero prime ideal $\mathcal{P}$ in $F(\mathcal{C})$ and, as usual, denote by $N(\mathcal{P})$ the order of the finite field $F(\mathcal{C})/\mathcal{P}$. We define the zeta function of $\mathcal{C}$ as the Euler product

$$\zeta_\mathcal{C}(s) \stackrel{\text{def}}{=} \prod_\mathcal{P} \frac{1}{1 - N(\mathcal{P})^{-s}}, \tag{2.36}$$

for $\Re(s) > 1$, where the product ranges over all nonzero prime ideals $\mathcal{P}$ in $F(\mathcal{C})$. Since $F(\mathcal{C})/\mathcal{P}$ is a finite field, $N(\mathcal{P})$ is a prime power p^m; we denote m by $\deg \mathcal{P}$. With the change of variables, $T = p^{-s}$, this allows us to write $\zeta_\mathcal{C}(s) = Z_\mathcal{C}(T)$, where

$$Z_\mathcal{C}(T) = \prod_\mathcal{P} \frac{1}{1 - T^{\deg \mathcal{P}}}. \tag{2.37}$$

With our assumption that $\mathcal{C}$ is an elliptic curve, we find (well, Artin did) that the zeta function is rational:

$$Z_\mathcal{C}(T) = \frac{1 + (N_p - p)T + pT^2}{1 - pT}, \tag{2.38}$$

where N_j is the number of points in $\mathcal{C}(\mathbf{F}_j)$. Dwork generalized this result to the zeta function of any algebraic set. Note that, for a projective curve, the function has an extra factor of $1/(1 - T)$ and N_p must be replaced by $N_p - 1$ to compensate for the extra point at infinity.

An important observation is that the Riemann hypothesis for the curve (ie, zeros on the line $\Re(s) = 1/2$) translates into a deviation bound on the number of zeros:

$$|N_p - p| \le 2\sqrt{p}. \tag{2.39}$$

This seems to indicate that the number of points varies randomly around p like a binomial distribution: We discuss this pseudorandom behavior and its relevance to computer science in our review of quadratic characters in §9.1.

To see the relation of this upper bound with the Riemann hypothesis, note that if $\zeta_\mathcal{C}(s)$ vanishes for s with $\Re(s) = 1/2$, then the roots of the numerator in (2.38) are of the form $T = p^{-1/2 - it}$. This rules out two distinct real roots and forces the discriminant of the quadratic equation in T to be nonpositive, ie, $(N_p - p)^2 - 4p \le 0$, and hence (2.39); again, remember to subtract one for the projective version of the bound, ie, $|N_p - p - 1| \le 2\sqrt{p}$. The Riemann hypothesis for elliptic curves was proven by Hasse in the 1930's and generalized to general curves by Weil in the late 1940's, and finally to algebraic varieties by Deligne in the 1970's.

One closing thought before we do an exercise. By formal manipulation involving taking derivatives and logarithms defined as power series (see details below), we derive

$$\ln Z_{\mathcal{C}}(T) = \sum_{m \geq 1} \frac{N_{p^m}}{m} T^m. \tag{2.40}$$

Since the expression for $Z_{\mathcal{C}}(T)$ in (2.38) depends only on N_p, what follows is particularly nice: Each of the N_{p^m}'s can be deduced from N_p alone! To do that, we read the power series in (2.40) as a Taylor expansion and recover each N_{p^m} by the usual formula, ie, setting $T = 0$ in

$$\frac{1}{(m-1)!} \frac{d^m}{dT^m} \ln Z_{\mathcal{C}}(T).$$

A Simple Example: To understand these concepts better, it might be good to work out a few of the calculations, and to prove the rationality of the zeta function for at least one simple object. We now regard $\mathcal{C}$ as an arbitrary affine plane curve. Again, fix some prime p.

Let $\overline{\mathbf{F}}_p$ denote an algebraic closure of $\mathbf{F}_p$. Given any point $\alpha \in \mathcal{C}(\mathbf{F}_{p^m})$, consider the smallest field $\mathbf{F}_{p^d} \subset \overline{\mathbf{F}}_p$ containing its coordinates. All the points

$$\alpha, \alpha^p, \alpha^{p^2}, \dots, \alpha^{p^{d-1}}$$

are also in $\mathcal{C}$ and they are all distinct; to raise a point to a power means to raise its coordinates to that power.[48] This set $\mathcal{D}$ of d points is called a *prime divisor* of $\mathcal{C}$ of degree d.

Let us count the number N_{p^m} of points of $\mathcal{C}(\mathbf{F}_{p^m})$. To do that, for each subfield F of $\mathbf{F}_{p^m}$, we count separately how many points lie in $\mathcal{C}(F)$ but not in any smaller subfield. As is well known, any subfield of $\mathbf{F}_{p^m}$ is of the form $\mathbf{F}_{p^d}$ for some d dividing m. So, obviously,

$$N_{p^m} = \sum_{d \mid m} d a_d,$$

---

[48] Why should α^p (and hence the other points) belong to $\mathcal{C}$? Within the field $\mathbf{F}_{p^d}$, we have the identities $(a + b)^p = a^p + b^p$ and (if $a \in \mathbf{F}_p$) $a^p = a$. This follows from, respectively, the characteristic being p and Fermat's theorem. This shows that if $x \in \mathbf{F}_{p^d}$ is a zero of a polynomial in $\mathbf{F}_p[X]$, then so is x^p (look at what happens to $f(x)^p$), and hence, any x^{p^i}.

where a_d is the number of prime divisors of degree d. Elementary algebraic geometry shows that the Euler product (2.37) can also be written as a product over all prime divisors, ie,

$$Z_C(T) = \prod_{D} \frac{1}{1 - T^{\deg D}} = \prod_{n \geq 1} \left(\frac{1}{1 - T^n} \right)^{a_n}.$$

Taking the logarithmic derivative[49] of $Z_C(T)$ yields

$$
\begin{aligned}
\frac{d}{dT} \ln Z_C(T) &= -\sum_{n \geq 1} a_n \frac{d}{dT} \ln(1 - T^n) = \frac{1}{T} \sum_{n \geq 1} \frac{n a_n T^n}{1 - T^n} \\
&= \frac{1}{T} \sum_{n \geq 1} n a_n T^n \sum_{k \geq 0} T^{kn} = \frac{1}{T} \sum_{m \geq 1} \left(\sum_{d \mid m} d a_d \right) T^m \\
&= \frac{1}{T} \sum_{m \geq 1} N_{p^m} T^m.
\end{aligned}
$$

Integrating term-wise gives us

$$Z_C(T) = \exp \left(\sum_{m \geq 1} \frac{N_{p^m}}{m} T^m \right).$$

This last expression is actually the standard way of defining the zeta function.[50] We avoided it because it seems to come out of nowhere, unlike the Euler product form which brings out the analogy with the Riemann zeta function.

To understand the rationality of the zeta function, we examine a very simple case. Instead of an elliptic curve, consider the affine line

$$x + y = 1.$$

Trivially, $N_{p^m} = p^m$. Using $-\ln(1 - x) = \sum_{m > 0} x^m / m$, we find that

$$\ln Z_C(T) = \sum_{m \geq 1} \frac{N_{p^m}}{m} T^m = \sum_{m \geq 1} \frac{(pT)^m}{m} = -\ln(1 - pT).$$

---

[49] Think of this operation algebraically as operating on the power series defining the logarithm and using term-wise derivation. Don't think analytically and worry about convergence.

[50] Now, one can understand our earlier comment about affine vs. projective. The difference in N_{p^m} is exactly one, so the change in $\ln Z_C(T)$ is a factor of $\sum_{m \geq 1} T^m / m$, which is $-\ln(1 - T)$, and the difference in $Z_C(T)$ is a factor of $1/(1 - T)$, just as we said.

Exponentiating leads to

$$Z_{\mathcal{C}}(T) = \frac{1}{1 - pT}.$$

Not only we see rationality right before our eyes, but we also recognize the bottom part of (2.38). This is telling us that this denominator has really nothing to do with the elliptic curve (or at least with the part that matters). In the case of an elliptic curve the numerator in (2.38) can be recovered by appealing to the Riemann-Roch theorem. Not that the reader really needs to know this, but the numerator has a deeper cohomological interpretation. The cohomological view was initiated by Weil and then elaborated greatly by Artin, Grothendieck and others via the notion of étale topology, leading to l-adic cohomology: These developments were motivated by the Weil conjectures and were, of course, directly relevant to Deligne's upper bound (2.30).

The Hasse-Weil L-Function

Of course, what we have done so far is very local: only one prime p at a time. The zeta functions bring into an Euler product form a sum over the number of points of the curve with coordinates in $\mathbf{F}_{p^k}$. We must now try to combine all of these functions together.

Take a "good" prime p (ie, one not giving a bad reduction). The Riemann hypothesis bounds the deviation of N_p from the number $p + 1$ of group elements (counting the point at infinity), so it makes sense to define the quantity $a_p = p + 1 - N_p$. We define the factor of the Euler product for prime p as

$$L_p(\mathcal{C}, s) = \frac{1}{1 - a_p\, p^{-s} + p^{1-2s}},$$

if p is good.[51] For the finite number of cases of bad reductions (ie, where p is bad) we set $a_p = -1, 0, 1$, depending on the type of singularity involved (happily there are only three cases), and we write

$$L_p(\mathcal{C}, s) = \frac{1}{1 - a_p\, p^{-s}}.$$

As expected, the L-function of the elliptic curve $\mathcal{C}$ is defined as the Euler

---

[51] Recall that in the projective case, the numerator of (2.38) is $1 + (N_p - 1 - p)T + pT^2$, which is precisely the denominator in $L_p(\mathcal{C}, s)$. Obviously, not a coincidence!

product

$$L(\mathcal{C}, s) = \prod_p L_p(\mathcal{C}, s).$$

Good things happen now: This Euler product converges for $\mathrm{Re}(s) > 3/2$ and is equal to the corresponding Dirichlet series, ie,

$$L(\mathcal{C}, s) = \sum_{n \geq 1} \frac{a_n}{n^s},$$

where a_n is extended to nonprimes by the multiplicative character formula explained earlier. This L-function is intimately related to the group structure on the curve. In fact, Faltings showed that two elliptic curves are isogenous (ie, one curve maps homomorphically to the other in a nontrivial way) if and only if their L-functions agree.

The Shimura-Taniyama Conjecture

Before we discuss the "grand unification" implied by the Shimura-Taniyama conjecture, we should show how to "read" an elliptic curve into a modular form. This will also, once again, reassure the reader about the soundness of the terminology: Modular forms are, indeed, differential forms. Take a cusp form f of weight 2 for $\Gamma_0(N)$ with Fourier coefficients in $\mathbf{Z}$.[52] Given $z_0 \in \mathcal{H}$, consider the integral

$$w_{z_0}(\gamma) \stackrel{\text{def}}{=} \int_{z_0}^{\gamma(z_0)} f(z)\, dz,$$

where $\gamma \in \Gamma_0(N)$. It is easy to see why the integral is independent of the path and the set

$$\{\, w_{z_0}(\gamma) \mid \gamma \in \Gamma_0(N) \,\}$$

is independent of z_0. Using the fact that the Fourier coefficients of the eigenform f are integral, it can be shown that this set forms a lattice Λ. As we saw earlier, it therefore defines an elliptic curve parameterized by the Weierstrass function for Λ.

Now, let us get on with the Shimura-Taniyama conjecture. Consider an elliptic curve $\mathcal{C}$ over $\mathbf{Q}$, and suppose that it has a parameterization $X_0(N) \mapsto \mathcal{C}$, with N as the *conductor* of the curve; the conductor is (roughly) the product of the bad primes for $\mathcal{C}$ raised to some powers determined by

---

[52] We actually need a few more conditions, such as being an eigenform and a *newform*, the latter implying that it does not arise from some lower level properly dividing N.

the type of singularity they correspond to. (Of all the candidate surfaces $X_0(N)$, focusing on the one where N is the conductor of the curve was a deep insight of Weil.) An important case, called *semistable*, is when the conductor is square-free. This essentially rules out cusplike singularities formed by triple roots.

It is known that if C can be parameterized by a pair of modular functions, then its Hasse-Weil L-function coincides with the L-function of a modular cusp form f of weight 2 for $\Gamma_0(N)$.[53] The Shimura-Taniyama conjecture claims precisely such a parameterization $X_0(N) \mapsto C$, and hence the modularity of C. Wiles' proof concerns the semistable case of the conjecture (which, as we mentioned, is not a terribly serious restriction). Before his result, the modularity of rational elliptic curves was known for those satisfying a property called *complex multiplication*, and it had been established by Shimura and others for many special cases.

In the event (of course, unlikely) that all of these technical words have somehow obscured the magic behind Wiles' result, let us again summarize the main point.

The punchline. *Take a rational projective elliptic curve with conductor N. For each prime p not dividing N, define a_p to be $p + 1$ minus the number of points of the curve with coordinates in the Galois field $\mathbf{F}_p$; set a_p to $-1, 0, 1$ for the other primes, and extend a_n to nonprimes by applying the multiplicative recipe discussed earlier. Finally, define the Fourier expansion*

$$f(z) = \sum_{n \geq 1} a_n e^{2\pi i n z}.$$

Lo and behold, $f(z)$ is a cusp form of weight 2 for the congruence subgroup $\Gamma_0(N)$.

So, there we started with an object encoding the "arithmetic" part of a cubic curve, and now we have an analytic object inheriting those beautiful symmetries of the hyperbolic plane. Moreover, its L-function $\sum a_n/n^s$ extends to the whole plane by meromorphic continuation and satisfies a functional equation. It's like complex analysis and number theory coming together![54]

Arithmetic algebraic geometry reveals a deep connection between the al-

---

[53] As we discussed earlier, the Fourier series expansion of f can be retrieved directly from the L-function by the inverse Mellin transform.

[54] All right, these are *just* cubic curves and not arbitrary polynomials. But, still, one has to recognize magic when one sees it.

gebraic and analytic properties of curves, which are shown to be the two sides of the same coin. Modular functions bring in analysis and invariance under certain discrete isometry groups. Dirichlet series and Euler products encode the numbers of points on the curve over various finite fields and attach analytic meaning to them as a whole. Finally, functional equations and analytic continuations provide the indispensable analytic glue. The apparent disparity between all of these notions is what makes the proofs so highly technical. Specific tools must be used to exploit the hidden symmetries behind these objects. One of the most powerful is the theory of Galois representations. This is because the modularity of a curve can be expressed as properties of certain objects related to Galois theory. We briefly explain.

Take an irreducible polynomial P with coefficients in some field $\mathbf{F}$ and let $\mathbf{G}$ denote the extension field that it defines (ie, the smallest field containing $\mathbf{F}$ and the roots of P in an algebraic closure of $\mathbf{F}$). We call $\mathbf{G}$ a *Galois extension* over $\mathbf{F}$. (Note that not all field extensions are Galois; any finite-degree extension, however, is contained in a Galois extension.) By looking at the symmetric polynomials we know that there are all sorts of relationships among the roots of P. We would like an algebraic structure, in fact, a group, to represent *all* such algebraic relations. Galois theory tells us that such a group does indeed exist. Denoted by $\mathrm{Gal}(\mathbf{G}/\mathbf{F})$, it can be defined from the set of root permutations that extend to automorphisms of $\mathbf{G}$ that leave $\mathbf{F}$ fixed. The reason why this group of automorphisms is so powerful is that all intermediate fields $\mathbf{F} \subseteq \mathbf{F}' \subseteq \mathbf{G}$ are in bijection with its subgroups H, with the correspondence such that H leaves $\mathbf{F}'$ fixed. Not only that, but if H is normal, then $\mathbf{F}'$ is Galois over $\mathbf{F}$ and its quotient group $\mathrm{Gal}(\mathbf{G}/\mathbf{F})/H$ is in fact the Galois group of $\mathbf{F}'$ over $\mathbf{F}$.

To put Galois theory to use with an elliptic curve $\mathcal{C}$, we first must try to define a Galois group $G = \mathrm{Gal}(\overline{\mathbf{Q}}/\mathbf{Q})$, where $\overline{\mathbf{Q}}$ is the algebraic closure of $\mathbf{Q}$, ie, the infinite field extension obtained by adjoining to $\mathbf{Q}$ all roots of all polynomials with coefficients in finite-degree extensions of $\mathbf{Q}$. Next, we study how G acts on various subgroups of $(\mathcal{C}(\mathbf{C}), +)$. In particular, we consider the elements of order p: This includes all the points A such that $pA = 0$. It forms a subgroup isomorphic to $(\mathbf{Z}/p\mathbf{Z})^2$. To tackle the formidable group G, we try our hand at simpler cases. We consider the Galois group $G_K = \mathrm{Gal}(\mathbf{K}/\mathbf{Q})$, where $\mathbf{K}$ is a finite-degree Galois extension of $\mathbf{Q}$. An element of G_K acts linearly on $\mathcal{C}(\mathbf{K})$, so it is like a linear transformation of $(\mathbf{Z}/p\mathbf{Z})^2$, which in turn gives us a *representation* of G_K in the group $\mathrm{GL}(2, \mathbf{Z}/p\mathbf{Z})$ of 2-by-2 invertible matrices with elements in $\mathbf{Z}/p\mathbf{Z}$.

From what we said earlier about Galois theory, we expect G_K to be a quotient of G by some normal subgroup. So we can repeat for all numbers

p^m and in this way create a representation of G using 2-by-2 matrices with elements over the p-adic numbers (ie, numbers expressed as power series in p with a finite number of negative exponents). The purpose of such representations is that (i) they encode a lot of information about the curve and (ii) we can actually work on them.

At this point, I am afraid we enter the exclusive province of the experts. The picture gets crowded with the cohomology of infinite Galois groups, Selmer groups, p-adic representations, Frobenius automorphisms, modular lifting, deformations, and so on. No, this is not an attempt to scare you away; it is only to guard you, the reader, against the temptation you might have to reward yourself with the belief that perhaps we are now halfway through understanding Wiles' proof. The truth is, we have not even begun. Nevertheless, it is my hope that this brief excursion into arithmetic algebraic geometry has helped the reader to make better sense of the orbital construction for placing points on the sphere by explaining its main ingredients and putting them in context.

2.7 The Laplacian and Optimum Principles *

As mentioned earlier, the main role played by the Laplacian in the proofs of Theorems 2.9 and 2.10 (page 58) is that, because it commutes with the operators of interest, its eigenfunctions can be used to decompose $L^2(S^2)$ into spaces that these operators leave invariant. This in turn allows us to use Fourier analysis over finite-dimensional spaces.

The Laplacian plays a major role in physics and applied mathematics. More to the point, it is central to the discrepancy method because of its relevance to expanders and pseudorandomness. It is useful to understand where the concept comes from. By going back to the Dirichlet principle, we show how the Laplacian ties together questions of equilibrium and optimization. The material is fairly classical and can be skipped by the reader with a background in applied mathematics. It is highly recommended to the others, however, because it attempts to explain the whys and wherefores of the Laplacian from a computer science perspective.

Consider a fluid in motion (in a two-dimensional environment), and assume that there is no viscosity and that the flow is irrotational (no vorticity, eg, no stirring with a spoon).[55] When the flow achieves its steady state

---

[55] The standard way to express the lack of rotation is to say that the circulation of the velocity is zero, ie, $\int v_x \, dx + v_y \, dy = 0$.

(ie, no longer changes over time), it can be shown that at any point the velocity of the fluid is the gradient of a potential function $u(x, y)$, ie,

$$v_x = \frac{\partial u}{\partial x} \quad \text{and} \quad v_y = \frac{\partial u}{\partial y}.$$

Furthermore, conservation of fluid (in the absence of sources or sinks) implies that (we'll see why later)

$$\frac{\partial v_x}{\partial x} + \frac{\partial v_y}{\partial y} = 0.$$

Eliminating the velocity vector gives

$$\frac{\partial^2 u}{\partial x^2} + \frac{\partial^2 u}{\partial y^2} = 0. \tag{2.41}$$

In other words, the divergence of the gradient of the potential must vanish. This equation, known as *Laplace's equation*, can be rewritten in "operator" notation,

$$\Delta u = 0,$$

where $\Delta = \partial^2/\partial x^2 + \partial^2/\partial y^2$ is called the *Laplacian*. Any function whose Laplacian vanishes is called *harmonic*.

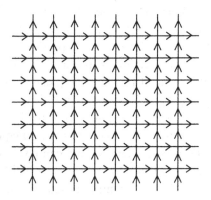

Fig. 2.14. Flow on a grid.

An intuitive understanding of this equation can be obtained by looking at a discrete model, which one might use in a computer simulation. Think of the medium as a fine square grid of pipes through which the fluid flows. Because the flow is irrotational, the grid consists of directed edges with no cycles. Actually, to simplify matters, assume that each horizontal (resp. vertical) edge is directed from left to right (resp. upward); see Figure 2.14.

Such a network can be represented by a matrix A, whose n columns and m rows correspond to the nodes and edges, respectively (Fig. 2.15). Specifically, if $e = (i, j)$ is an edge of the graph, then the row labeled e consists of 0's everywhere, except for -1 in column i and 1 in column j. Adding up all the columns together gives us a column of 0, so the rank of the matrix is less than n. It is easy to see that the rank is exactly $n - 1$. Indeed, as long as the network is connected, no proper subset of the columns can have both 1 and -1 in each of its rows, and thus no linear combination of the columns can cancel those "defective" rows.

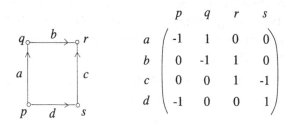

Fig. 2.15. A graph and its matrix A.

Each node i has a potential u_i, and the *flow* along the edge (i, j) is given by the difference $u_j - u_i$. The vector y of flows satisfies $Au = y$. Since flows involve only differences, it does not hurt to fix the potential of one node, say u_1, to 0 (thus "grounding" the network). This corresponds to removing one column from A, thus making it full-ranked. By the conservation of fluid (the analogue of Kirchhoff's law), as much flow enters a node as leaves it, which can be written as $A^T y = 0$. The two equations allow us to eliminate y and obtain

$$A^T Au = 0. \tag{2.42}$$

A comment is in order. In the continuous case, to say that the Laplacian of the potential is zero somewhere is a local statement. In the presence of sinks and sources, this need not be true everywhere, of course. Unfortunately, equation (2.42) makes a global statement and, as such, it is unrealistic. One should expect to have $A^T y = f$, where the left-hand side denotes the internal "forces" at work, and f the external ones. Away from sinks and sources, however, the equilibrium equations in $A^T Au = 0$ are still meaningful, and the analogy with Laplace's equation is valid.

The matrix $A^T A$ is called the *graph Laplacian* of the network. At each node, the potential is replaced by the sum of potential differences at each incident edge. It acts as an averaging operator. The clearest indication of the relevance of this notion to the connectivity structure of the network is that the quadratic form $u^T(A^T A)u$ is equal to $\sum_{i,j}(u_i - u_j)^2$, where the sum ranges over all directed edges (i,j). Predictably, relation (2.42) is precisely the discrete version of Laplace's equation. Indeed, if the grid is fine enough, any point can be thought of as a node. Its velocity vector is given by the flows of the outgoing edges. If y_i (resp. y_j) is the rightward (resp. upward) flow leaving a node (x,y) internal to the grid, then the velocity vector (v_x, v_y) is (y_i, y_j).

- The relation $Au = y$ is the discrete analogue of saying that the velocity is the gradient of a potential function. Indeed, notice that the ± 1 entries in A make it akin to the gradient operator $(\partial/\partial x, \partial/\partial y)$.

- The fluid conservation law $A^T y = 0$ is the discrete counterpart of divergence-free flow: $\partial v_x/\partial x + \partial v_y/\partial y = 0$.

Putting everything together, we see that $A^T Au = 0$ simply says that the gradient of the potential has zero divergence or, equivalently, that the potential is a harmonic function.

Our discussion of a fluid in motion can be applied to other physical systems (bed of springs, electrical networks, heat conduction, etc). Often, however, a few more bells and whistles are necessary. In the case of electrical networks, for example, u_i is the voltage at node i and y_j is the intensity of the current along edge j.

An edge might have a voltage source that affects the potential drop between its endpoints. Specifically, we assume that, along the directed edge (i,j), the voltage drop is $u_i - u_j + b_{i,j}$, instead of just $u_i - u_j$. In matrix form, this means that we must replace Au by $b - Au$; note the change of sign, which reflects the fact that the voltage drops and makes Au a negative vector. By Ohm's law, the equation $y = Au$ becomes $y = C(b - Au)$, where C is a diagonal matrix whose elements are the conductances of the edges. The main point is that it is a positive definite matrix C. Kirchhoff's law is expressed by $A^T y = g$ (the coordinate of g is zero if the corresponding node has no source of current on any incident edge). Thus, in general, the fundamental equations linking potentials and flows (or voltages and currents) are

$$C^{-1}y + Au = b \quad \text{and} \quad A^T y = g. \tag{2.43}$$

Eliminating y gives

$$A^T C A u = A^T C b - g. \tag{2.44}$$

If the matrix C is a simple scaling matrix (all of its entries are identical), then we have the discrete analogue of *Poisson's equation*, which is Laplace's equation with a nonzero right-hand side.

We now turn to the classical relation between equilibrium (Laplace or Poisson equation) and optimization (of the linear or quadratic sort). We shall discuss the discrete analogue of the Dirichlet principle, which seeks the minimum of the functional

$$\int_\Omega |\nabla u|^2 \stackrel{\text{def}}{=} \int\int_\Omega \left(\left(\frac{\partial u}{\partial x} \right)^2 + \left(\frac{\partial u}{\partial y} \right)^2 \right) dx\, dy, \tag{2.45}$$

with boundary conditions on $\partial\Omega$, among the functions u whose Laplacian vanishes inside Ω. Using nonstandard notation to make the intended connections more transparent, a typical linear program is of the form:

- Given a *cost vector* b and an m-by-n constraint *matrix* A, *minimize* $b^T y$, *subject to* $A^T y = g$ and $y \geq 0$.

This can be rewritten as

$$\min_{y \geq 0} \ \max_u \left\{ b^T y - u^T (A^T y - g) \right\}.$$

Indeed, note that the minimizer (which controls y) must set $A^T y - g$ to 0; otherwise, the maximizer (which controls u) is able to drive up the final cost to infinity. This being so, the maximizer is then rendered powerless (since it multiplies 0) and the minimizer is back to solving the original linear program. It is well known that min and max can be reversed in the expression above, which gives the equivalent formulation

$$\max_u \ \min_{y \geq 0} \left\{ b^T y - u^T (A^T y - g) \right\}. \tag{2.46}$$

(This can be shown by using Farkas' lemma.) Rewriting this as

$$\max_u \ \min_{y \geq 0} \left\{ g^T u - y^T (Au - b) \right\},$$

it is trivially equal to the maximum (over $u \in \mathbf{R}$) of $u^T g$, subject to the constraints $Au \leq b$. This is the duality theorem of linear programming, which says that the programs

$$\min_{y \geq 0} \left\{ b^T y \ : \ A^T y = g \right\}$$

and

$$\max\left\{\, g^T u \, : \, Au \le b \,\right\}$$

are equivalent. At the optimal (y, u) we have $Au - b \le 0$, and at those constraints $(Au - b)_j$ where the inequality is strict, the corresponding "weight" of the constraint, y_j is equal to 0. So the product $y_j(Au - b)_j$ is always zero. (This is known as the Kuhn-Tucker optimality condition.) The variables y and u are dual to each other. In the expression (2.46) it is natural to interpret u as a vector of Lagrange multipliers, because it is bringing the constraints and the optimization function together.[56] Setting the partial derivatives in u_i and y_j to 0, we obtain the two equations:

$$Au = b \quad \text{and} \quad A^T y = g,$$

which are the two fundamental equilibrium equations (2.43), where C is the conductance matrix with infinite values (the edges have no resistance).

The analogy is only half complete because the potentials and the flows do not interact in the same equation. To achieve this we turn to the classical quadratic programming problem:

- *Minimize $y^T B y - b^T y$, subject to the constraints $A^T y = g$.*

We assume that C is (real) symmetric positive definite. Write $B = \frac{1}{2}C^{-1}$. Introducing Lagrange multipliers in the form of a vector u, the question is identical to minimizing (over y) the maximum (over u) of

$$L(u, y) = \tfrac{1}{2} y^T C^{-1} y - b^T y + u^T (A^T y - g).$$

Setting the partial derivatives of L to 0 gives the system of equations,

$$C^{-1}y + Au = b \quad \text{and} \quad A^T y = g,$$

which are precisely the equations of equilibrium (2.43). Note that $y =$

---

[56] Recall that Lagrange multipliers are used to solve constrained optimization problems. For example, suppose that we wish to find the shortest distance from the origin in the plane to the unit-radius circle centered at $(1, 1)$. This can be written as $\min\{\,x^2 + y^2\,\}$, subject to $(x-1)^2 + (y-1)^2 = 1$. We introduce the multiplier λ and form the Lagrangian,

$$L(x, y, \lambda) \stackrel{\text{def}}{=} x^2 + y^2 + \lambda((x - 1)^2 + (y - 1)^2 - 1).$$

The answer is $\min_{x,y} \max_\lambda L(x, y, \lambda)$. (Indeed, the minimizer must satisfy the circle equation or else the maximizer can drive the Lagrangian to infinity.) The solution lies on a saddle point of the surface $z = L(x, y, \lambda)$. We find it by setting the three partials to 0. This gives $2x + 2\lambda(x - 1) = 0$ and $2y + 2\lambda(y - 1) = 0$, from which we derive the fact that $x = y$. The derivative in λ gives us back the circle equation, $(x - 1)^2 + (y - 1)^2 = 1$. It follows that the solution is $x = y = 1 - 1/\sqrt{2}$, as expected. (The solution $x = y = 1 + 1/\sqrt{2}$ solves the maximization problem.)

$C(b - Au)$; so by plugging back in we find that $L(u, y)$ has the value

$$-\tfrac{1}{2}(Au - b)^T C(Au - b) - u^T g.$$

This suggests the dual problem of maximizing the quadratic form above, or reversing signs:

- *Minimize $\tfrac{1}{2}(Au - b)^T C(Au - b) + u^T g$, subject to no constraints.*

Because C is positive definite, elementary algebra leads to the corresponding duality theorem, which states that the two minima are equal (at the saddle point of L). Potentials and flows are dual to each other. If we set C to be the identity and A to be the gradient matrix, then the potential vector obeys the discrete version of Poisson's equation (2.44). If we set $g = 0$, the problem becomes minimizing $\tfrac{1}{2}(Au)^T(Au)$. We recognize the discrete analogue of the Dirichlet problem (2.45), and the relation between Laplacian and optimization is now transparent. Nature, indeed, solves least-square problems all of the time! To do the same thing, however, we mere humans are stuck with having to solve partial differential equations derived from the Laplacian.

2.8 The Spanning Path Theorem

So far, our treatment of geometric discrepancy has been aimed at resolving the question: How well can a discrete measure approximate the Lebesgue measure? Volumes and areas can be regarded as discrete measures in the limit, so it is natural to ask how well a simple discrete measure can approximate a complicated one. We will show that it is possible to color n points in $\mathbf{R}^d$ red or blue, so that within no halfspace one color outnumbers the other by more than roughly $n^{1/2-1/2d}\sqrt{\log n}$. It is possible to remove the $\sqrt{\log n}$ factor and obtain a tight bound, but this requires a fair amount of work; so we shall satisfy ourselves with the weaker result.

What does this have to do with approximating discrete measures? If no color outnumbers the other by too much, then we can use the color with the fewer points as a good approximating measure. If we repeat this process k times, we end up with a sample of the original point set of size at most $n/2^k$ that induces a measure fairly similar to the original one. Red-blue discrepancy is thus a well-suited vehicle for the discrepancy method.

A red-blue coloring of a set P of n points in $\mathbf{R}^d$ is a partition of P into two sets, R and B, of "red" and "blue" points, respectively. In this section, all halfspaces are considered open. The discrepancy of a halfspace

τ is defined as

$$\Big| |R \cap \tau| - |B \cap \tau| \Big|.$$

Theorem 2.19 *Any set of n points in $\mathbf{R}^d$ can be two-colored in such a way that the maximum discrepancy of any halfspace is at most proportional to $n^{1/2-1/2d}\sqrt{\log n}$.*

We say that a hyperplane *cuts* a segment if it intersects the segment but does not contain either of its endpoints. Theorem 2.19 is a simple corollary of the following *spanning path theorem*:

Theorem 2.20 *Any set of n points in $\mathbf{R}^d$ can be ordered as $p_1, \ldots, p_n$, in such a way that no hyperplane cuts more than $cn^{1-1/d}$ segments of the form $p_i p_{i+1}$, for some constant $c > 0$.*

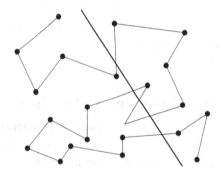

Fig. 2.16. Any collection of n points in $\mathbf{R}^d$ can be joined together by a polygonal line such that no hyperplane cuts more than $cn^{1-1/d}$ edges, for some constant $c > 0$. This is optimal in the worst case. Can you see why? Hint: take the points of a square grid.

First, we show why this implies Theorem 2.19. Given a halfspace τ, consider the segments of the form $p_{2i-1} p_{2i}$ that are cut by the hyperplane h bounding τ, for $1 \le i \le n/2$. Let $S(\tau)$ denote the set of their endpoints that lie in τ. The set system $\{\, S(\tau) \,|\, \text{halfspace}\, \tau \,\}$ contains $O(n^d)$ distinct sets. Color the points p_{2i} red or blue by applying the unbiased greedy algorithm of §1.1 to the set system. Each point of the form p_{2i-1} is colored the opposite of p_{2i} (or any color, if $2i > n$).

We thus have formed a perfect matching among the points of P (except for at most one point). Naturally, we can assume that h avoids all of the points of P because a small perturbation cannot decrease the number of edges that it cuts. Any edge of the matching that lies completely inside τ contributes zero discrepancy, while the edges straddling across h together contribute $O(\sqrt{K \log n})$, where K is the maximum size of any set $S(\tau)$. By the spanning path theorem (Theorem 2.20), $K \leq cn^{1-/1/d}$, and therefore Theorem 2.19 holds. $\square$

We now turn to the proof of the spanning path theorem, beginning with a brief, informal outline. Instead of finding a path spanning all n points, a simple recursive argument shows that it suffices to match a fraction of the points in pairs, while ensuring that no hyperplane cuts through too many of them. To do that, we join pairs of points that are not separated by too many of the $\binom{n}{d}$ canonical hyperplanes (passing through d points of P). In the next section, we show that such "close" pairs always exist. Pairing points in this fashion implies that the typical canonical hyperplane cuts through only a few pairs. To strengthen this statement and say that no canonical hyperplane cuts through many pairs, we use a weighting mechanism that we describe in a subsequent section.

It is easy to see that the theorem is optimal. Hint for the planar case: Place points at the vertices of $\sqrt{n} \times \sqrt{n}$ grid and count how many times the edges of the spanning path must cut the $2\sqrt{n}$ horizontal and vertical grid lines.

A Volume Argument

Let H be a collection of hyperplanes, $h_1, \ldots, h_m$, in $\mathbf{R}^d$ and let $\mathcal{A}(H)$ denote its arrangement (ie, the cell complex formed by the hyperplanes; see Appendix C). Given any two points p, q, we define their pseudodistance[57] $\Delta(p, q)$ to be the number of hyperplanes that cut pq. Recall that this does not count the hyperplanes passing through p or q. As we shall see, Δ behaves like the Euclidean metric in several ways. It satisfies the triangular inequality

$$\Delta(u, w) \leq \Delta(u, v) + \Delta(v, w),$$

provided that v does not lie on any hyperplane of H; otherwise, we must add the number of such hyperplanes to the right-hand side of the inequality as a corrective term. Perhaps less obvious is the fact that Δ shares some of the same packing properties as its Euclidean counterpart. The volume of a Euclidean ball of radius r is proportional to r^d. With the appropriate

---

[57] We use the term pseudodistance because $\Delta(p, q)$ might be zero even though $p \neq q$.

definition of volume, the same holds true of Δ (except that "proportional" becomes "at least proportional.") Assume for the time being that the hyperplanes of H are in general position. In this way, every vertex of $\mathcal{A}(H)$ is the intersection of exactly d hyperplanes. We define the ball $B(p,r)$ centered at point p to be the set of vertices v of $\mathcal{A}(H)$ such that $\Delta(p,v) \leq r$. The number of vertices in $B(p,r)$ is called the *volume* of the ball (Fig. 2.17). In the following we adopt the convention $\binom{a}{b} = 0$, if $a < b$.

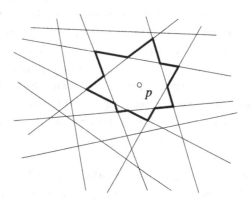

Fig. 2.17. The volume of $B(p,1)$ is 14.

Lemma 2.21 *Given any point $p \in \mathbf{R}^d$ and any nonnegative real $r \leq m = |H|$, the volume of $B(p,r)$ is at least $\binom{\lfloor r \rfloor}{d}/d!$.*

Proof: Let $V_d(m,r)$ be the minimum possible volume of $B(p,r)$ (over all p and H). We prove by induction on d that $V_d(m,r) \geq \binom{\lfloor r \rfloor}{d}/d!$. By a perturbation argument, it is clear that placing the center p of the ball on any hyperplane of H can only increase the volume of $B(p,r)$, so we assume that it is not the case. The case $d = 1$ follows from the fact that $V_1(m,r) \geq r$.

Assume that $d > 1$. Because of general position there exists a line L passing through p that intersects each hyperplane of H but not more than one at the same point. Let $q_1, \ldots, q_m$ be the sequence of intersections between L and the hyperplanes. If we choose the sequence so that the Euclidean distance between p and q_1, q_2, etc, is nondecreasing, then $\Delta(p,q_k) < k$. Let h_k be the hyperplane associated with q_k, and let $\mathcal{A}_k$ denote the arrangement formed by the $(d-2)$-flats $h_j \cap h_k$ ($j \neq k$). In $\mathbf{R}^{d-1}$, $\mathcal{A}_k$ appears as an arrangement of $m-1$ hyperplanes in general position and the restriction of Δ to h_k is itself a pseudodistance in $\mathbf{R}^{d-1}$ of the same type as Δ. Since

$r \leq m$, the points $q_1, \ldots, q_{\lfloor r \rfloor}$ are well defined. Every vertex of $\mathcal{A}(H)$ lies on exactly d hyperplanes, so we can use the triangular inequality to derive

$$V_d(m,r) \geq \frac{1}{d} \sum_{k=1}^{\lfloor r \rfloor} V_{d-1}(m-1, r-k+1) \geq \frac{1}{d} \sum_{k=1}^{\lfloor r \rfloor} \binom{\lfloor r-k \rfloor}{d-1} / (d-1)! .$$

We then immediately derive

$$V_d(m,r) \geq \frac{1}{d!} \sum_{k=0}^{\lfloor r \rfloor - 1} \binom{k}{d-1} = \frac{1}{d!} \binom{\lfloor r \rfloor}{d}.$$

$\square$

A simple corollary of the lemma is that, similar to the Euclidean case, a large collection of points enclosed in a small ball must have close pairs. We relax the assumption of general position on H.

Lemma 2.22 *Given n points $p_1, \ldots, p_n$ in $\mathbf{R}^d$, at least two of them, p_i, p_j ($i \neq j$), satisfy $\Delta(p_i, p_j) \leq b \lceil m/n^{1/d} \rceil$, for some constant $b > 0$.*

Proof: A small perturbation of H cannot decrease the Δ-distance between two fixed points; therefore, without loss of generality we can assume that H is in general position. Set $r = C \lceil m/n^{1/d} \rceil - d$, for some constant C large enough so that

$$n \binom{\lfloor r \rfloor}{d} / d! > \binom{m}{d}.$$

Since $\mathcal{A}(H)$ has exactly $\binom{m}{d}$ vertices (every d-tuple of hyperplanes provides one), it follows from Lemma 2.21 that the n balls $B(p_i, r)$ cannot be all disjoint. This implies the existence of a point q such that both $\Delta(p_i, q) \leq r$ and $\Delta(p_j, q) \leq r$, for some i, j. Because H is in general position, the triangular inequality shows that $\Delta(p_i, p_j) \leq 2r + d$, which completes the proof. $\square$

The Iterative Reweighting Method

We now complete the proof of the spanning path theorem (page 124). Recall that our goal is to connect n points into a path $p_1, \ldots, p_n$ in such a way that no hyperplane cuts too many edges. By definition, we should recall, a hyperplane cannot cut an edge through its endpoints. So, the hyperplane cutting the most edges in a path can always be chosen to avoid each of the n points. But in that case, we can perturb the points to make sure that they are in general position before computing a desired path. Having done so, it now suffices to ensure the low-cutting property for each of the $\binom{n}{d}$ canonical hyperplanes defined by d-tuples of points. Indeed, because the points are in general position, any hyperplane can be moved to a canonical position without changing the number of edges cut by more than an additive constant. Let H be the set of canonical hyperplanes; its size is $m = \binom{n}{d}$.

Let p_1, q_1 be the two closest points of P (measured with respect to Δ and H). Remove both points from P and *duplicate* each of the hyperplanes cutting p_1q_1; thus, from now on, we treat H as a multiset (Fig. 2.18). Iterate on this process $\lceil n/4 \rceil$ times, always selecting the closest pair p_i, q_i among the remaining points of P, and then deleting p_i and q_i and duplicating all of the hyperplanes of H cutting p_iq_i. Note that if a hyperplane appears k times in (the multiset) H, then duplicating it means bringing its multiplicity from k to $2k$.

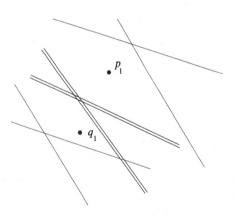

Fig. 2.18. We duplicate the hyperplanes for a repulsive effect.

By Lemma 2.22, at most $b\lceil m/n^{1/d} \rceil$ hyperplanes cut p_1q_1; therefore, the size of the multiset H after the first step is at most $(1 + 2b/n^{1/d})m$. By the same argument, the size of H after the k-th step is at most

$$m \prod_{0 \leq j < k} \left(1 + \frac{2b}{(n - 2j)^{1/d}} \right).$$

Notice that the term $-2j$ in the denominator reflects the removal of p_1, q_1, $\ldots, p_j, q_j$ from P.

Since $k \leq \lceil n/4 \rceil$, the size of H always remains at most

$$m\left(1 + \frac{b'}{n^{1/d}}\right)^{\lceil n/4 \rceil},$$

for some constant $b' > 0$. The pairs (p_1, q_1), (p_2, q_2), etc, produce a partial matching in P involving roughly half the points. The most remarkable fact about this matching is that no hyperplane can cut too many edges. Indeed, if a canonical hyperplane creates κ cuts, then it must be duplicated 2^κ times. The number of duplications cannot exceed the final size of H, so

$$2^\kappa \leq m\left(1 + \frac{b'}{n^{1/d}}\right)^{\lceil n/4 \rceil},$$

and hence $\kappa = O(n^{1-1/d})$; obviously we can assume that $d > 1$.

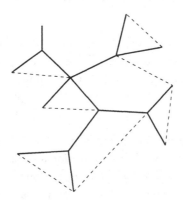

Fig. 2.19. Turning a tree into a path.

We can extend the partial matching by proceeding recursively. Choose one endpoint in each matching edge and remove it from P. Then, apply the construction recursively on the reduced set P. Each recursive call involves a set of size decreasing geometrically, so in the end we obtain a tree connecting all the points of P, with no hyperplane cutting more than

$$O\left(\sum_{i \geq 0} ((3/4)^i n)^{1-1/d}\right) = O\left(n^{1-1/d}\right)$$

edges. Finally, we must show how to turn this tree into a single path. Perform a depth-first traversal of the tree and connect the points in the order in which they are first encountered during the traversal (Fig. 2.19). This gives us the desired spanning path. Because any edge pq of the path that is not also an edge of the tree creates a cycle, a hyperplane that cuts pq must also cut an edge of the tree, which we can charge for that purpose. Since no tree edge can belong to more than two cycles, the charging scheme does not undercount by more than a factor of 2. Consequently, any hyperplane cuts $O(n^{1-1/d})$ edges of the path and Theorem 2.20 (page 124) is established. This also completes the proof of Theorem 2.19. □

2.9 Bibliographical Notes

Section 2.1: The bound on the error term in using sampling for numerical integration (Theorem 2.1, page 43) is due to Koksma [186]. Extensions can be found in Hlawka [161] and Niederreiter [240]. Applications of low-discrepancy point sets in computer graphics are described in Mitchell [231] and Shirley [286].

Section 2.2: The van der Corput sequence was introduced in [313, 314]. Halton-Hammersley points are named after Halton [156] and Hammersley [157]. Unfortunately, the $O(\log n)^{d-1}$ upper bound on the L^∞ norm of the discrepancy (Theorem 2.2, page 44) has been proven tight only in two dimensions. See Chapter 3 for a more detailed discussion of lower bounds.

Section 2.3: Arithmetic progressions modulo 1 have been extensively researched in number theory. Weyl's criterion and Lemma 2.5 date back to 1916 [321]. The L^2-norm bound of $O(\sqrt{\log n})$ on the discrepancy of axis-parallel boxes given in Theorem 2.4 (page 47) is from Davenport [105]. Earlier related work can be found in Behnke [41, 42]. More generally, an upper bound of $O(\log n)^{(d-1)/2}$ was established by Roth [264] for any fixed dimension $d > 0$. As discussed in Chapter 3, these bounds are tight.

Section 2.4: The jittered sampling construction for low-discrepancy rotated boxes (Theorem 2.7, page 54) is due to Beck; see Beck and Chen's book [37].

Sections 2.5,2.6: The orbital point set construction for the sphere and its analysis (Theorem 2.9, page 58, and Theorem 2.10, page 58) are due to Lubotzky, Phillips, and Sarnak [200]. The bound in the first theorem is not optimal, but the second one is (essentially). Schmidt [272] established the first nontrivial—and nearly tight—lower bound for spherical-cap discrepancy in S^d. His $\Omega(n^{1/2-1/2d-\varepsilon})$ bound was improved to $\Omega(n^{1/2-1/2d})$ by Beck [34]; the same bound can also be derived from later work by Alexander [10, 11]. The quasi-optimality for quadrature (Theorem 2.10) is proven in [200]. The Ramanujan bound is established in the companion paper by Lubotzky, Phillips, and Sarnak [201]; see also Sarnak's monograph [268]. The bound (2.30) on the Fourier coefficients for forms of even integral weight for $\Gamma(N)$ (one of the Ramanujan conjectures) was proven by Deligne [106, 107]. For background material on the spectral analysis of linear operators, we recommend Dunford and Schwartz [113] or Friedman [138]. A nice treatment of spherical harmonics can be found in Jones [173]. See also Stein and Weiss [297] for Fourier analysis on Euclidean spaces.

For an introduction to elliptic curves and modular forms, Koblitz [184], Zagier [334] (for modular forms), and Cassels [62], Silverman [287] (for elliptic curves) are excellent starting points. Ireland and Rosen's book [168] covers a wider front and also includes a nice discussion of zeta functions. Serre's classic [279] ends with a beautiful introduction to modular forms. McKean and Moll [206]'s book provides a highly readable introduction to elliptic curves. For elliptic and automorphic functions see Lang [194] and Shimura [285], respectively. As we showed in this chapter, the theory of modular forms has applications to other branches of mathematics: this is eloquently demonstrated in Sarnak's monograph [268].

A beautiful wide-audience (pre-Wiles) account of the Shimura-Taniyama conjecture is given by Mazur [223]. For Web fans we mention a few useful pages: http://www.fermigier.com/fermigier/elliptic.html.en (on elliptic curves), http://www.mbay.net/~cgd/flt/flt01.htm (on FLT), http://www.math.lsa.umich.edu/~jmilne/ (on elliptic curves and modular forms.) For an advanced treatment of contemporary issues on the subject, see [100]. Our example for the Eichler-Shimura theorem was taken from Silverman and Tate [288]. Wiles' proof of FLT is given in two papers [306, 323] (both for experts only), one of which is in collaboration with Taylor. For background in algebraic geometry, the texts by Hartshorne [159], Shafarevich [280], and especially Griffiths and Harris [150] are the classic points of entry.

As a closing note, we briefly mention the important role played recently by elliptic curves in cryptography. The operating theme is to trade the multiplicative groups used in modular arithmetic for the additive groups of elliptic curves over finite fields. Such substitutions can be beneficial for several reasons. One is that the discrete log problem seems more difficult for elliptic curves over $\mathbf{F}_{2^r}$ than for the multiplicative structure of the field $\mathbf{F}_{2^r}$ [246].[58] In other contexts, the advantage of elliptic curves is their generous supply of wildly different groups. Indeed, for a given prime p,

---

[58] Over the group $\mathcal{C}(\mathbf{F}_n)$, the discrete log problem is to solve the equation $xA = B$,

the field $\mathbf{F}_p$ provides us with a multiplicative group which may or may not have the property we want (for example, having its order $p - 1$ be a product of small primes). But by simply varying the coefficients of an elliptic curve $\mathcal{C}$, we get a whole family of distinct-looking groups $\mathcal{C}(\mathbf{F}_p)$ for a fixed p. Far-reaching applications of this idea have been given in public-key cryptography [185], primality testing [144], and integer factoring [195].

Section 2.7:   Our discussion of the Laplacian is fairly standard, and references can be found in many places: our choice of examples was inspired by the excellent text of Strang [299]; see also Biggs [45], Bollobás [53], Chung [91, 92], or the *Handbook of Combinatorics* [146] for a discussion of the Laplacian in graph theory. There is a large literature on the problem of placing points on a sphere, eg, Guralnik, Zemach, and Warnock [153], and Sloane [291].

Section 2.8:   The idea of coloring a low-cutting spanning path to achieve small red-blue discrepancy is due to Matoušek, Welzl, and Wernisch [222], who established Theorem 2.19 (page 124). As was mentioned in the text, a sharp bound of $O(n^{1/2-1/2d})$ was proven by Matoušek [215]. A weaker bound was obtained earlier by Beck in two dimensions [36]. The optimal cutting bound for spanning paths (Theorem 2.20, page 124) was established by Chazelle and Welzl [86], which improved on an earlier bound by Welzl [319].

---

where $x \in \mathbf{Z}$ is the unknown and $A, B \in \mathcal{C}(\mathbf{F}_n)$. Over $\mathbf{F}_n$, the problem is given by the equation $a^x = b$.

3

Lower Bound Techniques

roving lower bounds on the discrepancy of geometric set systems often involves looking at the L^2 norm of the discrepancy function. This makes sense given the wealth of techniques available for dealing with quadratic forms. A typical approach is to consider the incidence matrix A and bound the eigenvalues of $A^T A$ from below. Since the underlying set systems are often defined by "convolving" a shape with a set of points, one should expect Fourier transforms to be useful: After all, a characteristic feature of the Fourier transform is to diagonalize the convolution operator. Eigenvalue estimation can also be done by other methods, such as, in this case, wavelet transforms or finite differencing. We emphasize that the question is not to approximate eigenvalues numerically but to derive asymptotic bounds on them—a decidedly more difficult task.

This chapter samples the toolkit of available techniques: Haar wavelets, Riesz products, Fourier transforms and series, Bessel functions, Dirichlet and Fejér kernels, finite differencing, and so on. The emphasis of our discussion is on the methods rather than on the results themselves, so do not expect from this chapter a comprehensive coverage of the vast amount of knowledge on the subject. Instead, expect a wide assortment of powerful mathematical techniques for discrepancy lower bounds. In the spirit of the discrepancy method, it is important to master these techniques because of their importance in proving complexity bounds in later chapters.

Throughout this chapter we consider n points in the unit cube $[0, 1]^d$. Implicit in our discussion is the assumption that n is large enough. In §3.1 we consider axis-parallel boxes. The volume discrepancy of a box is the difference between the number of expected points in the box and the actual number. Using a general method based on orthogonal functions, we prove that some box has discrepancy $\Omega(\log n)^{(d-1)/2}$ in absolute value (Theorem 3.1); the bound is actually stronger since it holds in the L^2 sense. We show that it is possible to improve this bound to $\Omega(\log n)$ in two dimensions (Theorem 3.4). The result illustrates the use of Riesz products. This presents us with an unusual situation: In two dimensions, we have a tight bound of $\Omega(\log n)^{1/2}$ on the L^2 norm and a tight bound of $\Omega(\log n)$ on the L^∞ norm!

In §3.2 we examine what happens to the discrepancy when the boxes are allowed to be rotated. The two-dimensional case contains all of the ideas, so we confine our discussion to it. Using Fourier transforms as our main tool, we prove the surprising fact (Theorem 3.6) that the worst-case volume discrepancy shoots up to $\Omega(n^{1/4})$. We also discuss the discrepancy of disks, for which we derive the same lower bound (Theorem 3.7).

Finally, in §3.3 we introduce a different method, which is based on finite differencing and certain metric properties of the discrepancy. This yields tight lower bounds on the red-blue discrepancy of halfspaces (Theorem 3.9). We prove the two-dimensional case separately to showcase its intuitive geometric appeal.

3.1 The Method of Orthogonal Functions

We begin with one of the most beautiful proofs in all of discrepancy theory: without a doubt, a proof from *The Book*.

Given a point $q = (q_1, \ldots, q_d)$ in the unit cube $[0, 1]^d$, let B_q denote the box $[0, q_1) \times \cdots \times [0, q_d)$. Given a set P of n points in $[0, 1]^d$, the volume discrepancy $D(q)$ at a point $q \in [0, 1]^d$ is the difference between the expected number of points in the box B_q and the actual number (in other words, $D(q)$ is just the higher dimensional version of $D(B_q)$ given in §2.3):

$$D(q) = nq_1 \cdots q_d - |P \cap B_q|.$$

No matter how the points of P are placed, the average value of $D(q)^2$, over all $q \in [0, 1]^d$, is always $\Omega(\log n)^{d-1}$. Another way to say this is that[1] $\|D\|_2 \gg (\log n)^{(d-1)/2}$. The ideas of the proof are best illustrated in the two-dimensional case, so we prove the following theorem for $d = 2$. The case $d = 1$ is trivial, while the other cases involve an easy generalization of the two-dimensional result.

---

[1] Recall that $\ll$ and $\gg$ are shorthand for $O()$ and $\Omega()$, respectively.

Theorem 3.1 *Given n points in $[0,1]^d$, the mean-square discrepancy for axis-parallel boxes satisfies*

$$\int_{[0,1]^d} D(q)^2 \, dq > c(\log n)^{d-1},$$

for some constant $c = c(d) > 0$.

We begin with the complete proof and then discuss the method behind it and try to develop some intuition for it. In particular, we discuss a ubiquitous relationship between function orthogonality and probabilistic independence. The gist of the method is to choose a function $F : [0,1]^2 \mapsto \mathbf{R}$ whose mean-square $\int F^2$ is bounded above by a known quantity and whose inner product with D, ie, $\int FD$, can be easily estimated from below. (From now on, unless specified otherwise, the integration domain is assumed to be $[0,1]^2$ and variables are omitted when there is no ambiguity.) By Cauchy-Schwarz, we obtain the following lower bound:

$$\int D^2 \geq \left(\int FD \right)^2 \bigg/ \int F^2.$$

Geometrically, this is all quite trivial: We certify that a vector is long by showing that its projection on a conveniently chosen vector is long. The beauty of the technique is in the choice of the projection vector.

Haar Wavelets

The function F is defined by modifying the standard Rademacher functions, which are simply superpositions of Haar wavelets. Without loss of generality, we can assume that n is a power of 2, so let $n = 2^m$. (Indeed, we can always shrink the unit square to a square $[0,u]^2$ with the appropriate number of points inside.) We set

$$F = f_0 + \cdots + f_{m+1},$$

where each f_i is defined as follows. For any $0 \leq i \leq m+1$, let G_i be the grid obtained by dividing $[0,1]^2$ into $2n$ axis-parallel rectangles of size $2^{-i} \times 2^{i-m-1}$.

Note that to obtain a consistent partition of $[0,1)^2$, we should choose the grid cells as rectangles of the form $[x,y) \times [x',y')$. There are $2n$ cells but only n points; therefore, at least half the cells are empty. The function f_i is defined in terms of the "interaction" between P and each cell R of G_i:

- If $P \cap R \neq \emptyset$, set $f_i = 0$ over the entire cell.

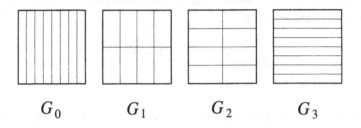

$$G_0 \qquad G_1 \qquad G_2 \qquad G_3$$

Fig. 3.1. The grids G_i.

- If $P \cap R = \emptyset$, subdivide R into four equal-size quadrants: set $f_i = 1$ over the northeast and southwest quadrants and $f_i = -1$ over the other two quadrants. (Figure 3.2 depicts a possible setting of f_i.)

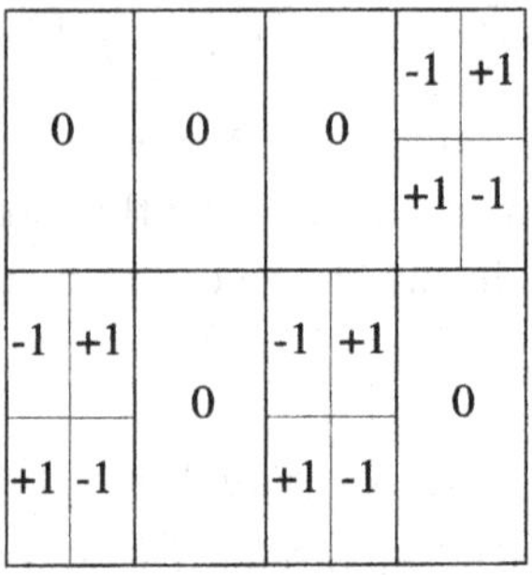

			-1 +1
0	0	0	+1 -1
-1 +1 +1 -1	0	-1 +1 +1 -1	0

Fig. 3.2. The function f_1.

It is not hard to prove directly that the family $\{f_i\}$ is orthogonal, meaning that, for any $i < j$,

$$\int f_i f_j = 0.$$

This is a particular consequence of a more general result that we will use extensively in the next section and which we prove now. Given two integers $0 \le a, b \le m + 1$, let $G_{a,b}$ be the grid obtained by subdividing $[0,1]^2$ into rectangles of size $2^{-a} \times 2^{-b}$. So, for example, G_i is really a shorthand for $G_{i,m+1-i}$. A function $f : [0,1]^2 \mapsto \mathbf{R}$ is called (a, b)-*checkered* if it satisfies the following conditions. For each cell R of $G_{a,b}$,

- If $P \cap R \neq \emptyset$, then $f = 0$ over R.

- If $P \cap R = \emptyset$, there exists $c(R) \in \{-1, 0, 1\}$ such that f is equal to $c(R)$ over the entire northeast and southwest quadrants of R and $-c(R)$ over the other two quadrants.

Note that f_i is $(i, m + 1 - i)$-checkered. An immediate property of an (a, b)-checkered function is that its integral over $[0, 1]^2$ vanishes. Thus, the orthogonality of the family $\{f_i\}$ follows directly from

Lemma 3.2 *If f is (a, b)-checkered and g is (a', b')-checkered, where $a' < a$ and $b < b'$, then fg is (a, b')-checkered.*

Proof: Over an empty cell R of $G_{a,b}$, $f(q)$ can be written as $c(R)(-1)^{r_1+r_2}$, where r_1 (resp. r_2) is 0 or 1, depending on whether q lies in the left or right (resp. upper or lower) half of R. Let S be a cell of $G_{a',b'}$ and assume that $T = R \cap S \neq \emptyset$. Then, by expressing $g(q)$ in the same manner as $f(q)$, we find that over T

$$fg = c(R)c(S)(-1)^{r_1+r_2+s_1+s_2}.$$

By construction of the grids, the vertical projections of R and S are dyadic intervals, and thus one lies completely inside the right or left half of the other; a similar fact holds for horizontal projections. Because $a' < a$ and $b < b'$, the vertical (resp. horizontal) projection of R encloses (resp. is enclosed by) that of S (Fig. 3.3).

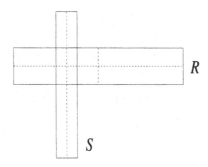

Fig. 3.3. The interaction between R and S.

It follows that (i) the functions r_1 and s_2 are constant over T, while (ii) r_2 and s_1 are equal to 1 on one side of their respective median line and -1 on the other side. This gives $fg = c(T)(-1)^{r_1+s_2}$ over T. Observe that the horizontal (resp. vertical) median line of R (resp. S) is also the median line of T. Thus, the functions r_1 and s_2 alternate between 1 and -1, as they should. Obviously, if $P \cap T$ is nonempty then neither is R nor S, and so $fg = 0$ over T. □

Since the L^2 norm of each f_i is at most 1 and $F = \sum_{i=0}^{m+1} f_i$, then, by orthogonality,

$$\int F^2 \leq m + 2. \qquad (3.1)$$

The second part of the method is to bound $\int FD$ from below. To do this we prove a more general result that will also be useful in the next section.

Lemma 3.3 *Given an (a, b)-checkered function f, let S be the set of mass centers of the northeast quadrants of the cells of $G_{a,b}$. Then,*

$$\int fD = \frac{n}{4^{a+b+2}} \sum_{p \in S} f(p).$$

Proof: Needless to say, there is nothing particular about mass centers: We just need a sample point in each northeast quadrant. We evaluate the integral over each cell R of $G_{a,b}$. If $P \cap R$ is nonempty, the integral over the cell is 0. Suppose that $P \cap R$ is empty. Then, for each point q in the northeast quadrant of R, let q^1, q^2, q^3 be the analogous points in the northwest, southwest, and southeast quadrants, respectively (Fig. 3.4). Let p be a point in R_0, the northeast quadrant of R. Then,

$$
\begin{aligned}
\int_{q \in R} f(q) D(q)\, dq &= \int_{q \in R_0} f(q) \Big(D(q) - D(q^1) + D(q^2) - D(q^3) \Big)\, dq \\
&= f(p) \int_{q \in R_0} n \cdot \text{area}\,[q, q^1, q^2, q^3]\, dq \\
&= n f(p) \Big(\frac{\text{area}\, R}{4} \Big)^2 = n f(p) 2^{-2a-2b-4},
\end{aligned}
$$

from which the lemma follows. □

Since in each grid G_i at least half the cells, ie, n of them, are empty, we have

$$\int f_i D = \frac{1}{2^6 n} \sum_{p \in S} f_i(p) \geq \frac{1}{2^6},$$

and hence

$$\int FD \geq 2^{-6} \log n. \qquad (3.2)$$

The two-dimensional restriction of Theorem 3.1 (page 135) follows imme-

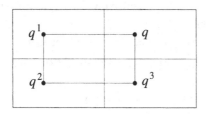

Fig. 3.4. The four points q, q^1, q^2, q^3.

diately from (3.1). As we said earlier, the case $d > 2$ is entirely similar. □

Riesz Products

Our notation and assumptions are the same as in the previous section. In particular, unless specified otherwise, the integration domain is still assumed to be $[0,1]^2$. Our goal is to establish a lower bound of $\Omega(\log n)$ on the L^∞ norm of the discrepancy. Unfortunately, the proof technique does not seem to generalize to higher dimensions.

Theorem 3.4 *Given n points in $[0,1]^2$, there exists a point $q \in [0,1]^2$, such that $|D(q)| = \Omega(\log n)$, where $D(q)$ is the discrepancy of the axis-parallel box $[0, q_x) \times [0, q_y)$.*

Being in the plane, we prefer to use (q_x, q_y) rather than (q_1, q_2) to denote the coordinates of q. As in the previous section, we bring in an auxiliary function $G(q)$. From the inequality

$$\left| \int GD \right| \le \int |GD| \le \|D\|_\infty \int |G|,$$

where $\|D\|_\infty$ is the supremum of $|D(q)|$ over $[0,1]^2$, we find that

$$\|D\|_\infty \ge \left| \int GD \right| \Big/ \int |G|.$$

Let γ be a small positive constant; we define the *Riesz product*

$$G(q) = -1 + \prod_{i=0}^{m+1} (1 + \gamma f_i(q)),$$

where, as we recall, $n = 2^m$. Intuitively, by expanding $G(q)$ we recognize the previous function F scaled by γ, together with all the "higher moments." This follows the standard approach of using higher moments for sharper estimates of a function's maximum. Obviously, it suffices to prove that

$$\int |G| \leq 2, \tag{3.3}$$

and

$$\left| \int GD \right| \gg \log n. \tag{3.4}$$

Note that since

$$\int |G| \leq 1 + \int \prod_{i=0}^{m+1} (1 + \gamma f_i),$$

expanding the product yields (3.3), provided that the following holds:

Lemma 3.5 *Given any $1 \leq j \leq m+2$ and any $0 \leq i_1 < \cdots < i_j \leq m+1$,*

$$\int f_{i_1} f_{i_2} \cdots f_{i_j} = 0.$$

Proof: We observed in the previous section that f_i is $(i, m + 1 - i)$-checkered. Repeated applications of Lemma 3.2 show that the product $f_{i_1} \cdots f_{i_j}$ is $(i_j, m + 1 - i_1)$-checkered. Since the integral over $[0,1]^2$ of an (a, b)-checkered function vanishes, the lemma is established. $\square$

All we have to do now is to establish (3.4). By the triangular inequality, expanding the product in G gives

$$\left| \int GD \right| \geq \gamma \left| \int G_1 D \right| - \sum_{j=2}^{m+2} \gamma^j \left| \int G_j D \right|, \tag{3.5}$$

where

$$G_j = \sum_{0 \leq i_1 < \cdots < i_j \leq m+1} f_{i_1} \cdots f_{i_j}.$$

Our first task is to find an upper bound on the second absolute value in the right-hand side of (3.5). Given $0 \leq i_1 < \cdots < i_j \leq m+1$, we just observed that the product $f_{i_1} \cdots f_{i_j}$ is $(i_j, m+1-i_1)$-checkered. Since $G_{i_j, m+1-i_1}$ consists of $2^{m+1+i_j-i_1}$ cells, Lemma 3.3 yields

$$\left| \int f_{i_1} \cdots f_{i_j} D \right| \leq \left(\frac{n}{4^{m+3+i_j-i_1}} \right) 2^{m+1+i_j-i_1} = \frac{1}{2^{i_j-i_1+5}}.$$

To tackle the integral $\int G_j D$ for $j \geq 2$, we sum over each possible value of $w \overset{\text{def}}{=} i_j - i_1$:

$$\left| \int G_j D \right| \leq \sum_{w=j-1}^{m+1} \frac{(m+1)}{2^{w+5}} \binom{w-1}{j-2}.$$

Therefore, by inverting the order of summation,

$$\sum_{j=2}^{m+2} \gamma^j \left| \int G_j D \right| \leq 2^{-6} \gamma^2 (m+1) \sum_{w=1}^{m+1} \frac{1}{2^{w-1}} \sum_{j=2}^{w+1} \binom{w-1}{j-2} \gamma^{j-2}$$

$$\leq 2^{-6} \gamma^2 (m+1) \sum_{w=1}^{m+1} \left(\frac{1+\gamma}{2} \right)^{w-1}$$

$$\leq 2^{-5} (m+1) \frac{\gamma^2}{1-\gamma}.$$

Note that $G_1 = F$ (see definition in previous section). So, from (3.2) and (3.5) we derive

$$\left| \int GD \right| \geq 2^{-6} \gamma \log n - 2^{-5} \frac{\gamma^2 (1 + \log n)}{1 - \gamma}.$$

Setting γ small enough ensures that $|\int GD| \gg \log n$. We now have established (3.3, 3.4), which proves Theorem 3.4 (page 139). □

Orthogonality and Independence *

We provide some intuition for the proof based on Haar wavelets (Theorem 3.1, page 135). This discussion is not necessary, strictly speaking, but it might help to relieve the reader of the sense of magic created by that (truly beautiful) proof.

Again, let q^1, q^2, q^3, q be the four corners of a rectangle of area $1/8n$ inside the box B_q (Fig. 3.5). For a fixed fraction of all placements of q, the rectangle is free of points, in which case its "discrepancy" is equal to $1/8$. This can be "picked up" by the inner product $\int fD$ by setting the function f to be 1 at q and q^2, and -1 at q^1 and q^3. In this way, the four points contribute

$$D(q) - D(q^1) + D(q^2) - D(q^3) \gg 1$$

to the integral. By observing that the same reasoning works for all *good* placements of q (which is a fixed fraction of all placements) we can set f as above at the relevant points and 0 elsewhere. With a bit of care, we can

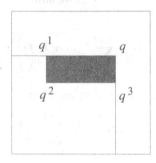

Fig. 3.5. Picking up a local discrepancy.

then derive a lower bound of the form

$$\int fD \gg \int_{\text{good } q} dq \gg 1.$$

We verify that $\int f^2 \ll 1$, so by Cauchy-Schwarz we have obtained, by nontrivial means, the trivial bound

$$\int D^2 \gg 1.$$

The next step is to show that local discrepancies can be added together to amplify the lower bound on $\int D^2$.

The hierarchy of grids G_i suggests considering about $\log n$ empty rectangles of the form q^1, q^2, q^3, q, each of area $1/8n$: Each such rectangle eats up a disjoint chunk of area at least $1/32n$ (the—intentionally off-scaled—dark boxes in Figure 3.6). Repeating the previous argument, we set F accordingly so as to "pick up" all those local discrepancies, ie, $F = \sum_{i=0}^{m+1} f_i(q)$. Integration is additive, so this yields the stronger lower bound,

$$\int FD \geq (\log n) \int_{\text{good } q} dq \gg \log n.$$

Now the punchline: The orthogonality of the f_i's ensures that the L^2 norm of F remains small. If we pick q randomly in $[0,1]^2$, then each $f_i(q)$ becomes a random variable x_i. Suppose that the x_i's were independent variables, or at least *behaved* likewise. Specifically, suppose that the expectation of any mixed product $x_i x_j$ ($i \neq j$) were the product of the

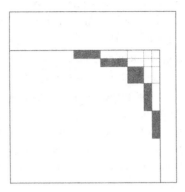

Fig. 3.6. Adding up local discrepancies.

expectations, ie, 0. It then would follow that

$$\mathbf{var} \sum x_i = \sum \mathbf{var}\, x_i \gg \log n.$$

But since the variance of $\sum x_i$ is the mean-square of $F = \sum f_i$, Theorem 3.1 follows immediately. The random variables x_i, now regarded as $f_i(q)$, are *orthogonal functions*. The connection between independence of random variables and orthogonality of functions is a key concept in discrepancy theory, which we shall encounter again in this chapter.

An intuitive understanding of Theorem 3.1 (page 135) that is helpful, although not entirely correct, is to view the discrepancy of B_q as the sum of all the local discrepancies in the dark boxes of Figure 3.6. These discrepancies now must be regarded as pairwise independent random variables uniformly distributed in some range $[-c, c]$, for some constant $c > 0$. Thus, $D(q)$ appears to be unbiased with a standard deviation of about $\sqrt{\log n}$. A random q should then place $D(q)$ within a fixed fraction of the standard deviation. The placement of the boxes along what is, in effect, a hyperbola ensures the sort of independence needed to make the argument carry through.

One final comment is that the functions f_i suggest finite difference operators. The relevance of finite differencing to discrepancy theory is the main focus of §3.3.

3.2 The Fourier Transform Method

Instead of restricting ourselves to axis-parallel boxes, let us now examine the discrepancy of arbitrary boxes, where rotations and translations are both allowed. By contrast with the previous scenario, where the discrepancy was polylogarithmic, we now can show that given n points in the unit cube $[0,1]^d$, there exists a box whose volume discrepancy is $\Omega(n)^{1/2-1/2d}$.

Again, all of the ideas in the proof are contained in the two-dimensional case, so this is where we confine our discussion. A box R is now understood as a rectangle $[0,a] \times [0,b]$ transformed by an arbitrary rigid proper motion (ie, translation/rotation). Given a set of n points in $[0,1]^2$, the discrepancy $D(R)$ is the difference between the expected number of points in R and the actual number points in R, ie,

$$D(R) = n \cdot \text{area}\left([0,1]^2 \cap R\right) - |P \cap R|.$$

Theorem 3.6 *Given n points in the unit square $[0,1]^2$, there exists a rotated box R such that $|D(R)| = \Omega(n^{1/4})$.*

It is worth mentioning that the high-discrepancy boxes of the theorem need not have large areas. Actually, one can always be found of area $O(1/\sqrt{n})$. Rotation plays a critical role in the discrepancy of boxes, for it allows us to choose the worst possible box slope for a given point set. Surprisingly, a disk "contains enough slopes around its boundary" to do the job all by itself.

Theorem 3.7 *Given n points in the unit square $[0,1]^2$, there exists a disk $\mathcal{D}$ such that $|D(\mathcal{D})| = \Omega(n^{1/4})$.*

A small technical point: We prove Theorem 3.7 for disks modulo $[0,1]^2$, ie, we "wrap around" the unit square as on a torus. A disk of radius at most $1/2$ might thus produce up to four disk sections in the unit square. Obviously, high discrepancy in the torus sense implies high discrepancy for at least one of the disk sections, so the theorem also holds when discrepancy is defined with respect to the intersection of a disk with the unit square (no wrap-around).

In both theorems the actual lower bound is even stronger than stated, since it holds in the L^2 sense as well. Both proofs appeal to harmonic analysis in an essential manner.[2] We include both proofs because, despite

---

[2] We refer the reader to Appendix B for a review of Fourier integrals and series. The term *Fourier transform method* is to be understood in a general sense, since on occasion

their common theme, they differ drastically. The first one illustrates the technique of discrepancy amplification due to Beck, while the second one, due to Montgomery, makes interesting use of Bessel functions and the Fejér kernel. As an introduction to the Fourier transform method we begin with an informal discussion of Theorem 3.6.

Beck's Amplification Method

To understand the relevance of the Fourier transform, observe that the discrepancy is a convolution of two functions. Indeed, let $R(q) \subset [0,1]^2$ be an axis-parallel square of side length r centered at q. The discrepancy of $R(q)$ can be expressed as

$$D(q) = \int_{R(q)} \mu(p)\, dp,$$

where $\mu(p)$ is $n - \sum_{p_i \in P} \delta(p - p_i)$ inside $[0,1]^2$ and 0 outside. We extend the integration domain to the whole plane by using a "mask" Π_r, which is 1 inside $R(O)$ and 0 outside. Then,

$$D(q) = \int_{\mathbf{R}^2} \Pi_r(p - q)\mu(p)\, dp.$$

Since $\Pi_r(p - q) = \Pi_r(q - p)$, the discrepancy $D(q)$ is the convolution $(\Pi_r \star \mu)(q)$. Taking Fourier transforms at this point seems a good idea for at least two reasons: By Parseval-Plancherel, the mean square of $D(q)$ is equal to the mean square of its Fourier transform $\widehat{D}(t)$, where $t \in \mathbf{R}^2$. Moreover, by the convolution theorem, $\widehat{D}(t)$ can be expressed quite simply in a manner where the geometric component $\widehat{\Pi}_r(t)$ is separated from the measure component $\widehat{\mu}(t)$:

$$\int_{\mathbf{R}^2} D(q)^2\, dq = \int_{\mathbf{R}^2} |\widehat{\Pi}_r(t)|^2\, |\widehat{\mu}(t)|^2\, dt.$$

As long as r is a little smaller than $1/\sqrt{n}$, $D(q)$ is bounded away from zero and we have the trivial lower bound

$$\int_{\mathbf{R}^2} D(q)^2\, dq \gg 1.$$

Suppose that we could show that at *any* value of t, $|\widehat{\Pi}_r(t)|^2$ grows at least linearly as a function of r (the *linear effect*). Then, by setting $r \approx 1$, we

---

we might choose to make the discrepancy function periodic and use Fourier series instead of Fourier transforms.

would immediately amplify the trivial lower bound to $\Omega(\sqrt{n})$. Unfortunately, there is no such linear effect. A simple calculation shows that

$$|\widehat{\Pi_r}(t)|^2 = \left(\frac{\sin(\pi t_x r)}{\pi t_x}\right)^2 \left(\frac{\sin(\pi t_y r)}{\pi t_x}\right)^2,$$

where $t = (t_x, t_y)$. One problem is that the function periodically touches its "upper" surface $z = 1/(\pi^2 t_x t_y)^2$. At those points, it obviously cannot grow with r at a fixed point t. One remedy is to average $\int D(q)^2\, dq$ over a small range of r, say, by considering

$$\frac{1}{r_0} \int_{r_0}^{2r_0} \int_{\mathbf{R}^2} D(q)^2 \, dq \, dr \,.$$

This should keep the function $|\widehat{\Pi_r}(t)|^2$ away from both its upper surface and 0. Another problem arises, however: As we will see later, the parameter r affects the value of $|\widehat{\Pi_r}(t)|^2$ only inside two narrow bands along the t_x- and t_y-axes. This means that outside these bands no amplification is to be expected as r increases. To fix this problem, we introduce rotations (as, obviously, we should). If $D_\theta(q)$ denotes the discrepancy of the square centered at the origin, rotated by θ and then translated by q, we consider

$$\int_0^{2\pi} \frac{1}{r_0} \int_{r_0}^{2r_0} \int_{\mathbf{R}^2} D_\theta(q)^2 \, dq \, dr \, d\theta \,.$$

A rotation in the q-plane corresponds to an inverse rotation in the frequency plane, so the net effect is that the contribution of a given t is now averaged out over the entire circle of radius[3] $|t|$ centered at O. Thus, even though t might be "bad" because it falls outside the bands, the circle of radius $|t|$ intersects the bands in (relatively long) arcs over which the linear effect takes place, so no frequency can now be entirely bad! This produces the desired amplification effect. This technique, due to Beck, can be extended to more complex shapes. In fact, what we present here is perhaps the simplest case.

In the discussion to come we consider rectangles and not just squares. This has the benefit of illustrating the fundamental principle that the discrepancy is a function of the boundary length of the body and not of its area.

---

[3]For notational convenience, we use $|t|$ to denote the L^2 norm of t.

THE PROOF

Let P be a set of n points in the unit square $[0,1]^2$. We define the rectangle $R(\alpha, \beta, \theta, q)$ by taking an axis-parallel $\alpha \times \beta$ rectangle centered at the origin, rotating it by an angle θ, and finally translating it by q (Fig. 3.7).

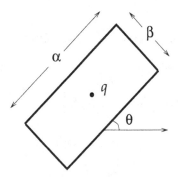

Fig. 3.7. The box $R(\alpha, \beta, \theta, q)$.

Fix two parameters α_0, β_0 such that

$$0 < \beta_0 \leq \alpha_0 \leq 1. \tag{3.6}$$

The discrepancy of the box $R(\alpha, \beta, \theta, q)$ is

$$D_{\alpha,\beta,\theta}(q) = n \cdot \text{area}\left(R(\alpha, \beta, \theta, q) \cap [0,1]^2\right) - |P \cap R(\alpha, \beta, \theta, q)|.$$

As explained earlier, to achieve the "linear effect," we average out $D_{\alpha,\beta,\theta}(q)^2$ not just over all q, but also over suitable intervals of α, β, and θ. We define

$$\Phi(\alpha_0, \beta_0) = \frac{1}{\alpha_0 \beta_0} \int_{\alpha_0}^{2\alpha_0} \int_{\beta_0}^{2\beta_0} \int_0^{2\pi} \int_{\mathbf{R}^2} D_{\alpha,\beta,\theta}(q)^2 \, dq \, d\theta \, d\alpha \, d\beta.$$

Any rectangle with side lengths between $1/4\sqrt{n}$ and $1/2\sqrt{n}$ has discrepancy bounded away from 0 (since n times its area is a number bounded away from an integer); therefore,

$$\Phi(n^{-1/2}/4, n^{-1/2}/4) \gg 1. \tag{3.7}$$

Our goal now is to amplify this trivial bound. Let $\Pi_{\alpha,\beta,\theta}$ denote the indicator function of $R(\alpha, \beta, \theta, O)$ and put

$$\mu(p) = \begin{cases} n - \sum_{p_i \in P} \delta(p - p_i) & \text{if } p \in [0,1]^2, \\ 0 & \text{else.} \end{cases}$$

We can express the discrepancy as

$$D_{\alpha,\beta,\theta}(q) = \int \Pi_{\alpha,\beta,\theta}(p-q)\mu(p)\,dp.$$

Because $\Pi_{\alpha,\beta,\theta}$ is even,

$$D_{\alpha,\beta,\theta} = \Pi_{\alpha,\beta,\theta} \star \mu,$$

and therefore

$$\widehat{D}_{\alpha,\beta,\theta}(t) = \widehat{\Pi}_{\alpha,\beta,\theta}(t) \cdot \widehat{\mu}(t),$$

from which it follows that

$$\Phi(\alpha_0,\beta_0) = \int \varphi_{\alpha_0,\beta_0}(t)|\widehat{\mu}(t)|^2\,dt, \tag{3.8}$$

where

$$\varphi_{\alpha_0,\beta_0}(t) = \int_0^{2\pi} \frac{1}{\alpha_0\beta_0} \int_{\alpha_0}^{2\alpha_0} \int_{\beta_0}^{2\beta_0} |\widehat{\Pi}_{\alpha,\beta,\theta}(t)|^2\,d\alpha\,d\beta\,d\theta.$$

Note that

$$\begin{aligned}
\widehat{\Pi}_{\alpha,\beta,\theta}(t) &= \int \Pi_{\alpha,\beta,\theta}(q)e^{-2\pi i\langle q,t\rangle}\,dq \\
&= \int \Pi_{\alpha,\beta,0}(p)e^{-2\pi i\langle p[\theta],t\rangle}\,dp,
\end{aligned}$$

where $p[\theta]$ denotes the vector p rotated by θ. In view of the fact that $\langle p[\theta],t\rangle = \langle p,t[-\theta]\rangle$,

$$\widehat{\Pi}_{\alpha,\beta,\theta}(t) = \widehat{\Pi}_{\alpha,\beta,0}(t[-\theta]).$$

We have

$$\begin{aligned}
\widehat{\Pi}_{\alpha,\beta,0}(t) &= \int_{-\alpha/2}^{\alpha/2} e^{-2\pi i x t_x}\,dx \int_{-\beta/2}^{\beta/2} e^{-2\pi i y t_y}\,dy \\
&= \frac{\sin \pi t_x \alpha}{\pi t_x} \cdot \frac{\sin \pi t_y \beta}{\pi t_y};
\end{aligned}$$

hence $\varphi_{\alpha_0,\beta_0}(t)$ is equal to

$$\int_0^{2\pi} \frac{1}{\alpha_0\beta_0} \int_{\alpha_0}^{2\alpha_0} \left(\frac{\sin(\pi\alpha t[-\theta]_x)}{\pi t[-\theta]_x}\right)^2 \int_{\beta_0}^{2\beta_0} \left(\frac{\sin(\pi\beta t[-\theta]_y)}{\pi t[-\theta]_y}\right)^2\,d\alpha\,d\beta\,d\theta.$$

Since the integration in θ is over $[0, 2\pi]$, the only relevant parameter about t is its L^2 norm, denoted by $|t|$, and so we can place t along the axis in the calculation. This shows that $\varphi_{\alpha_0,\beta_0}(t)$ is, in fact,

$$\int_0^{2\pi} \frac{1}{\alpha_0\beta_0} \int_{\alpha_0}^{2\alpha_0} \left(\frac{\sin(\pi\alpha|t|\cos\theta)}{\pi|t|\cos\theta}\right)^2 \int_{\beta_0}^{2\beta_0} \left(\frac{\sin(\pi\beta|t|\sin\theta)}{\pi|t|\sin\theta}\right)^2 d\alpha \, d\beta \, d\theta.$$

Suppose that $\alpha_0|t| \times |\cos\theta|$ remains larger than a small constant. Then, $\sin(\pi\alpha|t|\cos\theta)$ stays bounded away from 0 for any α over a fixed fraction of $[\alpha_0, 2\alpha_0]$, and so[4]

$$\frac{1}{\alpha_0} \int_{\alpha_0}^{2\alpha_0} \left(\frac{\sin(\pi\alpha|t|\cos\theta)}{\pi|t|\cos\theta}\right)^2 d\alpha \approx \frac{1}{|t|^2\cos^2\theta}.$$

If $\alpha_0|t| \cdot |\cos\theta|$ is small enough, however, a Taylor expansion shows that

$$\left|\frac{\sin(\pi\alpha|t|\cos\theta)}{\pi|t|\cos\theta}\right| \approx \alpha.$$

Thus, in all cases,

$$\frac{1}{\alpha_0} \int_{\alpha_0}^{2\alpha_0} \left(\frac{\sin(\pi\alpha|t|\cos\theta)}{\pi|t|\cos\theta}\right)^2 d\alpha \approx \min\left\{\alpha_0^2, \frac{1}{|t|^2\cos^2\theta}\right\}.$$

A similar derivation shows that

$$\frac{1}{\beta_0} \int_{\beta_0}^{2\beta_0} \left(\frac{\sin(\pi\beta|t|\sin\theta)}{\pi|t|\sin\theta}\right)^2 d\beta \approx \min\left\{\beta_0^2, \frac{1}{|t|^2\sin^2\theta}\right\}.$$

It then follows that

$$\begin{aligned}
\varphi_{\alpha_0,\beta_0}(t) &\approx \int_0^{2\pi} \min\left\{\alpha_0^2, \frac{1}{|t|^2\cos^2\theta}\right\} \min\left\{\beta_0^2, \frac{1}{|t|^2\sin^2\theta}\right\} d\theta \\
&\approx \int_0^{\pi/2} \min\left\{\alpha_0^2, \frac{1}{|t|^2\cos^2\theta}\right\} \min\left\{\beta_0^2, \frac{1}{|t|^2\sin^2\theta}\right\} d\theta \\
&\approx A + B,
\end{aligned}$$

where

$$A = \int_0^{\pi/4} \min\left\{\alpha_0^2, \frac{1}{|t|^2\cos^2\theta}\right\} \times \min\left\{\beta_0^2, \frac{1}{|t|^2\sin^2\theta}\right\} d\theta$$

and

$$B = \int_0^{\pi/4} \min\left\{\alpha_0^2, \frac{1}{|t|^2\sin^2\theta}\right\} \times \min\left\{\beta_0^2, \frac{1}{|t|^2\cos^2\theta}\right\} d\theta.$$

If $\beta_0 \geq 1/|t|\cos\theta$, then $B \geq A$ follows from the fact that $\alpha_0 \geq \beta_0$. On the other hand, if $\beta_0 < 1/|t|\cos\theta$, then we also have $\beta_0 \leq 1/|t|\sin\theta$, and hence

---

[4]Recall that the notation $\approx$ means both $\ll$ and $\gg$, ie, equal up to constant factors.

$B \geq A$. It follows that

$$\varphi_{\alpha_0,\beta_0}(t) \approx B$$

$$\approx \min\left\{\beta_0^2, \frac{1}{|t|^2}\right\} \int_0^{\pi/4} \min\left\{\alpha_0^2, \frac{1}{|t|^2\theta^2}\right\} d\theta$$

$$\approx \alpha_0 \min\{1, \alpha_0|t|\} \cdot \min\left\{\frac{\beta_0^2}{|t|}, \frac{1}{|t|^3}\right\}.$$

So, for $\alpha_1 \geq \alpha_0 > 0$,

$$\frac{\varphi_{\alpha_1,\beta_0}(t)}{\varphi_{\alpha_0,\beta_0}(t)} \approx \frac{\alpha_1 \cdot \min\{1, \alpha_1|t|\}}{\alpha_0 \cdot \min\{1, \alpha_0|t|\}} \gg \frac{\alpha_1}{\alpha_0}.$$

From (3.8) we derive that

$$\Phi(\alpha_1, \beta_0) \gg \left(\frac{\alpha_1}{\alpha_0}\right)\Phi(\alpha_0, \beta_0).$$

If we choose $\alpha_0 = \beta_0 = 1/4\sqrt{n}$, as in (3.7), and $\alpha_1 = 1$, we find that $\Phi(\alpha_1, \beta_0) \gg \sqrt{n}$. Because of our choice of α_1 the integrand in q in the definition of $\Phi(\alpha_1, \beta_0)$ is 0 outside a domain of constant area, and so there is a box R of discrepancy $\gg n^{1/4}$, which proves Theorem 3.6 (page 144).
$\square$

Note that by choosing different values of α_1, we derive the more general result that, for any $r \leq 1$, a rectangle of side lengths at most r and discrepancy $\Omega(\sqrt{r}\, n^{1/4})$ always exists.

Bessel Functions and the Fejér Kernel

Turning to the case of disks, we prove Theorem 3.7 (page 144). Let P be a set of n points in the unit square $[0,1]^2$. We define the discrepancy of the disk $\mathcal{D}(q, r)$ of radius r centered at q,

$$D_r(q) \stackrel{\text{def}}{=} n\pi r^2 - |P \cap \mathcal{D}(q, r)|.$$

To handle the case where the disk does not fit entirely within the unit square, we make a copy $P+u$ of P for each point u with integer coordinates. Equivalently, we operate on the torus obtained by identifying the opposite sides of the unit square. We wish to show that, for some $r \leq 1/2$,

$$\sup_q |D_r(q)| \gg n^{1/4}.$$

As usual, our first step is to express the discrepancy as a convolution. To do this, we introduce the indicator function

$$I_{r,q}(p) \overset{\text{def}}{=} \begin{cases} 1 & \text{if } p \text{ lies in } \mathcal{D}(q,r), \\ 0 & \text{else.} \end{cases}$$

Note that because we are on a torus, a point may lie in $\mathcal{D}(q,r)$ even though it is quite far from q in the Euclidean sense. Let

$$\mu(q) \overset{\text{def}}{=} n - \sum_{p \in P} \delta(q - p).$$

Note that

$$D_r(q) = \int_{[0,1]^2} \mu(p) I_{r,q}(p)\, dp,$$

or, if we write $I_r = I_{r,O}$,

$$D_r(q) = \int_{[0,1]^2} \mu(p) I_r(q - p)\, dp = (I_r \star \mu)(q).$$

Expanding the discrepancy in Fourier series, we find that

$$\widehat{D}_r(t) \overset{\text{def}}{=} \int_{[0,1]^2} D_r(q) e^{-2\pi i \langle t, q \rangle}\, dq = \widehat{I}_r(t)\, \widehat{\mu}(t),$$

where $t = (t_1, t_2) \in \mathbf{Z}^2$. By Parseval-Plancherel,

$$\|D_r\|_2^2 = \sum_{t \in \mathbf{Z}^2} |\widehat{D}_r(t)|^2 = \sum_{t \in \mathbf{Z}^2} |\widehat{I}_r(t)|^2 |\widehat{\mu}(t)|^2. \tag{3.9}$$

Our next step naturally is to derive a lower bound on $|\widehat{I}_r(t)|^2$. Unfortunately, the function vanishes infinitely often and no trivial lower bound exists. We circumvent this difficulty by considering the sum

$$|\widehat{I}_{1/2}(t)|^2 + |\widehat{I}_{1/4}(t)|^2$$

instead. We have

$$\widehat{I}_r(t) = \int_{[0,1]^2} I_r(q) e^{-2\pi i \langle t, q \rangle}\, dq = \int_{|q| \leq r} e^{-2\pi i \langle t, q \rangle}\, dq.$$

Again, we use $|q|$ as shorthand for $\|q\|_2$. In polar coordinates, $dq = dq_x dq_y = \rho\, d\rho\, d\theta$. The integrals above depend only on r and $|t|$, so we can assume that $t_1 = |t|$ and $t_2 = 0$. It follows that $\langle t, q \rangle = |t| \rho \cos \theta$, and therefore

$$\widehat{I}_r(t) = \int_0^r \rho\, d\rho \int_0^{2\pi} e^{-2\pi i |t| \rho \cos \theta}\, d\theta.$$

We introduce two Bessel functions of the first kind [2]:

$$J_0(x) = \frac{1}{2\pi} \int_0^{2\pi} e^{ix \cos \theta} \, d\theta$$

and

$$J_1(x) = \frac{1}{x} \int_0^x u J_0(u) \, du.$$

It is immediate that

$$\widehat{I}_r(t) = 2\pi \int_0^r \rho \, J_0(2\pi |t| \rho) \, d\rho,$$

and therefore

$$\widehat{I}_r(t) = \frac{r}{|t|} J_1(2\pi |t| r). \qquad (3.10)$$

From the classical approximation

$$J_1(x) = \sqrt{\frac{2}{\pi x}} \cos(x - 3\pi/4) + O(x^{-3/2}),$$

we derive

$$\frac{\pi x}{2} \left(J_1(x)^2 + J_1(2x)^2 \right) = \cos^2(x - 3\pi/4) + \cos^2(2x - 3\pi/4) + O(1/x).$$

Notice that $\cos^2(x - 3\pi/4) + \cos^2(2x - 3\pi/4)$ is bounded below by a positive constant, so for x large enough

$$\frac{\pi x}{2} \left(J_1(x)^2 + J_1(2x)^2 \right) \gg 1. \qquad (3.11)$$

Actually, by using finer properties of Bessel functions, one can show that (3.11) holds for *all* $x \geq 1$. It follows that

$$|\widehat{I}_{1/2}(t)|^2 + |\widehat{I}_{1/4}(t)|^2 \gg \frac{1}{|t|^3}.$$

From (3.9) we derive

$$\begin{aligned}
\|D_{1/2}\|_2^2 + \|D_{1/4}\|_2^2 &= \sum_{t \in \mathbf{Z}^2} \left(|\widehat{I}_{1/2}(t)|^2 + |\widehat{I}_{1/4}(t)|^2 \right) |\widehat{\mu}(t)|^2 \\
&\geq \sum_{|t| \geq 1} \left(|\widehat{I}_{1/2}(t)|^2 + |\widehat{I}_{1/4}(t)|^2 \right) |\widehat{\mu}(t)|^2 \\
&\gg \sum_{|t| \geq 1} \frac{1}{|t|^3} |\widehat{\mu}(t)|^2
\end{aligned}$$

and, hence,

$$\|D_{1/2}\|_2^2 + \|D_{1/4}\|_2^2 \gg \frac{1}{n^{3/2}} \sum_{\substack{t \neq 0 \\ |t_1|, |t_2| \leq \sqrt{2n}}} |\widehat{\mu}(t)|^2.$$

For $t \neq 0$,

$$\widehat{\mu}(t) = \int_{[0,1]^2} \mu(q) e^{-2\pi i \langle t, q \rangle} \, dq = -\sum_{p \in P} e^{-2\pi i \langle t, p \rangle} \, ;$$

therefore,

$$\|D_{1/2}\|_2^2 + \|D_{1/4}\|_2^2 \gg \frac{1}{n^{3/2}} \sum_{\substack{t \neq 0 \\ |t_1|, |t_2| \leq \sqrt{2n}}} \left| \sum_{p \in P} e^{-2\pi i \langle t, p \rangle} \right|^2.$$

The lower bound

$$\|D_r\|_2 \gg n^{1/4},$$

for $r = 1/2$ or $r = 1/4$, follows directly from Lemma 3.8 below. Indeed, setting $T_1 = T_2 = \lfloor \sqrt{2n} \rfloor$ in the lemma shows that

$$\|D_{1/2}\|_2^2 + \|D_{1/4}\|_2^2 \gg \frac{1}{n^{3/2}} \left((\sqrt{2n} - 1)^2 - n^2 \right) \gg \sqrt{n},$$

which implies that $\|D_r\|_2 \gg n^{1/4}$ for some $r \in \{1/4, 1/2\}$; hence Theorem 3.7 is proved (page 144). $\square$

Lemma 3.8 *For any set P of n points in the plane and any positive integers T_1, T_2,*

$$\sum_{\substack{t \neq 0 \\ |t_1| \leq T_1 \\ |t_2| \leq T_2}} \left| \sum_{p \in P} e^{-2\pi i \langle t, p \rangle} \right|^2 \geq n T_1 T_2 - n^2.$$

Proof: The sum

$$D_n(x) \stackrel{\text{def}}{=} \sum_{|k| \leq n} e^{2\pi i k x}$$

is called the *Dirichlet kernel*. It is easily observed that[5]

$$D_n(x) = \frac{\sin \pi (2n + 1) x}{\sin \pi x}.$$

---

[5] See Appendix B.

We also need the *Fejér kernel*:

$$F_n(x) \overset{\text{def}}{=} \frac{1}{n}(D_0(x) + \cdots + D_{n-1}(x))$$

$$= \frac{1}{n}\sum_{k=0}^{n-1} \frac{\sin \pi(2k+1)x}{\sin \pi x}$$

$$= \frac{1}{n}\left(\frac{\sin n\pi x}{\sin \pi x}\right)^2.$$

The last derivation follows simply from writing $\sin \pi(2k+1)x$ as

$$\frac{e^{i\pi(2k+1)x} - e^{-i\pi(2k+1)x}}{2i}$$

and reducing the resulting geometric sum. Note that this also handles the case $x = 0$, provided that we make the convention $\sin(Ax)/\sin(Bx) = A/B$. We now are ready to derive a lower bound on the left-hand side of the inequality in the lemma. Actually, we temporarily include the case $t = 0$. Define

$$S \overset{\text{def}}{=} \sum_{\substack{|t_1|\leq T_1 \\ |t_2|\leq T_2}} \left|\sum_{p\in P} e^{-2\pi i\langle t,p\rangle}\right|^2.$$

We have

$$S \geq \sum_{\substack{|t_1|\leq T_1 \\ |t_2|\leq T_2}} \left(1 - \frac{|t_1|}{T_1}\right)\left(1 - \frac{|t_2|}{T_2}\right)\left|\sum_{p\in P} e^{2\pi i\langle t,p\rangle}\right|^2$$

$$\geq \sum_{\substack{|t_1|\leq T_1 \\ |t_2|\leq T_2}} \left(1 - \frac{|t_1|}{T_1}\right)\left(1 - \frac{|t_2|}{T_2}\right)\sum_{p,q\in P} e^{2\pi i\langle t,p-q\rangle},$$

from which it follows that

$$S \geq \sum_{p,q\in P} \frac{1}{T_1 T_2} \sum_{\substack{|t_1|\leq T_1 \\ |t_2|\leq T_2}} \sum_{s_1=|t_1|}^{T_1-1} \sum_{s_2=|t_2|}^{T_2-1} e^{2\pi i\langle t,p-q\rangle}$$

$$\geq \sum_{p,q\in P} \frac{1}{T_1 T_2} \sum_{s_1=0}^{T_1-1} \sum_{s_2=0}^{T_2-1} \sum_{\substack{|t_1|\leq s_1 \\ |t_2|\leq s_2}} e^{2\pi i\langle t,p-q\rangle}.$$

Letting $p = (p_x, p_y)$ and $q = (q_x, q_y)$, we derive

$$
\begin{aligned}
S &\geq \sum_{p,q \in P} \frac{1}{T_1 T_2} \sum_{s_1=0}^{T_1-1} \sum_{s_2=0}^{T_2-1} \sum_{|t_1| \leq s_1} e^{2\pi i t_1 (p_x - q_x)} \sum_{|t_2| \leq s_2} e^{2\pi i t_2 (p_y - q_y)} \\
&\geq \sum_{p,q \in P} \frac{1}{T_1 T_2} \sum_{s_1=0}^{T_1-1} \sum_{s_2=0}^{T_2-1} D_{s_1}(p_x - q_x) D_{s_2}(p_y - q_y) \\
&\geq \sum_{p,q \in P} F_{T_1}(p_x - q_x) F_{T_2}(p_y - q_y) \\
&\geq \sum_{p,q \in P} \frac{1}{T_1 T_2} \left(\frac{\sin T_1 \pi (p_x - q_x) \sin T_2 \pi (p_y - q_y)}{\sin \pi (p_x - q_x) \sin \pi (p_y - q_y)} \right)^2.
\end{aligned}
$$

We obtain a lower bound by keeping only the diagonal terms $p = q$ in the last sum. This gives us

$$
S \geq \frac{1}{T_1 T_2} \sum_{p \in P} (T_1 T_2)^2 = n T_1 T_2,
$$

and therefore

$$
\sum_{\substack{t \neq 0 \\ |t_1| \leq T_1 \\ |t_2| \leq T_2}} \left| \sum_{p \in P} e^{-2\pi i \langle t, p \rangle} \right|^2 \geq n T_1 T_2 - n^2.
$$

$\square$

3.3 The Finite Differencing Method

Given n points on a d-dimensional cube grid, we show that no matter how we color them red and blue, there always exists a halfspace within which one color outnumbers the other by at least $\Omega(n^{1/2 - 1/2d})$. This lower bound also holds for any distribution where the ratio between the largest and smallest distances is at least on the order of $n^{1/d}$.

A *red-blue coloring* of a set P of n points in $\mathbf{R}^d$ is a partition of P into two sets, R and B, of "red" and "blue" points, respectively. Given a nonvertical[6] hyperplane h, we define

$$
D(h) = |R \cap h^+| - |B \cap h^+|
$$

to be the (red-blue) discrepancy of the closed halfspace h^+ above h. Let δ

---

[6] Ie, one whose normal is not perpendicular to, say, the x_d-axis.

and δ' be, respectively, the largest and smallest distances between any two distinct points in P. We say that P is *well-spread* if the ratio δ/δ' is less than $cn^{1/d}$, for some absolute constant $c > 0$.

Theorem 3.9 *Given any two-coloring of a well-spread set of n points in $\mathbf{R}^d$, there exists a hyperplane h such that $|D(h)| = \Omega(n^{1/2-1/2d})$.*

As mentioned in Chapter 2, the lower bound is tight: Any set of n points (well-spread or not) can be colored so that the discrepancy of any halfspace is bounded by $O(n^{1/2-1/2d})$. The proof of Theorem 3.9 uses two different kinds of tools. One is a remarkable integral-geometric identity that relates the mean-square discrepancy to metric properties of the point set. The other one is a finite differencing technique of which we caught a glimpse in the method of orthogonal functions discussed in §3.1. Finite differencing is useful because discrepancies are lower order phenomena. There is a beautiful illustration of this in two dimensions, where the technique reduces to the classical Buffon's needle experiment. We begin with this story and postpone the full saga (which is considerably more technical) for the general case.

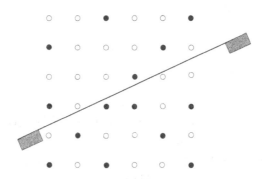

Fig. 3.8. The discrepancy is 2.

Buffon's Needle as a Discrepancy Tool

As usual, we prove a result stronger than Theorem 3.9 by establishing a lower bound on the L^2 norm of the red-blue discrepancy. To do this, our first task is to define the notion of a random halfplane or, equivalently, a random line. Consider the invariant measure of lines under rigid motions,

$d\omega(h) = d\rho d\theta$. (Up to a scaling factor, such a measure is unique.) By Cauchy's integral-geometric formula (also known as Crofton's formula), the measure of the set of lines crossing a convex region is equal to its perimeter. So, it is sensible to define a random line as one picked according to that measure among those crossing the square $\mathcal{U} = [0, 1/4]^2$. As we shall see, the proof's setup bears a striking resemblance to the classical Buffon's needle experiment.

Without loss of generality, assume that $n = m^2$ for some integer $m > 0$. We define P to be the set of n interior vertices of an $(m + 1)$-by-$(m + 1)$ grid covering $\mathcal{U}$, and we fix a red-blue coloring, $P = R \sqcup B$, once and for all. The case $d = 2$ of Theorem 3.9 follows from the L^2-norm lower bound on $D(h)$ below. (Well-spreadness is the only assumption on the point set necessary for the proof, so the result generalizes immediately from grid points to arbitrary well-spread points.) We use D below as shorthand for $D(h)$.

Lemma 3.10

$$\mathbf{E}_h \, D^2 \gg \sqrt{n}.$$

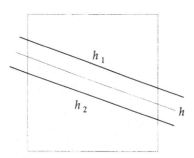

Fig. 3.9. The two slabs around h.

As usual, we expect the discrepancy to be "concentrated" around the boundary h, so we define the two lines h_1 and h_2 parallel to h at a distance $w/2$ from h, where $w = c_0/m$ for a small constant $c_0 > 0$. The line h is the medial axis of a slab of width w, denoted by slab(h), which is bounded by h_1 and h_2, with, say, h_1 above h_2 (Fig. 3.9). Define

$$D_w(h) \stackrel{\text{def}}{=} D(h_1) - 2D(h) + D(h_2).$$

This measures the difference in discrepancy between the halfslab between

h and h_2 and the one between h and h_1. We consider the expectation of $D_w(h)^2$ for a random h. By Cauchy-Schwarz,

$$D_w(h)^2 \le 6\Big(D(h_1)^2 + D(h)^2 + D(h_2)^2\Big),$$

and therefore[7]

$$\mathbf{E}_h\, D_w^2 \ll \mathbf{E}_h\, D^2. \tag{3.12}$$

So, it now suffices to establish a lower bound on the L^2 norm of D_w, which is an easier task, it turns out. Observe that

$$D_w(h)^2 = \Big(\sum_{R_1} 1 - \sum_{B_1} 1 - \sum_{R_2} 1 + \sum_{B_2} 1\Big)^2,$$

where R_1 is the set of red points in the upper halfslab, B_2 is the set of blue points in the lower halfslab, etc. If we expand the right-hand side, we see that companion pairs (ie, products of $\pm$ terms with themselves) amount to

$$|R_1| + |B_1| + |R_2| + |B_2| = |P \cap \text{slab}(h)|,$$

while mixed pairs contribute twice the number of segments $pq \subset \text{slab}(h)$ $(p \ne q)$ that are either monochromatic noncrossing or bichromatic crossing minus twice the number of segments that are bichromatic noncrossing or monochromatic crossing. Put differently,

$$D_w(h)^2 = |P \cap \text{slab}(h)| + 2(\sharp MN + \sharp BC - \sharp BN - \sharp MC),$$

where $\sharp MN$ denotes the number of positive-length segments $pq \subset \text{slab}(h)$ that avoid h and have monochromatic endpoints, etc. We can rewrite this as

$$D_w(h)^2 = |P \cap \text{slab}(h)| + \sum_{p \ne q} I_{pq}^{MN} + \sum_{p \ne q} I_{pq}^{BC} - \sum_{p \ne q} I_{pq}^{BN} - \sum_{p \ne q} I_{pq}^{MC},$$

where I_{pq}^{MN} is the indicator function equal to 1 if $pq \subset \text{slab}(h)$ avoids h and has monochromatic endpoints, and 0 else, etc. By using linearity of expectation, we find that

$$\mathbf{E}_h\, D_w^2 \ge \sum_{p \in P} \text{Prob}_h[p \in \text{slab}(h)] - \sum_{p \ne q} g(p, q),$$

---

[7]The random variables h_1 and h_2 do not come from precisely the same distribution as h, but this is a minor technicality that we can ignore. Indeed, the width of the slabs is much smaller than the distance from any point to the boundary of $\mathcal{U}$. As a result, over the domain where h_1 (or h_2) does not intersect $\mathcal{U}$, the corresponding h yields the same discrepancy as h_1, and so the expected value of $D(h_1)^2$ over that domain does not exceed $\mathbf{E}_h\, D^2$; therefore, conservatively, $\mathbf{E}_h\, D(h_1)^2 \le 2\,\mathbf{E}_h\, D^2$.

where

$$g(p,q) \;=\; \Big| \mathrm{Prob}_h[\, pq \subset \mathrm{slab}(h) \;\&\; \mathrm{crosses}\; h\,]$$
$$-\, \mathrm{Prob}_h[\, pq \subset \mathrm{slab}(h) \;\&\; \mathrm{avoids}\; h\,] \Big|.$$

By Cauchy's formula, a random line h is at a distance $w/2$ or less from any given point p with probability πw; therefore,[8]

$$\mathbf{E}_h\, D_w^2 \geq \pi w n - \sum_{p \neq q} \big| f(w, |p-q|) - 4f(w/2, |p-q|) \big|, \qquad (3.13)$$

where $f(x,y)$ is the probability that a random slab of width $x \in \{w/2, w\}$ contains a fixed segment $p_i p_j$ of length y. (Again, for notational convenience, we use $|p-q|$ to denote the L^2 norm of the vector $p-q$.)

Lemma 3.11 *For any $y \geq w$,*

$$\Big| f(w,y) - 4f(w/2, y) \Big| \leq \frac{w^4}{y^3}.$$

Proof: Because $x \leq w \leq y$, the maximum angle α between a segment of length y and the axis of an enclosing slab of width x is equal to $\arcsin(x/y)$; therefore,

$$f(x,y) = \int_{-\alpha}^{\alpha} (x - y\sin|\theta|)\, d\theta = 2\alpha x - 2(1 - \cos\alpha)y,$$

and so

$$f(w,y) - 4f(w/2, y) \;=\; 2w\Big(\arcsin\frac{w}{y} - 2\arcsin\frac{w}{2y}\Big)$$
$$+\, 2y\Big\{ 3 + \sqrt{1 - \Big(\frac{w}{y}\Big)^2} - 4\sqrt{1 - \Big(\frac{w}{2y}\Big)^2}\, \Big\}.$$

Because $y \geq 1/4(m+1)$ and $wm = c_0$ can be chosen as small as desired, a Taylor expansion around $w = 0$ gives

$$f(w,y) - 4f(w/2, y) = \frac{w^4}{16y^3} + O(w^6/y^5),$$

from which the proof follows. □

---

[8]Implicit to this argument, and the next one, is the fact that, by construction, p is more than $w/2$ away from the boundary of the square $\mathcal{U}$.

Combining (3.13) and Lemma 3.11, we obtain

$$\mathbf{E}_h D_w^2 \geq \pi w n - \sum_{p \neq q} \frac{w^4}{|p - q|^3}. \tag{3.14}$$

The edge length λ_0 of the grid is $1/4(m + 1)$, and there are $8k$ grid points at L^∞-distance $\lambda_0 k$ from a given grid point. It follows that

$$\sum_{p \neq q} \frac{1}{|p - q|^3} \ll \sum_{1 \leq k \leq m-1} \frac{n}{\lambda_0^3 k^2} \ll n^{5/2},$$

and hence

$$\mathbf{E}_h D_w^2 \geq \pi w n - O(w^4 n^{5/2}).$$

Keeping the constant $c_0 = wm$ small enough yields $\mathbf{E}_h D_w^2 \gg n^{1/2}$, which in view of (3.12) establishes Lemma 3.10 (page 157). □

A Probabilistic Interpretation *

The line of reasoning used in the two-dimensional proof provides an intuitive answer to the simple question: Why can't the discrepancy be 0 everywhere? Let h be a random line and suppose, for the sake of contradiction, that $D(h)$ is zero everywhere (of course, it is obvious that it must be at least ± 1 somewhere, so we should think of zero as meaning "very small"). It easily follows that within each of the two halfslabs in slab(h), the numbers of red and blue points match exactly,

$$R_h^+ = B_h^+ \quad \text{and} \quad R_h^- = B_h^-.$$

Now, let us perform the following experiment: Pick two random integers $1 \leq i < j \leq n$ and assume that both points p_i, p_j fall in slab(h). Note—this is crucial—that $i \neq j$. If the segment $p_i p_j$ crosses h, then obviously p_i and p_j are equally likely to be bichromatic or monochromatic; we write this as

$$BC = MC, \tag{3.15}$$

which reads: "probability of Bichromatic Crossing equals probability of Monochromatic Crossing." On the other hand, if $p_i p_j$ does not cross h, then

$$BN > MN, \tag{3.16}$$

meaning that the probability that $p_i p_j$ is Bichromatic Noncrossing exceeds that of being Monochromatic Noncrossing. Indeed, without loss of gener-

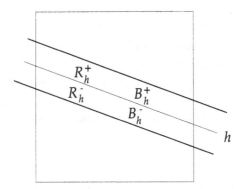

Fig. 3.10. Assume that there is no discrepancy.

ality, suppose that p_i is blue and lies in h^+. Then, p_j is picked randomly among R_h^+ red points and $B_h^+ - 1$ blue points (minus 1 because $i \neq j$). Because $B_h^+ = R_h^+$, the point p_j is thus ever so slightly more likely to be red than blue. Surprisingly, this rather trivial and innocent-looking fact is the key to the proof. To see this, add together (3.15) and (3.16). This gives $BN + MC > BC + MN$, and hence $BN > BC$ or $MC > MN$. Suppose that $BN > BC$ (a similar argument applies to the other case as well). Recall that h is random and so is the choice of $p_i p_j$ within slab(h). So, the meaning of the inequality is that a random bichromatic segment $p_i p_j$ (ie, with random i, j) that falls inside a random slab(h) is more likely to avoid h than to cross it. Let $\delta > 0$ be the difference in probability.

Now, fix $p_i p_j$ and consider a random slab(h) $\supset p_i p_j$. By changing our viewpoint, this is easily seen to be equivalent to fixing slab(h) and taking a random segment $s \subset$ slab(h) of length $|p_i p_j|$. (Note that the meaning of a random segment s is now with respect to the invariant measure for segments and not with respect to the indices i, j.) Throwing a random segment in a slab and counting how many times it crosses a line in the middle is essentially a repeat of the classical *Buffon needle experiment*. An elementary calculation shows that if s is "long," then it is almost as likely to cross the line h as to avoid it. But the inequality $BN > BC$ says that the two events must differ in probability by at least δ. Since p_i, p_j are grid points and the pair (i, j) is random, we must expect the segment $p_i p_j$ to be quite long, and so we have a contradiction.

At first it might seem doubtful that such a weak argument stands a chance of producing any kind of meaningful discrepancy, let alone a quasi-optimal bound. Indeed, the only reason that $|BC - BN|$ is nonzero is that the trivial segment $p_i p_i$ is forbidden from the randomized experiment. Obviously, even if we allowed such events, they still would be very unlikely to come up. Discrepancies are lower order effects, and however remarkable this might seem, ruling out the choice of $p_i p_i$ is all that it takes to produce the cancellations that are necessary to make the true discrepancy "bubble up to the surface." Quite an amazing mathematical phenomenon, wouldn't you say?

The Alexander-Stolarsky Formula

Let P be a well-spread set[9] of n points in $\mathbf{R}^d$. We fix a red-blue coloring of P, that is, a function $\chi : P \mapsto \{-1, 1\}$. The discrepancy of the closed halfspace h^+ bounded below by the (nonvertical) hyperplane h is

$$D(h) \stackrel{\text{def}}{=} \sum_{p \in P \cap h^+} \chi(p).$$

We want to show that $\mathbf{E}_h D^2 \gg n^{1-1/d}$. The underlying distribution is given by the motion-invariant measure ω for hyperplanes. To define a proper probability measure, we assume that a random hyperplane cuts a cube of unit boundary area. By Cauchy's formula, the measure of all such hyperplanes is 1. Furthermore, by appropriate scaling we can always assume that P lies comfortably within the unit cube. (The points should stay away from the boundary because we need some wiggling room.) The trick of using halfslabs is a special case of a general finite differencing technique, which we now discuss in detail. We begin with the indicator function

$$I_{pq}(h) = \begin{cases} 1 & \text{if } h \text{ separates } p \text{ and } q, \\ 0 & \text{else.} \end{cases}$$

For convenience we assume that P contains as many red points as blue ones. If this is not the case, we add new points of the deficient color. Note that if this requires the addition of too many points, say, more than $b_0 n^{1/2 - 1/2d}$, for some small positive constant b_0, then any halfspace that contains the cube has high discrepancy and we are done. Otherwise, if Theorem 3.9 holds for the new set of points, then it does for the old one as well, provided that b_0 is small enough. Thus, our assumption is not restrictive. We derive $D(h)^2 = -D(h)D^*(h)$, where $D^*(h)$ is the discrepancy

---

[9]This means that the ratio between the largest and smallest distances is $O(n^{1/d})$.

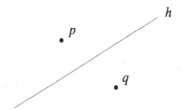

Fig. 3.11. $I_{pq}(h) = 1$.

of the halfspace below h. By pairing each term in $D(h)$ with each one in $D^*(h)$, we obtain[10]

$$D(h)^2 = - \sum_{p,q \in P} I_{pq}(h)\chi(p)\chi(q).$$

By Cauchy's formula,

$$\int I_{pq}(h)\, d\omega(h) = |p - q|,$$

from which the identity known as the *Alexander-Stolarsky formula* follows:

$$\mathbf{E}_h D^2 = - \sum_{p,q \in P} \chi(p)\chi(q)|p - q|. \tag{3.17}$$

Note that the fact that χ takes on values ± 1 was never used in establishing (3.17), and so it still holds for any real-valued function χ, as long as $\sum_\chi(p) = 0$.

The Discrepancy of Halfspaces

We now prove the discrepancy lower bound in its full generality (Theorem 3.9, page 156). We define the *forward differencing* operator Δ to act on a real function f as follows:

$$\Delta f(x) \stackrel{\text{def}}{=} f(x + 1) - f(x).$$

---

[10]Problems might seem to arise if p or q lies on h, but these events are of measure zero and can be ignored for integration purposes.

The t-fold iteration can be expressed as

$$\Delta^t f(x) = \sum_{i=0}^{t} (-1)^{t-i} \binom{t}{i} f(x+i).$$ (3.18)

For a function f of class $C^t(0,t)$, meaning that it has a continuous t-th derivative in $(0,t)$, there exists a number $\xi \in (0,t)$ such that

$$\Delta^t f(0) = f^{(t)}(\xi).$$ (3.19)

Let $v = (v_1, 0, \ldots, 0)$ denote a fixed vector in $\mathbf{R}^d$, where v_1 is a small positive real. We form the union of P with $t = \lceil d/2 \rceil + 1$ copies of it, each translated by a multiple of v:

$$P_v = \bigcup_{j=0}^{t} (P + jv).$$

For convenience, we assume that no two points in P are collinear with the x_1-axis. This way, we can easily extend the coloring of P to P_v by writing

$$\chi(p + jv) = (-1)^j \binom{t}{j} \chi(p).$$

We easily see that the discrepancy $D_v(h)$ for the new set P_v is given by

$$D_v(h) = \sum_{j=0}^{t} (-1)^j \binom{t}{j} D(h - jv).$$ (3.20)

We should point out the similarity with the previous section. Instead of making three copies of h and assigning the weights $1, -2, 1$, as we did earlier, we generalize this idea and use a total of $t+1$ copies of P to which we assign weights derived from the rows of Pascal's triangle: These are the signed coefficients for higher order finite differencing. The only real difference is that in two dimensions we dealt with slabs of fixed width, whereas now the slabs are of fixed x_1 width. By Cauchy-Schwarz,

$$D_v(h)^2 \le \sum_{j=0}^{t} \binom{t}{j}^2 \sum_{j=0}^{t} D(h - jv)^2,$$

and therefore

$$\mathbf{E}_h D_v^2 \ll_d \mathbf{E}_h D^2.$$ (3.21)

Again, as we did in two dimensions, we choose the width v_1 small enough so that the slight difference between the probabilistic space for h and $h - jv$ can be ignored in our calculations. We have a relation similar to the Alexander-

Stolarsky formula:

$$\mathbf{E}_h D_v^2 = -\sum_{p',q'\in P_v} \chi(p')\chi(q')|p'-q'|$$

$$= -\sum_{p,q\in P} \chi(p)\chi(q)G(p,q),$$

where

$$G(p,q) = \begin{cases} \sum_{j=-t}^{t}(-1)^j \binom{2t}{t+j}|p-q+jv| & \text{if } p\neq q, \\ -\binom{2t-2}{t-1}|v| & \text{if } p = q. \end{cases}$$

To verify these last two identities is straightforward. For example, in the case where $p\neq q$, the length $|p-q+jv|$ is weighted by the factor

$$(-1)^{2i+j}\sum_{i=0}^{t-j}\binom{t}{i}\binom{t}{i+j} = (-1)^j\sum_{i=0}^{t-j}\binom{t}{t-i}\binom{t}{i+j} = (-1)^j\binom{2t}{t+j}.$$

We now try to simplify the sum in the expression for $p\neq q$. For fixed p,q, the function

$$f(x) = |p-q+xv| = \sqrt{(p_1-q_1+xv_1)^2 + (p_2-q_2)^2 + \cdots + (p_d-q_d)^2}$$

is smooth locally around 0. In view of (3.18) and (3.19), we see that

$$G(p,q) = (\Delta^{2t}f)(-t),$$

so there exists $\xi \in (-t,t)$ such that

$$G(p,q) = f^{(2t)}(\xi),$$

from which we easily find that

$$|G(p,q)| \leq \frac{c_0 v_1^{2t}}{|p-q+\xi v|^{2t-1}} \leq \frac{c_0 v_1^{2t}}{|p-q|-tv_1},$$

for some constant $c_0 > 0$ (depending on d). We now can repeat the previous argument. At most $O(k^{d-1})$ points of P lie at distance roughly $k/n^{1/d}$ from any point, so as long as $|v|$ is much smaller than $n^{-1/d}$, for a fixed p,

$$\sum_{q\,;\,p\neq q}\frac{1}{|p-q+\xi v|^{2t-1}} \ll \sum_{k>0}\frac{k^{d-1}}{(k/n^{1/d})^{2t-1}} \ll n^{(2t-1)/d}.$$

It follows that

$$\sum_{p,q\,;\,p\neq q}|G(p,q)| \ll |v|^{2t}n^{1+(2t-1)/d},$$

and therefore

$$\mathbf{E}_h D_v^2 \;\gg\; -\sum_p G(p,p) - |v|^{2t} n^{1+(2t-1)/d}$$

$$\gg\; n\binom{2t-2}{t-1}|v| - |v|^{2t} n^{1+(2t-1)/d}.$$

Choosing $|v| = v_1 = c_1 n^{-1/d}$ for a small enough constant $c_1 > 0$ yields $\mathbf{E}_h D_v^2 \gg n^{1-1/d}$, and, by (3.21), $\mathbf{E}_h D^2 \gg n^{1-1/d}$. This establishes Theorem 3.9 (page 156). □

Approximate Diagonalization *

In Chapter 1 we discussed the relation between discrepancy and eigenvalues. In this chapter we have argued that the Fourier transform is the natural vehicle for diagonalization, as convolution often lurks behind the construction of the underlying set systems. It would be wrong to think that the finite differencing method is somehow fundamentally different. The twist is that, instead of seeking perfect diagonalization, finite differencing brings the original matrix in diagonally dominant form.[11] In a purely eigenvalue-based approach, a set system is represented by an $m \times n$ incidence matrix A, where each of the m rows of A is the characteristic vector of a distinct set in the set system. Given a coloring $x \in \{-1,1\}^n$, we estimate the L^2 norm of Ax or, equivalently, $x^T A^T Ax$. Because $B = A^T A$ is positive semidefinite, it can be diagonalized as

$$B = M^T \Lambda M,$$

where Λ is the diagonal matrix of eigenvalues. So,

$$\|Ax\|_2^2 = x^T Bx = y^T \Lambda y,$$

where $y = Mx$. Knowing the smallest eigenvalue of B immediately leads to a lower bound on the value of the quadratic form $y^T \Lambda y$, for $\|x\|_2 = \sqrt{n}$, and hence on the L^2-norm discrepancy $\|Ax\|_2$, for any $x \in \{-1,1\}^n$. Unlike its Fourier transform counterpart, the finite differencing method bypasses Λ. Instead, it manipulates A to bring B in (almost) diagonally dominant form. That finite differencing and a little bit of integral geometry added to the mix should do the trick is quite astonishing!

---

[11]Or, more generally, in a form that for vectors with ±1 coordinates has the same effect as if it were diagonally dominant.

Going back to our treatment of halfspace discrepancy, we simplify our discussion by discretizing the problem. We choose a large collection of representative halfspaces, and we replace the function $D(h)$ by the vector Ax, where x_i is the color of point p_i and the j-th row of A is the characteristic vector of the j-th halfspace in the collection. By analogy with D_v and the finite differencing identity (3.20), we form A_v by performing linear combinations on the rows of A. Inequality (3.21) then becomes

$$\|Ax\|_2 \gg \|A_v x\|_2.$$

Let B_v denote $A_v^T A_v$. Obviously,

$$\|A_v x\|_2^2 = x^T B_v x = \sum_{i,j} \left(\sum_k a_{ki} a_{kj} \right) x_i x_j,$$

where $A_v = (a_{ij})$. Now, observe that $-\sum_k a_{ki} a_{kj}$ is precisely the quantity $G(p, q)$ in the proof, where p is the i-th point and q is the j-th point. The remainder of the proof shows that, up to a constant factor, $x^T B_v x$ is at least $x^T B_v^* x$, if $x \in \{-1, 1\}^n$ and B_v^* is derived from B_v by zeroing out every nondiagonal element. This shows that, for diagonalization purposes, the full power of harmonic analysis is not always needed in discrepancy theory; in this case, it is replaced advantageously by finite differencing.

3.4 Bibliographical Notes

Section 3.1:   The elegant proof of Theorem 3.1 (page 135) is due to Roth [261]. In his paper, Roth proves only the two-dimensional case, but the generalization to d-space is not difficult. The reader interested in the connection between Rademacher functions and wavelets will turn to the classic text by Daubechies [104]. Recall that the $\Omega(\log n)^{(d-1)/2}$ lower bound on the L^2 norm of the discrepancy is matched by a corresponding upper bound; see Chapter 2. By using Hölder's inequality, instead of Cauchy-Schwarz, Schmidt [274] was able to strengthen Roth's result and show that $\Omega(\log n)^{(d-1)/2}$ is also a lower bound on the L^p norm of the discrepancy of axis-parallel boxes, for any fixed $p > 1$.

The lower bound of $\Omega(\log n)$ on the L^∞ norm of the discrepancy for axis-parallel boxes in two dimensions (Theorem 3.4, page 139) is due to Schmidt [273]. The proof we give, which is based on Riesz products, is due to Halász [155]. As discussed in Chapter 2, this lower bound is tight. Halász's paper also gives a lower bound of $\Omega(\log n)^{1/2}$ on the L^1 norm of the discrepancy in $\mathbf{R}^2$.

Section 3.2: The full power of the Fourier transform method was uncovered by Beck [33, 35] and Montgomery [232]. An earlier glimpse of the method was provided by Roth in his use of Fourier transforms for the discrepancy of arithmetic progressions (see §1.5). Baker also uses the basic underlying idea in [27]. Many applications of the method are given in Beck and Chen [37]. The lower bound of $\Omega(n^{1/4})$ on the discrepancy of rotated boxes (Theorem 3.6, page 144) is by Beck [35]. Schmidt [270] was the first to obtain a bound[12] close to $n^{1/4}$. Our discussion of the discrepancy of disks follows Montgomery [233], to whom Theorem 3.7 (page 144) is due [232]. An earlier bound of $\Omega(n^{1/2-1/2d-\varepsilon})$, for any fixed $\varepsilon > 0$, was obtained by Schmidt [270] for the toroidal discrepancy of balls in $[0,1]^d$. Schmidt also derived nontrivial results for the case where the ball is contained in the unit cube, but the near-optimal bound of $\Omega(n^{1/2-1/2d-\varepsilon})$ was achieved by Beck [35]. In that article, Beck proves the following, very general result: Given a d-dimensional convex body C in $[0,1]^d$, consider the family of all bodies obtained by similarity (ie, by translating, rotating, or homothetically transforming C); then, given any n points in $[0,1]^d$, the volume discrepancy is $\Omega(n^{1/2-1/2d-\varepsilon})$. For a more comprehensive account of the history behind all these results, see Beck and Chen [37], Drmota and Tichy [111], and Matoušek [219].

Section 3.3: The lower bound of $\Omega(n^{1/2-1/2d})$ on the red-blue discrepancy of halfspaces (Theorem 3.9, page 156) is due to Alexander [10]; it builds on earlier work by Alexander and Stolarsky, from which the Alexander-Stolarsky formula can be traced [298]. The lower bound is optimal, as discussed in Chapter 2. The slightly weaker bound of $\Omega(n^{1/4}\log^{-7/2}n)$ was obtained earlier by Beck [37] for Roth's *disc segment problem*, where points are placed in a disk; the bound was derived for the volume discrepancy, but it can be easily extended to the red-blue case as well.

The proof that we give in this text contains essential simplifications by Chazelle, Matoušek, and Sharir [82], who introduced the finite differencing method in discrepancy theory. Their paper also extends the technique to the volume discrepancy, providing an elementary alternative to Alexander's original proof [11]. For background material on integral geometry, the reader should consult Santaló's classical text [265].

---

[12]There are subtle variations depending on whether the box is taken modulo 1 (toroidal discrepancy) or not, and whether it is completely contained in the unit square or it may only overlap.

4

Sampling

his chapter is about extracting small representative samples from large data sets. In the process we develop a complete computational theory of geometric sampling, with an eye toward the derandomization applications that will be discussed in later chapters. It is difficult to overestimate the impact that this theory has had in computational geometry in the 1990's.

The combinatorial discrepancy of a set system indicates how well, relative to its constituent subsets, we can sample the ground set by selecting about half of it. It is natural to ask what happens for different sample sizes. At one extreme, we might wonder how well we can sample a set if we are allowed to pick only a constant number of elements. For example, given a finite collection of points in the plane, is it possible to choose a subset of constant size, such that any disk that encloses at least one percent of the points also includes at least one sample point? Surprisingly, the answer is yes.

In fact, something even stronger and stranger is true: Suppose that we want to estimate how many people live within 10 miles of a hospital in a given country. We can do this by sampling the population carefully, answering the question for the sample, and then scaling up appropriately. What is amazing is that, for a given relative error, the same sample size works just as well whether the country is Switzerland or China! Furthermore, we can change metrics and even lift the problem into higher dimensional space, and this still remains true. The magic lies in the concept of the Vapnik-Chervonenkis dimension (see §1.4). Because of its great generality, the result is best expressed within an abstract framework that unifies all of these specific cases.

Typically, sampling is used to estimate some parameter defined over a large population by looking at only a very small subset. We distinguish between two popular notions of sampling:

1. THRESHOLD-BASED SAMPLING: The sample should "hit" (ie, intersect) any large enough subset of the set system. This allows us to perform threshold tests. It is essentially a hypergraph cover problem, and we use ε-*nets* for that purpose.
2. COUNT-BASED SAMPLING: The relative size of a subset can be estimated within a small additive error; for this we use ε-*approximations*.

The chapter is organized as follows. In §4.1 we define ε-nets and ε-approximations, and we explore basic structural properties. Sampling algorithms for arbitrary set systems are given in §4.2. The important case of range spaces of bounded VC-dimension is treated in §4.3. In the same section, the concept of a product range space is also introduced as a tool for approximating more complex functions. This will be used extensively in the design of convex hull algorithms (Chapter 7). Finally, the related notion of weak ε-nets is briefly discussed in §4.4. This gives us an opportunity to take a quick excursion into hyperbolic geometry and show its usefulness for sampling.

4.1 ε-Nets and ε-Approximations

Let $(V, \mathcal{S})$ be a finite set system, where $|V| = n$ and $|\mathcal{S}| = m$. Given any $0 < \varepsilon < 1$, a set $N \subseteq V$ is called an ε-*net for* $(V, \mathcal{S})$ if $N \cap S \neq \emptyset$, for any $S \in \mathcal{S}$ with $|S|/|V| > \varepsilon$. A set $A \subseteq V$ is called an ε-*approximation* for $(V, \mathcal{S})$ if, for any $S \in \mathcal{S}$,

$$\left| \frac{|S|}{|V|} - \frac{|A \cap S|}{|A|} \right| \leq \varepsilon.$$

An equivalent formulation is that, given a random v uniformly distributed in V, for each $S \in \mathcal{S}$,

$$\left| \mathrm{Prob}[v \in S] - \mathrm{Prob}[v \in S \,|\, v \in A] \right| \leq \varepsilon.$$

Note that an ε-approximation is an ε-net but not the other way around. Whereas an ε-approximation can be used to estimate the size of any given set in $\mathcal{S}$, an ε-net is weaker and allows us only to certify sets larger than a certain threshold. The advantage of ε-nets is their smaller size. Like most threshold structures, however, they are mathematically awkward. On most other counts, ε-approximations are preferable. They enjoy nicer algebraic structures with useful compositional properties. Recall from Chapter 1 that, given $W \subseteq V$, the set system *induced* by W, denoted by $(W, \mathcal{S}|_W)$, is formed by the ground set W and the subsets $\{W \cap S \mid S \in \mathcal{S}\}$. The proofs of the two lemmas below are elementary and we omit them.

Lemma 4.1 *Let V_1, V_2 be disjoint subsets of V of the same size, and let A_i be an ε-approximation for the subsystem induced by V_i. If $|A_1| = |A_2|$, then $A_1 \cup A_2$ is an ε-approximation for the subsystem induced by $V_1 \cup V_2$.*

Lemma 4.2 *If A is an ε-approximation for $(V, \mathcal{S})$, then any ε'-approximation (resp. -net) for $(A, \mathcal{S}|_A)$ is also an $(\varepsilon + \varepsilon')$-approximation (resp. -net) for $(V, \mathcal{S})$.*

4.2 General Set Systems

Let $(V, \mathcal{S})$ be a set system, with $V = \{v_1, \ldots, v_n\}$ and $\mathcal{S} = \{S_1, \ldots, S_m\}$. Theorems 4.3 and 4.4 assert the existence of an ε-net of size $O(\varepsilon^{-1} \log m)$ and an ε-approximation of size $O(\varepsilon^{-2} \log m)$, respectively.

We begin our discussion with ε-nets. An equivalent statement of our goal is, given any integer $1 \leq r \leq n$, find a subset $N \subseteq V$ that intersects every S_i of size greater than n/r. Obviously, we can assume that $|S_i| > n/r$, for any i, and $m > 1$. Let $p = cr(\log m)/n$, for some large enough constant c. Sample the set V according to the binomial distribution $B(n, p)$, ie, form a set N by including in it each v_i with probability p. The probability that N fails to intersect some $S_i \in \mathcal{S}$ is less than $m(1 - p)^{n/r}$, which can be made smaller than any constant. Note that the sample N is of size $O(r \log m)$ with probability arbitrarily close to 1. Thus, with high probability, a random sample of size $O(r \log m)$ intersects every set of size $> n/r$. Checking that a given sample fits the bill takes $O(nm)$ time.

It is possible to derandomize the algorithm by using the method of conditional expectations discussed in Chapter 1. But there is a much simpler deterministic algorithm, which we describe next.

The Greedy Cover Algorithm

We follow a greedy approach. The idea is to keep selecting the element that belongs to the most sets. Specifically, initialize N to $\emptyset$. Then, iterate on the following process until the set $\mathcal{S}$ is empty:

STEP 1. Find the element v_i that is contained in the most sets of $\mathcal{S}$.

STEP 2. Remove v_i from V and add it to N. Then, remove from $\mathcal{S}$ every set that contains v_i.

When implemented properly, the algorithm takes $O(nm)$ time. It is simple, practical, and does not require storing large numbers. How big is N in the worst case? Note that the general problem of computing a minimum-size cover is NP-hard, so we should not expect N to be always minimum. But we can show that it is reasonably small. Let m_k be the number of sets left in $\mathcal{S}$ after k iterations. By definition, $m_0 = m$. After the k-th iteration, we have m_k sets left, each containing more than n/r elements. Thus, the next element to be removed belongs to more than $\frac{n/r}{n-k} \cdot m_k \geq m_k/r$ sets, and so $m_{k+1} < m_k(1 - 1/r)$; hence,

$$m_k \leq m\left(1 - \frac{1}{r}\right)^k.$$

For a large enough constant c, any $k \geq cr \log m$ is such that $m_k < 1$, and hence $m_k = 0$.

Theorem 4.3 *Given a set system* $(V, \mathcal{S})$, $|V| = n$ *and* $|\mathcal{S}| = m$, *and* $1 \leq r \leq n$, *the greedy cover algorithm finds a* $(1/r)$-*net for* $(V, \mathcal{S})$ *of size* $O(r \log m)$ *in time* $O(nm)$.

The Weighted Greedy Sampling Algorithm

To compute an ε-approximation, we use an approach similar in spirit to the unbiased greedy coloring algorithm of Chapter 1. Instead of mapping each $v_i \in V$ to ± 1, the "color" $\chi(v_i)$ will now take on values in $\{-q, 1-q\}$, for some fixed parameter $0 < q < 1$ to be specified later. For technical convenience, we add the set $S_0 = V$ into the set system. Given $S_i \in \mathcal{S}$, let $p_{i,k}$ (resp. $m_{i,k}$) be the number of $v_j \in S_i$ ($j \leq k$) such that $\chi(v_j) = 1 - q$ (resp. $\chi(v_j) = -q$). We fix a parameter $0 < \mu < 1$ and we define $H_q(i, k)$ as follows:

$$H_q(i, k) = (1 + (1-q)\mu)^{p_{i,k}}(1 - q\mu)^{m_{i,k}} + (1 + q\mu)^{m_{i,k}}(1 - (1-q)\mu)^{p_{i,k}}.$$

We introduce the weight function

$$H_q(k) = \sum_{0 \leq i \leq m} H_q(i, k).$$

Note that $H_q(0) = 2(m+1)$. The strategy is, for each $k = 0, 1, \ldots, n-1$, to choose the assignment of $\chi(v_{k+1}) \in \{-q, 1-q\}$ that produces the smaller value of $H_q(k+1)$. We define the set A as

$$A = \{\, v \in V \,:\, \chi(v) = 1 - q \,\}.$$

The algorithm takes time $O(nm)$. We easily check that we can afford a relative error of $1/n^{O(1)}$ for each operation, so all calculations can be carried out over $O(\log n)$-bit size words.

We now prove that A is a $(1/r)$-approximation. Given $S_i \in \mathcal{S}$, let $s_i = |S_i|$ and $\delta_i = |A \cap S_i| - q|S_i|$. We easily check that the strategy is unbiased. What we mean by this is that

$$H_q(i, k) = q H_q^{[1-q]}(i, k+1) + (1-q) H_q^{[-q]}(i, k+1),$$

where $H_q^{[j]}(i, k+1)$ is the value of $H_q(i, k+1)$ for the choice $\chi(v_{k+1}) = j$. It follows that $H_q(i, k+1) \leq H_q(i, k)$, and hence $H_q(k+1) \leq H_q(k)$. Therefore, $H_q(i, n) \leq H_q(n) \leq 2(m+1)$, and so, taking the first term in the sum for $H_q(i, n)$,

$$(1 + (1-q)\mu)^{qs_i}(1 - q\mu)^{(1-q)s_i}(1 + (1-q)\mu)^{\delta_i}(1 - q\mu)^{-\delta_i} \leq 2(m+1).$$

Taking logarithms and expanding in power series, we find that

$$\left(\mu - O(\mu^2)\right)\delta_i - \left(\mu^2 + O(\mu^3)\right)q(1-q)s_i \leq \ln(2m+2).$$

Setting

$$\mu = \sqrt{\frac{\ln(2m+2)}{q(1-q)n}}$$

gives

$$\delta_i \leq \left(1 + O(\mu)\right)\sqrt{4q(1-q)n \ln(2m+2)}.$$

By repeating this argument with respect to the second term in the sum for $H_q(i, n)$ and reversing the roles of q and $1 - q$, we find that $|\delta_i|$ has the same upper bound. Since $A = A \cap S_0$, we have

$$\frac{|A \cap S_i|}{|A|} = \frac{s_i}{n} \cdot \frac{1 + \delta_i/(qs_i)}{1 + \delta_0/(qn)};$$

therefore,

$$\left| \frac{|A \cap S_i|}{|A|} - \frac{s_i}{n} \right| = O\left(\frac{|\delta_i|}{qn} + \frac{s_i|\delta_0|}{qn^2}\right).$$

By setting $q = cr^2 (\ln m)/n$, for some constant c large enough, we find that

$$\left| \frac{|A \cap S_i|}{|A|} - \frac{s_i}{n} \right| < \frac{1}{r},$$

which establishes that A is a $(1/r)$-approximation for $(V, \mathcal{S})$. Its size is $qn + \delta_0$, which is $O(r^2 \log m)$. Recall that to ensure the validity of the argument, we must ensure that q and μ are small enough. This follows from the obvious fact that, in the theorem stated below, $1/r$ and $r/\sqrt{n/\log m}$ can always be assumed to be smaller than any fixed constant.

Theorem 4.4 *Given a set system $(V, \mathcal{S})$, $|V| = n$ and $|S| = m$, and $1 \leq r \leq n$, the weighted greedy sampling algorithm finds a $(1/r)$-approximation for $(V, \mathcal{S})$ of size $O(r^2 \log m)$ in time $O(nm)$.*

4.3 Sampling in Bounded VC-Dimension

Let $(X, \mathcal{R})$ be a range space of VC-dimension d; see definitions in Chapter 1. As usual, it is assumed that $(X, \mathcal{R})$ is actually a finite subsystem of an infinite range space of VC-dimension d. As it turns out, we rarely use the fact that d is the VC-dimension of $(X, \mathcal{R})$. Instead, we rely on its corollary (Lemma 1.6), stating that the (primal) shatter function $\pi_{\mathcal{R}}(m)$ is in $O(m^d)$. Thus, if $\pi_{\mathcal{R}}(m) = O(m^d)$ is the only fact we know about the range space, every theorem in this section still remains valid.

We prove the existence of a small ε-approximation for $(X, \mathcal{R})$ by describing a procedure that recursively computes ε-approximations for smaller subsystems and merges them together. The proof is inherently algorithmic, and it is quite easy to analyze its complexity. Next, we give three existence proofs for ε-nets. Why three? The first proof is a straightforward by-product of ε-approximations; the second one is an elegant application of discrepancy theory; finally, the last one—historically the first—uses an original probabilistic argument that is interesting in its own right. All three proofs are completely different and provide complementary perspectives on the problem.

In practice, it is often the case that range spaces are defined implicitly and are accessible via an *oracle*: a function that takes any $Y \subseteq X$ as input and returns the list of sets in $\mathcal{R}|_Y$ (each set represented explicitly). We assume that the time to complete this task is $O(|Y|^{d+1})$, which is linear in the maximum possible size of the oracle's output. As it turns out, the oracle is fairly realistic in practice. For example, in the case of points and balls in d-space, this assumes that, given n points, we can enumerate all subsets enclosed by a ball in time $O(n^{d+2})$. To do this, enumerate all k-tuples of points ($k \le d+1$) and, for each tuple, find which points lie inside the smallest ball enclosing the k points.

The existence of such an oracle is assumed in the following two theorems. For notational convenience we write $r = 1/\varepsilon$. As we indicated earlier, although stated in terms of VC-dimension, the results below hold just the same if all we know about the underlying range space is that its shatter function $\pi_{\mathcal{R}}(m)$ is in $O(m^d)$.

Theorem 4.5 *Let $(X, \mathcal{R})$ be a range space of VC-dimension d. Given any $r \ge 2$, a $(1/r)$-approximation for $(X, \mathcal{R})$ of size $O(dr^2 \log dr)$ can be computed in time $O(d)^{3d}(r^2 \log dr)^d |X|$.*

Theorem 4.6 *Let $(X, \mathcal{R})$ be a range space of VC-dimension d. Given any $r \ge 2$, a $(1/r)$-net for $(X, \mathcal{R})$ of size $O(dr \log dr)$ can be computed in time $O(d)^{3d}(r^2 \log dr)^d |X|$.*

The running time for computing a $(1/r)$-net can be reduced further to $O(d)^{3d}(r \log dr)^d |X|$, by using the notion of a *sensitive ε-approximation*. We omit this subject, which would take us too far afield, and instead refer the reader to the bibliographical notes at the end of this chapter for references.

Again, let us not miss the main point about these theorems. Remarkably, any bounded-VC dimensional range space $(X, \mathcal{R})$ admits ε-approximations and ε-nets of size *independent* of $|X|$. A superior sampling theory is where the notion of Vapnik-Chervonenkis dimension draws its computational power.

Building an ε-Approximation

Without loss of generality, assume that $n = |X|$ is a power of two.[1] We assume that the primal shatter function $\pi_{\mathcal{R}}(n)$ is in $O(n^d)$, and therefore that the number of sets $m = |\mathcal{R}|$ is in $O(n^d)$. For small values of ε, the problem is similar to two-coloring a set system for low discrepancy.

---

[1] To relax this assumption, we may pad the ground set with up to n artificial points and add the old set X to the new set system. It is immediate that decreasing ε by a constant factor and removing the artificial points at the end gives an ε-approximation for the original set system.

The Low-Error Case

Color the elements of X red or blue by applying the unbiased greedy algorithm of Chapter 1 to the set system $(X, \mathcal{R} \cup \{X\})$. By Theorem 1.2 (page 7), this guarantees that, within any $R \in \mathcal{R} \cup \{X\}$, no color outnumbers the other by more than $\Delta = \sqrt{2n \ln(2m + 2)}$. Take the more popular color in X, say red, and let A be an arbitrary set of $n/2$ red elements of X.

Lemma 4.7 *The set A is an ε-approximation for $(X, \mathcal{R})$ of size $n/2$, for $\varepsilon = 2\Delta/n$. It is computed in $O(n^{d+1})$ time.*

Proof: Let A_1 be the set of red elements in X. Because X is added to the set system, we have $0 \leq 2|A_1| - n \leq \Delta$. Dividing by 2 gives

$$\left| |A_1| - |A| \right| \leq \frac{\Delta}{2}.$$

On the other hand, given $R \in \mathcal{R}$, we have $|2|R \cap A_1| - |R|| \leq \Delta$, which shows that

$$
\begin{aligned}
|2|R \cap A| - |R|| \quad &\leq \quad \Delta + 2(|R \cap A_1| - |R \cap A|) \\
&\leq \quad \Delta + 2(|A_1| - |A|) \\
&\leq \quad 2\Delta.
\end{aligned}
$$

Dividing by n proves that A is an ε-approximation of size $n/2$ for the value of ε claimed in the lemma. $\square$

Using the fact that $m = O(n^d)$, we find that $(X, \mathcal{R})$ admits of an ε-approximation of size $n/2$ for $\varepsilon = f(n)$, where

$$f(x) = \sqrt{\frac{8d(\ln x + c_0)}{x}},$$

for some large enough constant c_0. (In the following the constants c_i are all assumed to be independent of d, r, n.)

The General Case

We need two pieces of terminology: (i) Given $Y, Z \subseteq X$, to form the union $Y \cup Z$ is called *merging* Y and Z; (ii) given an even-sized subset Y of X, computing an $f(|Y|)$-approximation for $(Y, \mathcal{R}|_Y)$ of size $Y/2$ in the manner just described is called *halving* Y. We now construct an ε-approximation A, given any $\varepsilon > f(n)$. Recall that $r = 1/\varepsilon$. For minor technical reasons, we assume that $r \geq 2$. We proceed in two stages (Fig. 4.1):

Fig. 4.1. The white and black dots represent the merging and halving steps, respectively.

STAGE 1. Partition X into arbitrary subsets of size 2^k, where $2^k = c_1 d^3 r^2 \log dr$, for some large enough constant c_1. (Recall that $n = |X|$ is a power of 2.) Form a binary tree with $n/2^k$ leaves, each associated with a distinct subset. Starting at the parents of the leaves, perform the following computation bottom-up until the root is reached: At node v, merge the two subsets associated with the children of v and halve the new set: The resulting set is now associated with v.

STAGE 2. Keep halving the set associated with the root of the tree until its size becomes $c_2 dr^2 \log dr$, for a large enough constant c_2. The final set is A.

The procedure just described almost works but not quite: We modify it by requiring that in Stage 1 the halving step should be skipped at every positive level[2] of the tree that is a multiple of $d + 2$. Why is the final set A produced in Stage 2 a valid $(1/r)$-approximation? First of all, notice that the size of A is $O(dr^2 \log dr)$, as desired.

Concerning the final error in the approximation provided by A, observe that only halving steps increase the error of the current partial approximations. Considering a whole level at a time, the union of the sets associated with its nodes form a δ-approximation for some "error" δ. At the first $(d + 1)$-st level, the current δ-approximation is formed as a disjoint union

---

[2]In this terminology, leaves are at level 0 and a pair merge/halve resides at a single level.

of sets of size 2^k (since alternating halving and merging steps keeps the size of the node sets constant). By Lemmas 4.1 and 4.2, the error δ of the current δ-approximation is $(d+1)f(2^k)$. Skipping the halving step at level $d+2$ doubles the size of the node sets; therefore, the next set of $d+1$ levels will add $(d+1)f(2^{k+1})$ to the current error. This produces a decreasing geometric sequence (with a rate independent of d), so that the total error at the end of Stage 1 is $O((d+1)f(2^k))$, with a constant factor independent of d. In sum, the error can be kept below $1/2r$ by choosing c_1 large enough.

By similar reasoning we see that the error in each halving step of Stage 2 increases geometrically, so the contribution of the last step is dominant. This gives an error of $O(f(c_2 dr^2 \log dr))$, which again is less than $1/2r$, if we choose c_2 large enough. By Lemma 4.2, the combined error is at most $1/2r + 1/2r$. So, the final set A is a $(1/r)$-approximation.

Lemma 4.8 *The set A is a $(1/r)$-approximation for $(X, \mathcal{R})$ of cardinality $O(dr^2 \log dr)$.*

Remark: Skipping a halving step once in a while is necessary to produce a geometric sequence and, thus, keep the error small. Failure to do so would add a logarithmic multiplicative factor to the final error. Our choice of $d+2$ in the skipping rate might seem arbitrary at this point. We justify it now.

Algorithmic Issues

By Lemma 4.7, to halve an n-element set takes $O(n^{d+1})$ time. It follows that, measured across the entire level, the first halving step of the algorithm takes time $O(n/2^k)(2^k)^{d+1}$, ie, $O(d)^{3d}r^{2d}(\log dr)^d n$. In the subsequent levels up to the first skipping level, the total number of sets decreases geometrically while their sizes remain the same, so the running time at each level also decreases geometrically. At each skipping level, the size of each node set doubles, which increases the time for halving them by 2^{d+1}. On the other hand, the $d+2$ previous merging steps up to the current level have reduced the total number of sets by 2^{d+2}, so the running time decreases geometrically between skipping levels. Stage 2 also entails geometrically decreasing costs. This shows that the total running time of the algorithm is $O(d)^{3d}r^{2d}(\log dr)^d n$, as stated in Theorem 4.5 (page 175). $\square$

As it turns out, an ε-approximation can be obtained very easily by a probabilistic algorithm. We do not prove the following result because we will give a very similar proof for ε-nets later.

Theorem 4.9 *Let $(X, \mathcal{R})$ be a range space of VC-dimension d and let $r \geq 2$. With probability close to 1, a random subset of X of size $cdr^2 \log dr$, for some large enough constant c, is a $(1/r)$-approximation for $(X, \mathcal{R})$.*

An Improved Construction

Suppose, for convenience, that $d > 1$. By Lemma 4.7, we know the existence of an ε_1-approximation for $(X, \mathcal{R})$ of size $\lceil n/2 \rceil$, for $\varepsilon_1 = 2\Delta(n)/n$, where $\Delta(n)$ is the discrepancy of $(X, \mathcal{R} \cup \{X\})$. By Theorem 1.8 (page 14), $\Delta(n) = O(n)^{1/2-1/2d}(\log n)^{1+1/2d}$. Apply the lemma iteratively on each new sample k times. By Lemma 4.2, the resulting sample is an ε_k-approximation for $(X, \mathcal{R})$ of size

$$\left\lceil \frac{1}{2} \left\lceil \cdots \left\lceil \frac{n}{2} \right\rceil \cdots \right\rceil \right\rceil = \left\lceil \frac{n}{2^k} \right\rceil,$$

where

$$\varepsilon_k \leq \sum_{i=0}^{k-1} \frac{2\Delta(n/2^i)}{n/2^i} \leq c \left(\frac{n}{2^k} \right)^{-1/2-1/2d} \left(\log \frac{n}{2^k} \right)^{1+1/2d},$$

for some constant $c > 0$. Setting k large enough so that $\varepsilon_k \leq 1/r$, we find:

Theorem 4.10 *Let $(X, \mathcal{R})$ be a range space of VC-dimension $d > 1$. Given any $r \geq 2$, there exists a $(1/r)$-approximation for $(X, \mathcal{R})$ of size at most proportional to $r^{2-2/(d+1)}(\log r)^{2-1/(d+1)}$.*

Again we mention that the theorem still holds as long as the primal shatter function is in $O(m^d)$, regardless of the actual value of the VC-dimension. By a similar argument, the discrepancy bound of Theorem 1.10 (page 17) yields the following result:

Theorem 4.11 *Let $(X, \mathcal{R})$ be a range space whose dual shatter function is in $O(m^d)$. Given any $r \geq 2$, there exists a $(1/r)$-approximation for $(X, \mathcal{R})$ of size at most proportional to $r^{2-2/(d+1)}(\log r)^{1-1/(d+1)}$.*

Three Ways to Build an ε-Net

Recall that $(X, \mathcal{R})$ is a subsystem of an infinite range space of VC-dimension d, with $n = |X|$ and $m = |\mathcal{R}|$, and that $\varepsilon = 1/r$. We give three distinct proofs of Theorem 4.6 (page 175).

Proof I: From ε-Approximations to ε-Nets

Let A be a $(1/2r)$-approximation for $(X, \mathcal{R})$. By Lemma 4.2, a $(1/2r)$-net N for $(A, \mathcal{R}|_A)$ is also a $(1/r)$-net for $(X, \mathcal{R})$. By Theorem 4.5 (page 175), a $(1/2r)$-approximation A of size at most the lesser of n and $O(dr^2 \log dr)$ can be computed in time $O(d)^{3d}(r^2 \log dr)^d n$ time, while by Theorem 4.3 (page 172) we can obtain a net N of size $O(dr \log dr)$ in $O(dr^2 \log dr)^{d+1}$ time. We easily verify that the latter cost is dominated by the upper bound on the cost for A. Thus, the total running time is $O(d)^{3d}(r^2 \log dr)^d n$, which establishes Theorem 4.6 (page 175). $\square$

Proof II: From Low Discrepancy to Nets

We color the elements of X red or blue by applying the unbiased greedy algorithm of Chapter 1 to the set system $(X, \mathcal{R} \cup \{X\})$. Let A_1 be the set of red elements. Next, we two-color the set system induced by A_1 and call A_2 the new set of red elements within A_1. We iterate on this coloring process k_0 times, where k_0 is defined by

$$2^{k_0} = \frac{n}{cdr \log dr},$$

for some large enough constant c.

Lemma 4.12 For $r \geq 2$, the set A_{k_0} is a $(1/r)$-net for $(X, \mathcal{R})$ of size $O(dr \log dr)$.

Proof: Take any $R \in \mathcal{R} \cup \{X\}$ of size greater than n/r. In the first stage, no color within R outnumbers the other by more than $\sqrt{2d|R|(\ln n + O(1))}$, so we have

$$\left| 2|R \cap A_1| - |R| \right| \leq \sqrt{2d|R|(\ln n + O(1))}.$$

Putting $A_0 = X$, we have the more general relation: For $k \geq 1$,

$$\left| 2|R \cap A_k| - |R \cap A_{k-1}| \right| \leq \sqrt{2d|R \cap A_{k-1}|(\ln |A_{k-1}| + O(1))}.$$

Using the fact that X is included as a subset of the set system, it follows from a tedious but easy inductive argument that,[3] for $k \geq 1$,

$$\left| |R \cap A_k| - 2^{-k}|R| \right| \ll \sqrt{\frac{d|R|}{2^k} \ln \frac{dn}{2^k}}. \tag{4.1}$$

---

[3] Recall that $\ll$ and $\gg$ are shorthand for $O()$ and $\Omega()$, respectively.

For c large enough, $2^{-k_0}|R|$ dominates

$$\sqrt{\frac{d|R|}{2^{k_0}} \ln \frac{dn}{2^{k_0}}}$$

by a large constant factor, so $R \cap A_{k_0} \neq \emptyset$, and A_{k_0} is a $(1/r)$-net. Choosing $R = X$, we find that

$$|A_{k_0}| = |R \cap A_{k_0}| < 2^{1-k_0}|R| = O(dr \log dr),$$

as claimed. $\square$

Proof III: A Probabilistic Argument

We prove a slightly more specific result by giving explicit constants. The proof is self-contained and does not rely on our previous discussion. It is simple but very ingenious, no doubt something Paul Erdős would have called a proof from *The Book*. In the following, the term *random s-sample* refers to a set of s elements drawn randomly in X without replacement, ie, every s-tuple of points is equally likely.

Lemma 4.13 *Given any integer $s \leq n$ large enough (in relation to d), a random s-sample of X constitutes an ε-net for $(X, \mathcal{R})$ with probability at least 2/3, where $\varepsilon = 7d(\log s)/s$.*

Proof: We rename $(X, \mathcal{R})$ to denote the set system obtained by removing from $\mathcal{R}$ all sets of size at most $7dn(\log s)/s$. We prove that, with probability at least 2/3, a random s-sample N intersects every set of $\mathcal{R}$. If it does not, we say that N *fails*. The probability that N is disjoint from a given $R \in \mathcal{R}$ is at most

$$\binom{n - \lceil 7dn(\log s)/s \rceil}{s} \Big/ \binom{n}{s} \leq \left(1 - \frac{7d \log s}{s}\right)^s$$

$$\leq \exp(-7d \log s) \leq 1/s^{7d}.$$

Thus, because $|\mathcal{R}| = O(n^d)$, the probability that N fails is $O(n^d/s^{7d})$. This probability can be made less than $1/3$ for, say, any $s \geq \sqrt{n}$.

So, assume now that $s < \sqrt{n}$. The key idea is to compare the behavior of s-samples and $2s$-samples. Given a $2s$-sample T we say that R is a *witness* for T if

$$2d \log s \leq |T \cap R| \leq s.$$

The term "witness" is used here to suggest good, normal behavior. Let

$f(s)$ be the probability that a random s-sample N fails. If N fails, we choose some $R \in \mathcal{R}$ such that $N \cap R = \emptyset$, which we denote by $R(N)$. Take a random s-sample N' in $X \setminus N$. With high probability, N' must "behave as expected," that is, make $R(N)$ a witness for $N \cup N'$. Let $p(s)$ be the probability that a random s-sample N fails and that $R(N)$ is a witness for $N \cup N'$, where N' is a random s-sample disjoint from N. We have

$$p(s) = f(s) \times \mathrm{Prob} \Big[|N' \cap R(N)| \geq 2d \log s \,|\, N \text{ fails} \Big].$$

Note that $(N \cup N') \cap R(N) = N' \cap R(N)$ and that the condition

$$\Big| (N \cup N') \cap R(N) \Big| \leq s$$

can be ignored since it follows automatically from the assumption that N is failing. To bound $p(s)$ from below, we can assume that $R(N)$ has the minimum allowed size of $u = \lceil 7dn(\log s)/s \rceil$. We obtain

$$p(s) \geq \Big(1 - \sum_{k=0}^{\lfloor 2d \log s \rfloor} h(u,k) \Big) f(s), \tag{4.2}$$

where

$$h(u,k) = \binom{u}{k} \binom{m-u}{s-k} \Big/ \binom{m}{s},$$

with $m = n - s$. It is a simple matter to show that

$$h(u,k) \leq s^{-d/3}, \tag{4.3}$$

for large enough $s < \sqrt{n}$ and $k \leq 2d \log s$ (see below for a proof). It follows from (4.2) that

$$p(s) \geq \Big(1 - (2d \log s + 1) s^{-d/3} \Big) f(s) \geq \frac{f(s)}{2}. \tag{4.4}$$

Let T be a $2s$-sample in X and let $p(T, s)$ be the probability that, given a random s-sample N in T, N fails and $R(N)$ is a witness for T. Let us now interpret $p(s)$ slightly differently. It is the probability that, given a random $2s$-sample T and a random s-sample $N \subseteq T$, N fails and $R(N)$ is a witness for T. So, we have

$$p(s) \leq \max \{ p(T, s) : T \subseteq X, \, |T| = 2s \}. \tag{4.5}$$

Now, consider a $2s$-sample T for which some fixed R is a witness. A random

s-sample in T will miss R entirely with probability at most

$$\binom{2s-2d\log s}{s} \Big/ \binom{2s}{s} \leq \left(1-\frac{d\log s}{s}\right)^s = o(1/s^{9d/8}).$$

But because the underlying range space has VC-dimension d,

$$\left|\{T\cap R : R\in\mathcal{R}\}\right| = O(s^d);$$

therefore $p(T,s) = o(1)$. It follows from (4.4) and (4.5) that $f(s) = o(1)$, and so, with high probability, a random s-sample does not fail, and the lemma is proven. $\square$

For the sake of completeness, here is a derivation of (4.3). From the approximation

$$\frac{(a-b)^b}{b!} \leq \binom{a}{b} \leq \frac{a^b}{b!},$$

we find that

$$h(u,k) \leq \frac{u^k(m-u)^{s-k}}{(m-s)^s}\binom{s}{k}.$$

By Stirling's approximation,

$$t^t e^{-t}\sqrt{2\pi t}e^{1/(12t+1)} < t! < t^t e^{-t}\sqrt{2\pi t}e^{1/(12t)},$$

it follows that

$$\binom{s}{k} \leq \frac{(s-k)^k}{k^k(1-k/s)^s} \leq \left(\frac{s}{k}\right)^k\left(1+\frac{2k}{s}\right)^s,$$

for any positive $k \leq s/2$. Using the fact that $s < m/2$, we derive

$$h(u,k) \leq \left(\frac{su}{mk}\right)^k\left(1-\frac{u}{m}\right)^{s-k}\left(1+\frac{2k}{s}\right)^s\left(1+\frac{2s}{m}\right)^s.$$

The function $x \mapsto (M/x)^x$, for $x > 0$, achieves its maximum at $x = M/e$, and so

$$h(u,k) \leq \exp\left(-\frac{u}{m}\left(s-k-\frac{s}{e}\right)+4d\log s+\frac{2s^2}{m}\right) \leq s^{-d/3},$$

which is (4.3).

Product Range Spaces

The tools we built in the previous section are useful in providing accurate estimates of geometric quantities. For example, given n hyperplanes in

d-space, an ε-approximation gives us an efficient way of estimating how many hyperplanes cut through an arbitrary segment, presented as a query. As we show below, the same data structure can be used to handle more complicated questions such as: Given a simplex, how many vertices of the arrangement of n hyperplanes lie within it (Fig. 4.2)? This will prove an indispensable tool for building ε-cuttings (Chapter 5) and convex hulls and Voronoi diagrams (Chapter 7). Again, the idea is simple: We look at the (presumably much smaller) arrangement formed by the ε-approximation and count how many vertices lie in the triangle. By scaling up that number appropriately we derive an accurate estimate on the desired number.

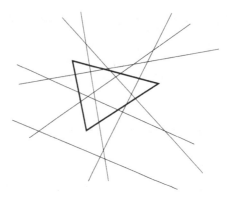

Fig. 4.2. How many vertices of the arrangement lie in the triangle? Answer: 6.

Rather than dealing with this particular geometric problem in an ad hoc fashion, we follow the same approach used earlier and provide a general framework of wide applicability. The central idea is to define a binary operation that allows us to combine range spaces to build more complicated ones. In the example above, we would combine range spaces involving lines to form a set system over pairs of lines, ie, vertices of the arrangement.

Given two finite range spaces $\Sigma_1 = (X_1, \mathcal{R}_1)$ and $\Sigma_2 = (X_2, \mathcal{R}_2)$, the *product range space* $\Sigma_1 \otimes \Sigma_2$ is defined as $(X_1 \times X_2, \mathcal{T})$, where $\mathcal{T}$ consists of all subsets $T \subseteq X_1 \times X_2$ such that each *cross-section*

$$T_{x_2}^1 = \{ x \in X_1 \mid (x, x_2) \in T \}$$

is a set of $\mathcal{R}_1$ and, similarly,

$$T_{x_1}^2 = \{ x \in X_2 \mid (x_1, x) \in T \}$$

belongs to $\mathcal{R}_2$.

By way of illustration, consider Σ_1 to be the range space induced by n blue lines in the plane and the set of all line segments: A range is the subset of blue lines intersected by a given segment. We define Σ_2 similarly, but we color its lines red. The product $\Sigma_1 \otimes \Sigma_2$ is a range space $(Z, \mathcal{T})$, where Z is the set of bichromatic pairs of lines, ie, red-blue vertices (assuming general position). A range is any subset T of Z such that, along any (blue or red) line ℓ, the vertices of T incident to ℓ (if any) appear consecutively among the red-blue vertices of ℓ.

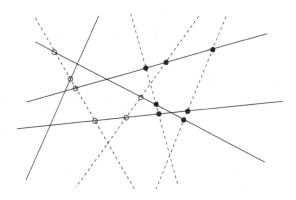

Fig. 4.3. A seven-point range.

Note, in particular, that the bichromatic intersections within any convex set constitute a range. This suggests that bounded VC-dimensionality might not be preserved under the product operation: Indeed, this can best be seen by observing that, in our example, any bichromatic matching of the lines gives a collection of n vertices, any of whose 2^n subsets is a valid range. Given two range spaces of bounded VC-dimension, the existence of a small-sized ε-approximation for the product might thus seem unlikely. And yet, it is indeed the case. Observe that the obvious counterargument, ie, the presence of large shattered subsets means no sampling, does not work here, because subspaces of product spaces are *not*, in general, product spaces themselves.

Theorem 4.14 *Given any* $0 \le \varepsilon_i \le 1$, *let* A_i *be an* ε_i-*approximation for a range space* Σ_i, *for* $i = 1, 2$. *Then, the Cartesian product* $A_1 \times A_2$ *is an* $(\varepsilon_1 + \varepsilon_2)$-*approximation for* $\Sigma_1 \otimes \Sigma_2$.

Proof: First, a little terminology: For $i = 1, 2$, put $\Sigma_i = (X_i, \mathcal{R}_i)$. Let $\mathrm{Prob}_{X_i}[R] = |R|/|X_i|$ be the probability that a random element x chosen uniformly in X_i lies in R, and let $\mathrm{Prob}_X[\,R\,|\,A_i\,]$ be the conditional probability that x is in R, given that it is in A_i. As we observed earlier, A_i is an ε_i-approximation for Σ_i if and only if, for each $R \in \mathcal{R}_i$,

$$\left| \mathrm{Prob}_{X_i}[\,R\,|\,A_i\,] - \mathrm{Prob}_{X_i}[R] \right| \le \varepsilon_i.$$

Let $Z = X_1 \times X_2$; by reversing the summation order we find that, for any $T \in \mathcal{T}$,

$$\mathrm{Prob}_Z[\,T\,|\,A_1 \times A_2\,] = \left\{ \begin{array}{l} \mathbf{E}_{X_1}\{\,\mathrm{Prob}_{X_2}[\,T^2_{x_1}\,|\,A_2\,]\,|\,A_1\,\}, \\ \mathbf{E}_{X_2}\{\,\mathrm{Prob}_{X_1}[\,T^1_{x_2}\,|\,A_1\,]\,|\,A_2\,\}. \end{array} \right.$$

Using similar identities, it follows that, for some δ_i $(i = 1, 2)$ such that $|\delta_i| \le \varepsilon_i$,

$$\begin{aligned} \mathrm{Prob}_Z[\,T\,|\,A_1 \times A_2\,] &= \mathbf{E}_{X_1}\{\,\mathrm{Prob}_{X_2}[\,T^2_{x_1}\,|\,A_2\,]\,|\,A_1\,\} \\ &= \mathbf{E}_{X_1}\{\,\mathrm{Prob}_{X_2}[\,T^2_{x_1}\,]\,|\,A_1\,\} + \delta_1 \\ &= \mathbf{E}_{X_2}\{\,\mathrm{Prob}_{X_1}[\,T^1_{x_2}\,|\,A_1\,]\,\} + \delta_1 \\ &= \mathbf{E}_{X_2}\{\,\mathrm{Prob}_{X_1}[\,T^1_{x_2}\,]\,\} + \delta_1 + \delta_2 \\ &= \mathrm{Prob}_Z[\,T\,] + \delta_1 + \delta_2, \end{aligned}$$

which completes the proof. $\square$

The product of range spaces is an associative operation, and we easily extend the theorem as follows:

Theorem 4.15 *Given an ε-approximation A for a range space Σ, the d-fold Cartesian product $A \times \cdots \times A$ is a $(d\varepsilon)$-approximation for the d-fold product $\Sigma \otimes \cdots \otimes \Sigma$.*

We close this section with an application of Theorem 4.15 to counting vertices in an arrangement of hyperplanes in d-space. We consider the range space $\Sigma = (H, \mathcal{R})$ formed by a set H of hyperplanes in $\mathbf{R}^d$, where each range $R \in \mathcal{R}$ is the subset of H intersected by an arbitrary line segment. Given a convex body σ (not necessarily full-dimensional), consider the arrangement formed by H within the affine span of σ. (See Appendix C for a definition of these terms.) Finally, let $V(H, \sigma)$ be the set of vertices of this arrangement that lie inside σ.

Theorem 4.16 *Given a set H of hyperplanes in $\mathbf{R}^d$ in general position, along with an ε-approximation A for $\Sigma = (H, \mathcal{R})$, for any convex body σ of dimension $k \le d$,*

$$\left| \frac{|V(H, \sigma)|}{|H|^k} - \frac{|V(A, \sigma)|}{|A|^k} \right| \le \varepsilon.$$

Proof: We can always reduce the dimensionality of the problem to that of σ, so we might as well assume that σ is full-dimensional. Let $\Sigma^d = (H^d, \mathcal{S})$ be the d-fold product of Σ, and let $H^{(d)}$ be the subset of H^d consisting of the d-tuples of *distinct* hyperplanes of H. By the general-position assumption, there exists a natural correspondence between $H^{(d)}$ and the vertex set $V(H)$ of the arrangement formed by H. Let $(h_1, \ldots, h_d)$ be a d-tuple in $H^{(d)}$, and let $f(h_1, \ldots, h_d) = h_1 \cap \cdots \cap h_d$ be the corresponding vertex of $V(H)$. The preimage $f^{-1}(v)$ of a vertex of $V(H)$ consists of $d!$ tuples. Given a convex body σ, it is easy to see that the inverse image, $T = f^{-1}(V(H, \sigma))$, is a range in Σ^d. Indeed, its cross-sections are obtained by fixing $d - 1$ hyperplanes $(h_1, \ldots, h_{i-1}, h_{i+1}, \ldots, h_d)$ and collecting the hyperplanes h such that the d-tuple $(h_1, \ldots, h_{i-1}, h, h_{i+1}, \ldots, h_d)$ is in T. But these are precisely the hyperplanes intersecting the line segment defined as the portion of the line $\cap_{j \neq i} h_j$ within σ. The cardinality of T is exactly $d!$ times the number of vertices of $V(H, \sigma)$. By Theorem 4.15, therefore,

$$\left| \frac{|V(H, \sigma)|}{|H|^d} - \frac{|V(A, \sigma)|}{|A|^d} \right| \leq \frac{d\varepsilon}{d!} \leq \varepsilon.$$

$\square$

4.4 Weak ε-Nets

Let P be a set of n points in $\mathbf{R}^d$. Given a parameter $\varepsilon > 0$, is it possible to select a sample of P of size *independent* of P, so that the convex hull of any subset of P of size at least εn contains at least one of the sample points? An ominous sign is that the range space formed by intersecting a set of points with convex regions has infinite VC-dimension. Indeed, if the points of P lie in convex position, then any subset is a valid range, and therefore the sample must be of size at least $(1 - \varepsilon)n$.

One reason for such a negative result is that we restrict ourselves to choosing the sample from within the set P itself. What if we relax this requirement and allow just any point in $\mathbf{R}^d$? A finite set $N \subseteq \mathbf{R}^d$ is called a *weak ε-net* (for convex sets, with P understood) if it intersects every convex region that contains at least εn points of P. Next, we give a very simple construction in two dimensions that produces a weak ε-net of size $O(1/\varepsilon^2)$. The construction can be generalized to higher dimensions, but it becomes more complicated and slightly less efficient.

Without much loss of generality we can assume that n is a power of 2 and that no three points are collinear. Let L be a vertical line splitting the set P into two equal-sized subsets. Let $q_1, q_2, \ldots$ be the intersections of L

with the segments formed by joining every pair of points on both sides of L. Note that there are exactly $(n/2)^2$ such intersection points; we assume that the list of q_i's is sorted along L. Let N_0 be the set obtained by taking every $\frac{1}{6}(\varepsilon n)^2$-th point in the list (Fig. 4.4). Now, recursively, compute a weak $(3\varepsilon/2)$-net of the points to the left (right) of L, and form the union N of N_0 with these two sets. If $3\varepsilon/2 \geq 1$, then, of course, no recursion is needed since any single point forms a valid net. We claim that the set N is a weak ε-net for P. Indeed, let C be a convex region that intersects P in at least εn points. We distinguish between two cases:

- If $C \cap P$ has at least a quarter of its points on each side of L, then these points alone contribute at least $3(\varepsilon n)^2/16$ points q_i's. The convex hull of these q_i's is an interval of L that lies entirely in C and contains at least one point of N_0. Thus, as desired, N intersects C.
- If $C \cap P$ has more than three-quarters of its points on one side of L, then C contains a fraction at least $3\varepsilon/2$ of the $n/2$ points on that side, and so by construction, C intersects the weak net built on the side in question.

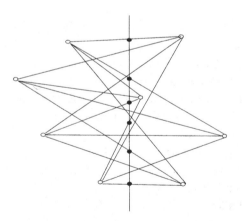

Fig. 4.4. The set N_0 consists of the filled dots.

Now that N has been shown to be a weak ε-net, it remains to see why it is of size $O(1/\varepsilon^2)$. Since N_0 contains $O(1/\varepsilon^2)$ points, the size $S(\varepsilon)$ of N follows the recurrence:

$$S(\varepsilon) = 2S(3\varepsilon/2) + O(1/\varepsilon^2),$$

where $S(\varepsilon) = 1$, if $\varepsilon \geq 1$. This gives $S(\varepsilon) = O(1/\varepsilon^2)$, as claimed.

The main merit of the construction is its simplicity. It is far from obvious, however, that it should be optimal. At this point, it seems natural to investigate what a "hard" distribution P might look like. Any convex region contains an ellipse that encloses at least a fraction of its area. Since ellipses produce range spaces of bounded VC-dimension, standard ε-nets constructions will basically work if the points of P are "uniformly" distributed within some convex domain. So, it is only natural to investigate the opposite extreme, which is points in convex position.

For concreteness we consider the case where the points of P are the vertices of a regular polygon, or at least are distributed on a circle evenly enough. We give a simple construction of a weak ε-net of size $O(1/\varepsilon)$, which obviously is optimal asymptotically. Rather surprisingly, a simple tiling of the hyperbolic plane is the main vehicle used in the construction. Although it might be tempting to translate the construction in Euclidean terms (which is not too difficult to do) and present it as such, this would be a mistake. Hyperbolic geometry is the native language of this particular sampling problem, and much of the poetry would get lost in translation. We begin with a review of the basic properties of the hyperbolic plane.

A Primer on Hyperbolic Geometry *

This section can be skipped by the reader familiar with the fundamentals of the hyperbolic plane. For additional information on the subject, however, one may consult [44, 101, 131, 207, 227, 307]. Euclid's fifth postulate states that, through a given point, exactly one line can be parallel to another given line. Hyperbolic geometry postulates that an infinite number of such lines exist. The hyperbolic plane $\mathbf{H}^2$ is typically modeled in one of three ways: the Klein model, the Poincaré model, or the halfplane model. We briefly discuss each of them. The hyperbolic plane can be defined as one sheet $\mathcal{H}^+$ (for $z > 0$) of the two-sheeted hyperboloid

$$\mathcal{H} : x^2 + y^2 - z^2 + 1 = 0$$

with the Riemannian metric

$$ds^2 = dx^2 + dy^2 - dz^2.$$

Note that this is like the Euclidean metric, where z is the imaginary axis: $ds^2 = dx^2 + dy^2 + (idz)^2$. (Similarly, up to rescaling, one will recognize in this metric on $\mathbf{R}^3$ the three-dimensional version of the four-dimensional Minkowski space-time of special relativity.) We can also view $\mathbf{H}^2$ in terms of the real projective plane by considering lines passing through the origin. In this way, a point of $\mathbf{H}^2$ is represented by a line in $\mathbf{R}^3$ passing through the origin and piercing the hyperboloid (Fig. 4.5).

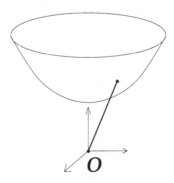

Fig. 4.5. The hyperboloid model.

To be able to view the hyperbolic plane as a plane and not as a set of lines, three models have been devised.

The Klein Model

This model, also known as the *projective model*, is obtained by projecting $\mathcal{H}^+$ centrally (ie, toward the origin) onto the plane $z = 1$: Specifically, the point $(x, y, z) \in \mathcal{H}^+$ maps to $(x/z, y/z)$ on the plane $z = 1$ (Fig. 4.5). The asymptotic cone $x^2 + y^2 - z^2 = 0$ intersects the plane $z = 1$ in the unit circle $\partial \mathcal{D}$, which corresponds to the points at infinity (and thus not in $\mathbf{H}^2$). A line from the origin that intersects the hyperboloid at p gives the point q on the unit disk at $z = 1$ (Fig. 4.6). Viewed in $\mathbf{R}^3$, a line of $\mathbf{H}^2$ is the intersection of the hyperboloid with the plane spanned by two lines passing through O. The intersection of that plane with $z = 1$ is a line; therefore, lines still look straight in the Klein model. The underlying metric, however, must be transferred from the hyperboloid to the unit disk $\mathcal{D}$, and so it is not the familiar Euclidean metric. In particular, angles and distances cannot be read off by simply looking at them in the Klein disk.

It is actually quite easy to express the metric of the Klein disk supplied with polar coordinates. Let (ρ, θ) be a point in the Klein disk with Euclidean polar coordinates (r, θ). The point is the image of a point $(t \cos \theta, t \sin \theta, z) \in \mathcal{H}^+$; therefore, $t^2 - z^2 = -1$. Since $r = t/z$, we find

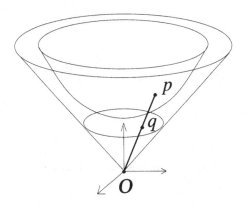

Fig. 4.6. The Klein model.

that $t = r(1 - r^2)^{-1/2}$, and hence

$$dt^2 = (1 - r^2)^{-3} \, dr^2.$$

From $z = t/r$, we derive that $z = (1 - r^2)^{-1/2}$, and hence

$$dz^2 = r^2(1 - r^2)^{-3} \, dr^2.$$

If we fix θ, we have

$$d\rho^2 = ds^2 = dx^2 + dy^2 - dz^2 = dt^2 - dz^2,$$

and therefore

$$d\rho = \sqrt{dt^2 - dz^2} = \frac{dr}{1 - r^2}.$$

A deviation of $d\theta$ produces an arc of length $t d\theta$, and so the area measure is $t d\rho d\theta$, which is

$$r(1 - r^2)^{-3/2} dr d\theta.$$

As mentioned earlier, this is the area density of the Klein model of $\mathbf{H}^2$.

The Klein model is useful for dealing with points and lines, but it does not help much in dealing with angles because it distorts them. To remedy this, we turn to *conformal* models, which preserve angles. In those models, angles can be read off directly, since they are the same as in the Euclidean case.

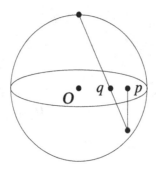

Fig. 4.7. From Klein to Poincaré.

The Poincaré Model

Let O stand now for the center of the Klein disk $\mathcal{D}$ and let $\mathcal{S}$ be the unit-radius sphere centered at O. A point p in the Klein disk is mapped to the point q in the Poincaré disk (which is also $\mathcal{D}$) as follows: Project p vertically down to the southern hemisphere of $\mathcal{S}$ and then centrally onto $\mathcal{D}$ toward the north pole of $\mathcal{S}$ (Fig. 4.7). Straight lines in the Poincaré disk model appear as circular arcs orthogonal to $\partial\mathcal{D}$. In Figure 4.8, the Poincaré disk is shaded and the line runs inside it from p to q.

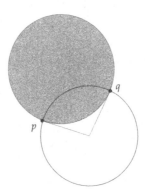

Fig. 4.8. A line in the Poincaré model.

In this model, isometries are generated by composing inversions in circles orthogonal to $\partial\mathcal{D}$ (with poles outside $\mathcal{D}$) and reflections through diameters of $\mathcal{D}$. As a function of the Euclidean metric $dx^2 + dy^2$, the Poincaré metric ds^2 satisfies

$$ds^2 = \frac{4}{(1 - r^2)^2}\,(dx^2 + dy^2),$$

where r is the Euclidean distance from the point (x,y) to O. Near the origin, Euclidean and hyperbolic distances are similar; but as we (isometrically) move a segment toward the boundary of $\mathcal{D}$, its Euclidean size shrinks by a factor roughly proportional to its (Euclidean) distance to the boundary.

By the relationship between projective transformations and isometries of $\mathbf{H}^2$, it is natural that the hyperbolic distance between two points p and q should be a function of the cross-ratio of (a,b,p,q), where a,b are the points of $\partial\mathcal{D}$ such that a,p,q,b appear in that order on a line (Fig. 4.9). Indeed, it is equal to

$$\frac{1}{2}\log\frac{|q-a||b-p|}{|p-a||b-q|}.$$

The logarithm is needed to make the distance function additive, while the factor $1/2$ is there to make the curvature -1. The notion of curvature is central to hyperbolic geometry: Indeed, in $\mathbf{H}^2$ the circumference of a circle of radius r is greater than $2\pi r$, which is similar to what happens (locally) in the Riemannian metric of a negatively curved surface, eg, a hyperbolic paraboloid. Note that, on the contrary, the circumference of a small circle taken on a positively curved surface, such as a sphere, is less than $2\pi r$.

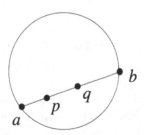

Fig. 4.9. The hyperbolic distance between p and q.

It is possible to construct negatively curved surfaces in $\mathbf{R}^3$ whose intrinsic metric is hyperbolic, but this can be only a local property because, by a theorem of Hilbert, no surface in $\mathbf{R}^3$ can be isometric to all of $\mathbf{H}^2$. Intuitively, there is not enough room in Euclidean space to accommodate the hyperbolic metric.

The Halfplane Model

The *halfplane model* is obtained by applying to the Poincaré disk an inversion whose pole lies on the unit circle. In other words, pick a point ω

on $\partial \mathcal{D}$ and map $\mathcal{D}$ through the inversion in the circle of radius 2 centered at ω. In Figure 4.10, p maps to q (both points are at infinity, and hence, strictly speaking, not in $\mathbf{H}^2$). Because inversions are conformal mappings, angles are still accurately represented in the halfplane model. The model is particularly convenient to work with because it can be thought of as the upper part of the complex plane. Pursuing this analogy, it can be shown that the group of isometries is precisely the group of linear fractional transformations (also known as Möbius transformations)

$$z \in \mathbf{C} \quad \mapsto \quad \frac{az+b}{cz+d},$$

with real coefficients and determinant $ad - bc \neq 0$.

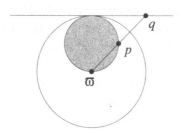

Fig. 4.10. From Poincaré to halfplane.

It is easy to verify that the area element is $dxdy/y^2$ at the complex point $z = x + iy$. This allows us to evaluate the area of a triangle very easily. First it can be shown that, unlike its Euclidean counterpart, a hyperbolic triangle is completely determined (up to congruence) by its three angles. In particular, all ideal triangles (those with vertices on $\partial \mathcal{D}$) are congruent because their three angles are all zero. To verify that their common area is π, consider the triangle with vertices $(-1,0)$, $(1,0)$, and $(0,\infty)$ in Figure 4.11. Its area is

$$\int_{-1}^{1} \int_{\sqrt{1-x^2}}^{\infty} \frac{1}{y^2} \, dy \, dx \ = \ \int_{-1}^{1} \frac{1}{\sqrt{1-x^2}} \, dx$$

$$= \ \int_{-\pi/2}^{\pi/2} \frac{\cos \theta}{\sqrt{1 - \sin^2 \theta}} \, d\theta = \pi.$$

We can treat the general case by a very simple argument due to Gauss. Since both the upper halfplane and Poincaré models are conformal, we

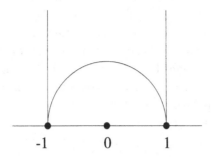

Fig. 4.11. All ideal triangles are congruent with area π.

can go back to the Poincaré model and freely reason about angles. Given $0 \leq \theta \leq \pi$, let $f(\theta)$ be the area of a triangle with two ideal vertices and one vertex with angle θ. For fixed θ, all of these triangles are congruent, so their area is indeed only a function of θ. By extending the edges of the triangles incident to the finite vertex, we easily derive

$$f(\alpha) + f(\pi - \alpha) = \pi.$$

We also leave it as an easy exercise that, by adding another edge toward the finite vertex, we obtain the identity

$$f(\alpha) + f(\beta) + f(2\pi - \alpha - \beta) = \pi.$$

Therefore,

$$f((\pi - \alpha) + (\pi - \beta)) = f(\pi - \alpha) + f(\pi - \beta) - \pi.$$

It follows that $f(x)$ is a linear polynomial in x and, more precisely, that $f(\alpha) = \pi - \alpha$. To deal with the general case, consider an arbitrary triangle with angles α, β, γ. Extend each side in one direction toward $\partial \mathcal{D}$ and form the ideal triangle with the three new vertices. In view of the previous result, the area A of the triangle satisfies

$$A + f(\pi - \alpha) + f(\pi - \beta) + f(\pi - \gamma) = \pi;$$

hence

$$A = \pi - (\alpha + \beta + \gamma),$$

as claimed earlier.

Hyperbolic Triangle Groups

Remarkably, *any* regular n-gon whose angles are $2\pi/k$, for some integer k, can be used to tile the whole hyperbolic plane. This shows how much more room $\mathbf{H}^2$ has compared to $\mathbf{E}^2$. Consider a regular n-gon centered at O. By triangulation, it immediately follows that its area is equal to $(n-2)\pi - n\alpha$, where α is its vertex angle. If the polygon is ideal, meaning that all of its vertices lie on $\partial\mathcal{D}$, then $\alpha = 0$. If we continuously decrease edge lengths and scale down the polygon toward O, however, its area decreases to 0, and consequently α tends to $(1-2/n)\pi$. If $n > 3$, this means that at some point during that process the angle α attains the value $2\pi/n$. At that point, let us draw the polygon and reflect it in its edges (which we can do without gaps or overlaps, since angles around the vertices sum up to 2π). Iterating these hyperbolic reflections everywhere yields a tiling of $\mathbf{H}^2$. Recall that from a Euclidean standpoint these reflections are circle inversions.

A useful class of tilings is obtained by reflecting triangles in their edges. It can be shown that given any positive integers l, m, n such that

$$\frac{1}{l} + \frac{1}{m} + \frac{1}{n} < 1,$$

the triangle with angles π/l, π/m, and π/n (which exists and is unique, up to congruence) can tile the entire hyperbolic plane. Let L (resp. M, N) denote the hyperbolic reflection in the edge of the triangle opposite the vertex with angle π/l (resp. $\pi/m, \pi/n$). The tiling is generated by the group, denoted by $T^*(l, m, n)$, with generators L, M, N and defined by the local relations

$$(LM)^n = (MN)^l = (NL)^m = 1,$$

$$L^2 = M^2 = N^2 = 1.$$

The first set of relations expresses the fact that reflected images of the triangle around a fixed vertex cycle back after exhausting an angle of 2π. The second set simply indicates that reflections are involutory.

The characterization of hyperbolic triangle groups given above immediately implies that any triangular tiling must consist of triangles whose area exceeds some absolute constant. Quite unlike the Euclidean case, the triangles involved in a tiling cannot be too small. We easily check that $T^*(2, 3, 7)$ is the tiling whose fundamental region has the smallest possible triangle (Fig. 4.12). Its area is $(1 - \frac{1}{2} - \frac{1}{3} - \frac{1}{7})\pi \approx 0.0748$.

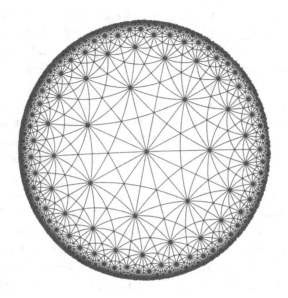

Fig. 4.12. The $(2, 3, 7)$-tiling. [Courtesy Daniel Huson]

Nets for Uniform Circular Distributions

We now have all the background needed to explain the construction of a weak ε-net for regular polygons.

Theorem 4.17 *Let P be the vertices of a regular polygon. There exists a weak ε-net for P of size $O(1/\varepsilon)$.*

We can slightly relax the conditions of the theorem and simply assume that the points of P lie on the boundary $\partial \mathcal{D}$ of the unit-radius disk $\mathcal{D}$, and that any arc of $\partial \mathcal{D}$ of length λ contains at most $\lceil c\lambda n \rceil$ points of P, for some constant $c > 0$. The following lemma trivially implies the existence of a weak ε-net for P of size $O(1/\varepsilon)$, which in particular proves Theorem 4.17. $\square$

Lemma 4.18 *Given any $\varepsilon > 0$, there exist $O(1/\varepsilon)$ points in the unit disk $\mathcal{D}$, such that any triangle of (Euclidean) side lengths $> \varepsilon$, whose vertices lie in $\partial \mathcal{D}$, encloses at least one of the points.*

To begin with, observe that the obvious approach of placing a fine square grid in $\mathcal{D}$ and using the grid points as our net fails. Indeed, this might succeed in hitting all the triangles of area $O(\varepsilon)$, but unfortunately some of the triangles to be hit may have an area as small as $O(\varepsilon^3)$. For example, take three points on $\partial\mathcal{D}$ as close together as allowed by the lemma. Such triangles are too small to always contain grid points. Intuitively, one would like to be able to change the underlying geometry so that all triangles with all three vertices on the circle $\partial\mathcal{D}$ should have the same area. By interpreting the disk $\mathcal{D}$ as the hyperbolic plane we are able to achieve precisely that.

As far as the distribution of points goes, instead of choosing a uniform grid we distribute the net points within $\mathcal{D}$ so that the density of the distribution increases as we approach the boundary $\partial\mathcal{D}$. Specifically, we use a nonuniform grid where the density of points at a distance r from the center is roughly $r(1 - r^2)^{-3/2}$. Of course, we recognize here the area density of the Klein model of $\mathbf{H}^2$. We begin with a simple technical lemma.

Lemma 4.19 *Let p and q be two points in the Poincaré disk, and assume that their Euclidean distance δ is equal to $\delta'/100$, where δ' is the Euclidean distance from $\partial\mathcal{D}$ to p or q, whichever is closer. Then the hyperbolic distance between p and q exceeds a positive constant (independent of p and q).*

Proof: As we saw earlier, in the Poincaré model the metric ds^2 is of the form $4(1 - r^2)^{-2}(dx^2 + dy^2)$. In view of the fact that δ is much smaller than δ', the hyperbolic distance between p and q can be approximated fairly accurately by integrating ds along the geodesic from p to q while keeping r set to the value $1 - \delta'$ in the expression for ds. Similarly, we can perform the integration along the Euclidean segment pq (instead of the geodesic) and lose only another constant factor in the final approximation. It follows that the hyperbolic distance between p and q is on the order of $\delta/\delta' = \frac{1}{100}$. $\square$

For concreteness, place the center O of $\mathcal{D}$ at a vertex of degree 14 in the tiling generated by $T^*(2, 3, 7)$, as in Figure 4.12. Fix $0 < r < 1$, and let $\mathcal{D}_r$ be the disk centered at O of Euclidean radius $r < 1$. From the hyperbolic metric of the Poincaré disk, we immediately find that the area of $\mathcal{D}_r$ is $O(1/(1 - r))$. Because the triangles of the $(2, 3, 7)$-tiling have bounded diameter, it easily follows that the number of triangles intersecting $\mathcal{D}_r$ is also $O(1/(1 - r))$.

We actually wish to have smaller triangles. By decomposing each triangle barycentrically (Fig. 4.13), and iterating in this fashion a constant number of times, we can bring down the hyperbolic diameter of every triangle below any desired positive constant. Of course, the total number of triangles among those intersecting $\mathcal{D}_r$ remains $O(1/(1-r))$. (Note that the new triangles are no longer congruent.) To summarize, we have constructed a triangulation of the Poincaré disk that consists of triangles of hyperbolic diameter below some arbitrary positive constant. Furthermore, given any $0 < r < 1$, the number of triangles intersecting the disk $\mathcal{D}_r$ of Euclidean radius r is $O(1/(1-r))$.

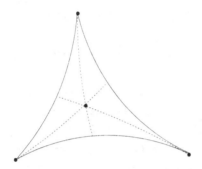

Fig. 4.13. A barycentric subdivision.

We are now ready to prove Lemma 4.18. Set $r = 1 - \varepsilon/10$, and choose a triangulation $\mathcal{T}$ of the Poincaré disk whose triangles are of hyperbolic diameter less than some suitably small constant $b > 0$. We now show that the set N consisting of the vertices of $\mathcal{T}$ within $\mathcal{D}_r$ satisfies the conditions of Lemma 4.18. Recall that the set contains $O(1/(1-r)) = O(1/\varepsilon)$ points, which is the desired number.

Next, consider an ideal triangle uvw in the Poincaré disk whose Euclidean side lengths exceed ε. Since a constant number of random points will hit any triangle with big enough sides, at the cost of adding a few new points to the net, we can certainly assume that the sides of uvw are fairly short. Now, let uvh be the triangle obtained as the intersection of the triangles uvw and $v'uv$, where v' is the reflection of v in Ou (Fig. 4.14).

We may assume that the segments uv and vw are congruent (in the Euclidean sense): Indeed, if w is further from v than u is, then we can slide w toward v until equidistance is achieved. The triangle uvh is where we will find a point of N, so sliding is not a problem because it causes only the triangle in question to shrink. Let λ be the Euclidean distance from u to v. A simple calculation shows that in the Euclidean plane the curved triangle uvh contains a disk B whose Euclidean radius and Euclidean distance to

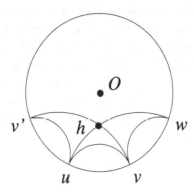

Fig. 4.14. Two ideal triangles.

$\partial \mathcal{D}$ are both greater than, say, $\lambda/10$. Because $\lambda > \varepsilon$, we have $\lambda > 10(1-r)$, and therefore B is entirely contained in $\mathcal{D}_r$. Suppose, by contradiction, that B does not contain any point of N. Then, the triangle of $\mathcal{T}$ that contains the (Euclidean) center p of B also contains a point at Euclidean distance at least $\lambda/10$ from p. Since the Euclidean distance from p to $\partial \mathcal{D}$ is less than 2λ, the triangle in question must contain a point q such that the pair p, q satisfies the conditions of Lemma 4.19. It follows that their hyperbolic distance exceeds a fixed positive constant. Thus, choosing b small enough leads to a contradiction.

The ideal triangle uvw has the same vertices in the Klein model; however, its edges are straight. Mapping the net N back to the Klein disk thus gives us a set of points that satisfies Lemma 4.18. This completes our discussion of weak ε-nets for points uniformly distributed on a circle, and completes the proof of Theorem 4.17 (page 197). $\square$

4.5 Bibliographical Notes

Section 4.2: The greedy cover algorithm leading to Theorem 4.3 (page 172) was proposed independently by Johnson [171] and Lovász [198]. The weighted greedy sampling algorithm of Theorem 4.4 (page 174) is due to the author and has not been published before.

Section 4.3: The randomized construction of ε-approximations for range spaces of bounded VC-dimension (Theorem 4.9, page 179) is due to Vapnik and Chervonenkis [317]. The deterministic construction given in the text (Theorem 4.5, page 175) is adapted from an algorithm proposed by Chazelle and Matoušek [81]. The latter is itself a simplification of an earlier algorithm presented by Matoušek [216] in an important paper that introduced the use of oracles and divide-and-conquer for effective sampling. The idea of refinement (ie, sampling from samples) originated in the derandomization paper of Chazelle and Friedman [78]. Earlier work by Matoušek [210, 211], although restricted to a geometric setting, nevertheless paved the way for subsequent developments. The improved constructions leading to Theorem 4.10 (page 179) and Theorem 4.11 (page 179) are due to Matoušek, Welzl, and Wernisch [222].

The first and third proofs for ε-nets are due to Matoušek [216] and Haussler and Welzl [160], respectively. The second proof has never been published; it was communicated to the author by Matoušek. Interestingly, the bound of $O(r \log r)$ for $(1/r)$-nets cannot be improved in general, as was shown by Komlós, Pach, and Woeginger [188]. Pach and Agarwal's book [247] has a nice discussion of this and related results. The running time for computing a $(1/r)$-net was improved to $O(d)^{3d}(r \log dr)^d |X|$ by Brönnimann, Chazelle, and Matoušek [57], using the notion of a *sensitive ε-approximation*.

Chazelle showed in [69] that ε-approximations could be used to estimate the number of vertices of an arrangement within a simplex. The technique was generalized by Brönnimann, Chazelle, and Matoušek [57], who introduced the notion of product range spaces and proved Theorem 4.16 (page 186).

Section 4.4: Alon et al. [14] initiated the study of weak ε-nets. They established the existence of weak ε-nets for convex sets in $\mathbf{R}^d$ of size independent of the cardinality of the point set. Specifically, they gave a construction of size $O(1/\varepsilon^2)$, for $d = 2$, and $O(1/\varepsilon)^{(d+1)(1-1/s_d)}$, where $s_d = (4d + 1)^{d+1}$, for general d.

The upper bound for $d \geq 3$ was improved by Chazelle et al. [77] to $O(\varepsilon^{-d} \log^{\beta_d} \varepsilon^{-1})$, where $\beta_2 = 0$, $\beta_3 = 1$, and

$$\beta_d \approx (\sqrt{e} - 1.5) \cdot 2^{d-1}(d - 1)! \approx 0.149 \cdot 2^{d-1}(d - 1)!.$$

These bounds are likely to be far from optimal. If the points are in convex position in the plane, then a net of size $O(\frac{1}{\varepsilon} \log^c \frac{1}{\varepsilon})$ can be found, where $c = \log_2 3 \approx 1.6$.

The optimal hyperbolic tiling construction used for the case of a regular polygon (Theorem 4.17, page 197) is also by Chazelle et al. [77]. A weaker bound of $2^{\log^*(1/\varepsilon)}O(1/\varepsilon)$ was obtained earlier by Capoyleas [61].

5

<hr style="border-top: 4px double #000;" />

Geometric Searching

o answer specific queries regarding a large collection of geometric data is what *geometric searching* is all about. The data could be a road map, the query could be a pair of coordinates obtained from a GPS navigational system, and the expected reply could be the name of the road that the car is on. With increasing frequency, such geometric databases can be found tucked into car dashboards, or at least tucked away in people's imagination of what a dashboard should look like. *Range searching* typically refers to more complex queries, eg, how many towns of more than 10,000 people can be found within 100 miles of Natchez, Mississippi? Or more challenging still: What is the library nearest you with a copy of this book?

This chapter highlights one of the finest vehicles for divide-and-conquer in computational geometry. It is a versatile data structure known as an ε-*cutting*. Suppose that we are given a set H of n hyperplanes in $\mathbf{R}^d$. We wish to subdivide $\mathbf{R}^d$ into a small number of simplices, so that none of them is cut by too many hyperplanes. Specifically, given a parameter $\varepsilon > 0$, a collection $\mathcal{C}$ of closed full-dimensional simplices is called an ε-cutting (Fig. 5.1) if:

(i) their interiors are pairwise disjoint, and together they cover $\mathbf{R}^d$ (hence, some are unbounded);[1]

(ii) the interior of any simplex of $\mathcal{C}$ is intersected by at most εn hyperplanes of H.

Intuitively, cuttings lend themselves to divide-and-conquer for the same reasons that graph separators do. They allow us to break down a problem

---

[1] If $d = 2$, for example, unbounded triangles consist of one or two edges. Alternatively, we might think of a projective model, where triangles "wrap around" infinity.

into independent subproblems of smaller size. Cuttings illustrate the typical discrepancy method pipeline: from low discrepancy to tools for divide-and-conquer via a bunch of intermediate steps (here, ε-approximations and ε-cuttings). Condition (ii) indicates the role of the parameter ε and the inherent tradeoff that it implies: The smaller ε, the easier the subproblems, but the more of them.

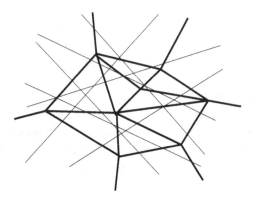

Fig. 5.1. An ε-cutting: No triangle is cut by more than εn lines.

The chapter is organized as follows: we build cuttings in §5.1, and we show how useful they are in §5.2. Extensions are also discussed. Section 5.3 addresses the following problem, known as simplex range searching: Given n points in d-space, find how many of them fall within a given simplex, presented as a query.

5.1 Optimal ε-Cuttings

For simplicity, we assume that the set H of n hyperplanes in $\mathbf{R}^d$ is in general position. This implies that the arrangement $\mathcal{A}(H)$ that it forms has exactly $\binom{n}{d}$ vertices. If need be, we can use the perturbation methods of [120, 332, 333] to deal with degeneracies. How large must a $(1/r)$-cutting $\mathcal{C}$ be? (For notational convenience, we prefer to write ε as $1/r$, for some $r > 0$.) Whereas the arrangement $\mathcal{A}(H)$ has on the order of n^d vertices, each $\sigma \in \mathcal{C}$ contains only $O(n/r)^d$ of them, so obviously $\mathcal{C}$ must consist of $\Omega(r^d)$ simplices. Is this lower bound tight? The answer, as we shall show, is yes.

In trying to prove this fact, we are quickly led to dismiss standard spatial decomposition techniques, such as grids, oct-trees, kd-trees, etc. These crude techniques are often computationally superior in practice, but mathematically they are inferior because they completely ignore the inner geometric structure of $\mathcal{A}(H)$. A better idea is to start with a $(1/r)$-net for H (Chapter 4).[2] Its VC-dimension is bounded, and so by Theorem 4.6 (page 175), a $(1/r)$-net N of size $O(r \log r)$ can be found in $nr^{O(1)}$ time.

Next, we define a *canonical triangulation* of $\mathcal{A}(N)$ by induction on the dimension d. The one-dimensional case is immediate, so we may assume that $d > 1$. We begin by ranking the vertices $\mathcal{A}(N)$ by the lexicographic order of their coordinate sequence. Then, inductively we form a canonical triangulation of the $(d-1)$-dimensional arrangement made by each hyperplane with respect to the $n - 1$ others. Finally, for each cell[3] σ of $\mathcal{A}(N)$, we lift toward its lowest ranked vertex each k-simplex ($k = 0, \ldots, d-2$) on the (triangulated) boundary of σ, provided that it does not lie in a $(d-1)$ face of $\mathcal{A}(N)$ that is incident to v.

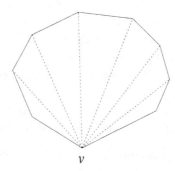

v

Fig. 5.2. The canonical triangulation in action: In this close-up of a two-dimensional face, v is the lowest ranked vertex of the face.

The *facial complexity* (ie, the size of the facial graph[4]) of $\mathcal{A}(N)$ is bounded by $O(r \log r)^d$. As is easily shown by induction, the same asymptotic bound applies to the triangulation of $\mathcal{A}(N)$. Therefore, the closures of its cells constitute a $(1/r)$-cutting for H of size $O(r \log r)^d$. This is good, but not good enough. By a slightly more subtle argument we can find cuttings that match the lower bound of $\Omega(r^d)$.

---

[2]Throughout this chapter, the underlying range space $(X, \mathcal{R})$ consists of a set X of hyperplanes and the collection $\mathcal{R}$ of subsets obtained by intersecting X with all possible open d-simplices.

[3]In this chapter, we use the word "cell" as shorthand for "full-dimensional cell."

[4]See Appendix C.

Theorem 5.1 *Given a collection H of n hyperplanes in $\mathbf{R}^d$, for any $r > 0$, there exists a $(1/r)$-cutting for H of optimal size $O(r^d)$. A full description of the cutting, including the list of hyperplanes intersecting the interior of each simplex, can be found deterministically in $O(nr^{d-1})$ time.*

To prove the theorem, we first sketch a simple strategy that almost works. We then show how to patch its holes.

A First Try

Suppose that we were able to solve the problem optimally for constant-size cuttings. Could we bootstrap the method to handle any size? In other words, assume that we have an efficient method for constructing a $(1/r_0)$-cutting of size $O(r_0^d)$, for some suitably large constant[5] r_0. To build a $(1/r)$-cutting, we might proceed as follows: Start out with a constant-size cutting, say the trivial cutting consisting of $\mathbf{R}^d$ as its sole "simplex," and then progressively refine the cutting, producing several generations of finer and finer cuttings, $\mathcal{C}_1, \mathcal{C}_2$, etc, where $\mathcal{C}_k$ is a $(1/r_0^k)$-cutting for H of size $O(r_0^{dk})$. That is the grand picture.

Now the details. Assume that we have recursively computed the cutting $\mathcal{C}_k$ for H. We further assume that, for each $\sigma \in \mathcal{C}_k$, the incidence list H_σ of hyperplanes intersecting the interior of σ is available. To produce the next-generation cutting $\mathcal{C}_{k+1}$, we could refine each σ in turn in this manner:

1. Compute a $(1/r_0)$-cutting for H_σ, using the construction assumed earlier.

2. Keep only the simplices that intersect σ and cut off their portion outside σ.

3. If the cut-off creates nonsimplicial cells, triangulate them by using the canonical method mentioned above (Fig. 5.3).

The collection of new simplices forms $\mathcal{C}_{k+1}$. One might think that we are done. For one thing, $\mathcal{C}_{k+1}$ certainly is an $(1/r_0^{k+1})$-cutting, as desired. The reason is that a simplex of $\mathcal{C}_{k+1}$ in σ is cut[6] by at most $|H_\sigma|/r_0$ hyperplanes of H_σ, and hence of H. By induction, this gives at most $(n/r_0^k)/r_0 = n/r_0^{k+1}$ cuts.

---

[5] Of course, since r_0 is a constant we could simply say "of size $O(1)$," but to use the bound $O(r_0^d)$ allows us to see the problems that would still arise from a bootstrapping approach, even if we could handle nonconstant values of r_0.

[6] Shorthand for: intersected in its interior.

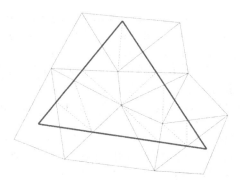

Fig. 5.3. The small triangles overlapping with the big triangle are clipped and then triangulated canonically.

What about the size of $\mathcal{C}_{k+1}$? The number of new simplices created within σ is at most cr_0^d, for some constant $c > 0$. From this, the size of $\mathcal{C}_{k+1}$ appears to be bounded by $cr_0^d \cdot O(r_0)^{dk}$, which might seem just fine; but it is not. Indeed, in the end, the size of the final cutting is $O(c'r_0)^j$, where $j = \lceil \log r / \log r_0 \rceil$ and $c' > 1$, which is not $O(r^d)$. The weakness of our method is that bounds for the current generation are based on bounds from the previous generation. This nesting leads to error buildup. To avoid this, we must use a global parameter to which the analysis constantly refers.

Which sort of parameter? Note that, in our analysis, bounding the number of new simplices within σ by cr_0^d is making the implicit assumption that most of the vertices in the arrangement formed by the newly chosen hyperplanes happen to fall within σ. If the number of σ's were to grow too fast from one generation to the next, obviously this could not remain true very long. Indeed, many numbers can grow from one generation to the next, but one that cannot is the total number of vertices in $\mathcal{A}(H)$. If too many simplices are created, then several of them must enclose relatively few vertices of $\mathcal{A}(H)$. Our effort will be aimed at reducing the growth of cuttings within those sparse simplices.

That this should be possible is suggested by the following observation. Suppose that the hyperplanes of H_σ (ie, those intersecting the interior of σ) create no vertices strictly inside σ. Then, for any subset K of H_σ, the portion of $\mathcal{A}(K)$ within σ has complexity $O(|K|^{d-1})$: This indicates that the size of a cutting within σ should have an exponent of $d-1$ and not d.

Getting it Right

We must be slightly more subtle in the way we refine a simplex $\sigma \in C_k$ into cells of generation $k+1$. We can assume that $|H_\sigma| > n/r_0^{k+1}$; otherwise, σ already satisfies the requirement of the next generation and needs no refining. We distinguish between two types of simplices, *full* and *sparse*. Given a set X of hyperplanes and a d-dimensional (closed) simplex σ, let $v(X, \sigma)$ be the number of vertices of $\mathcal{A}(X)$ in the interior of σ.

- A *full* simplex $\sigma \in C_k$ is one where $v(H, \sigma) \geq c_0 |H_\sigma|^d$, for some suitably small constant c_0, say $c_0 = 1/r_0^2$. We compute a $(1/r_0)$-net for H_σ. We triangulate the portion of the net's arrangement within σ to form a $(1/r_0)$-cutting of size $O(r_0 \log r_0)^d$. Its simplices form the elements of C_{k+1} that lie within σ.

- A *sparse* simplex σ is one that is not full. We treat it slightly differently. First, we compute a subset H_σ^o of H_σ that satisfies two conditions:

  (i) The canonically triangulated portion of $\mathcal{A}(H_\sigma^o)$ inside σ forms a collection C_σ^o of full-dimensional (closed) simplices of size at most $r_0^d/2$.

  (ii) The interior of each simplex of C_σ^o is intersected by at most $|H_\sigma|/r_0$ hyperplanes of H.

  The simplices of C_σ^o constitute the elements of C_{k+1} within σ.

It might not be obvious how all of these conditions can be met, let alone implemented efficiently or used profitably. But, assuming that full and sparse simplices can be processed as indicated, we now show that the net result is a new-generation cutting C_{k+1} with all the right characteristics.

Lemma 5.2 C_{k+1} *is a* $(1/r_0^{k+1})$*-cutting of size* $O(r_0^{d(k+1)})$.

Proof: First, the "cutting-number" condition. The case of a full simplex σ is trivial: Any of the new simplices into which σ is subdivided is cut by at most $|H_\sigma|/r_0$ hyperplanes and, by induction, $|H_\sigma| \leq n/r_0^k$. By (ii), the same argument applies to σ, even if it is sparse.

How large is the new cutting C_{k+1}? The number of full simplices cannot exceed $\binom{n}{d}/c_0|H_\sigma|^d$, and any one of them contributes $O(r_0 \log r_0)^d$ simplices to C_{k+1}. Since $n/r_0^{k+1} < |H_\sigma| \leq n/r_0^k$, this means that full simplices contribute $O(r_0^{d(k+1)})$ cells to C_{k+1}. Note that the constant behind the big-oh depends on d and c_0 but *not* on k.

We now assess the contribution of sparse simplices. The construction of simplices is hierarchical, so that it is natural to refer to σ as the parent simplex of the cells of C_{k+1} that lie within σ. A leaf of the hierarchy is a simplex cut by at most n/r_0^{k+1} hyperplanes. We use an accounting scheme to bound the number of sparse cells: Any sparse simplex charges its nearest full ancestor (or the root of the hierarchy if there is no such thing). In this way, a full simplex σ (at any generation, and not just the k-th one) is charged at most

$$O(r_0 \log r_0)^d \cdot \left(r_0^d/2 + (r_0^d/2)^2 + \cdots + (r_0^d/2)^i \right),$$

where i is the number of generations below σ, ie, i is the smallest integer such that $|H_\sigma|/r_0^i < n/r_0^{k+1}$. It follows that σ is charged at most $O(|H_\sigma| r_0^k/n)^d/2^i$, where the constant behind the big-oh depends on[7] r_0. Since σ is full (or is $\mathbf{R}^d$ in the case of the root), another way to phrase this is to say that each vertex of $\mathcal{A}(|H_\sigma|)$ within σ is charged $O(r_0^k/n)^d/2^i$. This implies that, overall, a vertex of $\mathcal{A}(H)$ is charged $O(r_0^k/n)^d$, and therefore the total number of sparse simplices is $O(r_0^k/n)^d \binom{n}{d}$, which is $O(r_0^{d(k+1)})$, as desired. $\square$

Our final task is to explain how to achieve conditions (i) and (ii) for sparse simplices and to bound the complexity of the algorithm. We begin with a simple technical fact:

Lemma 5.3 *The number of simplices created within the sparse simplex σ at generation $k+1$ is $O(v(H_\sigma^o, \sigma) + |H_\sigma^o|^{d-1})$.*

Proof: Prior to triangulation, the portion of $\mathcal{A}(H_\sigma^o)$ within σ has facial complexity in $O(v(H_\sigma^o, \sigma) + |H_\sigma^o|^{d-1})$. This is because every face is incident upon at least one vertex in the interior of σ or on its boundary. Furthermore, by our general-position assumption, vertices are incident upon at most a constant number of faces. The canonical triangulation itself can at most multiply the facial complexity by a constant factor. $\square$

Our first order of business is how to tell whether a simplex σ is full or sparse. By Theorem 4.5 (page 175), we can build a $(c_0/2)$-approximation A_σ for H_σ of constant size in $O(|H_\sigma|)$ time. We also know from Theorem 4.16 (page 186) that[8]

---

[7] One can now appreciate the importance of the dividing by two in condition (i) of a sparse simplex.

[8] For the application of the theorem, the range space should be defined by considering

$$\left| \frac{v(H,\sigma)}{|H_\sigma|^d} - \frac{v(A_\sigma,\sigma)|}{|A_\sigma|^d} \right| \le \frac{c_0}{2}, \tag{5.1}$$

and so we can estimate the value of $v(H,\sigma)$ in constant time with an error of at most $(c_0/2)|H_\sigma|^d$. As far as distinguishing between sparse and full goes, such an error is inconsequential (since no particular assumption has been made on the constant c_0 yet).

Next, we investigate the complexity of refining σ while satisfying the required conditions. The case of a full simplex is straightforward. By Theorem 4.6 (page 175), we can compute the $(1/r_0)$-net in $O(|H_\sigma|)$ time. Since its size is constant, computing the new set of simplices in σ, together with all of the incidence lists, can also be done in $O(|H_\sigma|)$ time.

To discuss the refinement of a sparse simplex, we begin with a probabilistic construction, which we then derandomize. We compute H^o_σ by choosing a random sample from A_σ of size $c_1 r_0 \log r_0$, for some constant c_1 large enough (independent of r_0). By Lemma 4.13, with probability at least $2/3$, the sample forms a $(1/2r_0)$-net for A_σ. By Lemma 4.2, it follows that H^o_σ is an $(c_0/2 + 1/2r_0)$-net for H_σ. By choosing c_0 smaller than $1/r_0$, we ensure that (ii) holds with probability at least $2/3$.

What about (i)? The probability that a vertex of $\mathcal{A}(A_\sigma) \cap \sigma$ turns up as a vertex of $\mathcal{A}(H^o_\sigma) \cap \sigma$ is equal to

$$\binom{|A_\sigma| - d}{|H^o_\sigma| - d} \Big/ \binom{|A_\sigma|}{|H^o_\sigma|} < \frac{1}{\sqrt{c_0}} \frac{|H^o_\sigma|^d}{|A_\sigma|^d},$$

using $\sqrt{1/c_0}$ in the role of a suitably large constant. From (5.1), it follows that the expected value of $v(H^o_\sigma,\sigma)$ is at most

$$\frac{1}{\sqrt{c_0}} \frac{|H^o_\sigma|^d}{|A_\sigma|^d} v(A_\sigma,\sigma) \le \frac{1}{\sqrt{c_0}} \frac{|H^o_\sigma|^d}{|H_\sigma|^d} v(H,\sigma) + \sqrt{c_0} \, |H^o_\sigma|^d.$$

Because σ is sparse, it follows that, for c_0 small enough,

$$\mathbf{E}\, v(H^o_\sigma,\sigma) \le 2\sqrt{c_0}\, (c_1 r_0 \log r_0)^d.$$

By Lemma 5.3, the number of simplices created within σ at generation $k+1$ is $O(v(H^o_\sigma,\sigma) + |H^o_\sigma|^{d-1})$; and so, with probability at least $2/3$, the

---

intersections of hyperplanes with line segments; the current range space, which is defined with respect to open d-simplices, is richer and works just as well, however.

number of child simplices derived from σ is

$$O(\sqrt{c_0})\,(c_1 r_0 \log r_0)^d + O(r_0 \log r_0)^{d-1}.$$

Recall that c_1 is an absolute constant that does not depend on r_0, so by choosing r_0 large enough and, say, $c_0 = 1/r_0^2$, we ensure that $c_0 < 1/r_0$ (as required), and we easily drive the number of child simplices below $r_0^d/2$, thus guaranteeing (i).

In summary, we have shown that, with probability at least $2/3$, the sample H_σ^o satisfies (i) and the same is true of (ii), so both conditions hold simultaneously with probability at least $1/3$. To derandomize the construction, we may simply test out every possible sample. Since A_σ is of constant size, this takes $O(1)$ time; of course, there are better ways of doing that, but this is just constant-factor finetuning, which we leave as an exercise. The running time for refining σ is thus $O(|H_\sigma|)$.

We conclude that the refinement of any simplex takes time proportional to the total size of the incidence lists produced in the process. By Lemma 5.2, this shows that the time for building generation $k+1$ is $O(r_0^{d(k+1)})\,n/r_0^{k+1}$, which is $O(nr_0^{(d-1)(k+1)})$. The construction proceeds until its reaches the first generation such that $r_0^k \geq r$. This brings the entire cost of building the final cutting to $O(nr^{d-1})$, which concludes the proof of Theorem 5.1 (page 206). $\square$

5.2 Cuttings in Action

There are many applications of cuttings in geometric searching. One of the most striking is simplex range searching, a problem for which cuttings (along with other tools) lead to a complete solution (§5.3). In Chapter 6, the solution is shown to be optimal in a very general model. This is one of the great success stories in theoretical computational geometry. As a warmup we mention a few simple applications of cuttings. Throughout our discussion, d is arbitrary but constant.

Point Location Among Hyperplanes

Preprocess an arrangement of n hyperplanes in $\mathbf{R}^d$ so that, given a query point, one can quickly find which face of the arrangement contains the point (Fig. 5.4). To simplify the discussion we assume that the hyperplanes are in general position and the point does not lie on any of the hyperplanes. Set $r = n$ in Theorem 5.1 (page 206) and observe that, from the nesting structure of C_1, C_2, etc, we can locate p in C_k (ie, find the cell that contains it) in constant time, once we know its location within C_{k-1}. Indeed, recall that a cell of C_{k-1} finds itself subdivided into $O(r_0 \log r_0)^d$ subcells in C_k. The algorithm is easily adapted to handle the case where the query point lies in a lower dimensional face of the arrangement. Thus, the construction of the cutting itself provides a data structure suitable for point location in an arrangement of n hyperplanes.

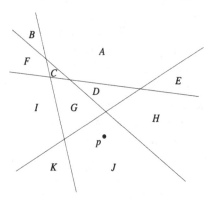

Fig. 5.4. The answer to the query p is the label J. Perhaps not the stuff of killer demos, but the source of many ingenious ideas over a quarter-century of research in computational geometry.

The query time is proportional to the depth of the hierarchy, which is $O(\log n)$. The space needed is that of the cutting itself, ie, $O(n^d)$. Finally, the preprocessing time is determined solely by the cutting construction. Of course, in addition, one would need the arrangement itself, with pointers from the terminal cells at the bottom of the hierarchy to their enclosing cells in the arrangement. It is an easy exercise to do all of the above in $O(n^d)$ time.

Theorem 5.4 *Point location among n hyperplanes can be done in $O(\log n)$ query time, using $O(n^d)$ preprocessing. This is the problem of preprocessing an arrangement of n hyperplanes in $\mathbf{R}^d$ so that, given a query point, the face of the arrangement that contains the point can be found quickly.*

If our goal is only to find out whether a query point lies on a hyperplane, we can use the same method, but the storage can be reduced a little. Indeed, we can do with a $(1/r)$-cutting for $r = n/\log n$. The cells at the bottom of the hierarchy are cut by only $O(\log n)$ hyperplanes; and so, once the cell enclosing the query point has been located, a naive $O(\log n)$-time incidence check finishes the job.

Theorem 5.5 *Given n hyperplanes in $\mathbf{R}^d$, it is possible to check whether a query point is incident to any of them in $O(\log n)$ time, using preprocessing time and space proportional to $n^d/(\log n)^{d-1}$.*

By duality[9] this also gives us the following result: Given n points in $\mathbf{R}^d$, it is possible to check whether a query hyperplane passes through any of them in $O(\log n)$ time, using preprocessing time and space proportional to $n^d/(\log n)^{d-1}$.

Hopcroft's Problem

The previous application is a good example of a geometric problem that gains in being rephrased in dual space. Of course, any dual-invariant problem is immune to such gains. A nice example of this was formulated by John E. Hopcroft. Given n points and n lines in the plane, check whether all the points lie outside all the lines. In other words, is the configuration free of any point/line incidence?

How hard should we expect this problem to be? There are good reasons to believe that $\Omega(n^{4/3})$ is a natural lower bound. Here is an intuitive, nonrigorous explanation why. By a classical construction of Erdős, there is an arrangement of n lines in the plane such that at least n of its vertices are each incident to $\Omega(n^{1/3})$ edges. A malicious adversary would choose these n lines as input to Hopcroft's problem. For the n points, he would choose one very near each of the high-degree points.

Intuitively, for the algorithm to decide whether a point actually touches its neighboring vertex or not, it should have to check in which wedge of the star formed around it the point lies. This would drive the total cost to be proportional to the total number of wedges over the n stars, ie, $\Omega(n^{4/3})$. Of course, that sort of pseudoargument should not be taken too seriously.

A simple solution to Hopcroft's problem goes like this. By Theorem 5.1 (page 206), we construct a $(1/r)$-cutting[10] for the n input lines, where $r = n^{1/3}/(\log n)^{2/3}$. Next, we locate each of the n points in their respective cells. Using the hierarchical nature of the construction, this can be done in

---

[9]See Appendix C.

[10]Obviously, one should not perturb lines when computing the cutting for fear of "perturbing" the outcome, too. Symbolic perturbation methods can easily be made to work, however.

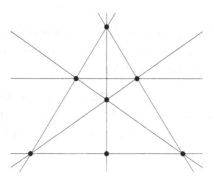

Fig. 5.5. By drawing n lines in a special way, we obtain an arrangement with at least n vertices of degree $\Omega(n^{1/3})$ each.

time $O(n \log n)$, while the cutting construction itself takes time $O(nr)$. If any point/line incidence is discovered at this time, we stop with the answer. Otherwise, we consider each cell (which is a triangle) of the cutting, and we break up the set of points inside it into subsets of size roughly n/r^2 or less. This gives us a total of $O(r^2)$ cells such that (i) only $O(n/r)$ lines cut through them, and (ii) only n/r^2 points lie inside. To find out whether there is any point/line incidence within any such cell, we dualize the lines into points and the points into lines, and ask the same question. To answer it, we appeal to Theorem 5.5, at a cost of $O(n/r) \log n + O(n/r^2)^2/\log n$ per cell. The total running time is $O(nr \log n + (n/r)^2/\log n)$, which, by our choice of r, is also $n^{4/3} \cdot O(\log n)^{1/3}$ time.

Theorem 5.6 *To decide whether n points and n lines in the plane are free of any incidence can be done in $n^{4/3} \cdot O(\log n)^{1/3}$ time.*

By a more careful balancing act among the various components of the algorithm, it is possible to reduce the complexity of the algorithm. The best time bound currently known is $n^{4/3} 2^{O(\log^* n)}$ (see bibliographical notes).

5.3 Simplex Range Searching

How should we preprocess a set P of n points in $\mathbf{R}^d$ so that, given a query (closed) simplex σ, the size of $P \cap \sigma$ can be evaluated very efficiently (Fig. 5.6)? This problem, commonly referred to as *simplex range*

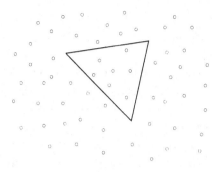

Fig. 5.6. The answer to the simplex range query is 7.

searching, is often considered in a slightly more general setting, in which weights are assigned to the points and the answer to a query is the sum of all of the weights within σ. For simplicity, we assume that weights belong to an additive group or semigroup, and that each weight can be stored in one unit of memory. The data structures that we discuss below are all easily adapted to the reporting case, where the answer to a query is the explicit listing of the points in $P \cap \sigma$.

Theorem 5.7 *Given n points in $\mathbf{R}^d$, there exists a data structure of size m (for any $n \le m \le n^d$), which allows simplex range searching to be done in time $O(n^{1+\varepsilon}/m^{1/d})$ per query, for any fixed $\varepsilon > 0$.*

Our discussion aims at explaining the main ideas behind this general theorem and not at providing a comprehensive treatment of the subject. In particular, we do not address issues such as preprocessing time. We also assume that the points are in general position. The proof touches on many ideas: VC-dimension, ε-nets, ε-approximations, ε-cuttings, reweighting, duality, etc. The theorem and its corresponding lower bound constitute a wonderful summary of the depth and variety of ideas in geometric searching. Indeed, as will be shown in the next chapter, the theorem is basically optimal over semigroups. The proof involves integral geometry, extremal graph theory, Heilbronn's problem, etc. Simplex range searching truly is one of the gold mines of computational geometry.

The following problem is an excellent vehicle for introducing several elegant, powerful ideas about geometric range searching: Our goal is to store a set P of n points in the plane in a linear-size data structure so that, given a query triangle σ, we can easily count or report the points in $P \cap \sigma$.

This is called *triangle range searching*. A natural approach is to construct a *partition tree*. This is a rooted tree $\mathcal{T}$ with the following characteristics: The root is associated with the point set P, which is partitioned into a number of subsets $P_1, \ldots, P_m$. Each P_i is associated with a distinct child v_i of the root. In addition, we define a convex open set R_i, called the *region* of v_i, that contains P_i. Unlike the point sets P_i, the regions R_i are not necessarily disjoint. If $|P_i| > 1$, the subtree rooted at that v_i is defined recursively with respect to P_i.

Suppose that we want to use a partition tree for triangle range counting (the reporting case being similar). Then we must only add one number to each node, specifying how many points of P lie in its region. Given a query triangle σ, we initialize a counter to 0, and we proceed to explore all children v_i of the root. We ask the following question: Does σ intersect the region R_i of v_i? Depending on the answer, we take one of three actions:

- Answer is yes, but the query triangle σ does not completely enclose the region R_i of v_i: We visit v_i and recurse.
- Answer is yes, but σ completely encloses R_i: We increment our counter by $|P_i|$, and we do not recurse at v_i.
- Answer is no: We do not recurse at v_i.

To make the discussion more concrete, we examine one of the earliest partition trees.

The Conjugation Tree

The set P consists of n points in the plane. We choose an arbitrary line L_1 that bisects P into two equal halves, P_1 and P_2 (for simplicity, assume that n is a power of 2). Having defined the root and its two children, we move on to the grandchildren. By the classical *ham-sandwich theorem*,[11] there exists another line L_2 that bisects each half into two quarters of equal size. This produces a second-generation partition of P into four sets of size $n/4$. At this point, we could simply recurse. But instead of treating each of the four wedges independently, we combine their treatment in the following way. Let P_1' and P_1'' be the two sets partitioning P_1. We use only one line to halve both sets: That we can do so again follows from the *ham-sandwich*

---

[11] In its discrete version, the ham-sandwich theorem states that, given any collection of d finite point sets in $\mathbf{R}^d$, each one can be simultaneously bisected by a single hyperplane. A hyperplane h is said to bisect a set of n points if at most $n/2$ of them lie in either one of the two open halfspaces bounded by h.

theorem. Similarly, we subdivide P_2' and P_2'' with a single cut, to produce a total of eight grandchildren. The construction proceeds recursively, always choosing the next bisecting line so as to halve two sets with a single cut. This defines a particular partition tree, called a *conjugation tree.*

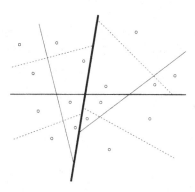

Fig. 5.7. A conjugation tree for 16 points.

The time for answering a query is proportional to the number of nodes visited in the process. This is at most twice the number of nodes *traversed,* which is our term for the nodes with visited children. The key observation is that, among the four grandchildren of a node, at most three of them can be traversed. It follows that the maximum number $T(n)$ of traversed nodes satisfies the recurrence relation:

$$T(n) \leq T(n/2) + T(n/4) + 2,$$

with the obvious boundary condition $T(1) = 1$. Writing $T(n) = f(\log n)$, we obtain a Fibonacci-like sequence,

$$f(k) \leq f(k-1) + f(k-2) + 2,$$

from which it follows that[12]

$$T(n) \ll \left(\frac{1+\sqrt{5}}{2}\right)^{\log n} = n^{\log(1+\sqrt{5})-1} \ll n^{0.695}.$$

The conjugation tree is an elegant construction that provides a linear-storage data structure for triangle range searching in $O(n^{0.695})$ time.

---

[12] Recall that $\ll$ and $\gg$ are shorthand for $O()$ and $\Omega()$, respectively.

The Spanning-Path Tree

Recall the spanning path theorem (Theorem 2.20, page 124). Given n points in $\mathbf{R}^d$, one can make them the vertices of a piecewise-linear curve $\mathcal{C}$, so that no hyperplane cuts more than $O(n^{1-1/d})$ edges of $\mathcal{C}$. This suggests building the following partition tree, called the *spanning-path tree* (Fig. 5.8):

- associate the root with the set of n points; then
- split the curve $\mathcal{C}$ into two curves with (roughly) the same number of vertices, and associate each of them with a distinct child of the root; finally,
- iterate as long as possible.

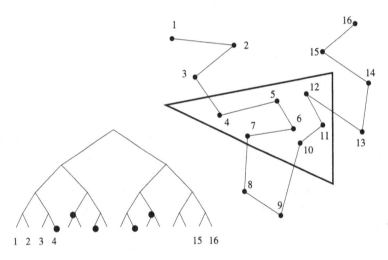

Fig. 5.8. The spanning path tree: To answer the triangle query, we must reach the filled nodes and add up the number of leaves (at or) descending from them, which gives 7.

What distinguishes this from a regular partition tree is that nodes are no longer associated with regions of space (only with subsets of the n points). This makes searching the tree hopelessly difficult in higher dimensions. If we knew which nodes to visit, however, the algorithm would be efficient because the number of nodes to be *traversed* (defined as the nodes whose children need to be visited) is $O(n^{1-1/d}\log n)$. To see why this is, think of how the curve $\mathcal{C}$ winds in and out of the query simplex σ. Since σ is bounded by only $d+1$ hyperplanes, the total number of entries and exits is $O(n^{1-1/d})$. This partitions $\mathcal{C}$ into curves, every one of which is either entirely inside or outside σ. Only the curves inside cause nodes to be traversed, and there can be at most $2 \log n$ such nodes per curve.

If we could quickly find the points of entry and exit, a solution would follow. But how to do that is unclear. Only low dimensions seem to offer a glimmer of hope. In the plane, for example, finding entry/exit points is an instance of the classical *ray-shooting problem*: Given a source point and a direction, find the first hit of a ray from the source along that direction. If C forms a simple polygonal curve, then this can be done in logarithmic time with linear storage (see bibliographical notes). We can always enforce simplicity by removing all edge intersections one at a time. When two edges ab and cd intersect, we remove them and reconnect the polygonal curve by adding ac, bd or ad, bc, whichever keeps C connected (Fig. 5.9). Iterating until no crossings are left produces a simple curve. Convergence is guaranteed because the Euclidean length of C decreases at every step and the process cannot cycle. Furthermore, edge flipping cannot increase the number of intersections with any given line, and so the number of cuts by a line remains $O(\sqrt{n})$.

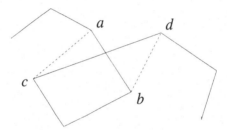

Fig. 5.9. Disentangling a nonsimple polygonal curve.

To summarize, this provides a linear-size data structure for triangle range searching with query time $O(\sqrt{n} \log n)$. In higher dimensions, the data structure is unusable but it proves the following: Given a set P of n points in $\mathbf{R}^d$, it is possible to form a collection S of $O(n)$ subsets of P such that, given any simplex σ, the intersection $P \cap \sigma$ can be expressed as the disjoint union of $O(n^{1-1/d} \log n)$ sets in S. Note that this statement is purely combinatorial, and says nothing about computing the subsets.

Simplicial Partitions

We return to the original goal of finding an efficient, "algorithmically usable" partition tree for a set of points P in $\mathbf{R}^d$. We slightly refine the

manner in which we distribute the points of P among the children of the
root. We say that the collection $\{(P_1, R_1), \ldots, (P_m, R_m)\}$ is a *simplicial
partition*, if (i) the P_i's partition P and (ii) each R_i is a relatively open
simplex enclosing P_i. Note that the R_i's can be of any dimension and need
not be disjoint. Also, P_i need not always be the same as $P \cap R_i$ (Fig. 5.10).
We say that a hyperplane *cuts* R_i if it intersects, but does not contain, it.
The maximum number of R_i's that a single hyperplane can cut is called
the *cutting number* of the simplicial partition. We postpone the proof of
the following fact:

Lemma 5.8 *Given a set P of n points in $\mathbf{R}^d$ ($d > 1$), for any integer
$1 < r \leq n/2$, there exists a simplicial partition of cutting number $O(r^{1-1/d})$
such that $n/r \leq |P_i| < 2n/r$ for each (P_i, R_i) in the partition.*

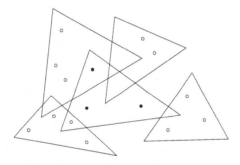

Fig. 5.10. A simplicial partition: Filled dots correspond to points belonging to
the set P_i of the middle triangle; note that P_i misses one of the points in the
triangle.

Using the lemma for a large enough constant r, we obtain a partition tree
whose number $T(n)$ of traversed nodes[13] satisfies the relations $T(O(1)) =
O(1)$ and

$$T(n) = O(r^{1-1/d}) \cdot T(2n/r),$$

from which it follows that $T(n) = O(n^{1-1/d+f(r)})$, where $f(r)$ tends to 0
as $r \to \infty$. This gives us a linear-size data structure for answering any
simplex range searching query in $O(n^{1-1/d+\varepsilon})$ time, for any fixed $\varepsilon > 0$.

---

[13] Recall that a node is traversed if its children are visited.

Theorem 5.9 *Given n points in* $\mathbf{R}^d$, *there exists a data structure of size* $O(n)$ *with which simplex range searching can be done in time* $O(n^{1-1/d+\varepsilon})$ *per query, for any fixed* $\varepsilon > 0$.

Lemma 5.8 can be viewed as a generalization of the spanning path theorem (Theorem 2.20, page 124). Its proof reflects this connection. Although the number of hyperplanes is infinite, a polynomial number of them are sufficient to test whether a simplicial partition has a small cutting number.[14] As it turns out, a well-chosen set of r hyperplanes is actually sufficient. In other words, we can find r magically chosen hyperplanes that are so representative of the others that any simplicial partition good for these hyperplanes is, in essence, good for all the others. The existence of such a small "test set" makes the construction of efficient simplicial partitions much easier.

Lemma 5.10 *Let P be a set of n points in* $\mathbf{R}^d$. *There exists a set* H_0 *of* r *hyperplanes such that, given a simplicial partition*

$$\Big\{ (P_1, R_1), \ldots, (P_m, R_m) \Big\}$$

for P, where each $|P_i| \geq n/r$, *the cutting number is* $(d+1)\kappa + O(r^{1-1/d})$, *where* κ *is the maximum number of* R_i's *cut by a single hyperplane of* H_0.

Proof: Let H be the set of hyperplanes dual[15] to the points of P. By Theorem 5.1 (page 206), there exists a $\Theta(r^{-1/d})$-cutting $\mathcal{C}$ for H with a number of simplices less than $r/(d+1)$. This implies that their combined set V of vertices is of size at most r. We now prove that choosing H_0 as the set of hyperplanes dual to V satisfies the lemma. Given a hyperplane h, let G be a simplex of $\mathcal{C}$ that contains the dual point of h. By definition, the dual hyperplane of any vertex of G cuts at most κ R_i's. This means that h can cut at most $(d+1)\kappa$ R_i's plus an undetermined number of R_i's not cut by any hyperplane dualizing to a point of V. By elementary properties of duality, any of the (at least n/r) points of P within such an R_i dualizes to a hyperplane that intersects the interior of G. Only $O(nr^{-1/d})$ hyperplanes can do that, so the number of such $R_i's$ is only $O(r^{1-1/d})$, bringing the total of cut $R_i's$ to $(d+1)\kappa + O(r^{1-1/d})$, as desired. $\square$

---

[14]This might seem rather obvious, but in fact it is not: Recall that the test set must work for "all" simplices R_i and that there is no obvious way to discretize the candidate simplices.

[15]See the discussion of duality in Appendix C.

We are now able to focus exclusively on the sparse set H_0, for which we customize a simplicial partition. We employ a weighting strategy similar to the argument used in Theorem 2.20 (page 124).

Lemma 5.11 *Let P be a set of n points in $\mathbf{R}^d$, and let H_0 be a set of hyperplanes. Given any integer $1 < r \leq n/2$, there exists a simplicial partition $\{(P_1, R_1), \ldots, (P_m, R_m)\}$ for P, such that (i) $n/r \leq |P_i| < 2n/r$, for each i, and (ii) the cutting number of any hyperplane of H_0 is $O(r^{1-1/d} + \log |H_0|)$.*

Proof: Suppose that $(P_1, R_1), \ldots, (P_i, R_i)$ have already been constructed (to get started we just apply the procedure below with $Q_0 = P$). If the leftover set $Q_i = P \setminus (P_1 \cup \cdots \cup P_i)$ is too small, ie, $|Q_i| < 2n/r$, we set $m = i + 1$, $P_m = Q_i$, $R_m = \mathbf{R}^d$, and the construction is complete. Assume now that $|Q_i| \geq 2n/r$. For each hyperplane $h \in H_0$, we define a *weight*,

$$w(h) = 2^{\kappa_i(h)},$$

where $\kappa_i(h)$ is the number of R_j's $(j \leq i)$ cut by h. Now, imagine duplicating each h of H_0 by making $w(h) - 1$ extra copies of it. Each new batch of $w(h)$ hyperplanes is slightly perturbed so that the resulting set, denoted by H_i, is in general position. Note that

$$|H_i| = \sum_{h \in H_0} 2^{\kappa_i(h)}. \tag{5.2}$$

By Theorem 5.1 (page 206), there exists an ε_i-cutting $\mathcal{C}$ for H_i whose simplices have, in total, fewer than $r|Q_i|/n$ faces of all dimensions, where $\varepsilon_i = \Omega(n/r|Q_i|)^{1/d}$. By the pigeonhole principle, one of these faces must contain at least n/r points of Q_i: We designate it by R_{i+1}. Note that the assumption $r \leq n/2$ implies that R_{i+1} is of dimension at least 1. We define P_{i+1} by choosing $\lceil n/r \rceil$ points of Q_i in R_{i+1} arbitrarily.

How large can the cutting number of $h \in H_0$ be? Intuitively, every time h is found to cut a new R_i, its weight is multiplied by 2. So, a large cutting number would imply a weight for the whole system that is exponentially larger. But to accumulate a large total weight is unlikely to happen because of our choice of R_i's: Each newly chosen R_i, being a simplex of a cutting, is cut by relatively few hyperplanes, and therefore its inclusion in the simplicial partition should increase the total weight by a relatively small amount. We flesh out this argument in the following.

If $\kappa_m(h)$ is the final cutting number of h, then the hyperplane must have $2^{\kappa_m(h)}$ copies of itself; therefore, by (5.2),

$$\kappa_m(h) \leq \log |H_m|. \tag{5.3}$$

We now bound the increase in size from H_i to H_{i+1}. At most $\varepsilon_i|H_i|$ hyperplanes can cut R_{i+1}: this is true by definition if R_{i+1} is full-dimensional; else it follows from the fact that cutting R_{i+1} implies intersecting the interior of any of its full-dimensional incident faces. These cutting hyperplanes are the only ones to be duplicated, so

$$|H_{i+1}| \leq (1 + \varepsilon_i)|H_i|.$$

It follows that, for some constant $c > 0$,

$$
\begin{aligned}
|H_m| \;&\leq\; |H_0| \cdot \prod_{i=0}^{m-1} (1 + \varepsilon_i) \\
&\leq\; |H_0| \cdot \prod_{i=0}^{m-1} \left(1 + \frac{cn^{1/d}}{r^{1/d}|Q_i|^{1/d}}\right) \\
&\ll\; |H_0| \cdot \prod_{k=1}^{r} \left(1 + \frac{c}{k^{1/d}}\right).
\end{aligned}
$$

Taking logarithms and using the fact that $\ln(1 + x) \leq x$,

$$\log|H_m| \ll \log|H_0| + \sum_{k=1}^{r} \frac{1}{k^{1/d}} \ll \log|H_0| + r^{1-1/d}.$$

The lemma follows from (5.3). $\square$

By combining Lemmas 5.10 and 5.11, we derive the existence of a simplicial partition $\{(P_1, R_1), \dots, (P_m, R_m)\}$ for P, such that (i) $n/r \leq |P_i| < 2n/r$, for each i, and (ii) the cutting number of any hyperplane is $(d + 1)\kappa + O(r^{1-1/d})$, where κ is $O(r^{1-1/d} + \log r)$. This bounds the maximum cutting number of any hyperplane by $O(r^{1-1/d})$, which completes the proof of Lemma 5.8. $\square$

The techniques we have used can be strengthened further. For example, treating each simplicial partition separately at each node is an obvious weakness. By using more global (and complex) arguments, the following can be shown.

Theorem 5.12 *Given n points in $\mathbf{R}^d$, there exists a data structure of size $O(n)$ with which simplex range searching can be done in time $O(n^{1-1/d})$ per query.*

Logarithmic Query Time

Having discussed one end of the spectrum, ie, minimal-storage solutions, we
turn to the other extreme: solutions providing minimal query time. First,
notice that if the query is a halfspace, the problem, called *halfspace range
searching*, is equivalent, by duality, to point location in an arrangement
of hyperplanes. Our choice of a dual transform is one known to preserve
above/below relations (see Appendix C):

$$(p_1,\ldots,p_d) \mapsto x_d = p_1 x_1 + \cdots + p_{d-1} x_{d-1} + p_d$$

and

$$x_d = p_1 x_1 + \cdots + p_{d-1} x_{d-1} + p_d \mapsto (-p_1,\ldots,-p_{d-1},p_d).$$

The term "above" refers here to the x_d-direction. We assume that none
of the n input points is at the origin. Halfspace range searching dualizes
to the problem of counting how many of n given hyperplanes lie above (or
below) a query point. Since each face of the arrangement formed by the n
hyperplanes corresponds to a single count, all we have to do in preprocessing is to compute these counts and set up the arrangement for fast point
location (Theorem 5.4, page 212). To answer a query becomes equivalent to
locating the query point in the arrangement. This can be done in $O(\log n)$
query time with $O(n^d)$ preprocessing.

The dual version of simplex range searching takes n hyperplanes as input,
and a query is a set of $d + 1$ labeled points $(p_0,\rho_0),\ldots,(p_d,\rho_d)$, where ρ_i
stands for "above," "through," or "below." The answer to the query is
obtained by counting the hyperplanes (or summing up their weights, if that
is how the problem is phrased) that satisfy each relation ρ_i with respect to
p_i, for $i = 0,\ldots,d$. We discuss only the counting case. We shall not use
subtraction, however, so that our method will also work in the case where
weights are chosen from an additive semigroup.

It helps to think of this problem in more general terms by considering
queries of the form $(p_0,\rho_0),\ldots,(p_k,\rho_k)$, for any k, and constructing a data
structure of *type* $(\rho_0,\ldots,\rho_k)$ to handle them. The idea is then to prepare
data structures of all possible types $(\rho_0,\ldots,\rho_k)$ to accommodate (in the
dual) for all possible query simplices (in the primal). If $k = 0$, the problem
is halfspace range searching and, as we just mentioned, a query can be
answered in $O(\log n)$ time using $O(n^d)$ storage.

Assume now that $k > 1$. We begin with a $(1/r)$-cutting $\mathcal{C}$, where $r = n^\varepsilon$,
for some small fixed $\varepsilon > 0$. Without loss of generality, suppose that ρ_0 says
"below." Then, for each cell c in $\mathcal{C}$, we identify the subset B_c of hyperplanes
passing below the interior of the cell. For each cell c, we iterate on this

process with respect to the set of hyperplanes that intersect the interior of c, using the same value of r at each iteration. This process can be modeled by a rooted tree of degree $O(r^d)$. At each node v of depth k, we have:

- A set H_v of hyperplanes of size at most n/r^k;
- A $(1/r)$-cutting for H_v, along with, for each cell c, (i) the subset B_c of hyperplanes in H_v passing below the interior of c, and (ii) a data structure of type $(\rho_1, \ldots, \rho_k)$ built recursively with respect to B_c.

To answer a query, we perform point location in the $(1/r)$-cutting stored at the root and thus identify the cell c enclosing p_0. Next, we recursively answer the query $(p_1, \rho_1), \ldots, (p_k, \rho_k)$ with respect to B_c. Finally, we move to the child v associated with c, and we recursively answer the original query $(p_0, \rho_0), \ldots, (p_k, \rho_k)$ with respect to H_v. The correctness of the algorithm is obvious.[16]

Because $r = n^\varepsilon$, the depth of the tree and, for that matter, of any of the auxiliary trees is constant; so the query time is at most proportional to the maximum time spent at any given node. Point location is the dominant cost at each node, so the query time is, indeed, $O(\log n)$, as desired.

The storage $S_k(n)$ follows the recurrence: $S_0(n) = O(n^d)$, $S_k(n) = O(1)$ for $n = O(1)$, and

$$S_k(n) = O(r^d) \cdot (S_k(n/r) + S_{k-1}(n)).$$

This gives $S_k(n) = O(n^{d+\varepsilon'})$, for some ε' tending to 0 as $\varepsilon \to 0$.

Theorem 5.13 *Given n points in $\mathbf{R}^d$, there exists a data structure of size $O(n^{d+\varepsilon})$, for any fixed $\varepsilon > 0$, that allows simplex range searching to be done in $O(\log n)$ time per query.*

Space-Time Tradeoffs

Suppose that $m \geq n$ units of storage are available. Plugging together the two previous algorithms provides a near-optimal solution to simplex range searching. The idea is to build a partition tree based on simplicial partitions but to prune the bottom of the tree. Specifically, as soon as the subproblem at a node involves fewer than n_0 points, for some parameter $n_0(n, m)$, we switch to the data structure of Theorem 5.13. The pruned

---

[16]We assume that the query point does not lie on the boundary of a cell of $\mathcal{C}$: To handle this particular case requires only minor modifications to the algorithm.

tree has only $O(n/n_0)$ leaves, each of which requires storage $O(n_0^{d+\varepsilon})$. The total amount of storage is therefore $O(nn_0^{d-1+\varepsilon})$, which is less than m if we set $n_0 = c(m/n)^{1/(d-1+\varepsilon)}$ for some small enough constant $c > 0$. The query time is equal to $t_1 t_2$, where t_1 (resp. t_2) is the time indicated in Theorem 5.9 (resp. Theorem 5.13) for a problem of size $O(n/n_0)$ (resp. n_0). This shows that the query time is bounded by

$$\left(\frac{n}{n_0}\right)^{1-1/d+\varepsilon} \log n_0,$$

which is $O(n^{1+\varepsilon'}/m^{1/d})$ for some ε' tending to 0 as $\varepsilon \to 0$. This completes the proof of Theorem 5.7 (page 215). $\square$

5.4 Bibliographical Notes

Section 5.1:   The optimal ε-cutting construction of Theorem 5.1 (page 206) is due to Chazelle [69]. A more complex argument was used earlier by Chazelle and Friedman to [78] to prove the same result, but with a higher preprocessing time. The idea of partitioning space into sparsely intersected simplices goes back to Clarkson [94] and Haussler and Welzl [160]. The term ε-cutting, and the definition given here, were introduced by Matoušek [211]. Near-optimal ε-cutting constructions were given by Agarwal [4, 5] (in two dimensions) and Matoušek [210, 211, 216] (in any dimension), and played an influential role in the development of the subject.

Section 5.2:   Point location and incidence detection are applications of cuttings discussed by Chazelle in [69], where Theorem 5.4 (page 212) and Theorem 5.5 (page 213) are established. Other applications not discussed here are given by Agarwal in [5]. The algorithm for Hopcroft's problem (Theorem 5.6, page 214) is also taken from [69]. The best bound currently known, due to Matoušek [213], is $n^{4/3} 2^{O(\log^* n)}$. Weaker bounds (though with the right exponent $4/3$) were obtained earlier by Edelsbrunner et al. [119] and Agarwal [4]. Our passing reference to a natural lower bound on Hopcroft's problem is substantiated by Erickson [125], who has established a lower bound of $\Omega(n^{4/3})$ in a restricted but fairly realistic model of computation.

Section 5.3:   The idea of using a partition tree for triangle range searching is due to Willard [324]. Edelsbrunner and Welzl [123] improved on the choice of partitioning lines to produce the conjugation tree. Generalizations to higher dimensions were also found, and many papers were written on the subject; see surveys [6, 214]. A breakthrough was achieved by Haussler and Welzl [160] in a highly influential paper that introduced ε-nets. They provided a linear-storage solution to simplex range searching in $\mathbf{R}^d$ with query time $O(n^\alpha)$, for $\alpha = 1 - \frac{1}{d(d-1)+1} + \varepsilon$ and any fixed $\varepsilon > 0$. In the planar case, $d = 2$, the query time was reduced to nearly optimal by Welzl in [319] by using low-cutting spanning paths. The construction was improved by Chazelle and Welzl [86]. The first quasi-optimal solution in any fixed dimension, including the space-time tradeoff of Theorem 5.7 (page 215), was given by Chazelle, Sharir, and Welzl [85]. The construction of partition trees based on simplicial partitions is due to Matoušek [212], as is Theorem 5.12 (page 223). The best space-time tradeoff currently known is due to Matoušek [213]. It represents a slight improvement over Theorem 5.7: Instead of $O(n^{1+\varepsilon}/m^{1/d})$, the bound is $O(n(\log m/n)^{d+1}/m^{1/d})$, for m/n large enough.

6

Complexity Lower Bounds

he discrepancy method is not just about designing algorithms. Techniques used for showing the necessity of disorder in complex structures can sometimes be recycled to prove the computational difficulty of solving certain problems. In this case, our aim is to translate high discrepancy into high complexity. To add a touch of irony, we will occasionally run into lower bound arguments that need highly uniform structures as auxiliary devices. So, expect low discrepancy to be part of the picture as well.

The arguments developed in this chapter are almost exclusively algebraic or Ramsey-type. The problems that they are trying to solve arise in the context of arithmetic circuits and geometric databases. They are all variations on the same "matrix complexity" theme: Let A be an n-by-n matrix with 0/1 elements. The goal is to assemble the matrix A by forming a sequence of column vectors $U_1, \ldots, U_s \in \mathbf{Z}^n$, where $s \geq n$ and

- $(U_1, \ldots, U_n)$ is the n-by-n identity matrix;
- $A = (U_{s-n+1}, \ldots, U_s)$;
- for any $i = n+1, \ldots, s$, there exist $j, k < i$ and $\alpha_i, \beta_i \in \mathbf{Z}$ such that $U_i = \alpha_i U_j + \beta_i U_k$.

The minimum length s of any sequence that satisfies these three conditions is called the *complexity* of A. We leave the following statements as warm-up exercises: All 0/1 matrices have complexity $O(n^2)$. A matrix does not have to be sparse to be trivial. For example, a triangular matrix with ones below

228

the diagonal and zeros elsewhere has linear complexity. A random $0/1$ matrix, on the other hand, has complexity $\Omega(n^2/\log n)$. Hint: Compare the number of possible sequences mod 2 and the number of matrices. What about classical matrices such as the $0/1$ matrix derived from Hadamard by adding 1 to its elements and dividing by 2 (Appendix B.1)? What about the complexity of geometric matrices, ie, incidence matrices of geometric set systems? Also, why restrict ourselves to coefficients in $\mathbf{Z}$? What about the discrete Fourier transform? Is the FFT optimal? Unfortunately, current technology is not up to the task. At the time of this writing, answers are unavailable.

So, in typical fashion, we restrict the model. First, we bound the size of the coefficients; then we distinguish between group and semigroup structures. In the *monotone* (resp. *nonmonotone*) model, all α_i, β_i's are confined to $\{0,1\}$ (resp. $\{-1,0,1\}$). Tools from extremal graph theory are powerful enough to elucidate most questions in the monotone case. Nonmonotonicity presents us with considerable difficulties, however, and more sophisticated machinery must be brought to bear. This tune has a familiar ring to it: The monotone vs. nonmonotone question echoes throughout contemporary complexity theory.

The problem of assembling matrices one column at a time raises intriguing complexity questions, and the simplicity of its statement is compelling. To make a good story even better, it has been the main vehicle for resolving the complexity of multidimensional searching. In that context, the problem is phrased in its dual, equivalent version, where the cast of characters consists of linear forms rather than vectors. So, purely for historical reasons, instead of linear combinations of column vectors, we speak of gates of linear circuits. The distinction between monotone and nonmonotone is the same. In fact, as will soon be obvious, the vector and circuit models are identical; only the language is different.

A *range searching problem* is specified by a set of n points in $\mathbf{R}^d$. Each point p_i is assigned a real number x_i, called its *weight*. A query is given by a *range* (ie, a region of space such as a box, a simplex, or a ball) and the answer is the sum of the weights of the points within the range. This is on-line range searching. There is also the off-line version, where all the queries are given ahead of time and are to be processed in batched mode. Figure 6.1 illustrates off-line triangle range searching. The input consists of five triangles R_i, five points, and five real numbers x_i (the weights of the points). The output is the sequence of numbers $(x_1, x_2 + x_3, x_3 + x_5, x_2 + x_4, x_5)$.

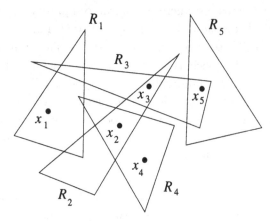

Fig. 6.1. Triangle range searching: The output is $(x_1, x_2+x_3, x_3+x_5, x_2+x_4, x_5)$.

The problem is neatly captured by the incidence matrix

$$A = \begin{pmatrix} 1 & 0 & 0 & 0 & 0 \\ 0 & 1 & 1 & 0 & 0 \\ 0 & 0 & 1 & 0 & 1 \\ 0 & 1 & 0 & 1 & 0 \\ 0 & 0 & 0 & 0 & 1 \end{pmatrix}.$$

Given $x = (x_1, \ldots, x_n)^T$, the task is to compute Ax. For that purpose, we use a circuit to encode A, whose input is x and output Ax. Roughly speaking, the nonmonotone model allows both addition and subtraction gates, while the monotone version allows only the former. This is the distinction in circuit complexity theory between monotone and nonmonotone computation. Nonmonotonicity is much harder to tackle because the computation can create unforeseen cancellations and take advantage of hidden symmetries more effectively. In fact, it is often hard to tell what a nonmonotone circuit *cannot* do. The monotone case raises mostly combinatorial questions related to the presence of complete subgraphs in certain geometric graphs. Quite differently, the nonmonotone case involves mainly algebraic issues, which we approach by investigating the spectrum of the corresponding incidence matrices.[1]

Monotone models are often derided as informing us less about complexity

---

[1] The term *spectrum* refers here to the set of singular values of the matrix A, ie, the square roots of the eigenvalues of $A^T A$.

theory than about our inability to prove interesting theorems. Let's face it: More often than not, the *only* reason we consider monotone models is to be able to prove something. This chapter is a happy exception. If anything, the natural habitat of range searching is a monotone model. This is because one might want to perform operations such as *max* that do not have inverses. Of course, to restrict the problem statement to operations with inverses is a logical step, but the reason is not that monotonicity is something to run away from, at least not in this case.

Discrepancy theory plays a major role in this chapter on lower bounds. In the nonmonotone case, high discrepancy implies high complexity. Why the ability of coloring evenly has anything to do with range searching is one of the wonders to behold in this chapter. The link between discrepancy and complexity is the spectrum of the incidence matrices. Interestingly, we also use data structures to establish upper bounds on the singular values of such matrices. This is an interesting twist where the theory of algorithms is used to prove purely combinatorial results. A fact that seems to have gone unnoticed is that much of the range searching literature (quad-trees, *kd*-trees, interval trees, segment trees, range trees, etc) is implicitly concerned with eigenvalue problems.

In the monotone case, the problem usually comes down to bounding the size of complete bipartite subgraphs. In the case of simplex range searching, we must solve a variant of Heilbronn's problem (place n points in the unit square forming no triangle of small area), a problem with a heavy discrepancy scent. We also use our old friends from Chapter 2, low-discrepancy point sets for boxes. This chapter introduces many different ideas. I have tried to keep the sections mostly independent of one another, with each one introducing at least one new idea.

A point of terminology: The incidence matrix $A = (A_{ij})$ of a set system defined by points and regions of space is called a *geometric matrix*. In our discussion, we restrict ourselves to square matrices defined by n points, $p_1, \ldots, p_n$, and n regions $R_1, \ldots, R_n \subseteq \mathbf{R}^d$, where

$$A_{ij} = \begin{cases} 1 & \text{if } p_j \in R_i, \\ 0 & \text{else.} \end{cases}$$

If the regions are boxes (resp. lines, triangles, simplices) we speak of box (resp. line, triangle, and simplex) matrices. In this chapter all boxes are understood to be axis-parallel.

- In §6.1 we investigate off-line range searching in the nonmonotone case. We prove a general *spectral lemma* relating the complexity of a problem instance to the spectrum of the incidence matrix of the set system associated with it. This gives us an opportunity to introduce a general entropy-based method of independent interest. We apply the spectral lemma to range searching with respect to boxes and triangles, successively. This entails charting out the spectrum of two important families of geometric matrices. We also use a second-moment variant of the spectral lemma, called the *trace lemma*, to investigate range searching with respect to lines and boxes in higher dimension. This section ties in nicely with the previous chapters on discrepancy theory by further exploring the algebraic properties of geometric set systems.

- In §6.2 we investigate the use of data structures to obtain upper bounds on the singular values of some geometric matrices. It must be understood that we are not concerned here with numerical computations of eigenvalues but with asymptotic bounds as a function of the size of the matrices, a much more difficult problem.

- We revisit the same range searching problems in §6.3, but this time in the monotone model. In other words, we investigate the monotone circuit complexity of box, line, and simplex matrices. We are able to prove much stronger results, in fact nearly optimal lower bounds. The main thrust of the proof techniques lies in extremal graph theory. We build geometric matrices containing no large rectangles of ones, ie, bipartite graphs with no large complete subgraphs. Again, discrepancy theory is used extensively through such devices as Halton-Hammersley point sets.

- In §6.4 we investigate space-time tradeoffs for on-line range searching in the monotone model. We consider the case of queries given as boxes and as simplices; these problems are commonly referred to as orthogonal range searching and simplex range searching, respectively.

$$x = (\,x_1\,,\dots,x_{\mathrm{n}}\,)^T$$

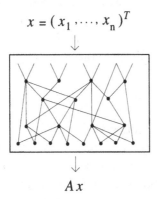

$$Ax$$

Fig. 6.2. The gates of an arithmetic circuit take two numbers as input and add or subtract them.

6.1 Arithmetic Circuits

Let A be an n-by-n $0/1$ matrix. We consider the problem of computing the vector $y = Ax$, where $x = (x_1, \dots, x_n)^T$, for any given $x \in \mathbf{R}^n$ and fixed A. The model of computation is a *linear arithmetic circuit* (Fig. 6.2). This is a directed acyclic graph whose nodes, the *gates*, have indegree 2. Each gate takes two numbers a, b as input and outputs $a + b$ or $a - b$. The size of the circuit is the number of edges. The *arithmetic complexity* of the matrix A is the size of the smallest circuit for computing $x \mapsto Ax$. Note that the circuit depends on A but *not* on x; it must work for any input $x \in \mathbf{R}^n$. The model is realistic inasmuch as it can describe all of the known algorithms for range searching.[2]

It is easy to prove that the size of the circuit is $\Omega(\log |\det A|)$. This is the *Morgenstern bound*. A rough proof sketch goes as follows: Each gate can be viewed as adding or subtracting two linear forms over the variables $x_1, \dots, x_n$, or, equivalently, adding or subtracting two vectors in $\mathbf{R}^n$. Regard the circuit as a directed acyclic graph and order the gates topologically. Going through the gates in sequence, we easily see that no gate can more than double the value of the biggest determinant formed by

---

[2]This excludes special range searching problems where one wishes to detect which ranges are empty, or report the points one by one. Ad hoc solutions sometimes exist for these problems. One could also allow more exotic operations at the gates (eg, multiplication, square roots). At present, nothing is known about such extensions.

any subset of n of the vectors computed so far. The lower bound follows immediately. Geometrically, this is saying that the circuit can "blow up" the unit sphere $\{ x : \|x\|_2 = 1 \}$ into the ellipsoid $\{ Ax : \|x\|_2 = 1 \}$ only so fast. The Morgenstern bound is useful for some highly structured linear maps, such as the discrete Fourier transform, but in general it suffers from serious weaknesses:

1. The matrix A needs to be of full rank.

2. To find good asymptotics for determinants can be exceedingly difficult. Recall that the Riemann hypothesis can be formulated as a bound on the determinant of the Redheffer matrix (page 32).

3. The Morgenstern bound can be expressed as $\Omega(n \log(\prod \lambda_i)^{1/n})$, where $\lambda_1, \ldots, \lambda_n$ are the eigenvalues of $A^T A$. In other words, it is a function of the *geometric mean* of the eigenvalues. There are many means to choose from: for lower bound purposes, the geometric mean is among the worst. A bound involving the arithmetic mean $\frac{1}{n} \sum \lambda_i$ would be much more favorable. Not only because it is never smaller than its geometric counterpart, but because the sum of the eigenvalues is the trace of $A^T A$, a much easier invariant to deal with than the determinant.

We establish a *trace lemma*, which shows that the complexity of $x \mapsto Ax$ is at least on the order of $n \log(\operatorname{tr} M/n - \varepsilon\sqrt{\operatorname{tr} M^2/n})$, where $M = A^T A$ and $\varepsilon > 0$ is any constant. The advantage of this formulation is that the trace of M (resp. M^2) has a simple combinatorial interpretation: it counts the number of ones (resp. rectangles of ones) in A. The lower bound follows from a more general result, called the *spectral lemma*, which bounds the circuit complexity by $\Omega(n \log \lambda_k)$, for any $k = \Theta(n)$.

To prove lower bounds for range searching, then we need either to estimate the traces of M and M^2 or, if that does not work, to find asymptotic estimates of the midrange of the spectrum of A. In both cases, tools from discrepancy theory come into play. Entropy considerations about the circuit close the pipeline from discrepancy theory to complexity lower bounds (Fig. 6.3).

The spectral lemma has the added advantage of accommodating up to a linear number of *help gates*. (Something the Morgenstern bound cannot do.) A *help gate* takes two inputs a, b and outputs $f(a, b) \in \mathbf{R}$, where f is *any* function. Help gates need not all use the same functions. One should not underestimate the power of help gates. Any Ax can be computed entirely with $2n - 1$ help gates. By using $n - 1$ gates arranged at the nodes

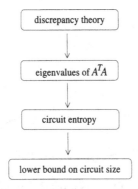

Fig. 6.3. The discrepancy method pipeline for bounding linear circuits.

of a tree we encode the vector $(x_1, \ldots, x_n)$ into a number. Then, with another n gates, we compute each coordinate of Ax (Fig. 6.4).

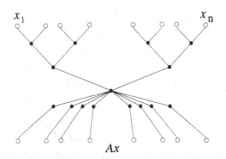

Fig. 6.4. The middle help gate encodes the vector x. The help gates below extract the relevant subset of x_i's and sum them up. They take only one input each but, to conform with the definition, a dummy one can always be added.

Obviously the Morgenstern bound cannot accommodate any help gates, since even a single one can be used to blow up the unit sphere to any size. On the other hand, the spectral lemma can be refined to show that any circuit for computing Ax has size $\Omega(k - 2h) \log \lambda_k$. where h is the number of help gates.

Entropy-Increasing Computation

We build a probability distribution on the input vector x and we monitor the entropy of the set of variables computed by the circuit. Assuming

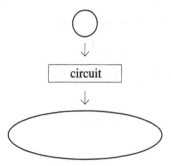

Fig. 6.5. The circuit turns a beach ball into a football.

that A is nonsingular, there is no loss of entropy, of course, and it is hard to argue anything interesting. So, we build a device through which, to a certain observer, the computation becomes entropy-increasing. Suppose that the computation of the circuit is visible at each gate, but that the observer is required to wear a certain type of glasses. Sadly for him, these glasses blur his vision, so that instead of showing the real number z, it shows $C\lfloor z/C \rfloor$ for a large fixed parameter C. To our observer's blurred eyes, the input vector x appears to have much less entropy than it has (because of the truncation). On the other hand, it can be shown that truncating has very little effect on the entropy of the output variables. In other words, the observer sees little entropy coming in and a lot of entropy getting out (note that for the observer, the circuit behaves nondeterministically, so there is no contradiction).

Intuitively, the reason why entropy is coming back into the system is that, if the map A has large singular values, it should stretch the input vector x, so that the various values of x that collide (with the glasses on) no longer collide after being transformed by A. From the observer's viewpoint, therefore, the circuit appears to reinject entropy into the system. The second step of the argument is to show that any gate can reinject at most a small amount of entropy, and so many of them are needed.

Blurring and computing are dual to each other. Their interaction manifests itself through losses and gains of information. So, it is intuitively natural to use entropy as the main monitoring parameter. It is not the only possible choice. It is the best, however, because the entropy function satisfies all sorts of identities and inequalites, and proves a wonderful vehicle for simplifying otherwise hopelessly messy calculations.

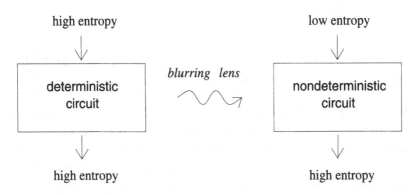

Fig. 6.6. A blurring lens makes the circuit appear nondeterministic. The computation becomes entropy-increasing.

The Spectral Lemma

The result below forms the main focus of this section. For those who view help gates as mere frills, we also state a weaker, bare-bones variant. Finally, we mention a useful corollary, which requires only a second-moment tail estimate on the distribution of eigenvalues, something we often can get at by purely combinatorial means.

Lemma 6.1 (THE SPECTRAL LEMMA) *Given an n-by-n 0/1 matrix A, any circuit for computing Ax has size at least $\Omega(k - 2h)\log \lambda_k$, for any $1 \leq k \leq n$, where h is the number of help gates and λ_k is the k-th largest eigenvalue of $A^T A$.*

Lemma 6.2 *Any circuit for computing Ax has size at least $\Omega(n \log \lambda_k)$, for any $k = \Theta(n)$.*

Lemma 6.3 (THE TRACE LEMMA) *Any circuit for computing Ax has size*

$$\Omega_\varepsilon \left(n \log \left(\operatorname{tr} M/n - \varepsilon \sqrt{\operatorname{tr} M^2/n} \right) \right),$$

where $M = A^T A$ and $\varepsilon > 0$ is an arbitrarily small constant.

Proof: An immediate corollary of the spectral lemma, together with Lemma 1.19 (page 34), which states that

$$\lambda_k \geq \operatorname{tr} M/n - \varepsilon \sqrt{\operatorname{tr} M^2/n},$$

for some $k = \Omega_\varepsilon(n)$. $\square$

Proof of the Spectral Lemma: To begin with, we specify a probability distribution on the input vectors $x \in \mathbf{R}^n$. We fix an arbitrary $1 \le k \le n$ and prove the lower bound of the spectral lemma for that particular k. Let K be the invariant subspace spanned by the eigenvectors for M associated with $\lambda_1, \ldots, \lambda_k$. (For convenience, we ensure that K is of dimension exactly k by dropping some of the eigenvectors for λ_k, in case of multiplicity.) Let $B_n(p, r)$ denote the Euclidean n-ball of radius r centered at p, and let $V_n(r)$ be its volume. Fixing a large parameter R, we consider the collection L of centers of the cubes of the form $\mathbf{Z}^n + [0, 1]^n$ that intersect $K \cap B_n(O, R)$. The underlying input distribution is defined by choosing x at random uniformly in L.

We show that it is possible to bucket the random vector Ax into a suitably large grid without losing too much entropy. For convenience, we scale the matrix to make the bucketing particularly simple. We bring the eigenvalue of interest into the interval $[1, 2]$; for this reason we must now consider matrices with rational elements and not simply those of the $0/1$ type.

Two pieces of notation: Given a random variable y, we let $H(y)$ denote the entropy of y (see Appendix A.3 for definitions and standard properties). Also, given $z = (z_1, \ldots, z_n)^T \in \mathbf{R}^n$, we use the shorthand $\lfloor z \rfloor$ to denote the vector $(\lfloor z_1 \rfloor, \ldots, \lfloor z_n \rfloor)^T$.

Lemma 6.4 *Let B denote an n-by-n matrix with rational elements and let $\beta_1 \ge \cdots \ge \beta_n$ be the eigenvalues of $B^T B$. If $1 \le \beta_k \le 2$, then*

$$H(\lfloor Bx \rfloor) \ge H(x) - \log V_n(5\sqrt{n}).$$

Proof: Note that the lemma refers to the parameter k that was fixed earlier. We must show that not too many integer points x can collide into the same unit cube via the transformation B. Let $x, x' \in L$ be such that $\lfloor Bx \rfloor = \lfloor Bx' \rfloor$. Write x as the direct sum $x_0 + u$, where $x_0 \in K$, $u \in K^\perp$, and do the same with x', ie, $x' = x_0' + u'$, where $x_0' \in K$, $u' \in K^\perp$. By abuse of terminology, we also use K to designate the space spanned by the eigenvectors associated with $\beta_1, \ldots, \beta_k$. Observe that

$$\|u - u'\|_2 \le \|u\|_2 + \|u'\|_2 \le \sqrt{n}.$$

By the variational characterization of eigenvalues,

$$\|B(x_0 - x_0')\|_2 \ge \sqrt{\beta_k}\, \|x_0 - x_0'\|_2.$$

Because $K^\perp$ contains $u - u'$ and is spanned by eigenvectors corresponding to $\beta_j \leq \beta_k$,

$$\|B(u - u')\|_2 \leq \sqrt{\beta_k}\,\|u - u'\|_2.$$

Because $\lfloor Bx \rfloor = \lfloor Bx' \rfloor$, we have $\|B(x - x')\|_2 \leq \sqrt{n}$, and hence (for $1 \leq \beta_k \leq 2$)

$$
\begin{aligned}
\|x - x'\|_2 &\leq \|x_0 - x_0'\|_2 + \|u - u'\|_2 \\
&\leq \|B(x_0 - x_0')\|_2 + \|u - u'\|_2 \\
&\leq \|B(x - x')\|_2 + \|B(u - u')\|_2 + \|u - u'\|_2 \\
&\leq \sqrt{n} + 3\|u - u'\|_2 \leq 4\sqrt{n}.
\end{aligned}
$$

Thus, the preimage of a fixed $z \in \mathbf{R}^n$ under $x \in L \mapsto \lfloor Bx \rfloor$ lies entirely in a ball $B_n(x, 4\sqrt{n})$, for some $x \in L$. It follows that the uniform distribution within that preimage has entropy at most $\log V_n(5\sqrt{n})$. Indeed, the unit cubes centered at the points of the preimage all lie in $B_n(x, 5\sqrt{n})$. Standard identities on the entropy of joint distributions (see Appendix A.3), namely,

$$H(x) = H(x, \lfloor Bx \rfloor) = H(\lfloor Bx \rfloor) + H(x \mid \lfloor Bx \rfloor),$$

complete the proof. $\square$

Let $z = (z_1, \ldots, z_s)^T$ be the vector of $\mathbf{R}^n$ whose coordinates are the intermediate variables computed by the nodes; we append the input variables at the beginning of the list ($z_j = x_j$, for $1 \leq j \leq n$) and the output variables at the end ($z_{s-n+j} = y_j$, for $1 \leq j \leq n$). We assume that the list corresponds to a topological ordering of the circuit's nodes, meaning that, for any $j > n$,

$$z_j = \alpha_j z_{f(j)} + \beta_j z_{g(j)},$$

where $f(j) \leq g(j) < j$ and $\alpha_j, \beta_j \in \{-1, 0, 1\}$. If the node corresponds to a help gate, then z_j is an arbitrary real function of $z_{f(j)}$ and $z_{g(j)}$. For convenience we use the notation

$$\mu = \lfloor \sqrt{\lambda_k} \rfloor.$$

The input x to the circuit is chosen so that $\tilde{x} = \mu x$ is a random variable uniformly distributed in L. We shall assume that λ_k is large enough, something the spectral lemma certainly allows us to do.

For our unfortunate observer wearing blurring glasses, bucketing the input x appears to cause a substantial loss of entropy (Lemma 6.5). On the other hand, bucketing the entire vector z causes almost no entropy loss (Lemma 6.6). Consequently, the gates of the circuit must appear to the observer as restoring lost entropy. Lemma 6.7 shows that a single gate can contribute only that much to the restoring process, and therefore many gates are required to do the job.

Lemma 6.5

$$H(\lfloor x \rfloor) \le 2n + \log V_k(R/\mu).$$

Proof: Let C be the set of cubes of the form $(\mu \mathbf{Z})^n + [0, \mu]^n$. Any cube of C that contains a point of L contains the entire unit cube centered at that point, and so it intersects $K \cap B_n(O, R)$. The number of such cubes is easily shown to be bounded by $3^n V_k(R/\mu)$, so that

$$
\begin{aligned}
H(\lfloor x \rfloor) &= H(\lfloor \tilde{x}_1/\mu \rfloor, \ldots, \lfloor \tilde{x}_n/\mu \rfloor) \\
&\le 2n + \log V_k(R/\mu).
\end{aligned}
$$

□

Lemma 6.6

$$H(\lfloor z \rfloor) \ge \log V_k(R) - \log V_k(\sqrt{n}) - \log V_n(5\sqrt{n}).$$

Proof: The unit-side length cubes whose centers are in L cover the ball $B_k(O, R)$ embedded in K. The intersection of K with each of these cubes fits into a k-ball of radius $\sqrt{n}/2$, so $|L| \ge V_k(R)/V_k(\sqrt{n})$, and hence

$$H(\tilde{x}) = \log |L| \ge \log V_k(R) - \log V_k(\sqrt{n}).$$

Because A is a linear map, the circuit outputs $B\tilde{x}$, where

$$B \overset{\text{def}}{=} \left(\frac{1}{\mu}\right) A$$

obviously satisfies the conditions of Lemma 6.4. Thus,

$$H(\lfloor z \rfloor) \ge H(\lfloor B\tilde{x} \rfloor) \ge H(\tilde{x}) - \log V_n(5\sqrt{n}),$$

which proves the lemma. □

As the circuit computes Ax, the vector $\lfloor z \rfloor$ gets modified by gaining one more coordinate at each step. We argue that its entropy cannot increase too much at any step. We distinguish between regular (ie, nonhelp) gates, where the increase is shown not to exceed 3, and help gates, where the entropy increase can be more substantial.

Without help gates, it is clear that the output of any gate is a variable that can be expressed as a linear form over the input variables $x_1, \ldots, x_n$.

Let us call the output of any help gate a help variable. It is now equally clear that any gate variable can be expressed uniquely as a linear form over the input and help variables. Note that no attempt is made to express help variables in terms of other variables. Let us break down the expression for z_j by writing $z_j = z'_j + z''_j$, where z'_j (resp. z''_j) denotes the linear form over the input (resp. help) variables. Our observer will monitor the information contents of the pairs

$$\mathbf{z}_j \overset{\text{def}}{=} (\lfloor z'_j \rfloor, z''_j).$$

We denote by $\mathbf{z}$ the vector $(\mathbf{z}_1, \ldots, \mathbf{z}_s)$. The relevant conditional entropies can be bounded as follows:

Lemma 6.7 *Given* $n < j \leq s$,

$$H(\mathbf{z}_j \mid \mathbf{z}_{f(j)}, \mathbf{z}_{g(j)}) \leq \begin{cases} 3 & \text{for any regular variable } z_j, \\ 2(\log \mu + 1) & \text{for any help variable } z_j. \end{cases}$$

Proof: Suppose that z_j is a regular variable. Obviously,

$$H(z''_j \mid z''_{f(j)}, z''_{g(j)}) = 0, \tag{6.1}$$

because z''_j is completely specified by $\mathbf{z}_{f(j)}$ and $\mathbf{z}_{g(j)}$. To deal with z'_j, we use simple inequalities about entropy (Appendix A.3), such as subadditivity and $H(A \mid B) \leq H(A \mid C) + H(C \mid B)$:

$$H(\lfloor z'_j \rfloor \mid \lfloor z'_{f(j)} \rfloor, \lfloor z'_{g(j)} \rfloor) \leq H(\lfloor z'_j \rfloor \mid \lfloor \alpha_j z'_{f(j)} \rfloor, \lfloor \beta_j z'_{g(j)} \rfloor)$$

$$+ H(\lfloor \alpha_j z'_{f(j)} \rfloor \mid \lfloor z'_{f(j)} \rfloor) + H(\lfloor \beta_j z'_{g(j)} \rfloor \mid \lfloor z'_{g(j)} \rfloor).$$

Given two random variables ξ, ξ' arbitrarily distributed in $\mathbf{R}$,

$$H(\lfloor \xi + \xi' \rfloor \mid \lfloor \xi \rfloor, \lfloor \xi' \rfloor) \leq 1.$$

Indeed, the only information missing is a one-bit carry. Similarly, given a real random variable ξ, we have $H(\lfloor -\xi \rfloor \mid \lfloor \xi \rfloor) \leq 1$. Since $z_j = z_{f(j)} \pm z_{g(j)}$, it follows that

$$H(\lfloor z'_j \rfloor \mid \lfloor z'_{f(j)} \rfloor, \lfloor z'_{g(j)} \rfloor) \leq 3.$$

By (6.1), we have

$$H(\mathbf{z}_j \mid \mathbf{z}_{f(j)}, \mathbf{z}_{g(j)}) \leq H(\lfloor z'_j \rfloor \mid \lfloor z'_{f(j)} \rfloor, \lfloor z'_{g(j)} \rfloor),$$

which concludes the first case.

If z_j is a help variable, then $\mathbf{z}_j = z_j''$, where z_j'' is an arbitrary function of $z_{f(j)}$ and $z_{g(j)}$. Regarding $z_{f(j)} = z_{f(j)}' + z_{f(j)}''$, the only information we have at our disposal is $\mathbf{z}_{f(j)} = (\lfloor z_{f(j)}' \rfloor, z_{f(j)}'')$. There is no loss of information in the second component, $z_{f(j)}''$. The same is not true of the first one, however. The key observation is that $z_{f(j)}'$ is a linear form over $x_1, \ldots, x_n$ with integer coefficients. Thus, since $2\mu x_i$ is itself integral, so is $2\mu z_{f(j)}'$. It follows that the fractional part of $z_{f(j)}'$ can be one of only 2μ possible values, and hence $H(z_{f(j)}' \mid \lfloor z_{f(j)}' \rfloor) \leq \log \mu + 1$. By using standard properties of the entropy function, in particular, subadditivity and the nonincreasing effect of conditioning, we easily derive (skipping a few intermediate steps)

$$
\begin{aligned}
H(\mathbf{z}_j \mid \mathbf{z}_{f(j)}, \mathbf{z}_{g(j)}) &\leq H((z_{f(j)}, z_{g(j)} \mid \mathbf{z}_{f(j)}, \mathbf{z}_{g(j)}) \\
&\leq H(z_{f(j)}' \mid \lfloor z_{f(j)}' \rfloor) + H(z_{g(j)}' \mid \lfloor z_{g(j)}' \rfloor) \\
&\leq 2(\log \mu + 1).
\end{aligned}
$$

$\square$

By Lemmas 6.5 and 6.7 and standard facts about entropy,

$$
\begin{aligned}
H(\mathbf{z}) &= H(\lfloor x \rfloor) + \sum_{n+1 \leq j \leq s} H(\mathbf{z}_j \mid \mathbf{z}_1, \ldots, \mathbf{z}_{j-1}) \\
&\leq H(\lfloor x \rfloor) + \sum_{n+1 \leq j \leq s} H(\mathbf{z}_j \mid \mathbf{z}_{f(j)}, \mathbf{z}_{g(j)}) \\
&\leq 2n + \log V_k(R/\mu) + 3(s - n - h) + 2h(\log \mu + 1) \\
&\leq 3s - n + \log V_k(R/\mu) + 2h \log \mu,
\end{aligned}
$$

and therefore

$$
\begin{aligned}
H(\lfloor z \rfloor) &\leq H(\mathbf{z}, \lfloor z \rfloor) = H(\mathbf{z}) + H(\lfloor z \rfloor \mid \mathbf{z}) \\
&\leq H(\mathbf{z}) + \sum_{j=n+1}^{s} H(\lfloor z_j' + z_j'' \rfloor \mid \lfloor z_j' \rfloor, z_j'') \\
&\leq 4s + 2h \log \mu + \log V_k(R/\mu).
\end{aligned}
$$

Bringing the lower bound of Lemma 6.6 to bear, we derive

$$
\begin{aligned}
4s \geq\ &-2h \log \mu - \log V_k(R/\mu) + \log V_k(R) - \log V_k(\sqrt{n}) \\
&- \log V_n(5\sqrt{n}).
\end{aligned}
$$

Using for $V_k(r)$ the approximation

$$
\frac{1}{\sqrt{\pi k}} \left(\frac{2e\pi}{k} \right)^{k/2} r^k,
$$

we find that

$$4s \geq -2h \log \mu + k \log \mu - \log V_k(\sqrt{n}) - \log V_n(5\sqrt{n}).$$

The last two terms add up to $O(n)$ in absolute value. Since λ_k is large enough, it follows that

$$s \geq \tfrac{1}{8}(k - 2h) \log \lambda_k - O(n).$$

The size of the circuit is the number of edges, which is at least $n + s$. By scaling down the constant c in the spectral lemma accordingly, its proof is now complete (page 237). $\square$

A Wavelet Argument for Box Matrices

Our first application of the spectral lemma is to establish a lower bound for off-line range searching with respect to axis-parallel boxes. By using a wavelet approach reminiscent of something we did in Chapter 3, we provide an eloquent example of the discrepancy method in action.

Theorem 6.8 *There are n-by-n box matrices of arithmetic complexity $\Omega(n \log \log n)$. This remains true in the presence of up to $n/9$ help gates.*

We build a large $N \times n$ set system B, where $N = n^{O(1)}$, from which we extract an $n \times n$ set system A that satisfies the lower bound. (This is *not* the same B as the one used in Lemma 6.4.) To construct B, as in Chapter 2, we use a set P of n points corresponding to a two-dimensional Halton-Hammersley point set; for boxes we take the southwest quadrants cornered at the N points of a fine square grid. Let $\mu_1 \geq \cdots \geq \mu_n$ be the eigenvalues of $B^T B$. (From now on we identify a set system with its incidence matrix.) The proof of the lower bound consists of three steps:

- STEP 1: Show that $\mu_k \gg (n - k)N(\log n)/n^2$, for any $1 \leq k \leq n$ (Lemma 6.10).[3]
- STEP 2: Prove the existence of an n-by-n submatrix A of B such that $\det A^T A = \Omega(\log n)^n$ (Lemma 6.11).
- STEP 3: Show that the k-th largest eigenvalue λ_k of $A^T A$ is at least $(\log n)^{\Omega(1)}$, for any k up to roughly $n/4$ (Lemma 6.12).

Combining the lower bound of this last step with the spectral lemma proves Theorem 6.8. $\square$

---

[3] Recall that $\ll$ and $\gg$ denote $O()$ and $\Omega()$, respectively.

We now discuss each step separately. Let m be a large enough power of 2, and let $n = m/8$. The set P is a subset of the bit-reversal point set (Halton-Hammersley in two dimensions):

$$Q = \left\{ \left(\frac{1}{2m} + c(k), \frac{1}{2m} + \frac{k}{m} \right) \;\middle|\; 0 \le k < m \right\},$$

where $c(k) = \sum_{i \ge 0} b(i)/2^{i+1}$ and $\{b(i)\}$ is the binary expression for the running index k, ie, $k = \sum_{i \ge 0} b(i)2^i$. To illustrate this rather abstruse definition, we take the simple example $m = 4$. The set Q consists of the points shown in Figure 6.7. Their coordinates in binary are

$$\left\{ (0.001, 0.001), (0.101, 0.011), (0.011, 0.101), (0.111, 0.111) \right\}.$$

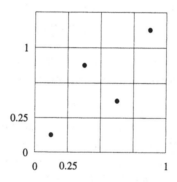

Fig. 6.7. The set Q for $m = 4$.

For any $1 \le k \le \log m$, let $\mathcal{G}_k$ be the grid obtained by dividing $[0, 1]^2$ into m axis-parallel rectangles of size $2^{-k} \times (2^k/m)$. See Figure 6.8. It is easy to see that each cell σ of $\mathcal{G}_k$ is a rectangle of area $1/m$ that contains exactly one point q of Q. We say that q is *well-centered* for $\mathcal{G}_k$ if is lies near the center c_σ of σ, specifically, within the box $(\sigma + c_\sigma)/2$. At least a quarter of the points in Q are well-centered for $\mathcal{G}_k$. We omit the proof of this simple fact.[4] It follows that at least $m/8$ points of Q are each well-centered for

---

[4] A hint: Subdivide each cell of $\mathcal{G}_k$ into m squares of area $1/m^2$. For each cell, exactly one of the squares is occupied by a point of Q. From cell to cell the point in question occupies a different square; since a quarter of a cell is well-centered, the point in question is well-centered a quarter of the time.

at least $(\log m)/8$ grids X_k. We define P to consist of $n = m/8$ of these points.

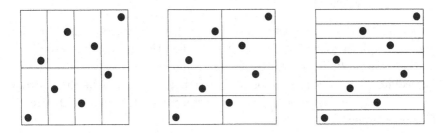

Fig. 6.8. The three grids for $m = 8$.

Let $\mathcal{G}$ be the $(\sqrt{N} - 1) \times (\sqrt{N} - 1)$ square grid covering $[0, 1]^2$, where $N = (m^2 + 1)^2$. Each column of the N-by-n matrix B corresponds to a distinct point of P. Each row of B is associated with the southwest quadrant cornered at a distinct grid point; in other words, for each grid point (x, y) there is a distinct row in B that is the characteristic vector of the set $P \cap (-\infty, x] \times (-\infty, y]$. Note that the N rows are not all distinct. Next, we show that the set system B contains a hard subsystem (for range searching). Our first task is to bound the L^2 norm of Bx in terms of the L^1 norm of x.

Lemma 6.9 *For any $x \in \mathbf{R}^n$,*

$$\|Bx\|_2 \gg \frac{1}{n}\sqrt{N \log n}\, \|x\|_1.$$

Proof: Fix $x = (x_1, \ldots, x_n)^T \in \mathbf{R}^n$. Each x_i corresponds to a distinct point of P, so by abuse of terminology x_i will also be used to refer to its corresponding point. By reversing signs if necessary, we can always assume that

$$\|x\|_1 \leq 2 \sum_{x_i > 0} x_i. \tag{6.2}$$

Given $1 \leq k \leq \log m$, we say that a cell σ of $\mathcal{G}_k$ is *k-heavy* if it contains a well-centered point x_i and $x_i > 0$. We assign a weight to each grid point q of $\mathcal{G}$ as follows: Let σ be any cell of $\mathcal{G}_k$ that contains q.

- If σ is not uniquely defined (because q lies on its boundary), or if σ is not heavy, then assign q a weight of 0.

- Else, subdivide σ into four equal-size quadrants (similar to σ): Assign q a weight of 1 if it lies in the interior of the northeast or southwest quadrant; assign a weight of -1 if it lies in the interior of the northwest or southeast quadrant. If q lies elsewhere, assign it a weight of 0.

One might recognize a discrete version of the Haar wavelets used in the method of orthogonal functions (Chapter 3). Using the same ordering as the rows of B, let $g_k \in \mathbf{R}^N$ be the column vector of weights. It is immediate that the $\log m$ vectors g_k are orthogonal. Let G be the matrix $(g_1, \ldots, g_{\log m})$ and let u be the column vector made of $\log m$ ones. Because G is orthogonal,

$$\|Gu\|_2^2 = \sum_{k=1}^{\log m} \|g_k\|_2^2 \le N \log m.$$

By summing separately over each k-heavy cell σ, we obtain

$$g_k^T Bx \gg \frac{N}{m} \sum_i \{\, x_i \in \ k\text{-heavy cell of } \mathcal{G}_k \,\}.$$

Why this is true is a little subtle, and an example might help. Figure 6.9 illustrates how, in a k-heavy cell, the weights of 1 at q_1, q_3 and -1 at q_2, q_4 produce the cancellations that contribute x_i to $g_k^T Bx$, for any q_1 higher and to the right of x_i. Since the cell in question is heavy, the set of such q_1's covers at least a fraction of the cell.

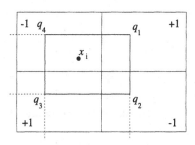

Fig. 6.9. The four discrepancies at the corners of the rectangle $q_1 q_2 q_3 q_4$ cancel out to produce the discrepancy within the rectangle, ie, x_i.

Because each $x_i > 0$ is well-centered for at least a fraction of the grids $\mathcal{G}_k$, it follows from (6.2) and $m = 8n$ that

$$(Gu)^T Bx \gg \frac{N \log n}{n} \|x\|_1.$$

Finally, by Cauchy-Schwarz,

$$\frac{N \log n}{n} \|x\|_1 \ll (Gu)^T Bx \le \|Gu\|_2 \cdot \|Bx\|_2 \le \sqrt{N \log n} \, \|Bx\|_2.$$

$\square$

Lemma 6.10 *For any $1 \le k \le n$, the k-th largest eigenvalue of $B^T B$ satisfies*

$$\mu_k \gg \frac{(n - k + 1) N \log n}{n^2}.$$

Proof: Let $\{v_i\}$ be an orthonormal eigenbasis for $B^T B$, where v_i is a unit eigenvector associated with μ_i. If $Q = (q_{ij})$ denotes the orthonormal matrix whose rows are the eigenvectors v_i, then the column vector ξ obtained by expressing x in the basis $\{v_i\}$ satisfies $\xi = Qx$. The difficulty in applying Lemma 6.9 is that the L^2 norm of Bx is bounded in terms of the L^1 norm of x. To get over this hurdle, we show that the invariant subspace spanned by $\{v_k, \ldots, v_n\}$ contains a unit vector x whose L^1 norm is as large as $\sqrt{n - k + 1}$. By the variational characterization of eigenvalues, the lower bound will then follow.

We use a probabilistic argument. Let $R = (r_{ij})$ be the matrix obtained by replacing each of the first $k - 1$ rows of Q by a row of zeros. If $y = (y_1, \ldots, y_n)$ is a random vector chosen uniformly in $\{-1, 1\}^n$, then

$$
\begin{aligned}
\mathbf{E} \|Ry\|_2^2 &= \sum_{i=1}^n \mathbf{E} \left(\sum_{j=1}^n r_{ij} y_j \right)^2 \\
&= \sum_{i \ge k} \sum_{j=1}^n q_{ij}^2 \, \mathbf{E} \, y_j^2 + \sum_{i \ge k} \sum_{j \ne j'} q_{ij} q_{ij'} \, \mathbf{E} \, y_j y_{j'} \\
&= \sum_{i \ge k} \sum_{j=1}^n q_{ij}^2 = n - k + 1.
\end{aligned}
$$

This implies the existence of a vector $y \in \{-1, 1\}^n$ such that

$$\|Ry\|_2^2 \ge n - k + 1,$$

and therefore the $(n - k)$-flat defined by the equations

$$
\begin{cases}
\xi_i = 0 & (1 \le i \le k - 1), \\
(Qy)^T \xi = \sqrt{n - k + 1},
\end{cases}
$$

intersects the unit-radius ball centered at the origin. Indeed, in the subspace spanned by $\{v_k, \ldots, v_n\}$, the distance of the flat specified above to the origin is equal to $\sqrt{n-k+1}/\|Ry\|_2$. If x is a point of the intersection, the relation $\xi = Qx$ shows that

$$\|x\|_1 \geq y^T x = (Qy)^T \xi = \sqrt{n-k+1},$$

and, from Lemma 6.9,

$$\mu_k \geq \mu_k \|x\|_2^2 \geq \sum_{i=1}^{n} \mu_i \xi_i^2 = \|Bx\|_2^2 \gg \frac{N \log n}{n^2} \|x\|_1^2 \geq \frac{(n-k+1)N \log n}{n^2}.$$

$\square$

Since the determinant of $B^T B$ is the product of the eigenvalues,

$$\det B^T B = \Omega\left(\frac{N \log n}{n^2}\right)^n n! . \tag{6.3}$$

Lemma 6.11 *There exist n points and n southwest quadrants whose incidence matrix A satisfies*

$$\det A^T A = \Omega(\log n)^n.$$

Proof: By the Binet-Cauchy formula,

$$\det B^T B = \sum_{1 \leq j_1 < \cdots < j_n \leq N} \left| \det B \begin{pmatrix} j_1 & j_2 & \cdots & j_n \\ 1 & 2 & \cdots & n \end{pmatrix} \right|^2 .$$

Therefore, by (6.3) there exists an n-by-n submatrix A of B such that

$$
\begin{aligned}
\det A^T A &= \left| \det B \begin{pmatrix} j_1 & j_2 & \cdots & j_n \\ 1 & 2 & \cdots & n \end{pmatrix} \right|^2 \\
&\geq \binom{N}{n}^{-1} \det B^T B \\
&= \Omega(1)^n \left(\frac{n}{eN}\right)^n \left(\frac{n}{e}\right)^n \left(\frac{N \log n}{n^2}\right)^n = \Omega(\log n)^n.
\end{aligned}
$$

$\square$

This completes the second step of the proof. Let $\lambda_1 \geq \cdots \geq \lambda_n \geq 0$ be the eigenvalues of $A^T A$. Moving on to the last step, we find that most of the work left in establishing a lower bound on λ_k involves first proving upper bounds on these eigenvalues. In view of the spectral lemma (page 237), Theorem 6.8 (page 243) follows directly from the lower bound below. $\square$

Lemma 6.12 *For any $k \leq n/4.1$, we have $\lambda_k = (\log n)^{\Omega(1)}$.*

Proof: Of course, we can assume that n is large enough. The proof follows easily from Lemma 6.17, proven in the next section, which states that

$$\lambda_j \leq \frac{5n^2 \log^4 n}{j^2}.$$

Indeed, by Lemma 6.11,

$$
\begin{aligned}
(n-k)\log \lambda_k \quad &\geq \quad \log \det A^T A - \sum_{j=1}^{k-1} \log \lambda_j \\
&\geq \quad n \log \log n - O(n) \\
&\qquad\qquad -k(2\log n - 2\log k + 4\log \log n).
\end{aligned}
$$

Choosing $k = n/4 - \varepsilon n$, for any fixed $\varepsilon > 0$, gives $\lambda_k = (\log n)^{\Omega(1)}$ and completes the proof. $\square$

Triangle Matrices via Buffon's Needle

For our second application of the spectral lemma we derive a lower bound for off-line range searching with respect to triangles. The proof is also a nice illustration of the discrepancy method, since it hinges on Alexander's work on the discrepancy of halfspaces [10] and, hence, as explained earlier, Buffon's needle experiment.

Theorem 6.13 *There are n-by-n triangle matrices of arithmetic complexity* $\Omega(n \log n)$. *This remains true in the presence of up to $n/5$ help gates.*

As before, for notational convenience, we do not distinguish between the input points of P and their assigned weights $x_1, \ldots, x_n$. Our goal is to exhibit n halfplanes $h_1, \ldots, h_n$ and prove that the geometric set system $\{P \cap h_i\}$ has high spectrum and, hence, is hard for range searching. (Recall that the spectrum refers to the singular values of the associated incidence matrix A.) A construction described in §1.5 provides such a set system. By (1.11) on page 31, we know that

$$\det A^T A = \Omega(n)^{n/2}. \tag{6.4}$$

Following the approach of the previous section, we bound the singular values of A from both above and below. We show that the k-th largest eigenvalue λ_k of $A^T A$ is at least $n^{\Omega(1)}$ for any k up to roughly $n/2$ (Lemma 6.14). Together with the spectral lemma (page 237), this establishes Theorem 6.13 (page 249). $\square$

Lemma 6.14 *For any fixed $\varepsilon > 0$ and any $k \le n/2 - \varepsilon n$, we have $\lambda_k = n^{\Omega(1)}$.*

Proof: The lemma follows easily from the upper bound proven in Lemma 6.19:

$$\lambda_j \le \frac{cn^2 \log(j+1)}{j},$$

for some constant $c > 0$. Indeed,

$$
\begin{aligned}
\log \det A^T A \ &= \ \sum_{i=1}^{n} \log \lambda_i \\
&\le \ (n-k) \log \lambda_k + \sum_{j=1}^{k} \log \frac{cn^2 \log(j+1)}{j} \\
&\le \ (n-k) \log \lambda_k + k(2 \log n \\
&\qquad - \log k + \log \log k + c'),
\end{aligned}
$$

for some constant $c' > 0$. By (6.4), it immediately follows that the lower bound of the lemma holds for any $k \le n/2 - \varepsilon n$ and any fixed $\varepsilon > 0$. $\square$

Applications of the Trace Lemma

The main appeal of the trace lemma (page 237) is that it does not involve individual eigenvalues but only first and second moments. The traces of M and M^2 have simple combinatorial interpretations, which makes their asymptotic estimation sometimes very easy. We give two examples below. As we observed in Chapter 1, $\operatorname{tr} M$ is the number of ones in A, while $\operatorname{tr} M^2$ is the number of rectangles of ones in A.

Lemma 6.25 (page 263) asserts the existence of n points and n lines in the plane, all of them distinct, such that each point belongs to $\Theta(n^{1/3})$ lines and each line contains $\Theta(n^{1/3})$ points. The trace of M is the number of incidences, which is $\Theta(n^{4/3})$. The matrix A is square-free, so the number of rectangles of ones is $\Theta(n(n^{1/3})^2)$ and the trace of M^2 is $O(n^{5/3})$. By the trace lemma, we derive this result, which can be viewed as a strengthening of Theorem 6.13.

Theorem 6.15 *There are n-by-n line matrices of arithmetic complexity $\Omega(n \log n)$.*

In the proof of Theorem 1.21 (page 36), we exhibit a set of n points and n boxes in dimension $\Theta(\log n)$ such that, for $c \approx 1.0955$, $\operatorname{tr} M = \Theta(n^c)$ and $\operatorname{tr} M^2 = O(n^{2c-1})$. The trace lemma gives us the following lower bound.

Theorem 6.16 *There are n-by-n box matrices in dimension $\Theta(\log n)$ of arithmetic complexity $\Omega(n \log n)$.*

6.2 Data Structures and Eigenvalues

We show in this section how data structures for multidimensional searching can help in bounding the singular values of geometric matrices.[5] Once again, this shows how eigenvalues are the bridge between discrepancy and algorithms. Let A be an n-by-n matrix with $0/1$ elements, and let

$$\lambda_1 \geq \cdots \geq \lambda_n \geq 0$$

be the eigenvalues of $A^T A$. By the Courant-Fischer characterization of eigenvalues, we have

$$\lambda_k = \min_{\dim F = n-k+1} \; \max_{\substack{x \in F \\ x \neq 0}} \frac{x^T A^T A x}{x^T x}. \tag{6.5}$$

So, to find an upper bound on λ_k, we can choose any subspace F of codimension $k-1$, ie, any set of $k-1$ homogeneous linear constraints, which we call F-*constraints*, and bound the maximum value of the Rayleigh quotient above. Our goal in choosing the F-constraints is to "decimate" as many ones as possible in the matrix A, the rationale being that a $0/1$ matrix with few ones has small singular values. We explain the logic of the approach on a small example. Suppose that $n = 4$, $k = 3$, and

$$A = \begin{pmatrix} 1 & 1 & 0 & 1 \\ 1 & 1 & 1 & 1 \\ 0 & 0 & 1 & 0 \\ 0 & 1 & 1 & 0 \end{pmatrix}.$$

We now must specify the F-constraints. Being entitled to $k - 1 = 2$ of them, we choose $x_1 + x_2 + x_4 = 0$ and $x_3 = 0$. Then, over F the map specified by the matrix A coincides with the one specified by

---

[5]Recall that a matrix A is called geometric if it is the incidence matrix of a geometric set system. The singular values of A are the square roots of the eigenvalues of $A^T A$.

$$C = \begin{pmatrix} 0 & 0 & 0 & 0 \\ 0 & 0 & 0 & 0 \\ 0 & 0 & 0 & 0 \\ 0 & 1 & 0 & 0 \end{pmatrix}.$$

No eigenvalue of $C^T C$ can exceed 1; and so, $\lambda_k \leq 1$ and the same is true of the k-th largest singular value of A. Granted, this example was built to impress. In general, what is a judicious choice of F-constraints? When A is a geometric matrix, the answer is to be found in data structures for the corresponding range searching problem. Magically, their design suggests how to form the constraints in F. Isn't it interesting that, unbeknownst to them, algorithm designers have been solving eigenvalue problems in disguise all of these years? We explain this below by considering two cases: box and triangle matrices.

The Spectrum of Rectangle Matrices

Let A be an n-by-n incidence matrix of a set system formed by n points and n rectangles (ie, axis-parallel boxes in $\mathbf{R}^2$): A_{ij} is 1 if and only if rectangle i contains point j. Again, let $\lambda_1 \geq \cdots \geq \lambda_n$ be the eigenvalues of $A^T A$.

Lemma 6.17 *For any $1 \leq k \leq n$,*

$$\lambda_k \leq \frac{5n^2 \log^4 n}{k^2}.$$

Proof: Going from left to right, place the n points in bijection with the leaves of a complete binary tree. Each node v of the tree is naturally associated with the vertically sorted list N_v of the points in the leaves at or below v. This forms what is known as a *range tree*. Given a southwest quadrant, a binary search for its x-coordinate leads to a decomposition of the quadrant into rectangles, every one of which is associated with a distinct node of the tree (Fig. 6.10). Thus, any set specified by a row of A can be partitioned into at most $\nu = \log n + 1$ subsets:

(i) each one is a prefix of a list N_v;
(ii) all of the relevant lists N_v are on different levels of the tree.

This allows us to break up A into ν separate set systems, one for each level of the tree. Their n-by-n incidence matrices $A_1, \ldots, A_\nu$ satisfy $A = \sum_i A_i$. For example, if row j of matrix A encodes $\{a, d, e, f\}$ (Fig. 6.10), then row

j of matrix A_i, where i denotes the third level from the bottom, encodes $\{a, d\}$.

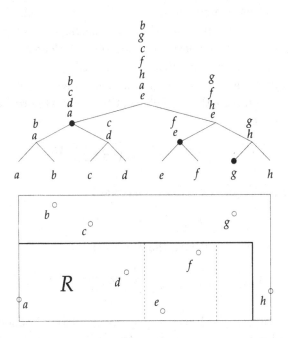

Fig. 6.10. The rectangle R is decomposed into three canonical rectangles, each one associated with a node of the tree. The lists N_v appear next to each node.

To define the F-constraints, we break up the lists N_v into sublists. For the calibration of these lists, we assign integer weights to the points in N_v. We do this in two passes. Initially, every weight is equal to 1. Let i be the level of v. Among the rows of A_i corresponding to prefixes of N_v, we check whether any of them are identical. For each group of identical rows, we take the last element (ie, the point with the highest y-coordinate) in the prefix and assign that point a weight equal to the size of that group. Note that the total weight of all the elements at level i in the tree is at most $2n$. Next, consider each list N_v (after the previous preprocessing) and subdivide it into contiguous sublists of weighted size r (or less); r is an integer parameter to be specified later.

To summarize, at each level we have a collection of at most $2n/r$ sublists of weighted size exactly r along with a number of other sublists of size less than r. Any subset specified by a row of A_i can thus be written as a union of sublists of weight r and a *remainder* set of size less than r.

Recall that point i is associated with the i-th coordinate of the vector $x = (x_1, \ldots, x_n)^T$. For each sublist of weight r, write the linear constraint expressing that the x_i's in that sublist sum up to 0. This gives us a total of at most $2\nu n/r$ F-constraints. Assume that all are satisfied. Then, A_i can be expressed more concisely as an n-by-n 0/1 matrix, denoted by C_i, whose rows each have fewer than r ones. In other words, as long as x satisfies the given constraints, $A_i x = C_i x$. Because of the calibration of the sublists, note that similarly no column of C_i can have more than r ones. It is a standard result in matrix theory [193] that the spectral norm[6] of a matrix $M = (M_{ij})$ satisfies

$$\|M\|_s^2 \leq \left(\max_i \sum_j |M_{ij}| \right) \left(\max_j \sum_i |M_{ij}| \right),$$

and therefore $\|C_i\|_s^2 \leq r^2$. Using the fact that

$$\|C_i x\|_2 \leq \|C_i\|_s \|x\|_2 \leq r \|x\|_2,$$

we find that, for any x satisfying the F-constraints,

$$\|Ax\|_2 \leq \sum_{i=1}^{\nu} \|A_i x\|_2 = \sum_{i=1}^{\nu} \|C_i x\|_2 \leq r\nu \|x\|_2.$$

It follows that over F the Rayleigh quotient in (6.5) is bounded by $r^2\nu^2$. Since $\lambda_1 = \|A\|_s^2 \leq n^2$, the lemma is trivial if k is $O(\nu)$, so we assume that it is not the case. There are at most $2\nu n/r$ F-constraints, so setting $\lfloor 2\nu n/r \rfloor = k - 1$ completes the proof. $\square$

The Spectrum of Triangle Matrices

Let A be an n-by-n incidence matrix of a set system formed by n points and n triangles: A_{ij} is 1 if and only if triangle i contains point j. Again, $\lambda_1 \geq \cdots \geq \lambda_n$ denote the eigenvalues of $A^T A$. We derive a crude upper bound very simply by using simplicial partitions.

Lemma 6.18 *For any $1 \leq k \leq n$, $\lambda_k = O(n^2/\sqrt{k})$.*

Proof: Recall from Chapter 5 that, given the set P of n points corresponding to the columns of A, a collection $\{(P_1, R_1), \ldots, (P_m, R_m)\}$ is a simplicial partition, if

---

[6]The spectral norm of M is defined as the maximum value of $\|Mx\|_2$ over all unit vectors x.

- the P_i's partition P and

- each R_i is a relatively open triangle enclosing P_i.

The maximum number of R_i's that a single line can cut is called the cutting number. Replace r by r^2 in Lemma 5.8; then, for any positive integer $r \leq \sqrt{n/2}$, there exists a simplicial partition of cutting number at most cr, for some constant $c > 0$, such that $n/r^2 \leq |P_i| < 2n/r^2$ for each P_i.

Recall that point j is associated with the j-th coordinate of the vector $x = (x_1, \ldots, x_n)^T$. For each triangle R_i, write the linear constraint expressing that the sum of the x_j's whose corresponding points fall inside the triangle is null. This gives us the set of F-constraints; there are at most r^2 of them, so we set $r = \lfloor \sqrt{k-1} \rfloor$ and assume that $k > 1$. Over F, the map defined by A can be specified by means of a sparse 0/1 matrix C, where no row of C contains more than $\rho = \lfloor 6cn/r \rfloor$ ones: the number of edges (ie, 3) times the cutting number (ie, cr) times an upper bound on $|P_i|$ (ie, $2n/r^2$). The spectral norm of C satisfies

$$\|C\|_s^2 \leq \left(\max_i \sum_j |C_{ij}| \right) \left(\max_j \sum_i |C_{ij}| \right) \leq n\rho,$$

and so, by (6.5), $\lambda_k = O(n\rho)$, which proves the lemma. Obviously, the bound remains true for $k = 1$, since $\lambda_1 = \|A\|_s^2 \leq n^2$. $\square$

In the proof of the arithmetic circuit lower bound for triangle matrices we had the additional assumption that the points were on a grid. This is a case where we can do better with little effort.

Lemma 6.19 *If the n points are the vertices of a square grid, then, for any set of n triangles and any $1 \leq k \leq n$,*

$$\lambda_k = O \left(\frac{n^2 \log(k+1)}{k} \right).$$

Proof: Let $\mathcal{L}$ be the set of lines passing through the edges of the triangles. We break up the square enclosing the grid into triangles with few of the n points in any of them and few lines of $\mathcal{L}$ crossing them. To do that, we first subdivide the square into a regular $r \times r$ grid of lines, for some integer parameter r large enough. In addition, we throw in a random sample of r lines chosen among $\mathcal{L}$. Next, we form the arrangement of these $3r + 4$ lines (counting the boundary of the square) and triangulate it (Fig. 6.11). For a constant c taken large enough for future purposes, with high probability no triangle is cut by more than $cn(\log r)/r$ lines of $\mathcal{L}$ and none contains more than cn/r^2 points.[7]

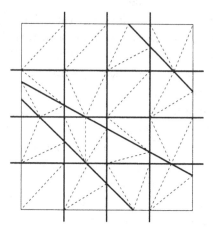

Fig. 6.11. The square is subdivided into triangles with few points and few lines cutting through them.

The point labeled j is associated with the j-th coordinate of the vector $x = (x_1, \ldots, x_n)^T$. For each triangle, write the linear constraint expressing that the sum of the x_j's whose corresponding points fall inside of it is null. This gives us a set of at most cr^2 linear F-constraints. If all are satisfied, then A can be rewritten in a simpler form by means of a sparse matrix C. Specifically, by the classical "zone theorem" for line arrangements [122], we know that no line can cut more than cr triangles, and so the sum of the x_j's within any triangle is a linear form over at most $3cr$ variables. Therefore, within the restriction to the constraint space F, each row of A corresponds to a linear form with at most $3c^2n/r$ nonzero coefficients. Let $C = (C_{ij})$ be the new matrix formed by these coefficients. Note that no column (resp. row) of C contains more than $cn(\log r)/r$ (resp. $3c^2n/r$) ones. The spectral norm of C satisfies

$$\|C\|_s^2 \le \left(\max_i \sum_j |C_{ij}|\right)\left(\max_j \sum_i |C_{ij}|\right) \le \frac{3c^3n^2\log r}{r^2}.$$

---

[7]The theory of ε-nets developed in Chapter 4 gives a comprehensive treatment of this sampling technique and explains why not too many lines can cut any triangle. The bound on the number of points is where we use the property of the grid.

By (6.5), this shows that

$$\lambda_k = O\left(\frac{n^2 \log r}{r^2}\right),$$

from which the lemma follows for k large enough and the setting $k - 1 = \lfloor cr^2 \rfloor$. If $k = O(1)$, we use the fact that $\lambda_1 = \|A\|_s^2 \le n^2$ to conclude. $\square$

6.3 Monotone Circuits

Arithmetic complexity in the monotone case disallows subtractions. Formally, this means that we are confined to the semigroup $(\mathbf{R}, +)$, and the product Ax must be computed by using additions only. Remember that the matrix is $0/1$, so this is possible to do with $O(n^2)$ additions. As before, the model of computation is a circuit or, equivalently, a straight-line program: Each gate (or step) performs an operation of the form

$$z \leftarrow x + y,$$

where x and y are previously computed variables or input weights. The circuit depends on the linear map A, but it must work for any assignment of the input variables (ie, the weights of the points) in the additive semigroup of reals.

The *monotone arithmetic complexity* of the matrix A is the size of the smallest monotone circuit for computing $x \mapsto Ax$. We prove three lower bounds on the complexity of geometric matrices. Recall that a box (resp. line, simplex) matrix is the incidence matrix of a set system defined by points and axis-parallel boxes (resp. lines, simplices). For any n large enough, we have:

Theorem 6.20 *In any dimension $d \ge 1$, there are n-by-n box matrices of monotone arithmetic complexity $\Omega(n(\log n / \log \log n)^{d-1})$.*

Theorem 6.21 *There are n-by-n line matrices of monotone arithmetic complexity $\Omega(n^{4/3})$.*

Theorem 6.22 *In any dimension $d > 2$, there are n-by-n simplex matrices of monotone arithmetic complexity[8] $\widetilde{\Omega}(n)^{2-2/(d+1)}$.*

---

[8]The notation $\widetilde{\Omega}(f(n))$ means $\Omega(f(n))/(\log n)^{O(1)}$.

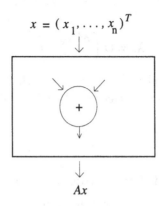

$$x = (x_1, \ldots, x_n)^T$$

$+$

Ax

Fig. 6.12. A monotone circuit only has addition gates.

Recall that, in all three cases, the input consists of n points, n reals (the weights), and n ranges (ie, boxes, lines, simplices). The output is the list of numbers obtained by summing the weights of the points in each range. All three lower bounds are essentially optimal, which means that (at least in the monotone case) off-line range searching with respect to boxes and simplices is a solved problem. Theorem 6.22 brings unexpected news: If d is large, the bound is close to n^2 and the naive approach of comparing all points against all simplices is, in practice, the method of choice.

All three proofs are applications of the following lemma. The ways in which the lemma is applied are very different, however; each one offers an interesting new perspective on the complexity of geometric matrices.

Let $A = (a_{ij})$ denote the n-by-n incidence matrix of the underlying range searching problem.

Lemma 6.23 *If A is an n-by-n incidence matrix with no p-by-q submatrix of ones, then the complexity of computing Ax on a monotone circuit is at least on the order of*

$$\frac{1}{pq}\left(\sum_{i,j} a_{ij}\right) - \frac{n}{p}.$$

Proof: Obviously, each input to a gate is a linear form over the n input variables, and the output is the sum of the two corresponding linear forms. A gate is called *heavy* if in the linear form it outputs, $\sum_j \alpha_j x_j$ ($\alpha_j \in \mathbf{N}^+$), at least q variables x_j are involved. Given row i of the matrix A, let $S_i = \sum_j a_{ij}$ be the sum of its elements. The output gate g computing the form $\sum_j a_{ij} x_j$ is connected to S_i input variables x_{i_1}, x_{i_2}, etc. The circuit forms a directed acyclic graph. Let T_i be a subtree of the graph rooted at g with x_{i_1}, x_{i_2}, etc, at its leaves (all edges directed from child to parent). A subtree of T_i with at most q leaves is called maximal if its root either has no parent or has one with at least q leaves descending from it.[9]

The maximal subtrees of T_i are all mutually disjoint. Each such subtree has one fewer two-child nodes (ie, nodes with two children) than leaves. There are at least S_i/q such subtrees, so together they account for no more than $S_i - S_i/q$ two-child nodes. It follows that at least $S_i/q - 1$ internal nodes of T_i coincide with heavy gates. Obviously, no heavy gate can provide a node for p (or more) trees of the form T_i (over all output gates g), since this would create a p-by-q submatrix of ones in A. The lower bound then follows. $\square$

Box Matrices and Chinese Remaindering

We exhibit n points and n axis-parallel boxes in $\mathbf{R}^d$ whose corresponding range searching problem requires $\Omega(n(\log n/\log\log n)^{d-1})$ additions in the monotone model (Theorem 6.20, page 257). Trivially, we can assume that $d > 1$. The set of input points is obtained from a *Halton-Hammersley* construction. For completeness, we repeat the construction used in §2.2. Let $p_1 < p_2 < \cdots < p_{d-1}$ be consecutive primes. Any integer m has a unique decomposition in base p_k: $m = \sum_{i \geq 0} b_k(i) p_k^i$. We consider the function

$$x_k(m) = \sum_{i \geq 0} \frac{b_k(i)}{p_k^{i+1}}.$$

This allows us to construct the input point set in $[0,1)^d$:

$$P = \left\{ \left(x_1(m), \ldots, x_{d-1}(m), \frac{m}{n} \right) \,\middle|\, 0 \leq m < n \right\}.$$

We define an interval of *type* $\langle k, j \rangle$ to be of the form

$$\left[\frac{M}{p_k^j}, \frac{M+1}{p_k^j} \right),$$

where M is a nonnegative integer. Finally, we call a box B *special* if it is of the form $I_1 \times \cdots \times I_d \subseteq [0,1)^d$ and the three additional conditions below hold:

---

[9] Descending means going against the direction of the edges.

- $I_1, \ldots, I_{d-1}$ are intervals of type $\langle 1, j_1 \rangle, \ldots, \langle d-1, j_{d-1} \rangle$, respectively, for some integers $j_1, \ldots, j_{d-1} \geq 0$.
- I_d is of the form $[Mp_1Q/n, (M+1)p_1Q/n)$, where $Q = \prod_{0<k<d} p_k^{j_k}$ and M is an integer.
- $p_1 Q \leq n$ (Fig. 6.13).

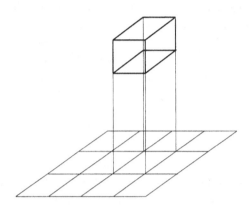

Fig. 6.13. A special box.

Lemma 6.24 *A special box contains exactly p_1 points of P.*

Proof: Let $B = I_1 \times \cdots \times I_d$ be a special box indexed by $(j_1, \ldots, j_{d-1})$, with $I_k = [m_k/p_k^{j_k}, (m_k + 1)/p_k^{j_k})$; $k < d$. By the Chinese remainder theorem, the system of congruences

$$m \equiv m_k \pmod{p_k^{j_k}},$$

for $0 < k < d$, has a unique solution m^* modulo Q. Every other solution is of the form $m^* + lQ$. It is immediate that each point

$$\Big(x_1(m^* + lQ), \ldots, x_{d-1}(m^* + lQ) \Big)$$

lies in $I_1 \times \cdots \times I_{d-1}$. Since, by assumption,

$$I_d = \Big[\frac{Mp_1Q}{n}, \frac{(M+1)p_1Q}{n} \Big)$$

lies entirely in $[0, 1)$, exactly p_1 values of $(m^* + lQ)/n$ lie in I_d, and therefore $|B \cap P| = p_1$. $\square$

Next, we count the number N of special boxes:

$$N = \sum_{j_1,\ldots,j_{d-1} \geq 0} \left\lfloor \frac{n}{p_1 \prod_{0<k<d} p_k^{j_k}} \right\rfloor \prod_{0<k<d} p_k^{j_k}$$

$$\geq \sum \left\{ \frac{n}{p_1} - \prod_{0<k<d} p_k^{j_k} \,\middle|\, j_1,\ldots,j_{d-1} \geq 0 \text{ and } \prod_{0<k<d} p_k^{j_k} \leq \frac{n}{2p_1} \right\}$$

$$\geq \sum_{t=0}^{\lfloor \log n / \log p_{d-1} - 2 \rfloor} \sum_{j_1+\cdots+j_{d-1}=t} \frac{n}{2p_1}.$$

We conclude that

$$N \gg \frac{n}{p_1} \left(\frac{\log n}{\log p_{d-1}} \right)^{d-1}.$$

By choosing p_1 around $(\log n)^{d-1}$, which implies that all of the other p_j's are in $O(\log n)^{d-1}$, we then can find about $n/(\log \log n)^{d-1}$ boxes that define a set system whose incidence matrix A contains at least $n(\log n / \log \log n)^{d-1}$ ones. By adding boxes that contain no points, we can make the set system n-by-n. It remains for us to show that A is square-free (ie, has no two-by-two submatrix of ones).

Consider the intersection of two distinct special boxes B and B', with parameters $(j_1,\ldots,j_{d-1})$ and $(j_1',\ldots,j_{d-1}')$, respectively. We may assume that $Q < Q'$ (since equality gives an empty intersection). The intersection of an interval of type $\langle k,j \rangle$ with one of type $\langle k,j' \rangle$, for $j' < j$, is either empty or of type $\langle k,j \rangle$. This shows that $B \cap B'$ is a box $J_1 \times \cdots \times J_d$, where each J_k $(k < d)$ is an interval of type $\langle k, \max\{j_k, j_k'\} \rangle$ and J_d has length $\leq p_1 Q/n$.

Assume that the box $B \cap B'$ intersects P, and let m be the index of a point in the intersection. By the Chinese remainder theorem, m is entirely specified modulo $\prod p_k^{\max\{j_k, j_k'\}}$. Because $Q < Q'$, this determines m uniquely modulo $p_i Q$, for some $0 < i < d$. The point's d-th coordinate m/n lies in an interval J_d of length $\leq p_1 Q/n$. Since in addition the residue class of m modulo $p_i Q$ is specified and $p_i \geq p_1$, the value of m is uniquely determined. Thus, two special boxes intersect in at most one point of P. As a result, A is square-free and Theorem 6.20 (page 257) follows directly from Lemma 6.23. $\square$

Line Matrices and the Euler Totient Function

It is remarkably easy to prove a lower bound of $\Omega(n^{4/3})$ on the monotone complexity of line range searching. The input consists of n weighted points and n lines, and the problem is to compute the sum of the weights along each line. In other words, we seek the monotone complexity of a line matrix whose rows correspond to lines and columns to points. Note that by thickening the lines ever so slightly and clipping them, we can turn them into long, thin triangles. Any lower bound for line matrices thus applies to triangle matrices as well.

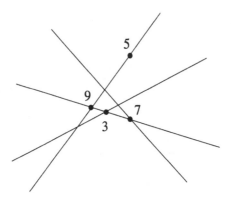

Fig. 6.14. The output is $\{14, 19, 7, 3\}$.

Let m be a large enough integer. We specify the set of lines first. Each line

$$u(X - i) = v(Y - j)$$

is parameterized by u, v, i, j, where

- $1 \leq u \leq v < m^{1/3}$ and $\mathrm{GCD}\,(u, v) = 1$;
- $1 \leq i \leq v$ and $1 \leq j \leq m$.

The first thing to notice is that all the lines are distinct. Indeed, because u and v are relatively prime, the slope u/v is distinct for each pair (u, v). Similarly, given u, v, the affine term $vj - ui$ is distinct for each pair (i, j); else v would have to divide a term of the form $0 < i - i' < v$ (the case $i = i'$ implies that $j = j'$, and thus is ruled out). Since

$$\sum_{k=1}^{N} \varphi(k) = \frac{3N^2}{\pi^2} + O(N \log N),$$

where $\varphi(k)$ is the Euler totient function,[10] the number of lines is on the order of

$$\sum_{1 \leq v \leq m^{1/3}} mv\varphi(v) \approx m^2.$$

A given line passes through integer points of the form $(i + vl, j + ul)$. Since $0 < i, u, v < m^{1/3}$ and $0 < j \leq m$, at least $m^{2/3}$ of these points lie in $[1, 2m]^2$. If we choose the integer points

$$\left\{ (x, y) \mid 1 \leq x, y \leq 2m \right\},$$

then we end up with m^2 points and m^2 lines (up to constant factors): every line (resp. point) is incident to roughly $m^{2/3}$ points (resp. lines).

Lemma 6.25 *There exist n points and n lines in the plane, all of them distinct, such that each point belongs to $\Theta(n^{1/3})$ lines and each line contains $\Theta(n^{1/3})$ points.*

Thus, we have an n-by-n set system of points and lines whose incidence matrix contains at least $\Omega(n^{4/3})$ ones. The lines are distinct and two of them intersect in at most one point, so the matrix is square-free, and Theorem 6.21 (page 257) follows from Lemma 6.23.

Simplex Matrices and Heilbronn's Problem

We can generalize the previous result and derive a general lower bound on the monotone complexity of range searching with respect to simplices in any dimension $d > 2$. The exponent in the expression of the complexity increases from $4/3$ to $2 - 2/(d + 1)$; for technical reasons, we lose a small polylogarithmic factor in the process.

In the proof of Theorem 6.22 (page 257), we assume that n is large enough. We construct a set P of n points in $\mathbf{R}^d$ together with a collection $\{S_q\}$ of n slabs: (i) Each slab contains roughly $n^{1-2/(d+1)}$ points; (ii) the intersection of any $k \geq c \log n$ slabs, for some constant $c > 0$, contains at most a polylogarithmic number of points. We apply Lemma 6.23 and conclude. The details follow.

Given $q \in \mathbf{R}^d$ distinct from the origin, let H_q denote the hyperplane normal to Oq passing through q; its equation is $\langle p, q \rangle - \|q\|_2^2 = 0$. We define S_q to be the slab of width w consisting of all the points at most

---

[10] $\varphi(k)$ is the number of integers $\leq k$ relatively prime to k.

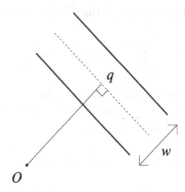

Fig. 6.15. The slab S_q.

$w/2$ away from H_q (Fig. 6.15). The parameter w will be set to its proper value below. To specify the collection of slabs $\{S_q\}$, it thus suffices to provide a set Q of n points q. We show that if any d of the points of Q are sufficiently "spread apart," then the d corresponding slabs have a small common intersection. We use a construction related to Heilbronn's problem to exhibit a suitable set Q. Finally, by throwing in random points in the unit cube we form the set P and complete the construction of the set system.

Building the Slabs

We use the notation $\mathrm{vol}_d(A)$ and $\mathrm{conv}(A)$ to refer to the d-dimensional volume and the convex hull of A, respectively.

Lemma 6.26 *Let $q_1,\ldots,q_d$ be d points in $[0,1]^d$, and assume that the central projection q_i' of each q_i on the hyperplane $x_1 = 1$ also lies in $[0,1]^d$. Then*

$$\mathrm{vol}_d \bigcap_{i=1}^{d} S_{q_i} \ll w^d \Big/ \mathrm{vol}_{d-1}\Big(\mathrm{conv}\,\{q_1',\ldots,q_d'\}\Big).$$

Proof: In the two-dimensional case illustrated in Figure 6.16, the lemma formalizes the intuitive notion that the longer the segment $q_1' q_2'$ the smaller the intersection between the two slabs S_{q_1} and S_{q_2}. Let $[u_1,\ldots,u_d]$ denote the matrix whose columns are the vectors u_i spanning the parallelepiped $\bigcap S_{q_i}$. By convention, each u_i has the direction specified by the intersection of hyperplanes bounding the slabs S_{q_j}, for all $j \neq i$. Since each u_i is normal to any q_j $(j \neq i)$,

$$\det\left([u_1, \ldots, u_d]^T [q_1, \ldots, q_d]\right) = \prod_{i=1}^{d} \|q_i\|_2 \cdot \|u_i\|_2 \cos(q_i, u_i)$$

$$= w^d \prod_{i=1}^{d} \|q_i\|_2,$$

and hence

$$\operatorname{vol}_d \bigcap_{i=1}^{d} S_{q_i} = |\det[u_1, \ldots, u_d]|$$

$$= w^d \left(\prod_i \|q_i\|_2\right) / |\det[q_1, \ldots, q_d]|$$

$$= w^d \left(\prod_i \|q_i'\|_2\right) / |\det[q_1', \ldots, q_d']|$$

$$\ll w^d / |\det[q_1', \ldots, q_d']|,$$

from which the lemma easily follows. $\square$

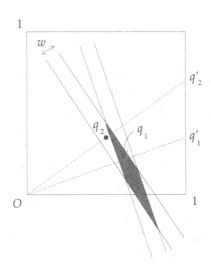

Fig. 6.16. The intersection of the slabs shrinks as the convex hull of $\{q_1', q_2'\}$ expands.

The point set P should be well "spread out" in the unit cube. Specifically, the convex hull of any $d+1$ of them should have volume $\Omega(1/n)$. Except in the trivial case, $d = 1$, this turns out to be impossible to achieve. A slight relaxation of the goal is within reach, however. We show in Lemma 6.27 that if P is random and $k/\log n$ is large enough, the convex hull of any k points has volume $\Omega(k/n)$.

This type of question is known as *Heilbronn's problem*: Given n points in the unit cube, what is the smallest volume of the simplex formed by any $d + 1$ of them? The goal is to choose the points so as to maximize this number. As shown by Roth [263], the intuitive answer of $O(1/n)$ turns out to be wrong. In fact, any set of n points in the unit square always contains a triangle of area $O(1/n^c)$, for some constant $c > 1$. Weaker results were obtained earlier by Roth [260] in 1950 and then by Schmidt [271] over twenty years later. The best current bound of $O(1/n^{8/7-\varepsilon})$, for any $\varepsilon > 0$, is due to Komlós, Pintz, and Szemerédi [189]. From the other end, the same authors [190] have shown[11] the existence of point sets with all $\binom{n}{3}$ triangles of area $\Omega\big((\log n)/n^2\big)$. A nice chronological account of the problem is given by Moser [235].

For our purposes here, a set of n points in $[0, 1]^d$ is said to satisfy the *Heilbronn property* if, for some fixed $c > 0$, the convex hull of any subset of $k \geq c \log n$ points has d-dimensional volume $\Omega(k/n)$.

Lemma 6.27 *A random set of n points uniformly distributed in $[0, 1]^d$ satisfies the Heilbronn property with probability tending to 1 as n goes to infinity.*

Proof: We begin with a simple approximation result. Fix a large enough constant a, and let $\mathcal{E}$ be the set of ellipsoids of volume at least $1/n$ entirely contained in the cube $[-a, a]^d$. For some constant $b = b(a)$, there exists a set $\mathcal{E}_0$ of $O(n^b)$ ellipsoids with the following property: Given any ellipsoid $E \in \mathcal{E}$, there is some $E_0 \in \mathcal{E}_0$ such that $E \subseteq E_0$ and $\mathrm{vol}_d E_0 \leq 2\,\mathrm{vol}_d E$.

Here are a few words of explanation why this should be true. First, we observe that an ellipsoid can be specified by its center and its principal vectors. In the case of $\mathcal{E}$, any such parameter point lies in $[-a, a]^{d(d+1)}$. Round the $d(d+1)$ coordinates to their ℓ-th decimal position, where $\ell/\log n$ is a large enough constant. The resulting set of rounded-off parameter points creates a polynomial-size collection of ellipsoids that can be used as $\mathcal{E}_0$. Indeed, given $E \in \mathcal{E}$, find the nearest $E_1 \in \mathcal{E}_0$, where nearest is defined with respect to the Euclidean distance of their respective parameter points.

---

[11] As noted by Erdős, to achieve $\Omega(1/n^2)$ is immediate: Consider the set of points $\frac{1}{n}(x, x^2 \bmod n)$, for $x = 0, 1, \ldots, n - 1$. The area of any triangle can be expressed, up to a constant factor, as a 3-by-3 Vandermonde determinant that is trivially $\gg 1/n^2$.

The two ellipsoids E and E_1 have a Hausdorff distance[12] at most $1/n^\alpha$, for an arbitrarily large constant $\alpha = \alpha(\ell)$ (Fig. 6.17). Keep the center of E_1 fixed and scale up its principal vectors (while staying in $\mathcal{E}_0$) to ensure that the resulting ellipsoid E_0 contains all of E. Because the volume of E is not too small, neither are the lengths of its principal axes. It easily follows that the volume of E_0 need not exceed, say, twice that of E to ensure inclusion. This establishes our claim.

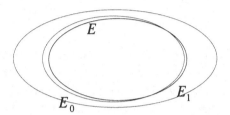

Fig. 6.17. The ellipsoid E_0 encloses E and is at most twice its volume.

Throw n points into $[0,1]^d$ at random uniformly and independently. Given $E_0 \in \mathcal{E}_0$, the expected number of points falling in E_0 is equal to $n \, \mathrm{vol}_d E_0 \cap [0,1]^d$. Suppose that the volume of E_0 is at most $c(\log n)/2n$. Then the expected number of points falling in E_0 does not exceed $c(\log n)/2$. Furthermore, by Chernoff's tail estimate (Lemma A.3),[13] the probability that the number of points is at least $c \log n$ is less than $1/n^{c'}$, for some constant c' growing monotonically to infinity with c. By choosing c large enough, we can thus ensure that, with probability arbitrarily close to one, each $E_0 \in \mathcal{E}_0$ of volume at most $(\log n)/n$ contains fewer than $c \log n$ points.

We now can verify that the point set satisfies the lemma. Consider a subset of size $k \geq c \log n$ and suppose, by contradiction, that the volume of its convex hull is less than $k/(c^3 n)$. Subdivide the convex hull into $\lfloor k/(c \log n) \rfloor$ parallel slices of the same volume. By the pigeonhole principle, one of them, C, contains at least $c \log n$ points, while its volume is less than $(\log n)/cn$. From the existence of the Löwner-John ellipsoid[14] it follows

---

[12]The Hausdorff distance between two compact bodies A and B is defined as

$$\max\{ \max_{x \in A} d(x, B), \max_{x \in B} d(x, A) \},$$

where $d(x, A)$ is the minimum Euclidean distance from x to any point of A.

[13]In the application of Lemma A.3, one can always assume that the probability of success is *exactly* $c(\log n)/2n$, since any lower probability can only produce fewer hits.

[14]For any convex body K in $\mathbf{R}^d$, there exist a point p and a linear transformation f

that C can be enclosed by an ellipsoid E of volume at most $(\log n)/2n$. Furthermore, this Löwner-John ellipsoid can be made to fit within C by a constant-factor scale-down, and therefore $E \subset [-a/2, a/2]^d$ (just choose a big enough). Of course, we have enough wiggle room to inflate E within $[-a, a]^d$, if necessary, to make it into an ellipse in $[-a, a]^d$ of volume at least $1/n$, and hence an ellipse $E \in \mathcal{E}_0$. By our previous discussion, we know that E is itself enclosed by some $E_0 \in \mathcal{E}_0$ of volume at most $(\log n)/n$. The ellipsoid E_0 contains at least $c \log n$ points, which gives a contradiction. □

Choose an integer $m = \lfloor c_0 n w \rfloor$, for some constant $c_0 > 0$. By Lemma 6.27, we can place m points in $(1, 0, \ldots, 0) + [0, 1]^{d-1}$ so that the convex hull of any $k \geq c \log m$ points has $(d-1)$-dimensional volume at least $\Omega(k/m)$. For each such point q', place points on the segment Oq' at intervals of length $2w$. This produces $\Theta(c_0 n)$ points q, and if w is small enough, then, for at least a constant fraction of them, the slab S_q intersects the cube $[0, 1]^d$ in a polytope of volume $\Omega(w)$. By choosing c_0 large enough, we can find a set Q of n such points. To summarize, Q consists of n points such that, for any $q \in Q$,

$$\text{vol}_d\, S_q \cap [0, 1]^d \gg w. \tag{6.6}$$

For any distinct $q_1, \ldots, q_k \in Q$, with $k \geq c \log m$, either at least two q_i's have the same central projection q_i', in which case $\text{vol}_d \bigcap_{i=1}^{k} S_{q_i} = 0$, or else

$$\text{vol}_{d-1}\left(\text{conv}\,\{q_1', \ldots, q_k'\}\right) \gg \frac{k}{m}.$$

By triangulating the convex hull of $q_1', \ldots, q_k'$, using $O(k^{\lfloor (d-1)/2 \rfloor})$ simplices,[15] we derive the existence of d points, say, $q_1', \ldots, q_d'$, whose convex hull has volume $\gg k^{2 - \lceil d/2 \rceil}/m$. By Lemma 6.26, this shows that, in all cases, $k \geq c \log m$ implies that

$$\text{vol}_d \bigcap_{i=1}^{k} S_{q_i} \leq \text{vol}_d \bigcap_{i=1}^{d} S_{q_i} \ll n w^{d+1} k^{\lceil d/2 \rceil - 2}. \tag{6.7}$$

In summary, we have constructed n slabs of width w, denoted by $S_{q_1}, \ldots, S_{q_n}$, that satisfy (6.7).

---

such that K is contained in $p + df(B^d)$ and contains $p + f(B^d)$, where B is the Euclidean unit ball ([151], page 654).

[15] See Appendix C.

Placing the Points

As we said earlier, we construct the point set P by choosing n points in $[0,1]^d$ at random uniformly and independently. Let $w = n^{-2/(d+1)}$ and set $k = \lceil c \log n \rceil$. Any intersection of k distinct slabs has volume $(\log n)^{O(1)}/n$, and therefore, by Chernoff's tail estimate, it contains $(\log n)^{O(1)}$ points with probability $1/n^b$, where b can be chosen (conservatively) as an arbitrarily large constant. There are many k-wise intersections, but by simple examination of (6.7) we see that only $\binom{n}{d}$ of them matter. By choosing, say, $b = d$, we immediately derive the existence of a point set for which any k-wise intersection of slabs encloses only $(\log n)^{O(1)}$ points.

Similarly, by (6.6) a given slab S_{q_i} contains $\Omega(wn)$ points on average. Chernoff's bound again shows that the probability of falling much beyond wn (ie, by a large constant factor) is exponentially small in n. So we can easily ensure that each of them contains $\Omega(wn)$ points.

The point set P and the slabs S_q form a set system whose incidence matrix A has no p-by-q submatrix of ones, for $p, q > (\log n)^\beta$, with constant $\beta > 0$. It follows from Lemma 6.23 that the monotone complexity of the map $x \mapsto Ax$ is $\Omega(wn^2)/(\log n)^{O(1)}$, which proves Theorem 6.22 (page 257). $\square$

6.4 Geometric Databases

Assume that the class of range queries and the underlying point set are fixed once and for all, but that queries are specified on-line. For example, in the case of simplex range searching in two dimensions, a database will store a collection of weighted points in the plane so that, given an arbitrary query triangle, the sum of the weights of the points within the triangle can be computed efficiently. Intuitively, the more storage the database system is equipped with, the faster the queries should be answerable. We provide lower bounds on space-time tradeoffs to support this intuition. The model is particularly simple. It is especially relevant for lower bounds and, if anything, it underestimates the true costs of a real system, since it ignores the cost of memory access (which is good, not bad).

The results are stunningly pessimistic. For example, to answer a simplex range searching query in dimension 20 requires roughly $n^{0.9}$ additions in the worst case on a database with as much as n^2 storage. Since any range searching query can be answered in linear time, using no extra storage and a 20-line piece of code, the moral of the story is clear: In higher dimension, naive is best! True speed-ups require unrealistically high amounts of storage, especially in higher dimension, and the naive algorithms are unbeatable.[16]

Consider the range space $(P, \mathcal{R})$, where $\mathcal{R}$ is the collection of subsets of the form $P \cap R$, for all allowable ranges R (eg, simplices or boxes). Given an integer $m > 0$ (the size of the database), we wish to find the smallest integer $t = t(m)$ such that any query can be answered with fewer than t additions. There is a nice set-theoretical way to model this: We require the existence of m subsets $P_1, \ldots, P_m \subseteq P$ such that, given any $R \in \mathcal{R}$, $P \cap R$ can be expressed as the union of t of them, ie,

$$P \cap R = P_{i_1} \cup \cdots \cup P_{i_t}, \tag{6.8}$$

for some indices $i_1 \leq \cdots \leq i_t$. The t indices need not be distinct, so this framework allows the (likely) possibility that some $P \cap R$ might be expressible by fewer than t subsets.

We dispense with a formal discussion of the model and its relevance. The intuition is fairly obvious, however: The weights of the points in each P_i have been added up and stored in the database (hence, the size m). A query is answered by collecting an appropriate, preferably small, collection of P_i's and adding up their weights. The precomputed weights are used as shortcuts to speed up the computation. Proving lower bounds becomes an extremal problem on set systems. The following two results provide lower bounds on $t(m)$ for two basic problems. In both cases, the number m can be chosen arbitrarily.

Theorem 6.28 *Given n points in $\mathbf{R}^d$, range searching with respect to simplices requires $\widetilde{\Omega}(n/m^{1/d})$ additions on a database of size m.*

Theorem 6.29 *Given n points in $\mathbf{R}^d$, range searching with respect to axis-parallel boxes requires $\Omega(\log n/ \log(2m/n))^{d-1}$ additions on a database of size m.*

Recall that the notation $\widetilde{\Omega}(f(n))$ means $\Omega(f(n))/(\log n)^{O(1)}$. The two proofs are quite different and give two new interesting perspectives on extremal geometric graph theory. Both of them are textbook examples of the discrepancy method.

---

[16]See how useful lower bounds can be: the ultimate show-stoppers.

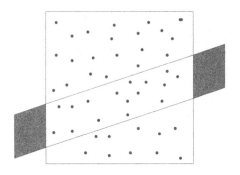

Fig. 6.18. Counting points in slabs.

Simplex Queries: An Isoperimetric Inequality

We consider the set system $(P, \mathcal{R})$ formed by a set P of n points in $[0, 1]^d$ and sets $\{R \in \mathcal{R}\}$, each one defined as the intersection of P and a closed slab of width $w = 1/cm^{1/d}$, for some large enough constant c. Throughout the proof we make the following (obviously nonrestrictive) assumptions:[17] $d > 1$ and $cn \leq m \leq n^d / \log^c n$. The idea behind the proof of Theorem 6.28 is to show that no large P_i can be used for "too many" queries.

Lemma 6.30 *If the incidence matrix of a set system $(P, \mathcal{R})$ has at least pN ones, where $N = |\mathcal{R}|$, and it does not contain any y-by-$\lfloor pN/2my \rfloor$ submatrix of ones, for any integer $y \leq y_0$, then $t(m) > y_0 m/N$.*

Proof: By contradiction, suppose that $t = t(m) \leq y_0 m/N$. Among the precomputed sets P_i, consider the *fat* ones, ie, those of size at least $pN/2my_0$. Of the t sets involved in answering a query $R \in \mathcal{R}$, the nonfat ones cover fewer than $tpN/2my_0 \leq p/2$ points. Summing over all R's, we find that, counting multiplicity, fat P_i's cover more than $pN/2$ points of P. But, as we show below, no fat P_i can contribute as much as $pN/2m$ to this count; so the number of fat P_i's must exceed m, which is a contradiction.

Let y be the number of queries for which a given fat P_i is used. We cannot have $y \geq y_0$, since this would produce a y_0-by-$\lfloor pN/2my_0 \rfloor$ submatrix of ones. But $y < y_0$ implies that $|P_i|$ is less than $pN/2my$, and hence the contribution of P_i to the count is $y|P_i| < pN/2m$. $\square$

---

[17] Note that m must be at least n or it is impossible to answer certain queries, so the lower bound is trivial in the case $d = 1$.

Theorem 6.28 follows directly from the next lemma; recall that $w = 1/cm^{1/d}$. The polylogarithmic factor hidden in the $\widetilde{\Omega}$ notation is in fact $1/\log n$. It is possible to remove it in two dimensions, but the higher dimensional case is still open. (See bibliographical notes.) □

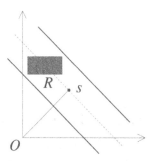

Fig. 6.19. The slab of width w indexed by s containing the box R.

Lemma 6.31 *There exist a set of n points in $\mathbf{R}^d$ and a collection of $N = n^b$ slabs, for some constant $b > 1$, that together form a set system whose incidence matrix has pN ones and no y-by-$\lfloor pN/2my \rfloor$ submatrix of ones, for any $y \le w^{d+1}nN/\log n$.*

What follows is the proof of the lemma. It is broken down into auxiliary lemmas. As usual, we assume that n is large enough. We postpone the discussion of the set P and instead address the selection of slabs right away. All of the slabs have the common width $w = 1/cm^{1/d}$, so each one is specified entirely by its bisecting hyperplane. We consider the set of slabs whose bisecting hyperplanes intersect the unit cube. A standard result in integral geometry says that there is a unique probability measure over such hyperplanes (and hence slabs) that is invariant under rigid motions [265]. Since our method is based on integration, we may assume that no slab ever passes through the origin. Thus, a slab can be specified by s, the point on its bisecting hyperplane nearest to the origin. Up to a constant factor (which, for our purposes, we can simply ignore), the differential element at the slab indexed by $s = (s_1, \ldots, s_d)$ is

$$d\omega(s) = \frac{ds_1 \wedge \cdots \wedge ds_d}{\|s\|_2^{d-1}} .$$

The heart of the lower bound proof rests in the following lemma. It is an isoperimetric inequality that bounds the probability that a random slab encloses a box as a function of the volume of the box. The relation with the forbidden submatrix of ones in Lemma 6.31 is simple: The probability will give the number of rows and the volume the number of columns.

Lemma 6.32 *Given a box $R \subseteq [0,1]^d$, the probability that a random slab encloses R is $O(w^{d+1}/\mathrm{vol}_d R)$.*

Proof: The probability, denoted by π, that a random slab encloses R is at most $\int d\omega(s)$, where the integral is taken over all the slabs (indexed by) s that enclose R. (We may not have equality because the bisecting hyperplane of an enclosing slab may not always intersect the unit cube.) To upper-bound this integral, we can always assume, by motion invariance, that the box is of the form $\prod_{1 \le i \le d}[1, r_i]$, where each $r_i > 1$. Placing the box away from the origin in this way has one advantage: We can assume that the vector Os is not too short. Before we show why this is a good thing, let us prove that it is, indeed, the case.

By symmetry we can restrict the domain of integration to $s \ge 0$ (ie, each $s_i \ge 0$). To see why is easy: The hyperplanes normal to the axes passing through the center of R define d reflections generating a subgroup of symmetry of order 2^d. Any slab indexed by an s with at least one negative coordinate can thus be mapped uniquely to a point $s' \ge 0$ (Fig. 6.20). So, up to a factor of 2^d, we can upper-bound the probability π by integrating $d\omega(s)$ over slabs indexed by $s \ge 0$. Assuming from now on that $s \ge 0$, we prove that

$$\tfrac{1}{2} < \|s\|_2 < 2d. \tag{6.9}$$

The slab must contain $(1, \ldots, 1)$, so

$$\left| \sum_{1 \le i \le d} s_i - \|s\|_2^2 \right| \le \frac{w}{2}\|s\|_2, \tag{6.10}$$

and since $x < 1 + x^2/2$ for all x,

$$\|s\|_2^2 - \frac{w}{2}\|s\|_2 \le \sum_{1 \le i \le d} s_i < d + \frac{\|s\|_2^2}{2}.$$

It follows that

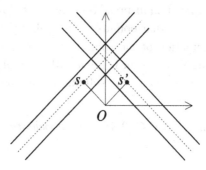

Fig. 6.20. Reflecting a slab about the vertical axis to make its parameter coordinates nonnegative: $s = (s_1, s_2) \mapsto s' = (-s_1, s_2)$.

$$\|s\|_2 < \frac{w}{2} + \sqrt{w^2/4 + 2d},$$

and with w small enough, $\|s\|_2 < 2d$. Similarly, by (6.10) and $s \geq 0$,

$$\|s\|_2^2 + \frac{w}{2}\|s\|_2 \geq \sum_{1 \leq i \leq d} s_i \geq \|s\|_2,$$

and therefore $\|s\|_2 \geq 1 - w/2 > 1/2$, which establishes (6.9). It follows from it that one of the coordinates must exceed $1/2\sqrt{d}$. By symmetry we can assume that the coordinate in question is s_1, which will cause the loss of at most a factor of d. If Δ is the set of $s \geq 0$ $(s_1 > 1/2\sqrt{d})$ whose associated slabs enclose R, then

$$\pi \ll \int_\Delta d\omega(s). \tag{6.11}$$

To tackle this integral, we perform the following change of variables: $u_1 = \|s\|_2$ and, for $i > 1$, let $u_i = s_i/u_1$. The transformation acts bijectively between

$$\left\{ (s_1, \ldots, s_d) \in \mathbf{R}^d \setminus \{O\} \,|\, s_1 \geq 0 \right\}$$

and

$$\left\{ (u_1, \ldots, u_d) \,|\, u_1 > 0 \text{ and } \sum_{2 \leq i \leq d} u_i^2 \leq 1 \right\}.$$

To compute its Jacobian J_u, we notice that

$$\frac{\partial u_i}{\partial s_j} = \begin{cases} s_j/u_1 & \text{if } i = 1; \\ 1/u_1 - s_i^2/u_1^3 & \text{if } i > 1 \text{ and } i = j; \\ -s_i s_j/u_1^3 & \text{if } i > 1 \text{ and } i \neq j. \end{cases}$$

It follows that

$$\det J_u = \begin{vmatrix} s_1/u_1 & s_2/u_1 & \cdots & s_d/u_1 \\ -s_2 s_1/u_1^3 & 1/u_1 - s_2^2/u_1^3 & \cdots & -s_2 s_d/u_1^3 \\ \vdots & \vdots & \ddots & \vdots \\ -s_d s_1/u_1^3 & -s_d s_2/u_1^3 & \cdots & 1/u_1 - s_d^2/u_1^3 \end{vmatrix}.$$

We derive:

$$\det J_u = \frac{\left(\prod_{1 \leq i \leq d} s_i\right)^2}{s_1 u_1^{3d-2}} \times \begin{vmatrix} 1 & 1 & \cdots & 1 \\ -1 & (u_1/s_2)^2 - 1 & \cdots & -1 \\ \vdots & \vdots & \ddots & \vdots \\ -1 & -1 & \cdots & (u_1/s_d)^2 - 1 \end{vmatrix}.$$

The determinant above is made triangular by subtracting the first column from the others, which gives $\det J_u = s_1/u_1^d$. Let $s(u)$ denote the point s in bijection with u. By (6.9, 6.11),

$$\int_{s(u) \in \Delta} du_1 \wedge \cdots \wedge du_d = \int_{s \in \Delta} \frac{s_1}{u_1^d} ds_1 \wedge \cdots \wedge ds_d$$

$$\gg \int_{s \in \Delta} ds_1 \wedge \cdots \wedge ds_d \gg \pi.$$

Note that $|\langle p, s \rangle - \|s\|_2^2| \leq (w/2)\|s\|_2$ for the two corners of the box R: $p = (1, \ldots, 1)$ and $p = (r_1, \ldots, r_d)$. Since $s \geq 0$, this shows that

$$-\frac{w}{2} \|s\|_2 \leq \sum_{1 \leq i \leq d} s_i - \|s\|_2^2 \leq \sum_{1 \leq i \leq d} r_i s_i - \|s\|_2^2 \leq \frac{w}{2} \|s\|_2,$$

and, by (6.9),

$$\sum_{1 < i < d} (r_i - 1)s_i \leq w\|s\|_2 < 2dw.$$

Because $s \geq 0$ and $r_i > 1$,

$$0 \leq u_i \leq \frac{w}{r_i - 1}$$

for each $i \geq 2$, and

$$0 \leq s_i < \frac{2dw}{r_i - 1} \qquad (6.12)$$

for each $i \geq 1$. When $u_2, \ldots, u_d$ are fixed, the direction of the vector s is fixed, and so $u_1 = \|s\|_2$ can only vary in an interval of length no greater than w. It easily follows that

$$\int_{s(u) \in \Delta} du_1 \wedge \cdots \wedge du_d \leq \frac{w^d}{\prod_{2 \leq i \leq d} (r_i - 1)}.$$

We know from (6.12) that $s_1 < 2dw/(r_1 - 1)$; therefore,

$$r_1 - 1 < 4d\sqrt{d}\,w,$$

and consequently

$$\pi \ll \frac{w^{d+1}}{\prod_{1 \leq i \leq d} (r_i - 1)}.$$

$\square$

We now have the tools for proving Lemma 6.31. The previous lemma presents us with an infinite set of slabs. We can discretize it, however, by restricting the points s specifying the slabs to lie in a polynomial-size low-discrepancy sample. (One way to do this is to sample the parameter space according to the motion-invariant measure. We did this sort of thing on page 30 already, so we skip the details.) It is easy to make the number N of slabs equal to n^b while at the same time ensuring that the probability, over the continuous uniform distribution, that a random slab contains a box differs from its counterpart over the discrete uniform distribution by at most $1/n^{b'}$; here, both b and $b' = b'(b)$ are arbitrarily large constants. In this way, Lemma 6.32 still holds verbatim, since $w^{d+1}/\mathrm{vol}_d R$ is not too small, ie, it exceeds $1/n^{2d}$.

The set P is obtained by taking n random points uniformly distributed independently in $[0, 1]^d$. With high probability, the number of ones in the incidence matrix of the set system is at least $c_0 wnN$, for a constant $c_0 > 0$ independent of c. To see why, observe that, because w tends to 0 as n goes to infinity, we can assume that over 99% of the slabs "eat up" a chunk of the unit cube of volume $\Omega(w)$. With high probability, any such slab contains $\Omega(wn)$ points. The proof is elementary: Up to a constant factor, the expected number of points in a given slab is $\Omega(wn) \gg \log^2 n$; so by Chernoff's bound (see Appendix A) the number is $\Omega(wn)$ with probability at least $1 - 1/n^{2b}$ for one slab (conservatively), and hence $1 - 1/n^b$ for all slabs.

To summarize, we have exhibited a set system $(P, \mathcal{R})$ formed by n points and N slabs, such that (i) its incidence matrix has pN ones, where $p \geq c_0 wn$, and (ii) P satisfies the Heilbronn property (see Lemma 6.27, page 266).[18] To establish Lemma 6.31, we assume, by contradiction, that some $S \subseteq P$ is a common subset of at least y sets of $(P, \mathcal{R})$, where $y \leq w^{d+1} nN/\log n$ and $|S| = \lfloor pN/2my \rfloor$. Because $d > 1$, we have $|S| \geq \lfloor \frac{1}{2} c_0 c^d \log n \rfloor > c \log n$, and so, by the Heilbronn property, the convex hull of S has volume $\gg |S|/n$. This convex hull contains a box R of volume at least proportional to it. Indeed, we know that this is true of the largest enclosed ellipsoid,[19] and hence of the largest (arbitrarily rotated) enclosed box. By Lemma 6.32 (and our previous observation about the discrete approximation of the measure on slabs), the total number of slabs enclosing R is

$$O(Nw^{d+1}/\text{vol}_d R) = O(y/c^d) < y,$$

which gives a contradiction, and proves Lemma 6.31. $\square$

Box Queries: The Hyperbolic Boundary

We prove Theorem 6.29 (page 270) in the two-dimensional case only. Extensions to higher dimension are possible, but they present technical difficulties, none of which is particularly germane to the discrepancy method. The input set P consists of n (carefully chosen) points in the unit square $[0, 1]^2$. Given m subsets $P_i \subseteq P$, we want to exhibit a "hard" axis-parallel box B, ie, one whose associated set $P \cap B$ cannot be expressed as the union of too few P_i's.

The key idea is to focus, not on the precomputed sets P_i themselves, but on the upper right corner p_i of their smallest enclosing box. Specifically, let $P_i \subseteq P \subseteq [0, 1]^2$ be one of the m sets whose weighted sum has been precomputed. We define p_i to be the upper right vertex of the smallest axis-parallel box enclosing P_i (Fig. 6.21).

As a matter of terminology, we say that a point $p = (p_x, p_y)$ *dominates* q

---

[18] Note that the lemma implies an absolute lower bound for the constant c. Of course, any higher value for c (like the one we are using here) works just the same.

[19] See footnote on page 268.

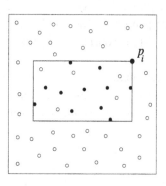

Fig. 6.21. The set P_i consists of the filled dots.

if $p_x \geq q_x$ and $p_y \geq q_y$. For the purpose of the lower bound, we restrict our queries to axis-parallel boxes that contain the origin. Given such a box B, we assume that the set $P \cap B$ can be written as the union of t of the sets P_i; see (6.8) on page 270. Obviously, the t corners p_i derived from these P_i's *collectively dominate* $P \cap B$. By this we mean that each point in $P \cap B$ is dominated by at least one of the t corners.

So, we can think of the problem quite differently. Given n blue points and m red points in the unit square, we define the *cost* of a box $B_q = [0, q_x] \times [0, q_y]$, where $q = (q_x, q_y) \in [0, 1]^2$, to be the minimum number of red points in B_q that collectively dominate the blue points in B_q. Note that to avoid infinite costs we must have $m \geq n$. We show that with "only" m red points at our disposal and a low-discrepancy blue set, some boxes do not have any small dominating red set.

The lower bound construction is built around the idea of a *room* R_q (Fig. 6.22): This is a box within $[0, 1]^2$ with upper right corner q, divided up into four smaller boxes that share a common corner along the diagonal (of positive slope) of R_q. Of the four, we focus on the upper right (resp. lower left) box, which we call the *NE-chamber* (resp. *SW-chamber*). The room and its subdivision are calibrated so that the NE-chamber (resp. SW-chamber) is of area $1/8m$, (resp. $16/n$). We easily verify that the area of R_q is

$$\left(\frac{1}{\sqrt{8m}} + \frac{4}{\sqrt{n}} \right)^2 < \frac{20}{n}.$$

Note that this area is independent of the actual shape of the room.

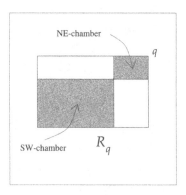

Fig. 6.22. The room R_q with its two chambers.

Lemma 6.33 *If $q \in [1/2, 1]^2$, the box $[0, q_x] \times [0, q_y]$ contains a collection of $\Omega(\log n / \log(2m/n))$ rooms R_q such that the intersection of any two of them is disjoint from both their SW-chambers. This collection can be chosen to be the same for all q (up to translation).*

Proof: Figure 6.23 illustrates the lemma. As we just observed, all rooms have the same area A, irrespective of their aspect ratio. We build the first room R_q by choosing the width $1/2$ and the height $2A$. We easily check that the ratio, width of NE-chamber over width of R_q, is equal to

$$\sqrt{\frac{\text{area of NE-chamber}}{\text{area of } R_q}} = \frac{1}{1 + 8\sqrt{2m/n}}.$$

Note that this ratio is independent of the particular room R_q. This means that, to carve out rooms with the disjointness condition of the lemma, it suffices to pick a sequence of widths, beginning at $1/2$, that is decreasing geometrically with a ratio of $1 + 8\sqrt{2m/n}$. The heights increase by the same ratio. Elementary geometry shows that neither SW-chamber overlaps with the other room. Since the rooms must stay within the unit square, we safely bound heights and widths by $1/2$, which gives us a number of such rooms at least proportional to

$$\frac{\log n}{\log(1 + 8\sqrt{2m/n})}.$$

$\square$

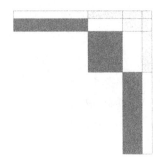

Fig. 6.23. Three rooms R_q with disjoint SW-chambers.

Remark: The placement of rooms R_q suggests a hyperbola: Look where the SW-corners lie (Fig. 6.23). In the metric where the distance between two points is the area of the smallest enclosing box, a ball has hyperbolic branches as its boundary (hence the subtitle of this section). Should one be surprised that so much attention is given to the region around a hyperbola near the NE-corner of B_q? Of course not. The discrepancy theory of axis-parallel boxes (§3.1) shows that the "action" is to be found near the upper right corner of a random box, specifically, in the cells of the Haar wavelet grids incident to that corner. Such cells lie right above a hyperbola. The function for a hyperbola, $f(x) = 1/x$, has indefinite integral $\ln x$, which is the reason why logarithms turn up in the discrepancy of boxes as well as in Lemma 6.33.

We conclude the proof of Theorem 6.29 (page 270). Without loss of generality, we assume that n is a power of two. The set P is formed by the bit-reversal permutation (Halton-Hammersley in two dimensions), as explained in §6.1 (page 244); see also §2.2. As was observed, the grid obtained by subdividing $[0, 1]^2$ into n boxes of size $2^{-k} \times (2^k/n)$, for any fixed $0 \leq k \leq \log n$, has the property that every cell contains exactly one blue point (ie, point of P). Any interval in $[0, 1]$ contains a dyadic interval, ie, one of the form $[K/2^{-j}, (K + 1)2^{-j}]$, at least one-fourth its length; so any box of area $\geq 16/n$ encloses at least one box of area $\geq 1/n$ that is the Cartesian product of two dyadic intervals. Such a box necessarily contains a dyadic product of area exactly $1/n$, ie, a grid cell (for some k). It follows that the SW-chamber of any room contains at least one blue point.

Given $q \in [1/2, 1]^2$, let $\mathcal{S}_q$ denote the collection of rooms in Lemma 6.33. Choose a point q randomly, uniformly in $[1/2, 1]^2$. Each of the SW-chambers in the rooms of S_q contains at least one blue point. Since q is random over an area $1/4$, a given NE-chamber, being of area of $1/8m$, is free of red points with probability at least $1/2$. By linearity of expectation, it follows that the expected number of red-free NE-chambers among the rooms of S_q is at least half the total, and so there exists some S_q at least half of whose NE-chambers are free of red points. The corresponding SW-chambers, called *good*, have the following property: Each one contains a blue point, but its corresponding NE-chamber has no red point.

Because of the disjointness condition of the rooms in S_q, SW-chambers are naturally ordered from left to right and from top to bottom, and any point that dominates three points in distinct SW-chambers dominates the middle chamber entirely. Thus, no red point in B_q can dominate blue points in more than two distinct good SW-chambers. It follows that the cost of B_q, ie, the minimum number of red points in B_q needed to dominate collectively the blue points in B_q, is at least half the number of good SW-chambers, ie, $\Omega(\log n / \log(2m/n))$. This proves the two-dimensional case of Theorem 6.29 (page 270). $\square$

6.5 Bibliographical Notes

Section 6.1:   The Morgenstern bound, named after its discoverer, was established in [234]. The spectral lemma is due to Chazelle [74]. The lower bounds for box and triangle range searching in the general arithmetic model (Theorem 6.8 on page 243 and Theorem 6.13 on page 249) were also established by Chazelle in [71] and [74], respectively. The trace lemma was introduced and used by Chazelle and Lvov [79, 80] to prove Theorems 6.15 and 6.16 (page 251).

For the interested reader, we should also mention the existence of an intriguing combinatorial approach due to Valiant [312]: The idea is to relate circuit size to *matrix rigidity*, which measures how many entries must be changed in order to reduce the rank. Unfortunately, this has proven a rather difficult quantity to evaluate. See also [16] for a study of linear circuits over $\mathbf{F}_2$.

Section 6.2:   The relationship between eigenvalues and data structures for range searching was discovered and investigated by Chazelle [71, 74].

Section 6.3:   In the monotone circuit model, the lower bounds for axis-parallel boxes (Theorem 6.20, page 257) and for simplices (Theorem 6.22, page 257) were proven by Chazelle [71]; they are essentially optimal (within polylog or polyloglog factors). For the case of triangles, a dynamic version of Theorem 6.21 was proven by Fredman [134]; the lower bound requires the use of inserts and deletes. Lemma 6.25 (page 263) is attributed to Erdős [118].

Section 6.4:   Chazelle [65] established the lower bound for range searching with simplices (Theorem 6.28, page 270). As we saw in Theorem 5.7 (page 215), this bound is essentially optimal. The lower bound for axis-parallel boxes (Theorem 6.29, page 270) was also proven by Chazelle in [67]. The proof given here, being in two dimensions, is much simpler. The lower bound is tight for any $m \gg n(\log n)^{d-1+\varepsilon}$, with any fixed $\varepsilon > 0$.

Other lower bound proofs in the arithmetic model are given in [58, 83, 133, 311, 331]. More restrictive lower bounds for pointer machines or partition-based methods can be found in [66, 84, 125, 126]. There is a huge literature on range searching, in particular in the on-line (database) model. A good starting place is the survey article by Agarwal and Erickson [6] or the one by Matoušek [214].

7

Convex Hulls and Voronoi Diagrams

![drop cap T] he reader looking here for working codes may be disappointed. The problem of how to compute the convex hull of n points in optimal time is viewed mostly through a theoretical lens. Optimality is understood here in a worst-case setting: Given n points in $\mathbf{R}^d$, the convex hull is a polytope with $O(n^{\lfloor d/2 \rfloor})$ faces, and possibly as many as that. It is well known that computing the polytope entails sorting, so the complexity we are aiming for is $O(n \log n + n^{\lfloor d/2 \rfloor})$. To approach this complexity is reasonably easy, but to design an optimal deterministic algorithm is surprisingly challenging. In this chapter we do just that.

The algorithm is to this day the most sophisticated example of derandomization in computational geometry. It is also the most unlikely, considering how hopeless its basic line of attack might seem at first. To overcome such odds, the whole kitchen sink of sampling technology developed in the previous chapters is called into action. Interestingly, sampling is used for two very different purposes: One is to provide a divide-and-conquer mechanism (as we've seen in Chapter 5), and the other is to evaluate complicated potential functions approximately very fast (as we've seen nowhere yet). No other (deterministic) algorithm has yet been found for computing convex hulls optimally. So, perhaps more than any other, this chapter shows how deep and uniquely powerful the discrepancy method is in the area of algorithm design.

The algorithm brings together many concepts and ideas. Fortunately, a good deal of them have already been introduced in previous chapters. As explained in Appendix C, the problem of computing the convex hull of n points in $\mathbf{R}^d$ reduces, by duality, to computing the intersection of n halfspaces (Fig. 7.1). Furthermore, computing the Voronoi diagram (or, equivalently, the Delaunay triangulation) of a finite set of points[1] in Euclidean d-space can be reduced to a convex hull problem in $(d + 1)$-space. It thus appears that the halfspace intersection problem holds the key to both the convex hull and the Voronoi diagram problem.

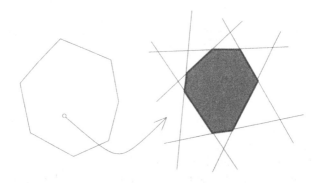

Fig. 7.1. A convex hull is, by duality, an intersection of halfspaces.

The intersection of n halfspaces is a convex polyhedron of combinatorial complexity $O(n^{\lfloor d/2 \rfloor})$. We show that, in any fixed dimension d, its facial graph can be computed deterministically in time $O(n \log n + n^{\lfloor d/2 \rfloor})$. Because the asymptotic bound of $O(n^{\lfloor d/2 \rfloor})$ on the combinatorial complexity is tight for certain inputs, the complexity of the algorithm is optimal in the worst case. By the reductions mentioned above, we derive the immediate corollary:

Theorem 7.1 *The convex hull of a set of n points in $\mathbf{R}^d$ can be computed deterministically in $O(n \log n + n^{\lfloor d/2 \rfloor})$ time, for any fixed $d > 1$.*

Theorem 7.2 *The Voronoi diagram of a set of n points in $\mathbf{E}^d$ can be computed deterministically in $O(n \log n + n^{\lceil d/2 \rceil})$ time, for any fixed $d > 1$.*

In some sense, computing the intersection of halfspaces is as simple as one could hope for: We insert each halfspace one after the other and maintain the current intersection at all times. Although such a naive approach might not always be optimal, it can be shown that, if the insertion sequence corresponds to a random permutation of the halfspaces, then the

---

[1] For Delaunay triangulations, some general-position assumptions are necessary: It is assumed that no $d + 1$ points lie in a common hyperplane and no $d + 2$ points lie on a common $(d - 1)$-sphere.

expected complexity of the algorithm is optimal. Of course, one needs a supporting data structure to allow for efficient updating of the intersection. Surprisingly, a simple *conflict-graph* structure works. This consists of a triangulation of the current intersection polyhedron, together with a bipartite graph indicating which halfspace intersects which cell of the triangulation. This leads to an optimal probabilistic algorithm. We show how to derandomize the method to produce an optimal deterministic algorithm. The idea is to compute deterministically a "random-looking" permutation of the halfspaces that behaves as well as the random one used in the probabilistic algorithm. This is where most of the effort goes; to maintain the conflict graph is the easy part. Our main result below implies Theorems 7.1 and 7.2.

Theorem 7.3 *The facial graph of the polyhedron formed by the intersection of n halfspaces in $\mathbf{R}^d$ can be computed in $O(n \log n + n^{\lfloor d/2 \rfloor})$ time.*

7.1 Geode and Conflict Lists

Let H be a set of n hyperplanes in $\mathbf{R}^d$ ($d > 1$). We assume that none of them passes through the origin O and that the hyperplanes are in general position. Since the theorem aims only at a worst-case bound, anyway, we can always use symbolic perturbation methods [120] to achieve the effect of general position. Let $H^\cap$ denote the closure of the full-dimensional cell enclosing O in the arrangement formed by H. Our goal is to compute a full facial description of $H^\cap$. Some notation:

- Given a simplex[2] s and $R \subseteq H$, let $R_{|s}$ designate the subset of hyperplanes of R that intersect but do not contain s.
- Let $V(R)$ denote the set of vertices of the arrangement of $R \subseteq H$. If P is a polyhedron, then $V(P)$ refers to its vertex set. If s is a simplex (not necessarily full-dimensional), $V(R, s)$ denotes the set of vertices within s in the arrangement formed by R and the affine span of s. Given a vertex v of $V(H)$, the *conflict list* of v, denoted by $H_{|Ov}$, is the set of hyperplanes of H separating O from v, ie, intersecting the relative interior of the segment (Fig. 7.2). The size of the conflict list of v is denoted by n_v. By extension,

---

[2] In this chapter, the term "simplex" refers to the relative interior of the convex hull of at most $d + 1$ points. The dimension of its affine span is between 0 and d.

we define the conflict list $H_{|Os}$ of a simplex s as the union of the conflict lists of its vertices; equivalently, it includes exactly the hyperplanes that intersect the relative interior of the convex hull of $s \cup \{O\}$. Similarly, we set $n_s = |H_{|Os}|$.

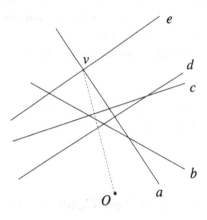

Fig. 7.2. The conflict list $H_{|Ov} = \{b, c, d\}$.

Given $R \subseteq H$, we define a canonical triangulation of $R^\cap$, called the *geode* of R and denoted by $\mathcal{G}(R)$ (Fig. 7.3). This is similar to the triangulation defined in Chapter 5. For completeness, we describe it again. We assume that $R^\cap$ is bounded. We can always add a fictitious box large enough to enclose all the vertices of $\mathcal{G}(R)$, if necessary. For $k = 1, 2, \ldots, d$, in that order, triangulate each k-face f of $R^\cap$ recursively as follows. The case $k = 1$ is obvious, so assume that $1 < k < d$. Let the *apex* v of f be its vertex with minimum n_v. (Break ties by taking the vertex with the lexicographically smallest coordinate vector.) Lift centrally toward v the triangulation of each j-face ($j < k$) of $R^\cap$ that lies within the boundary of f but is not incident upon v. By our choice of v this produces a triangulation of f that is consistent with that of its boundary. (This is because v is also the vertex chosen for triangulating the faces on the boundary of f that are incident upon v.) For $k = d$, simply lift toward O the triangulation of $\partial R^\cap$ just obtained. It follows easily from the Upper Bound Theorem that the size of the geode is $O(|R|^{\lfloor d/2 \rfloor})$. The next result motivates our choice of triangulation.

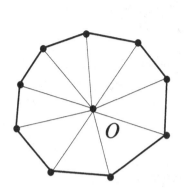 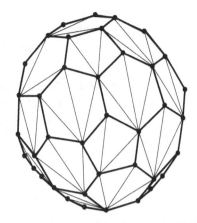

Fig. 7.3. A geode in $\mathbf{R}^2$ (inside view) and $\mathbf{R}^3$ (outside view).

Lemma 7.4 *Given any constant $c \geq 1$ and any $R \subseteq H$, there exists a constant C (dependent on c) such that*

$$\sum_{s \in \mathcal{G}(R)} n_s^c \leq C \sum_{v \in V(R^\cap)} n_v^c.$$

Proof: We prove by induction that, for any k-face f of $R^\cap$, the sum $S_f = \sum n_s^c$, over all faces s of $\mathcal{G}(R)$ that lie within the closure $\mathrm{cl} f$ of f, satisfies

$$S_f \leq k!(2^c + 1)^k \sum_{v \in V(R^\cap) \cap \mathrm{cl} f} n_v^c,$$

where $0! = 1$. Plugging in $k = d$, where f is the interior of $R^\cap$, proves the lemma.

Vertices are 0-faces, so the case $k = 0$ is obvious. Assume now that $k > 0$. Any simplex s of the geode within $\mathrm{cl} f$ is obtained by lifting a simplex s' incident to f. By definition of the apex v, we have

$$n_s \leq n_{s'} + n_v \leq 2n_{s'}.$$

It follows that

$$S_f \leq (2^c + 1) \sum_g S_g,$$

where g ranges over all the $(k-1)$-faces of $R^\cap$ that are incident to f. (The term 1 comes from the contribution of the geode faces incident to f, while the term 2^c accounts for the simplices within f itself.) Because of general position, no vertex can belong to more than k faces of dimension $k-1$, and thus is counted at most k times in the sum $\sum_g S_g$. Substituting into the induction-hypothesis bound completes the proof. $\square$

In the next section, we set the stage for a probabilistic solution to the problem of computing $H^\cap$. We show how the method can be efficiently derandomized in the subsequent section to yield an optimal deterministic algorithm.

7.2 A Probabilistic Algorithm

To compute $H^\cap$, we insert the hyperplanes of H one at a time, breaking down the process into successive rounds. In the first round, a constant number c of input hyperplanes are inserted.[3] These hyperplanes need not be randomly chosen. Any will do; for convenience one might want to add the bounding hyperplanes of a large fictitious box, to ensure that the initial geode is bounded. This (optional) step is meant to simplify the maintenance of the geode, or at least our discussion of it.

Moving now to the i-th round, we let R denote the subset of hyperplanes of H that have been inserted in the $i-1$ previous rounds. We assume a standard facial-graph representation for $\mathcal{G}(R)$ that allows us to navigate across adjacent faces of the geode. We also keep the conflict list of each simplex of the geode. To open the i-th round, we begin by selecting a probability

$$p = \frac{2r}{3(n-r)},$$

where $r = |R|$. Next, we form a set S by randomly picking each hyperplane of $H \setminus R$, independently, with probability p. For each simplex $s \in \mathcal{G}(R)$, we use its conflict list to compute the portion of the arrangement of S within s, and from it we extract the intersection between s and $(R \cup S)^\cap$. Putting all of the pieces together, we derive the facial graph of $(R \cup S)^\cap$ and set up the conflict lists of the faces of the new geode $\mathcal{G}(R \cup S)$.

Observe that the average number of hyperplanes finding their way into $R \cup S$ is $5r/3$, and so we should expect the number of hyperplanes inserted to grow exponentially at each round, and hence the number of rounds to be logarithmic. After each round, R is updated to denote the new set $R \cup S$ of

---

[3]To simplify the notation, we use b and c to denote large enough constants, with b sufficiently larger than c. Both constants depend on d.

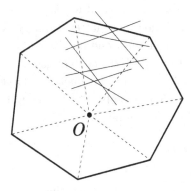

Fig. 7.4. Each simplex of the geode has its conflict list.

inserted hyperplanes; as soon as the size of R exceeds n/c, we insert all of the remaining hyperplanes by using the same method. By identifying the hyperplanes of S in the conflict list of s in time $O(n_s + 1)$, the arrangement of S within s is then easily computed in $O(|S_{|Os}|^d)$ time. Thus, it is routine to ensure that the work spent in round i is at most on the order of

$$\sum_{s \in \mathcal{G}(R)} \left(|S_{|Os}|^d + n_s + 1 \right). \tag{7.1}$$

Since the hyperplanes are chosen randomly, one should anticipate that, for a typical $s \in \mathcal{G}(R)$, the value of n_s is on the order of n/r. Of the n_s "cutting" hyperplanes, only a constant of them should end up in S, and so $|S_{|Os}|$ is expected to be $O(1)$. Unfortunately, this is not quite true of every simplex, but only of an "average" one. By using higher moments, however, we can capture this fact concisely in a manner that minimizes the negative effect of the "deviant" simplices. Some additional notation:

$$\begin{cases} q_s &= (r/n)n_s + 1, \\ r_s &= |S_{|Os}| + 1. \end{cases}$$

Note that although q_s and r_s might occasionally be large, for all practical purposes they will act as constants in our analysis and thus should be intuitively thought of as such. By the Upper Bound Theorem, $R^\cap$ has $O(r^{\lfloor d/2 \rfloor})$ vertices, and a typical vertex v should have $n_v = O(n/r)$. Therefore, a geode that conforms to one's expectation of a random choice of R should be a *semicutting*, which is our way of saying that it should satisfy

$$\sum_{v \in V(R^\cap)} n_v^c \leq N \stackrel{\text{def}}{=} br^{\lfloor d/2 \rfloor} \left(\frac{n}{r} \right)^c.$$

The sum is taken over the vertices, and not the simplices of the geode, to simplify the analysis. This does not matter since, by Lemma 7.4, an upper bound on the vertex sum implies one on the simplex sum. We assume inductively that, prior to round i, the current geode is a semicutting. (The case $i = 1$ is satisfied by choosing b sufficiently larger than c.) Having chosen R, we now pick S randomly. As we show below, the following conditions hold with high probability:

INV1 $r/2 \leq |S| \leq r$;

INV2 $\sum_{s \in \mathcal{G}(R)} r_s^c \leq b^2 r^{\lfloor d/2 \rfloor}$;

INV3   The geode of $R \cup S$ is a semicutting.

To see why these properties hold, it is convenient to define a function associated with each of them:

$$
\begin{cases}
A_1(S) &= \frac{1}{4r}(|S| - 2r/3)^2 , \\[2mm]
A_2(S) &= \frac{1}{b^2 r^{\lfloor d/2 \rfloor}} \sum_{s \in \mathcal{G}(R)} r_s^c , \\[2mm]
A_3(S) &= \frac{1}{N} \sum_{v \in V((R \cup S)^\cap)} n_v^c .
\end{cases}
$$

Note that R is understood and does not appear as an argument. Finally, we denote the expected values of these quantities by:

$$\mathcal{E}_j = \mathbf{E} A_j(S),$$

for $1 \leq j \leq 3$, and we define

$$\mathcal{E} = \mathcal{E}_1 + \mathcal{E}_2 + \mathcal{E}_3 .$$

These expectations refer to a random choice of S in round i, using the distribution mentioned earlier, conditioned upon the earlier choice of R, which is now fixed. By analogy with a thermodynamic system that cools down through an energy-minimizing process,[4] we refer to these expectations as *energies*. Next we show that, as anticipated, the energies are bounded by constants. It is worth noting that, of the three lemmas below, only one requires that $\mathcal{G}(R)$ should be a semicutting.

Lemma 7.5 *We have $\mathcal{E}_1 < 1/6$.*

---

[4]In this analogy, the entropy of the system, ie, the amount of randomness left in it, decreases over time.

Proof: The energy $\mathcal{E}_1$ is equal to $(\mathbf{var}|S|)/4r$. For independent variables, the variance behaves linearly, and so

$$\mathcal{E}_1 = \left(\frac{n-r}{4r}\right)p(1-p) < \frac{1}{6}.$$

□

Lemma 7.6 *If the geode of R is a semicutting, then $\mathcal{E}_2 = O(1/b)$.*

Proof: A hyperplane in $H_{|Os}$ contributes 1 to $|S_{|Os}|$ with probability p. By Lemma A.1, it follows that

$$\mathbf{E}\,|S_{|Os}|^c \le (c + p|H_{|Os}|)^c = (c + pn_s)^c = O(q_s^c).$$

Next, we sum up over all simplices s in $\mathcal{G}(R)$ and apply the fact that $\mathcal{G}(R)$ is a semicutting. By Lemma 7.4,

$$\mathbf{E}\left\{\sum_{s\in\mathcal{G}(R)} r_s^c\right\} = O\left(\sum_{s\in\mathcal{G}(R)} q_s^c\right) = O(br^{\lfloor d/2\rfloor}),$$

from which the lemma follows. □

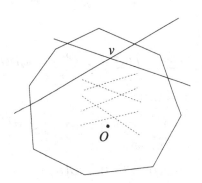

Fig. 7.5. The probability that v ends up a vertex of $(R\cup S)^\cap$ is $p^2(1-p)^5$.

Lemma 7.7 *We have $\mathcal{E}_3 = O(1/b)$.*

Proof: The probability of a given vertex $v \in V(H) \cap R^\cap$ turning up in $V((R\cup S)^\cap)$ is $p^{d_v}(1-p)^{n_v}$, where d_v counts the number of hyperplanes of $H\setminus R$ passing through v; recall that n_v is the number of hyperplanes separating v from O. Therefore,

$$N\mathcal{E}_3 = \mathbf{E}\left\{\sum_{v\in V((R\cup S)^\cap)} n_v^c\right\} = \sum_{v\in V(H)\cap R^\cap} p^{d_v}(1-p)^{n_v}\,n_v^c, \qquad (7.2)$$

where N is, as we recall, $br^{\lfloor d/2 \rfloor}(n/r)^c$. The sum above would be easily handled, were it not for that pesky factor n_v^c. To take care of it, we use a nice little trick. Consider another random sample T chosen in $H \setminus R$, where each hyperplane is now picked with probability $t = p/2$. By definition, the expected number Q of vertices of $(R \cup T)^{\cap}$ is

$$Q \overset{\text{def}}{=} \sum_{v \in V(H) \cap R^{\cap}} t^{d_v}(1-t)^{n_v}.$$

By the Upper Bound Theorem and Lemma A.1, it follows that[5]

$$Q \ll \mathbf{E}\left\{|R \cup T|^{\lfloor d/2 \rfloor}\right\} = O(r^{\lfloor d/2 \rfloor}).$$

Using the derivations

$$\frac{1-t}{1-p} \geq 1 + \frac{p}{2} \geq e^{p/4} \geq e^{r/8n},$$

we infer that

$$\begin{aligned}
Q &= \sum_v p^{d_v} 2^{-d_v}(1-p)^{n_v}\left(\frac{1-t}{1-p}\right)^{n_v} \\
&\geq 2^{-d}\sum_v p^{d_v}(1-p)^{n_v} e^{n_v r/8n}.
\end{aligned} \tag{7.3}$$

Obviously,

$$n_v^c = \left(\frac{8cn}{r}\right)^c \left(\frac{n_v r}{8cn}\right)^c \leq \left(\frac{8cn}{r}\right)^c e^{n_v r/8n}.$$

By (7.2) and (7.3) we find that

$$\begin{aligned}
N\mathcal{E}_3 &= \sum_{v \in V(H) \cap R^{\cap}} p^{d_v}(1-p)^{n_v} n_v^c \leq 2^d \left(\frac{8cn}{r}\right)^c Q \\
&= O\left((n/r)^c r^{\lfloor d/2 \rfloor}\right) = O(N/b).
\end{aligned}$$

$\square$

In the last two lemmas, the big-oh notation hides a constant that does not depend on b, but does depend on c. For b large enough, we thus have:

$$\mathcal{E} < \tfrac{1}{3}. \tag{7.4}$$

By Markov's inequality, it follows that, with probability at least $1/2$,

$$A_1(S) + A_2(S) + A_3(S) < 1,$$

---

[5] Recall that $\ll$ is our notation for $O()$.

which implies that conditions INV1–INV3 are satisfied. In particular, the geode of $R \cup S$ is a semicutting. It is worth noting that, for this to be true, Lemma 7.7 does not require that the geode of R itself should be a semicutting. This robustness property is crucial. Indeed, the derandomization procedure introduces errors into the computation, whose build-up and propagation between rounds must be avoided at all costs. Lemma 7.7 circumvents this potential difficulty.

Suppose that INV1–INV3 are satisfied at each round. It follows easily that each round i runs in optimal time. Why is that? Recall from (7.1) that the execution of that round takes time on the order of

$$\sum_{s \in \mathcal{G}(R)} \left(|S_{|Os}|^d + 1 + n_s \right).$$

By INV2, the first two terms of the sum add up to $O(r^{\lfloor d/2 \rfloor})$. To bound the third term, we use the fact that the geode of R is a semicutting, and therefore

$$\sum_{v \in V(R^\cap)} n_v^c \le b r^{\lfloor d/2 \rfloor} \left(\frac{n}{r} \right)^c.$$

By Hölder's inequality[6] and the Upper Bound Theorem, the inequality remains valid if we substitute 1 for c. It follows from Lemma 7.4 that

$$\sum_{s \in \mathcal{G}(R)} n_s \le \sum_{v \in V(R^\cap)} b n_v \le b^2 r^{\lfloor d/2 \rfloor} \left(\frac{n}{r} \right),$$

and thus the total amount of work at round i is $O(r^{\lfloor d/2 \rfloor} + n r^{\lfloor d/2 \rfloor - 1})$. By INV1, the size of each new sample grows (roughly) geometrically between rounds. So, if all three conditions are always satisfied, the total running time of the algorithm is then $O(n \log n + r^{\lfloor d/2 \rfloor})$, which is optimal. The deterministic algorithm that we describe below maintains all three conditions at each round. Note that this would be gross overkill if our target were an optimal probabilistic algorithm. By using linearity of expectation, we could do with simply bounding the expected completion time of each round—a much easier task.

---

[6] $\sum u_i v_i \le \left(\sum u_i^p \right)^{1/p} \left(\sum v_i^q \right)^{1/q}$, for any $u_i, v_i \ge 0$, $p, q > 1$, where $1/p + 1/q = 1$.

7.3 Derandomization

It seems natural to derandomize the probabilistic approach by using the method of conditional expectations discussed in Chapter 1. The idea is sound, although, as we shall see, it runs into a number of technical difficulties. The chief obstacle is that the expectations needed are too costly to compute, and we can only hope to estimate them with high enough accuracy.

Again we consider the algorithm at the opening of round i, assuming that the geode of R is a valid semicutting. Our next step is to find a new sample $S \subseteq H \setminus R$ such that $A_1(S) + A_2(S) + A_3(S) < 1$, with an eye toward satisfying the three conditions INV1–INV3. We order the hyperplanes of $H \setminus R$ arbitrarily, $h_1, \ldots, h_{n-r}$, and for each of them in turn we decide whether to accept it or reject it, ie, include it in S or not. Suppose that we have already decided on the status of $h_1, \ldots, h_k$. Let $S^{(k)}$ be the set of accepted hyperplanes among them. We define the energy $\mathcal{E}_j^{(k)}$ to be the conditional expectation

$$\mathcal{E}_j^{(k)} = \mathbf{E}\left[A_j(S) \,\middle|\, S^{(k)} = S \cap \{h_1, \ldots, h_k\} \right],$$

and again we denote their sum $\mathcal{E}_1^{(k)} + \mathcal{E}_2^{(k)} + \mathcal{E}_3^{(k)}$ by $\mathcal{E}^{(k)}$, which is the total energy at the k-th step. Finally, we let

$$\mathcal{E}_j^{(k|in)} = \mathbf{E}\left[A_j(S) \,\middle|\, S^{(k)} = S \cap \{h_1, \ldots, h_k\} \text{ and } h_{k+1} \in S \right]$$

and

$$\mathcal{E}_j^{(k|out)} = \mathbf{E}\left[A_j(S) \,\middle|\, S^{(k)} = S \cap \{h_1, \ldots, h_k\} \text{ and } h_{k+1} \notin S \right]$$

designate the corresponding energy after accepting and rejecting h_{k+1}, respectively. Recall that the method of conditional expectations stipulates that the hyperplane h_{k+1} should be accepted or rejected, depending on which outcome yields the lesser total energy $\mathcal{E}^{(k+1)}$. Because of the identity

$$\mathcal{E}^{(k)} = p\mathcal{E}^{(k|in)} + (1-p)\mathcal{E}^{(k|out)}, \qquad (7.5)$$

such a selection rule ensures that

$$\mathcal{E}^{(k+1)} \leq \mathcal{E}^{(k)},$$

for each $0 \leq k < n - r$, where $\mathcal{E}^{(0)}$ is the expectation $\mathcal{E}$ at the outset of round i, when no hyperplane has yet been committed to S. By (7.4), this indicates that, once S has been entirely selected, the energy is still less than 1/3. Since no randomization is left, we have, in effect,

$$\sum_{j=1}^{3} A_j(S^{(n-r)}) < \frac{1}{3},$$

and the sample S thus satisfies all three conditions. To turn this into an algorithm we must be able to compute the energies as we assemble S. The case of $\mathcal{E}_1^{(k)}$ or $\mathcal{E}_2^{(k)}$ is easy. Unfortunately, the same cannot be said of $\mathcal{E}_3^{(k)}$. So, instead of evaluating $\mathcal{E}^{(k)}$ exactly, we estimate it by computing the *approximate energy*

$$\mathcal{AE}^{(k)} = \mathcal{E}_1^{(k)} + \mathcal{E}_2^{(k)} + \mathcal{AE}_3^{(k)}.$$

The trick is to ensure that the approximate energy $\mathcal{AE}^{(k)}$ still follows a relation similar to (7.5), and tbus keep the basic derandomization approach valid.

Sharp Energy Estimation

Recall from (7.2) that, at the beginning of round i, the third energy component is of the form

$$\mathcal{E}_3^{(0)} = \frac{1}{N} \sum_{v \in V(H) \cap R^\cap} p^{d_v} (1-p)^{n_v} n_v^c,$$

where d_v is the number of hyperplanes of $H \backslash R$ passing through v. Breaking down the sum along the facial structure of the geode of R, we obtain

$$\mathcal{E}_3^{(0)} = \frac{1}{N} \sum_{s \in \mathcal{G}(R)} \sum_{v \in V(H) \cap s} p^{d_v} (1-p)^{n_v} n_v^c.$$

With the assumption of general position, a vertex v of $V(H)$ that lies within a j-simplex s of $\mathcal{G}(R)$ is contained in $d - j$ hyperplanes of R and j hyperplanes of $H \backslash R$, and so $d_v = j = \dim s$. It follows that $V(H) \cap s = V(H, s)$, and hence

$$\mathcal{E}_3^{(0)} = \frac{1}{N} \sum_{s \in \mathcal{G}(R)} \sum_{v \in V(H,s)} p^{d_v} (1-p)^{n_v} n_v^c. \tag{7.6}$$

For the time being, we assume the existence of an oracle that, given a j-simplex $s \in \mathcal{G}(R)$ as input, returns an approximation $\mathcal{O}(s)$ of the inner

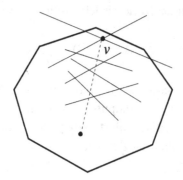

Fig. 7.6. The contribution of v to the third energy component is $p^2(1-p)^5 \cdot 5^c$, up to a scaling factor of $1/N$.

sum above. Specifically, we require that[7]

$$\left| \mathcal{O}(s) - \sum_{v \in V(H,s)} p^j (1-p)^{n_v} n_v^c \right| \leq E_s \stackrel{\text{def}}{=} \frac{n_s^c}{b^2 q_s^{\sqrt{c}}}. \qquad (7.7)$$

Initially, the approximate energy is

$$\mathcal{AE}_3^{(0)} \stackrel{\text{def}}{=} \frac{1}{N} \sum_{s \in \mathcal{G}(R)} \mathcal{O}(s).$$

The geode of R is assumed to be a semicutting, so by (7.6, 7.7) and the usual mix of Lemma 7.4 and the Upper Bound Theorem, we find that

$$\left| \mathcal{E}_3^{(0)} - \mathcal{AE}_3^{(0)} \right| \leq \frac{1}{N} \sum_{s \in \mathcal{G}(R)} E_s \leq \frac{1}{b^3 r^{\lfloor d/2 \rfloor}} \sum_{s \in \mathcal{G}(R)} q_s^{c-\sqrt{c}} = O(1/b^2). \quad (7.8)$$

This proves the case $k = 0$ of the following, more general result:

Lemma 7.8 We have $|\mathcal{E}_3^{(k)} - \mathcal{AE}_3^{(k)}| < 1/3$, for any $0 \leq k \leq n-r$.

Proof: Of course, for this lemma to make any sense, we need to define the approximate energy for arbitrary $k > 0$. Given a vertex $v \in (R \cup S^{(k)})^\cap$, let m_v be the number of hyperplanes among $\{h_{k+1}, \ldots, h_{n-r}\}$ in its conflict list (note that m_v depends on k). We now define d_v as the number of

---

[7]The reader should not be frightened by the exponent $\sqrt{c}$. This is just a convenient mechanism for using constants with implied order relations without introducing new notation. As hard to believe as it may be, the reader will come to appreciate this.

hyperplanes among $\{h_{k+1}, \ldots, h_{n-r}\}$ passing through v and write

$$\mathcal{E}_3^{(k)} = \frac{1}{N} \sum_v p^{d_v} (1-p)^{m_v} n_v^c, \tag{7.9}$$

where the summation extends over the vertices v of the arrangement of $R \cup S^{(k)} \cup \{h_{k+1}, \ldots, h_{n-r}\}$ that lie within the polyhedron $(R \cup S^{(k)})^\cap$; this excludes vertices lying on rejected hyperplanes.

We generalize the oracle $\mathcal{O}$ to provide good estimates on the contribution of a given face to the sum above. The input is a j-dimensional polytope σ that lies entirely within a single simplex $s \in \mathcal{G}(R)$, and whose affine span is (i) $\mathcal{R}^d$, or (ii) a hyperplane of H, or (iii) the intersection of several of them.[8] The oracle returns an approximation $\mathcal{O}^{(k)}(\sigma)$ such that

$$\left| \mathcal{O}^{(k)}(\sigma) - \sum_{v \in V(\{h_{k+1}, \ldots, h_{n-r}\}, \sigma)} p^j (1-p)^{m_v} n_v^c \right| \leq E_s, \tag{7.10}$$

with E_s defined as in (7.7). One might be tempted to approximate the energy $\mathcal{E}_3$ by considering each simplex s of the geode of R separately and, using the oracle, estimating the contribution of each face in the portion of the arrangement formed by $S^{(k)}$ within s. The problem with that approach is that the errors would grow to be too large. Instead, we use the oracle—the source of error—only to approximate the difference in energy between accepting or rejecting the new hyperplane.

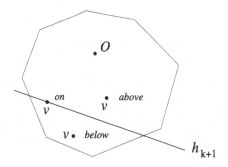

Fig. 7.7. What happens to vertex v if we accept h_{k+1}?

Consider the contribution of a given vertex to $\mathcal{E}_3$, depending on whether

---

[8] This excludes polytopes within the "walls" of $\mathcal{G}(R)$, ie, within faces of dimension less than that of the faces of $R^\cap$ in which they live. For correctness, we should extend the general-position assumption to rule out vertices within these walls.

the new hyperplane h_{k+1} is accepted or rejected. In the accepting case, the contribution of the vertices on the same side of h_{k+1} as the origin remains unchanged; for lack of a better word, we say that these vertices are *above* h_{k+1}. The contribution of the vertices *below* h_{k+1} is zero. For a vertex v that lies on h_{k+1}, the parameter d_v decreases by one, which has the effect of multiplying its energy contribution by $1/p$. Let $\mathcal{E}_{on}^{(k)}$ be the contribution of the vertices on h_{k+1} to the sum (7.9), and let $\mathcal{E}_{below}^{(k)}$ be the contribution of the vertices below the hyperplane. Obviously,

$$\mathcal{E}_3^{(k|in)} = \mathcal{E}_3^{(k)} - \mathcal{E}_{below}^{(k)} + \left(\frac{1}{p} - 1\right)\mathcal{E}_{on}^{(k)}.$$

Let Σ_{on} be the union, for all nonwall $s \in \mathcal{G}(R)$, of the collection of polytopes σ obtained by intersecting s with each face of the polytope $h_{k+1} \cap (R \cup S^{(k)})^{\cap}$; we include the polytope itself as a face. We define

$$\mathcal{AE}_{on}^{(k)} \overset{\text{def}}{=} \frac{p}{N} \sum_{\sigma \in \Sigma_{on}} O^{(k)}(\sigma). \tag{7.11}$$

Why the factor p? Because the oracle includes the factor $p^{\dim \sigma}$, whereas as far as $\mathcal{E}^{(k)}$, and hence $\mathcal{AE}_{on}^{(k)}$, are concerned, the desired factor is $p^{\dim \sigma+1}$.

We define Σ_{below} similarly to Σ_{on}, except that now s is clipped within the portion of $(R \cup S^{(k)})^{\cap}$ below h_{k+1}. This leads to

$$\mathcal{AE}_{below}^{(k)} \overset{\text{def}}{=} \frac{1}{N} \sum_{\sigma \in \Sigma_{below}} O^{(k)}(\sigma), \tag{7.12}$$

and, finally,

$$\mathcal{AE}_3^{(k+1)} \overset{\text{def}}{=} \mathcal{AE}_3^{(k|in)} = \mathcal{AE}_3^{(k)} - \mathcal{AE}_{below}^{(k)} + \left(\frac{1}{p} - 1\right)\mathcal{AE}_{on}^{(k)}.$$

The case where h_{k+1} is rejected is similar to the accepting case. The vertices lying in h_{k+1} no longer contribute anything to the energy, and for any vertex v below h_{k+1} the parameter m_v decreases by one, so their contribution to the energy is multiplied by $1/(1-p)$. This justifies introducing the following definition:

$$\mathcal{AE}_3^{(k+1)} \overset{\text{def}}{=} \mathcal{AE}_3^{(k|out)} = \mathcal{AE}_3^{(k)} + \left(\frac{1}{1-p} - 1\right)\mathcal{AE}_{below}^{(k)} - \mathcal{AE}_{on}^{(k)}.$$

From this follows an identity analogous to (7.5),

$$\mathcal{AE}_3^{(k)} = p\mathcal{AE}_3^{(k|in)} + (1-p)\mathcal{AE}_3^{(k|out)},$$

and hence

$$\mathcal{AE}^{(k)} = p\mathcal{AE}^{(k|in)} + (1-p)\mathcal{AE}^{(k|out)},$$

with the obvious meaning of $\mathcal{AE}^{(k|in/out)}$. As expected, h_{k+1} is accepted if $\mathcal{AE}^{(k|in)} < \mathcal{AE}^{(k|out)}$, and rejected otherwise. With this rule, we ensure that

$$\mathcal{AE}^{(n-r)} \leq \mathcal{AE}^{(0)}.$$

By (7.4), we know that $\mathcal{E}^{(0)} < 1/3$, and so, by (7.8),

$$\mathcal{AE}^{(n-r)} \leq \mathcal{E}_1^{(0)} + \mathcal{E}_2^{(0)} + \mathcal{AE}_3^{(0)} < \tfrac{1}{3} + \mathcal{AE}_3^{(0)} - \mathcal{E}_3^{(0)} < \tfrac{1}{2}. \qquad (7.13)$$

This immediately implies that

$$\mathcal{E}_1^{(n-r)} + \mathcal{E}_2^{(n-r)} < \tfrac{1}{2}.$$

Since no randomness is left at stage $k = n - r$,

$$\sum_{j=1,2} A_j(S^{(n-r)}) < \tfrac{1}{2},$$

and so the final sample $S = S^{(n-r)}$ satisfies conditions INV1 and INV2.

To complete the proof, we must perform the error analysis relative to the approximations used in the algorithm, and show that the upper bound that we have on $\mathcal{AE}_3^{(k)}$ is basically valid for $\mathcal{E}_3^{(k)}$ as well. By using the identities relating $\mathcal{AE}_3^{(k+1)}$ and $\mathcal{AE}_3^{(k)}$ in the accepting and rejecting cases, and combining them with the analog identities between $\mathcal{E}_3^{(k+1)}$ and $\mathcal{E}_3^{(k)}$, we derive

$$\left| \mathcal{E}_3^{(k)} - \mathcal{AE}_3^{(k)} \right| \leq \left| \mathcal{E}_3^{(0)} - \mathcal{AE}_3^{(0)} \right|$$

$$+ \sum_{i:\, h_i \in S^{(k)}} \left(\left| \mathcal{E}_{below}^{(i-1)} - \mathcal{AE}_{below}^{(i-1)} \right| + \frac{1-p}{p} \left| \mathcal{E}_{on}^{(i-1)} - \mathcal{AE}_{on}^{(i-1)} \right| \right)$$

$$+ \sum_{i \in \{1,\ldots,k\}:\, h_i \notin S^{(k)}} \left(\frac{p}{1-p} \left| \mathcal{E}_{below}^{(i-1)} - \mathcal{AE}_{below}^{(i-1)} \right| + \left| \mathcal{E}_{on}^{(i-1)} - \mathcal{AE}_{on}^{(i-1)} \right| \right).$$

By (7.8), the first term is bounded by $O(1/b^2)$. Turning now to the contribution of a single simplex $s \in \mathcal{G}(R)$ to the other terms, we distinguish between the accepted and the rejected hyperplanes (ie, the second and third terms above). Using a conservative estimate, the number of faces that s contributes to Σ_{on} or Σ_{below} is $O(r_s^d)$; of course, r_s is to be understood here with respect to the final sample, ie, $r_s = |S_{|Os}^{(n-r)}| + 1$. Also, there are no more than r_s accepted hyperplanes cutting s or separating it from O, so the total number of polytopes within s for which the oracle is called is $O(r_s^{d+1})$. For an accepted hyperplane h_i, the definitions of $\mathcal{AE}_{below}^{(k)}$ in (7.12) and $\mathcal{AE}_{on}^{(k)}$ in (7.11) show that the oracle's error bound of (7.10) must be multiplied by, respectively, $1/N$ and $(1-p)/N \le 1/N$ to provide upper bounds for the middle sum above. So, the total error contribution for the accepted hyperplanes is $O(r_s^{d+1}E_s/N)$.

As regards a rejected hyperplane h_i, the same argument leads to multiplying the oracle's error bound by $p/(1-p)N \le 2p/N$ for the faces of Σ_{below} and p/N for those of Σ_{on}. Therefore, the contribution of s to the third term of the sum above is $O(n_s r_s^d p E_s/N) = O(q_s r_s^d E_s/N)$, and so the last two terms together account for $O(q_s r_s^{d+1} E_s/N)$ and the first one for $O(1/b^2)$. This gives

$$\left| \mathcal{E}_3^{(k)} - \mathcal{AE}_3^{(k)} \right| \ll \frac{1}{b^2} + \frac{1}{b^3 r^{\lfloor d/2 \rfloor}} \sum_{s \in \mathcal{G}(R)} q_s^{c-\sqrt{c}+1} r_s^{d+1},$$

where the constant behind the notation $\ll$ does not depend on b. We already have bounds on the powers of r_s and q_s. To deal with products of powers, we use the inequality $xy \le x^u + y^v$, which is easily shown to hold for any $1 < u \le v$ such that $1/u + 1/v = 1$. In this case, we set

$$\begin{cases} x &= r_s^{d+1}, \\ y &= q_s^{c-\sqrt{c}+1} \le q_s^{c-\sqrt{c}/2}, \\ u &= 2\sqrt{c}, \\ v &= \frac{1}{1-1/u} = \frac{c}{c-\sqrt{c}/2}. \end{cases}$$

We derive

$$r_s^{d+1} q_s^{c-\sqrt{c}+1} \le r_s^{2(d+1)\sqrt{c}} + q_s^c,$$

and therefore

$$\left| \mathcal{E}_3^{(k)} - \mathcal{AE}_3^{(k)} \right| \ll \frac{1}{b^2} + \frac{1}{b^3 r^{\lfloor d/2 \rfloor}} \left(\sum_{s \in \mathcal{G}(R)} r_s^{2(d+1)\sqrt{c}} + \sum_{s \in \mathcal{G}(R)} q_s^c \right).$$

By INV2 (which, earlier, was shown to hold true), the first sum does not exceed $b^2 r^{\lfloor d/2 \rfloor}$. Because the geode of R is a semicutting, the second sum is bounded by $O(b r^{\lfloor d/2 \rfloor})$. Therefore, the total error is $O(1/b)$, which completes the proof of Lemma 7.8. $\square$

Lemma 7.9 *Assuming that the geode $\mathcal{G}(R)$ produced in round $i-1$ is a semicutting, then round i computes a subset S that satisfies all three invariants* INV1–INV3.

Proof: By (7.13) and the previous lemma, at the end of round i, we have:

$$\begin{aligned}
\mathcal{E}^{(n-r)} &= \mathcal{E}_1^{(n-r)} + \mathcal{E}_2^{(n-r)} + \mathcal{E}_3^{(n-r)} \\
&< \mathcal{E}_1^{(n-r)} + \mathcal{E}_2^{(n-r)} + A\mathcal{E}_3^{(n-r)} + \tfrac{1}{3} \\
&< A\mathcal{E}^{(n-r)} + \tfrac{1}{3} \le \tfrac{5}{6} .
\end{aligned}$$

Again, because no randomization is left at the end of the round,

$$\mathcal{E}^{(n-r)} = \sum_{j=1}^{3} A_j(S^{(n-r)}) < \tfrac{5}{6} ,$$

and the final sample $S = S^{(n-r)}$ satisfies the three invariants. $\square$

The Oracle

We describe an implementation of the oracle $\mathcal{O}^{(k)}$ used in the proof of Lemma 7.8 with the following characteristics: Given an input polytope σ contained in a simplex $s \in \mathcal{G}(R)$, the oracle returns its answer in $O(|\sigma|q_s^{a\sqrt{c}})$ time, for some absolute constant a, where $|\sigma|$ denotes the combinatorial complexity of σ, ie, its total number of faces of all dimensions. The output is an approximation $\mathcal{O}^{(k)}(\sigma)$ that satisfies (7.10). The total time needed for updating the data structure for s during the round is bounded by[9] $O(n_s r_s^d q_s^{a\sqrt{c}})$.

For notational convenience, we introduce another parameter ρ_s, which, like q_s and r_s, behaves "mostly" like a constant:

$$\rho_s \stackrel{\text{def}}{=} b^{3d+6} q_s^{4d\sqrt{c}}.$$

Let $H_s^{(k)}$ denote the set of hyperplanes yet to be processed, among those in the conflict list of s. Specifically,

$$H_s^{(k)} = \{h_{k+1}, \dots, h_{n-r}\}_{|O_s} .$$

We maintain an ε-approximation $A_s^{(k)}$ for the set $H_s^{(k)}$ as it evolves dynamically. (A full treatment of the concept of an ε-approximation is given in Chapter 4.) The underlying range space is defined with respect to line segments. To maintain such an ε-approximation under deletion of hyperplanes, we can start by using, say, a $(1/2\rho_s)$-approximation $A_s^{(0)}$ and recompute a brand-new $(1/2\rho_s)$-approximation after every $n_s/2\rho_s$ deletions.

---

[9]The big-oh notation, when used for expressing running times, hides constants that might depend on b.

By Theorem 4.5 (page 175), each such ε-approximation can be computed in $O(\rho_s^{2d+1} n_s)$ time, so the total maintenance time for all the ε-approximations is $O(n_s q_s^{a\sqrt{c}})$, as desired. The size of any $A_s^{(k)}$ is $O(\rho_s^2 \log \rho_s)$.

Recall that to answer a call to the oracle $\mathcal{O}^{(k)}$ is to provide a sharp estimate on the value of

$$\sum_{v \in V(H^{(k)}, \sigma)} p^{\dim \sigma} (1 - p)^{m_v} n_v^c .$$

To do this, we define the output of the oracle to be

$$\mathcal{O}^{(k)}(\sigma) \overset{\text{def}}{=} \sum_{v \in V(A_s^{(k)}, \sigma)} (\alpha_s p)^{\dim \sigma} (1 - p)^{m_v} n_v^c , \qquad (7.14)$$

where

$$\alpha_s \overset{\text{def}}{=} |H_s^{(k)}| \Big/ |A_s^{(k)}| .$$

The parameters m_v and n_v can be computed for any vertex $v \in V(A_s^{(k)}, \sigma)$ in time $O(\rho_s^{3d} n_s)$. Since $O(\rho_s)$ distinct approximations are needed, this brings up the cost to $O(\rho_s^{3d+1} n_s)$. With this information handy, given σ, we can compute the output of the oracle in $O(|\sigma| \cdot |A_s^{(k)}|^d)$ time, which is $O(|\sigma| q_s^{a\sqrt{c}})$ time, for some suitably large constant a. The total combinatorial complexity of all the polytopes σ in the sets Σ_{below} and Σ_{on} introduced while processing a hyperplane h_i is in $O(r_s^d)$, and this bound also covers the cost of computing them. Thus, the overall time spent within s while processing a given hyperplane is $O(r_s^d q_s^{a\sqrt{c}})$. There are n_s hyperplanes to process, so we have the following result.

Lemma 7.10 *The work spent on any simplex $s \in \mathcal{G}(R)$ during round i takes $O(n_s r_s^d q_s^{a\sqrt{c}})$ time.*

To conclude the proof of our opening claims, we now must show that the oracle is accurate enough. For notational convenience, we deal only with the case $k = 0$, the other cases being entirely similar; this allows us to omit the superscript (k) from our notation, and to simplify our discussion by observing that $m_v = n_v$. As was shown in §4.3, when suitably scaled, the number of vertices falling within a polytope that are formed by the ε-approximation approximates the number of vertices of H_s within that polytope with relative accuracy ε.

By analogy with the finite-element method, our strategy is to subdivide σ into a small number of cells, so that the variation of the summand within each cell is small. Within such a cell the summand in question is estimated (erroneously, of course) by a single number. We begin with a discussion of the error caused within a given cell. Then, we explain how to subdivide σ into cells. The function f to be summed is

$$f(x) = p^j (1-p)^x x^c,$$

where $j = \dim \sigma$. We define its *modulus of continuity* over a set T as the function denoted by $M_{f,T}(h)$, where

$$M_{f,T}(h) = \sup_{\substack{x_1, x_2 \in T \\ |x_2 - x_1| \leq h}} |f(x_2) - f(x_1)|.$$

Given a j-dimensional cell $\xi \subseteq \sigma$ (Fig. 7.8), we use the shorthand

$$\Sigma_\xi f = \sum_{v \in V(H, \xi)} f(n_v) \quad \text{and} \quad \Sigma_\xi^* f = \sum_{v \in V(A_s, \xi)} \alpha_s^j f(n_v).$$

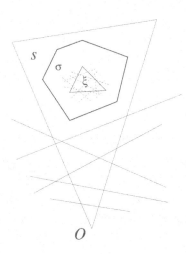

Fig. 7.8. Assuming $j = d$, the cell ξ lies within $\sigma \subseteq s$.

To bound the error $|\Sigma_\xi f - \Sigma_\xi^* f|$ over the cell ξ, we define

$$\begin{cases} T_\xi &= \{\, n_v : v \in V(H, \xi) \,\}, \\ f_\xi^{min} &= \min\{\, f(x) : x \in T_\xi \,\}, \\ f_\xi^{max} &= \max\{\, f(x) : x \in T_\xi \,\}, \\ \Delta_\xi(f) &= f_\xi^{max} - f_\xi^{min}. \end{cases}$$

It is immediate that

$$|V(H, \xi)| \cdot f_\xi^{min} \leq \Sigma_\xi f \leq |V(H, \xi)| \cdot f_\xi^{max}$$

and

$$\alpha_s^j |V(A_s, \xi)| \cdot f_\xi^{min} \leq \Sigma_\xi^* f \leq \alpha_s^j |V(A_s, \xi)| \cdot f_\xi^{max},$$

and therefore

$$\left| \Sigma_\xi f - \Sigma_\xi^* f \right| \leq \left| \alpha_s^j |V(A_s, \xi)| - |V(H, \xi)| \right| \cdot f_\xi^{max} + |V(H, \xi)| \Delta_\xi(f).$$

By Theorem 4.16 (page 186), for any j-dimensional cell ξ within $\sigma \subseteq s$,

$$\left| \alpha_s^j |V(A_s, \xi)| - |V(H, \xi)| \right| \leq \frac{n_s^j}{\rho_s}, \tag{7.15}$$

from which it follows that

$$\left| \Sigma_\xi f - \Sigma_\xi^* f \right| \leq \frac{n_s^j}{\rho_s} \cdot f_\xi^{max} + |V(H, \xi)| \Delta_\xi(f). \tag{7.16}$$

Note that, technically speaking, inequality (7.15) holds, but not for the reasons we mentioned. Indeed, recall that we do not actually use a $(1/\rho_s)$-approximation, but a sequence of $(1/2\rho_s)$-approximations, which we compute after every $n_s/2\rho_s$ deletions. This means that, to the standard bound of $n_s^j/2\rho_s$, we must add the overcount of vertices lying on rejected hyperplanes, ie, $\binom{n_s}{j-1} n_s/2\rho_s$; this makes the previous bound of n_s^j/ρ_s still valid.

Lemma 7.11 *For any* $T \subseteq [0, \tau]$, $M_{f,T}(h) \leq h \, p^j \tau^{c-1} (c - \tau \log(1 - p))$.

Proof: By the mean value theorem, for any $x_1, x_2 \in T$,

$$|f(x_2) - f(x_1)| \leq |x_2 - x_1| \times \sup_{x \in [x_1, x_2]} |f'(x)|.$$

Because $f'(x) = p^j (1 - p)^x (cx^{c-1} + x^c \log(1 - p))$ and $\log(1 - p) < 0$, it follows that $|f'(x)| \leq cp^j \tau^{c-1} - p^j \tau^c \log(1 - p)$, which implies the lemma. $\square$

To estimate the total error within the cell σ, we subdivide it into smaller subcells. Setting

$$\nu_s \stackrel{\text{def}}{=} b^3 q_s^{2\sqrt{c}},$$

we choose a $(1/\nu_s)$-net $\mathcal{N} \subseteq H_{|Os}$ (for the same underlying range space) of size $O(\nu_s \log \nu_s)$. Note that this is a device used only for the error analysis, which means no extra work for the algorithm. Let Υ be the portion of the arrangement of $\mathcal{N}$ within σ; because of general position, we may restrict our attention to j-dimensional cells, so that Υ is a subdivision of σ into $|\Upsilon| = O((\nu_s \log \nu_s)^j) = O(\nu_s^{d+1})$ cells. For any $\xi \in \Upsilon$, we have $\sup T_\xi \leq n_s$; so by virtue of $\mathcal{N}$ being a $(1/\nu_s)$-net and the fact that $-\log(1 - p) \ll r/n$, it follows from Lemma 7.11 that

$$\Delta_\xi(f) \leq M_{f,T_\xi}\left(\frac{n_s}{\nu_s}\right) \ll \frac{n_s p^j}{\nu_s}\left(1 + \frac{rn_s}{n}\right)n_s^{c-1} \ll \frac{q_s p^j n_s^c}{\nu_s}. \qquad (7.17)$$

Putting together (7.16) and (7.17), we obtain

$$\begin{aligned}|\Sigma_\sigma f - \Sigma_\sigma^* f| &\leq \sum_{\xi \in \Upsilon}|\Sigma_\xi f - \Sigma_\xi^* f| \ll \nu_s^{d+1}\frac{n_s^j}{\rho_s}f_\sigma^{max} + n_s^j\frac{q_s p^j n_s^c}{\nu_s}\\ &= O\left(\frac{\nu_s^{d+1}q_s^d}{\rho_s} + \frac{q_s^{d+1}}{\nu_s}\right)n_s^c = O(E_s/b),\end{aligned}$$

which establishes (7.7) and (7.10), and proves our claim about the oracle's performance.

Complexity Analysis

By Lemma 7.10, we derive the complexity of a round by summing up the given time bound over all simplices of the geode. Using the inequality $xy \leq x^2 + y^2$,

$$\sum_{s \in \mathcal{G}(R)} n_s r_s^d q_s^{a\sqrt{c}} \leq \left(\frac{n}{r}\right)\sum_{s \in \mathcal{G}(R)}r_s^d q_s^{a\sqrt{c}+1} \leq \left(\frac{n}{r}\right)\sum_{s \in \mathcal{G}(R)}\left(r_s^{2d} + q_s^{2a\sqrt{c}+2}\right).$$

As we observed in the proof of Lemma 7.8, INV2 holds. We also assumed that the geode of R is a semicutting; so, by Lemma 7.4 and the fact that c is arbitrarily larger than a, the sum above is $O(nr^{\lfloor d/2\rfloor-1})$. This applies to all of the rounds except the first and last ones.

The first one takes $O(n)$ time, while for the last round we use a naive $O(n_s^d)$ algorithm for each simplex s of the geode $\mathcal{G}(R)$ computed in the previous round. By Lemma 7.4, the semicutting property of that geode, and Hölder's inequality, the costs sum up to $O(r^{\lfloor d/2\rfloor})(n/r)^d$. Since $r > n/c$, this gives $O(n^{\lfloor d/2\rfloor})$. From the geometric growth of the sample size at each round, we conclude that the running time of the convex hull algorithm is $O(n\log n + n^{\lfloor d/2\rfloor})$, which completes the proof of Theorem 7.3 (page 285) and, hence, of Theorems 7.1 and 7.2 as well. $\square$

A closing observation concerns the issue of finite precision. Since the algorithm is polynomial, it suffices to represent numbers over $c\log n$ bits, for c large enough, to ensure that the numerical (relative) errors added at each operation are small enough, eg, $O(1/n^d)$. It is easily checked that such errors are then inconsequential in the decision process.

7.4 Bibliographical Notes

Optimal deterministic convex hull algorithms in dimensions 2 and 3 were found by Graham [145] and Preparata and Hong [252]. For the case of two-dimensional Voronoi diagrams, an optimal divide-and-conquer solution was proposed by Shamos and Hoey [281], and a plane-sweep method was developed later by Fortune [132].

Convex hulls are considerably more complicated to compute in dimension higher than 3. Seidel [276] gave an optimal $O(n^{\lfloor d/2 \rfloor})$-time algorithm in any even dimension $d > 2$, and a suboptimal $O(n^{\lfloor d/2 \rfloor} \log n)$ solution in any dimension $d > 1$ [277]. The problem was finally settled by Chazelle [70], who discovered an optimal $O(n \log n + n^{\lfloor d/2 \rfloor})$ deterministic algorithm in any dimension d. The algorithm was subsequently simplified by Brönnimann, Chazelle, and Matoušek [57]. Our presentation here follows the latter treatment. The algorithm is but one example of derandomization in computational geometry. A nice survey of the subject is given by Matoušek [217].

On the probabilistic front, the main breakthrough was the randomized convex hull algorithm of Clarkson and Shor [97]. Their algorithm is of the Las Vegas type. It always computes the right answer and its expected running time is optimal. A nice, simpler variant was later proposed by Seidel [278]. See also the textbooks by Mulmuley [237], Boissonnat and Yvinec [52], and the survey by Clarkson [95] for a good coverage of randomized geometric algorithms.

8

Linear Programming and Extensions

 eometrically, linear programming is the problem of minimizing a linear functional (the cost function) over a convex polyhedron. Algebraically, one of the many ways of stating the problem is

$$\text{Minimize} \quad c^T x,$$

subject to the constraints

$$Ax \leq b \quad \text{and} \quad x \geq 0,$$

where b and c are column vectors in $\mathbf{R}^n$ and $\mathbf{R}^d$, respectively, and A is an n-by-d real matrix. The number d of variables is the dimension of the ambient space: It can be arbitrarily large, but in our context it will be considered constant. In other words, we are primarily interested in the dependency of the running time on n rather than on d. The main purpose of this chapter is to use the discrepancy method to derive a deterministic algorithm for linear programming that is linear in n and singly exponential in d.

That linear programming can be solved in time linear in the number of constraints is quite stunning. How can such a problem be easier than, say, sorting? Furthermore, the algorithm is a variant of the well-known (dual) simplex algorithm, and so these bounds hold in the unit-cost RAM model. Unlike the classical algorithms of Khachiyan or Karmarkar, no assumption is needed on bit encodings.

The underlying method is sufficiently general that it can be used for a broad class of optimization problems. Remarkably, what happened with VC-dimension theory will repeat itself. The problems can be treated combinatorially with no reference to geometry or linear algebra whatsoever. This generality allows us to derive linear-time algorithms for nonlinear optimization problems, eg, finding the smallest ball or even the smallest ellipsoid enclosing n points in $\mathbf{R}^d$. If linear-time LP seems stunning, what about the following? Given n points in $\mathbf{R}^{100}$, the smallest enclosing ellipsoid (which is unique) can be found in $O(n)$ time!

8.1 LP-Type Problems

We define a class of combinatorial problems, called *LP-type*.[1] Any problem in that class is characterized by a pair (H, w), where H is a finite set and w is a "cost" function mapping certain subsets of H (not necessarily all of 2^H) to a totally ordered universe $(W, \leq)$. The elements of H are called the *constraints* of the problem. We say that $h \in H$ *violates* a subset of constraints $G \subseteq H$ if $w(G) < w(G \cup \{h\})$. A *basis* B of $G \subseteq H$ is a minimal set of constraints with the same cost as G. Specifically, we must have $w(B) = w(G)$ and $w(C) < w(B)$ for any proper subset C of B. We define the *combinatorial dimension of* (H, w), denoted by δ, to be the maximum size of any basis (of any subset of H). To solve the problem (H, w) is to find a basis of H. We need certain combinatorial and computational assumptions to make this an efficient process.

The Four Axioms

The first two axioms concern the behavior of the function w; the others provide computational oracles for the algorithm:

1. MONOTONICITY. Given any $F \subseteq G \subseteq H$, $w(F) \leq w(G)$.

2. LOCALITY. If $h \in H$ violates $G \subseteq H$, then it violates any basis of G.

Note that the converse of the locality assumption (the violation of a basis implies the violation of G) follows from the monotonicity assumption. Before we go any further, it is useful to make a simple observation.

Lemma 8.1 *A basis B of any subset of H is also a basis of $B \cup N$, where N is any collection of nonviolating constraints for B.*

Proof: We prove this by induction on the size of N. The case $|N| = 0$ is trivial. For $|N| > 0$, consider any $h \in N$. Because B is a basis of

---

[1] Yes, LP stands for what you think.

$B \cup (N \setminus \{h\})$, by the locality assumption, h cannot violate $B \cup (N \setminus \{h\})$, and therefore

$$w(B \cup N) = w(B \cup (N \setminus \{h\})) = w(B).$$

□

Given a basis B (of some subset of H), let $V(B)$ be the set of violating constraints:

$$V(B) = \left\{ h \in H \,\middle|\, w(B) < w(B \cup \{h\}) \right\}.$$

We define a range space $(H, \mathcal{R})$, where $\mathcal{R}$ is the collection of sets $V(B)$, for all bases B. We make the assumption that $(H, \mathcal{R})$ has finite VC-dimension. Let γ be a bound on the exponent in its shatter function, ie, we assume that $\pi_{\mathcal{R}}(m) = O(m^\gamma)$. In practice, γ is either equal or larger than δ. We need to make two computational assumptions: One is that violations can be detected effectively; the other is an oracle for $(H, \mathcal{R})$ of the kind described in Chapter 4.

3. VIOLATION TEST. Given any basis B and a constraint $h \in H$, check whether h violates B. (Return an error message if B is not a basis.)

4. ORACLE. Given any subset $Y \subseteq H$, compute the set $\mathcal{R}|_Y$ in time $O(|Y|^{\gamma+1})$.

Linear Programming as an LP-Type Problem

How does all of that relate to linear programming? By rotation, the cost function can be assumed to be of the form $(1, 0, \ldots, 0)^T x$. Because of the nonnegativity constraints on the coordinates ($x \geq 0$) this implies that, if the system is feasible, it must have a bounded optimal solution. Such a solution can be made unique by choosing the one whose coordinate vector is lexicographically smallest. For the time being, we assume that the system is feasible. (We shall remove that assumption later.) The correspondence with an LP-type problem is straightforward:

- H is the set of n closed halfspaces formed by the inequalities $Ax \leq b$; note that we do not include the inequalities $x \geq 0$ in H.
- $W = \mathbf{R}^d$, ordered lexicographically.
- Given $G \subseteq H$, $w(G)$ is the unique (lexicographically) minimal point with nonnegative coordinates in the halfspaces of G. Its existence follows from the feasibility of the original linear program.

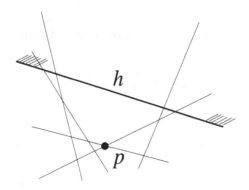

Fig. 8.1. The halfspace h violates G.

A halfspace $h \in H$ violates $G \subseteq H$ if $w(G) < w(G \cup \{h\})$, ie, adding h to G would strictly increase the cost of the optimal solution. Geometrically, this means that the hyperplane corresponding to h cuts off the old solution from the new feasible set. A basis consists of at most d halfspaces, and its combinatorial dimension is d. The monotonicity assumption expresses the fact that adding more constraints cannot improve the optimal solution. The locality assumption asserts that the violation of a set of constraints can always be witnessed locally by looking at any one of its bases. Why is that the case? The optimal is achieved at a unique point p in space, which is unambiguously specified by any basis. (Note that the point p is the intersection of the hyperplanes in B, possibly together with some of the coordinate hyperplanes.) If a new halfspace h causes a violation, then, as we observed, its bounding hyperplane cuts away the point p from the new feasible set. In doing so, it also cuts p away from the feasible set defined by h and any basis (Fig. 8.1).

Regarding the computational assumptions, a violation test can be performed in $O(d^3)$ time by Gaussian elimination. If the vertex (specified by the constraints of the basis and the nonnegativity constraints) is available explicitly—as it often will be—then the time is reduced to $O(d)$. To implement the oracle, we consider the arrangement formed by the hyperplanes bounding the halfspaces $x \geq 0$ and those in Y. Each basis for a subset of Y corresponds to a vertex in that arrangement (but often not the other way around). We form $V(B)$ by checking which of the halfspaces of Y violate that basis, using the fact that violation means that the halfspace does not contain the vertex associated with the basis B. This is done naively in time at most proportional to

$$d^3 m \binom{m+d}{d},$$

where $m = |Y|$. The running time of the oracle is therefore $O(|Y|^{d+1})$.

Finally, what is to be done with the linear program if it has no feasible solution? As will soon be evident, the algorithm will automatically detect unfeasibility during Step 1 (or its recursive incarnations).

A Deterministic Solution

We give a recursive procedure that takes a set H of constraints as input and returns a basis of H. Let $D = \max\{\delta, \gamma\}$, and let $|H| = n$.

The Algorithm

STEP 1. If $n \leq cD^4 \log D$, for some large enough constant c, compute a basis of H directly by examining all possible j-tuples of constraints in the order $j = 1, \ldots, \delta$, and returning the first one that is not violated by any constraint of H.

STEP 2. Compute a $(1/4D^2)$-net N for $(H, \mathcal{R})$.

STEP 3. Compute a basis B of N recursively. Let V be the set of all constraints in H violating the basis. If V is empty, then return B and stop; otherwise, add all of the violating constraints to the set N and repeat Step 3.

Correctness

Why is the algorithm correct? In fact, why does it even terminate? It is easy to see that the algorithm can iterate through Step 3 at most δ times. Indeed, assume that V is not empty after the first iteration. Then, by the monotonicity assumption, the original basis B is such that $w(B) < w(H)$. Let B^* be any basis of H. By Lemma 8.1, if none of the constraints in B^* violated B, then B would also be a basis for $B \cup B^*$. But then, by monotonicity, $w(B) = w(B \cup B^*) = w(H)$, which is a contradiction. This means that, after the first iteration, N contains at least one constraint from any basis for H. These constraints can never again violate N in later iterations. Thus, the same reasoning shows that after each iteration at least one additional *new* constraint from any basis of H joins N. After δ steps, therefore, the process stops with $V = \emptyset$.

Note that an alternative to Step 1 is to try all possible j-tuples J, where $1 \leq j \leq \delta$, and report the one that maximizes $w(J)$. This is correct because of the monotonicity assumption. Of course, this assumes that w is computable, which, interestingly enough, is not part of our assumptions. The correctness of the output follows from Lemma 8.1, since any basis that has no violation is also a basis of H.

Complexity

Step 1 requires at most

$$n\binom{n}{1} + \cdots + n\binom{n}{D} = O(n/D + 1)^D n$$

violation tests. Assuming that each test costs time t_v, this amounts to $O(D)^{3D+4}(\log D)^{D+1} t_v$ time. By Theorem 4.6 (page 175), a $(1/r)$-net of size $O(Dr \log(Dr))$ can be computed in time $O(D)^{3D} nr^{2D} \log^D(Dr)$; therefore, Step 2 takes time $O(D^7 \log D)^D n$. The size of no set V in Step 3 can ever exceed $n/4D^2$. It follows that, at all times during the course of the algorithm, $|N| \leq n/4D$. Therefore, each recursive call involves a problem of size at most $n/4D + O(D^3 \log D)$, which is less than $n/2D$ if c is chosen large enough. The violation costs in Step 3 are at most nt_v. If $T(n)$ denotes the running time of the algorithm, we have the recurrence:

$$T(n) = \begin{cases} O(D)^{3D+4}(\log D)^{D+1} t_v & \text{if } n \leq cD^4 \log D, \\ O(D^7 \log D)^D n + nt_v + DT(n/2D) & \text{else.} \end{cases}$$

Assuming that t_v is at most $O(D)^{4(D-1)}/\log D$, which is the case in our applications, we find that $T(n) = O(D^7 \log D)^D n$.

Theorem 8.2 *An LP-type problem (H, w) satisfying assumptions 1–4 can be solved in time $|H| \cdot O(D^7 \log D)^D$. This assumes that a violation test can be done in time $O(D)^{4(D-1)}/\log D$. The parameter D is the combinatorial dimension or the exponent in the complexity of the range space oracle, whichever is larger.*

Note that the complexity can be slightly improved by using the more complex (but faster) ε-net construction mentioned in §4.3.

8.2 Linear Programming in Linear Time

We continue our previous discussion of linear programming in light of Theorem 8.2. The combinatorial dimension δ is equal to d. From what we said earlier, the oracle time is $O(|Y|^{d+1})$; therefore, we can set γ to d, and hence $D = d$. The time t_v to test a violation is $O(d^3)$, which is certainly $O(D)^{4(D-1)}/\log D$ (for $d > 1$). From Theorem 8.2 we derive

Theorem 8.3 *Linear programming with n constraints and d variables can be solved in $d^{O(d)}n$ time.*

8.3 Computing Löwner-John Ellipsoids

Given a set H of n points in $\mathbf{R}^d$, the smallest ellipsoid enclosing all n points is unique. It is called the *Löwner-John* ellipsoid of H. How fast can we compute it? Fortunately, it is not too difficult to see that we have an LP-type problem in disguise. Each of the n points is a constraint. Adding points cannot decrease the volume of the Löwner-John ellipsoid, so the monotonicity assumption holds. In two dimensions, it takes at most five points to specify an ellipse, and this upper bound is tight. Setting the problem as convex programming easily shows that the combinatorial dimension is 5. In $\mathbf{R}^d$, an ellipsoid is specified by an inequality of the form

$$(x - c)^T Q(x - c) \leq 1,$$

where c is the center of the ellipsoid and Q is a symmetric positive definite d-by-d matrix. Again, the combinatorial dimension δ can be shown to be at most the degree-of-freedom (DOF) number[2] of an ellipsoid, which is $d(d + 3)/2$: d for the choice of c plus $d(d + 1)/2$ for the choice of Q. The locality assumption follows from the uniqueness of the Löwner-John ellipsoid.

The violation test addresses the following question: Given a basis B and a point h, verify whether h lies outside the Löwner-John ellipsoid of B. We use a characterization of Juhnke which says that the point h lies in the Löwner-John ellipsoid for B if and only if the real matrix Q is symmetric positive definite and there exist reals $\alpha \geq 0$, $\lambda(p) > 0$ ($p \in B$), such that

$$
\begin{aligned}
(h - c)^T Q(h - c) &\leq 1 \\
(p - c)^T Q(p - c) &= 1, \quad p \in B \\
\sum_{p \in B} \lambda(p) \cdot p &= \left(\sum_{p \in B} \lambda(p)\right) c \\
(\alpha \det Q)U &= \sum_{p \in B} \lambda(p)Q(p - c)(p - c)^T,
\end{aligned}
$$

---

[2]DOF considerations alone might not be sufficient to bound the combinatorial dimension. A nice illustration of this was indicated to the author by Jirka Matoušek. Consider the narrowest strip enclosing n points in general position in the plane. A strip has three degrees of freedom and, sure enough, the narrowest one is specified by three points. The combinatorial dimension can be as high as n, however. Can you see why?

where U denotes the d-by-d matrix with ones everywhere. Note that positive definiteness can be enforced by requiring that all of the upper left submatrices of Q have positive determinants. To check the feasibility of such a system of polynomial equalities and inequalities, we use a known result concerning the complexity of the existential theory of the reals: A system of m equalities and inequalities of maximum degree b in r variables can be decided in $(mb)^{O(r)}$ time (in the unit-cost RAM model of computation). Thus, it appears that a violation test can be performed in time $d^{O(d^2)}$.

The last point to check is the range space oracle. Given a set Y of points in $\mathbf{R}^d$, the problem can be solved by enumerating all possible subsets of Y of the form $Y \cap F$, where F is the complement of an ellipsoid, and then—the easy part—keeping those defined by actual bases. The enumeration is easily done by lifting the problem into higher dimensions. This is intended to linearize the inequality

$$(x - c)^T Q(x - c) \leq 1$$

which expresses the membership of x in the ellipsoid $\mathcal{E}(Q, c)$. We map $x = (x_1, \ldots, x_d)^T$ to

$$f(x) = (x_1, \ldots, x_d, x_1^2, x_1 x_2, x_1 x_3, \ldots, x_1 x_d, x_2^2, x_2 x_3, \ldots, x_d^2).$$

In this way, x belongs to $\mathcal{E}(Q, c)$ if and only if $f(x)$ lies in a halfspace $h(Q, c)$ in $\mathbf{R}^{d(d+3)/2}$. This shows that the subsets $Y \cap F$ can be enumerated by trying out all $O(|Y|)^{d(d+3)/2}$ halfspaces whose bounding hyperplanes pass through $d(d+3)/2$ points. Checking whether a point lies above or below a hyperplane in $\mathbf{R}^{d(d+3)/2}$ takes $O(d^6)$ time by Gaussian elimination, so the total reporting time is $O(|Y|)^{d(d+3)/2+1} d^6$. A simple variation of the analysis in the proof of Theorem 8.2 gives:

Theorem 8.4 *The ellipsoid of minimum volume that encloses a set of n points in $\mathbf{R}^d$ can be computed in time $d^{O(d^2)} n$.*

8.4 Bibliographical Notes

The first polynomial-time algorithm for linear programming was discovered by Khachiyan [181]. However, it is not a practical alternative to the simplex algorithm of Dantzig [102], even though the latter is exponential in the worst case. A polynomial algorithm shown to be efficient in practice was given by Karmarkar [178]. The complexity of both Khachiyan's and

Karmarkar's algorithms depends on the bit-complexity of the input. By contrast, all of the algorithms discussed in this chapter, being dual simplex algorithms in disguise, run in time that depends only on the numbers of variables and constraints (assuming a unit-cost RAM model).

The first algorithm for linear programming with a running time linear in the number of constraints was found by Megiddo [224, 225]. Its complexity of $2^{2^d} \cdot O(n)$ is, unfortunately, highly dependent on the number of variables. The multiplicative factor was reduced to 3^{d^2} by Dyer [114] and Clarkson [93]. Randomized algorithms with even lower dependency on d were found by Dyer and Frieze [116], Clarkson [96], and Seidel [278]. Among those, the lowest asymptotic complexity is achieved by Clarkson's algorithm [96], with a running time of $O(d^2 n + d^{d/2+O(1)} \log n)$. Kalai [175, 176] and, independently, Matoušek, Sharir, and Welzl [220] broke the exponential barrier. In combination with Clarkson's algorithm, this led to a randomized algorithm for linear programming with expected running time

$$O(d^2 n) + e^{O(\sqrt{d \ln d})}.$$

Extensions to certain cases of convex programming were provided by Gärtner [141]. No deterministic algorithm can match this bound at present.

Section 8.1:   The LP-type formalism was developed by Sharir and Welzl [283]; see also [220]. The linear deterministic algorithm of Theorem 8.2 (page 312) was discovered by Chazelle and Matoušek [81]. Extensions of the LP-type formalism are discussed by Gärtner and Welzl [142].

Section 8.2:   The bound for linear programming (Theorem 8.3, page 313) was derived by Chazelle and Matoušek [81]. The running time of $d^{O(d)} n$ is the best known to date; however, it still falls short of the bounds for randomized algorithms mentioned above.

Section 8.3:   The linear algorithm for the smallest enclosing-ellipsoid problem (Theorem 8.4, page 314) is due to Chazelle and Matoušek [81]. An earlier linear algorithm was found by Dyer [115] with a higher dependency on d. The problem was also investigated by Post [251] and Welzl [320]. The related problem of computing the smallest enclosing ball was solved by Megiddo [224, 225] deterministically, and later by Welzl [320] probabilistically. Here are references for the various results on the Löwner-John ellipsoid that we mentioned without proof: Uniqueness is proven in [103]; the number of degrees of freedom of an ellipsoid is given in [103, 174]; and the result on the existential theory of the reals that we used in the proof of Theorem 8.4 is due to Renegar [259].

9

Pseudorandomness

s we saw in Chapter 4, the difficulty of sampling geometric spaces directly reflects their discrepancy. In the presence of unbounded VC-dimension, we have no combinatorial structure to hang on to, and naive randomization is often the preferred route. The trouble is that the underlying probability spaces are usually of exponential size and straightforward derandomization is intractable. This chapter shows that, by sampling sparse low-discrepancy subsets of the probability spaces, we can often considerably reduce the amount of randomness needed. The connection is intuitively obvious: A low-discrepancy subset should be mostly indistinguishable from the whole set. So, by sampling from it we should be able to fool the casual observer into thinking that we are actually sampling from the whole set. Of course, "casual observer" is our euphemism for "polynomially bounded algorithm." Thus, if this chapter needed a wordy subtitle, it could be: How designers of probabilistic algorithms can limit the amount of randomness they need through the judicious use of the discrepancy method.

Suppose that we wish to find a random sample S of size s in a universe with n elements. For concreteness, the universe can be thought of as $\{0, \ldots, n-1\}$. The quality of the random sample is measured by its discrepancy relative to any subset. In other words, imagine that we fix a certain $F \subseteq \{0, \ldots, n-1\}$. When picking S at random, it is desired that the discrepancy

$$\left| \frac{|F|}{n} - \frac{|F \cap S|}{|S|} \right|$$

be small. (By analogy with the ε-approximations defined in Chapter 4, we use a relative discrepancy measure.) Note the ordering of the statement. First we choose F, then S. Naturally, we cannot hope for a small discrepancy for *all* F's once S has been chosen. This is the major difference with set systems of finite VC-dimension. Standard calculations can tell us how large a discrepancy we should expect with random sampling. This is not the issue.

What is at issue is the number of bits required for sampling. If s is sufficiently smaller than n, the number of bits is about $\log\binom{n}{s} \approx s\log n$. In other words, each sampled item requires $\log n$ bits. This might seem natural, given the fact that simply to write down an item takes that many bits. But, how do we know that all of these bits need to be random? After all, a random item in the universe may require $\log n$ bits, but to indicate whether it is in F or not should require only one bit, so perhaps one could hope to reduce the total number of (truly) random bits to only $O(1)$ per sampled item. Indeed, one can.

There are many applications of this result. We discuss one of the most celebrated: amplification of success probability for **BPP**.[1] Recall that a probabilistic algorithm is a deterministic algorithm that takes two inputs: One is the standard input x to the problem at hand; the other one is a random m-bit string R. The class **BPP** includes all the languages L that can be recognized by a randomized polynomial-time algorithm with the following characteristics:

- If a string x is in the language L, then the algorithm accepts the string with probability at least $2/3$.
- If x is not in L, then the algorithm rejects the string with probability at least $2/3$.

Fix an input x once and for all. A **BPP** algorithm[2] does the "right thing" with probability at least $2/3$. There is nothing magical about $2/3$. Any constant bounded away from below by $1/2$, say, 0.5001 or 0.999, would work just as well.[3] The reason is that there is a straightforward way to *amplify* the success probability. Simply run the algorithm k times, and take a majority vote on all of the k outcomes. We expect the runs to give us the right answer at least $2k/3$ times. By ensuring independence among the random strings, we can use Chernoff's bound to show that taking a majority vote fails to provide the right answer with probability at most $1/2^{\Omega(k)}$.

---

[1] BPP= Bounded-error Probabilistic Polynomial.

[2] By abuse of terminology, we call an algorithm of the type above **BPP**.

[3] Of course, replacing $2/3$ by $1/2$ would be disastrous, since we could just flip a coin and answer yes if the outcome is heads.

This means that reducing the error probability to $1/2^k$ requires on the order of km random bits. Is it really the case that we need m random bits per run? Let F be the set of m-bit strings R that cause the algorithm to fail on input x. Intuitively, after the first run, the only "entropy" used up is the amount of randomness needed to tell whether R is in F or not. One bit should be enough for that. This is an indication that $m - 1$ or, say, at least $m - O(1)$ bits of the string R should still be usable for the next runs. Wishful thinking? The answer is, beautifully, no.

Theorem 9.1 *Given a* **BPP** *algorithm that uses m random bits per run, the probability of failure can be reduced to $1/2^k$ by using no more than $O(m + k)$ random bits and $O(k)$ runs.*

The connection to discrepancy theory is hard to miss. Any set of m-bit strings has a natural measure, ie, its relative size within the set of all m-bit strings. Fix an input x and consider the measure of the set F of bad strings (ie, those causing the algorithm to fail on the input x). Now pick k (truly) random m-bit strings. The proportion of such strings within F provides a good approximation on the size of F. Indeed, Chernoff's bound shows exponentially decaying deviation from the true answer. In other words, a random collection of k m-bit numbers provides a low-discrepancy set for F (in the sense of an ε-approximation). The theorem says that we can restrict the choice of a random collection to a much smaller universe and still have discrepancy that is just as low. Specifically, the size can be cut down from 2^{km} to $2^{O(k+m)}$ without adding much discrepancy.

The theorem is proven in §9.4 by using random walks on expanders. We prove two weaker results in §9.2 and §9.3. This gives us a vehicle for introducing two fundamental techniques in pseudorandomness: pairwise independence and universal hash functions. The latter have proven very useful in complexity theory. In this text we use them for hashing randomly with small likelihood of collision. An important dual use of universal hash functions is to provide pseudorandom colorings, where each preimage defines a color. We will not discuss this application here beyond mentioning its relevance to the distinction between public and private coins.

In the following sections, we revisit the tight connection between low discrepancy and small Fourier coefficients to build pseudorandom sequences. In §9.5 we use quadratic characters to construct a pseudorandom m-bit string, using much fewer than m (truly) random bits, while in §9.6 we build sparse low-discrepancy sets for arithmetic progressions, and we show how to use them for polynomial interpolation.

9.1 Finite Fields and Character Sums *

We begin with a review of basic facts from algebra [24] that the mathematically literate will be able to skip. First, finite fields. All of them are commutative, and all of those with the same number of elements are isomorphic. This number is always a prime power p^m, and the field in question is denoted by $\mathbf{F}_{p^m}$ (or $\mathrm{GF}(p^m)$, for Galois field, in honor of its discoverer). The two operations are called addition and multiplication, with unit elements denoted by 0 and 1, respectively. The characteristic of $\mathbf{F}_{p^m}$ is p, which means that adding the same element to itself p times always gives 0. The case $m = 1$ is trivial, for $\mathbf{F}_p$ is simply the set of integers mod p (which is a field because p is a prime).

For $m > 1$, the field $\mathbf{F}_{p^m}$ is *not* the set of integers modulo p^m. However, the multiplicative group $\mathbf{F}_{p^m} \setminus \{0\}$, denoted by $\mathbf{F}_{p^m}^*$, is cyclic. This means that all of its elements are powers of a single one, called a *primitive element*. The order of any element in a group divides its cardinality, so $x^{p^m-1} = 1$ for any $x \in \mathbf{F}_{p^m}^*$ (Fermat's theorem). This shows that the elements of $\mathbf{F}_{p^m}$ are all roots of the polynomial $x^{p^m} - x$. In fact, by a counting argument, we see that these are precisely the roots. So, it appears that $\mathbf{F}_{p^m}$ is an algebraic extension field of $\mathbf{F}_p$ in which the polynomial $x^{p^m} - x$ resolves into linear factors.

That fact alone is not too useful for carrying out field operations. For that, we choose some irreducible polynomial $f(x)$ of degree m with coefficients in $\mathbf{F}_p$, and we identify $\mathbf{F}_{p^m}$ with $\mathbf{F}_p[x]/(f(x))$, ie, the ring of polynomials with coefficients in $\mathbf{F}_p$ taken modulo (the ideal generated by) $f(x)$. What makes this a field is that every nonzero element g has an inverse. If you have forgotten why, just compute the GCD of f and g by Euclid's algorithm. This allows you to express this GCD as $pf + qg$, where p, q are polynomials of degree less than m. The GCD is constant; in fact, it can always be chosen to be 1. So by reducing modulo f, you now find that q is the inverse of g. In this way, we can add, subtract, multiply, and divide field elements quite easily.

We can do better still. The multiplicative group formed by the nonzero polynomials in $\mathbf{F}_p[x]/(f(x))$ maps isomorphically to the cyclic group $\mathbf{F}_{p^m}^*$. Wouldn't it be nice to ensure that the monomial x maps to a primitive element α? In this way, all field elements could be obtained directly as powers of x. For this to happen we must choose f as the minimal polynomial of α (ie, monic polynomial of least degree with α as a root). The reason for this is simple. If x is to map to α isomorphically, then $f(x)$ must map to $f(\alpha)$. But in the quotient ring, $f(x)$ is identically zero; therefore, $f(\alpha)$ must be

0, too. Our claim now follows from the fact that the degree of f cannot be less than m (else how could α generate the whole multiplicative group?). So, it is exactly m, and $f(x)$, being irreducible, can be used to form the field $\mathbf{F}_{p^m}$. Note that not all irreducible polynomials of degree m have this property. For example, over $\mathbf{F}_2$, both $x^4 + x + 1$ and $x^4 + x^3 + x^2 + x + 1$ are irreducible; but, while x corresponds to a primitive element of $\mathbf{F}_{2^4}$ in the first case, it does not in the second (hint: $x^6 = x$).

The advantage of such a "primitive polynomial" representation is clear. We can set up a table $0, 1, x, x^2, \ldots, x^{p^m-2}$ listing all of the field elements. In each entry $i \geq 0$, we write x^i as a polynomial $Q_i(x)$ of degree less than m. For consistency, define $Q_{-1}(x) = 0$. To add two field elements is trivial: Just add the two corresponding polynomials $Q_i(x)$ and $Q_j(x)$ term by term, performing addition of coefficients mod p. To multiply $Q_i(x)$ and $Q_j(x)$ is even easier. If i or j is -1, then we get 0; else, observe that

$$Q_i(x) \times Q_j(x) = x^i \times x^j = x^{i+j} = Q_{i+j}(x),$$

where $i + j$ is understood mod $p^m - 1$. So, a simple table lookup gives the answer; what we have is, in effect, a logarithm table.

Our discussion now will move to quadratic characters. Fix an odd prime p. We say that $x \in \mathbf{F}_p$ is a *quadratic residue* if it is of the form y^2, for some $y \in \mathbf{F}_p$. For reasons we will explain below, we define

$$\chi_p(x) = \begin{cases} 0 & \text{if } x = 0, \\ 1 & \text{if } x \text{ is a quadratic residue,} \\ -1 & \text{else.} \end{cases}$$

The function $\chi_p(x)$ is extended to all of $\mathbf{Z}$ by making it periodic, with period p. It is easy to see that χ_p is multiplicative, ie,

$$\chi_p(xy) = \chi_p(x)\chi_p(y).$$

In fact, χ_p is nothing but a multiplicative character of the sort encountered in §2.6. We explain why, and we show which particular character it is. Pick a primitive element α in the multiplicative group $\mathbf{F}_p^*$. Every element m in the group gives rise to its own distinct character

$$\chi_p^{[m]}(\alpha^t) \stackrel{\text{def}}{=} e^{2\pi i m t/(p-1)},$$

for $0 \leq t \leq p-2$. A quadratic character is one whose square is identically 1. Of course, the case $m = (p-1)/2$ fits the bill. Now, let us check that, for that value of m, $\chi_p^{[m]}$ agrees over $\mathbf{F}_p^*$ with the function χ_p defined above.

We must show that

$$\chi_p(\alpha^t) = e^{\pi i t}. \tag{9.1}$$

Obviously, the quadratic residuosity of $x = \alpha^t$ depends only on the parity of t (its discrete log): x is a quadratic residue if and only if t is even; hence (9.1).

By Fermat's theorem, the polynomial x^{p-1} is equal to 1 over $\mathbf{F}_p^*$. This shows that

$$(x^{(p-1)/2} - 1)(x^{(p-1)/2} + 1) = 0;$$

therefore, either $x^{(p-1)/2} = 1$ or $x^{(p-1)/2} = -1$ over $\mathbf{F}_p^*$. If x is a quadratic residue y^2, then

$$x^{(p-1)/2} = y^{p-1} = 1.$$

If x is a nonresidue, as we just saw, $x = \alpha^t$, for odd t, and so $x = \alpha y$, where y is a quadratic residue. It follows that

$$x^{(p-1)/2} = \alpha^{(p-1)/2} = -1.$$

The reason for the -1 is that if it were 1, then the powers of α would cycle past $(p-1)/2$ and would generate at most half the group: So much for a primitive element! This implies the classical characterization of quadratic characters as:

$$\chi_p(x) = x^{(p-1)/2}.$$

From the Riemann hypothesis for curves over finite fields (proven by Weil), we can bound how far the number of points of a curve with coordinates in the field deviates from the field order. This has bearing on how "randomly" $\chi_p(x)$ switches from 1 to -1. (The word "random" is not to be taken in a complexity-theoretic sense here but in a statistical one.) Weil showed that, given any polynomial f with coefficients in $\mathbf{F}_p$, if f is not a square and has exactly n distinct zeros, then (see [275], page 43)

$$\left| \sum_{x \in \mathbf{F}_p} \chi_p(f(x)) \right| \leq (n-1)\sqrt{p}. \tag{9.2}$$

Here is a simple example to illustrate how this relates to the Riemann hypothesis and to randomness. Take a cubic polynomial $f(x)$ with distinct roots in an algebraic closure of $\mathbf{F}_p$. Let $\mathcal{C}$ denote the affine elliptic curve of equation $y^2 = f(x)$, and consider its number N of points (x, y) with coordinates in $\mathbf{F}_p$. For any $x \in \mathbf{F}_p$, how many points with abscissa x lie on the curve C? If

$$f(x) = \begin{cases} 0, \text{ the answer is 1: } (x,0); \\ y^2, \text{ then it is 2: } (x,y) \text{ and } (x,-y); \\ \text{a quadratic nonresidue, the answer is 0.} \end{cases}$$

So, we have

$$\begin{aligned} N &= \sum_{x \in \mathbf{F}_p} \Big(\chi_p(f(x)) + 1 \Big) \\ &= p + \sum_{x \in \mathbf{F}_p} \chi_p(f(x)). \end{aligned}$$

By the Riemann hypothesis for elliptic curves over finite fields,[4] we know that

$$|N - p| \leq 2\sqrt{p}.$$

See (2.39) on page 110. This proves (9.2) for $n = 3$. Up to a constant factor, the upper bound is the standard deviation of the binomial distribution $B(p, 1/2)$. As x runs through the field, it looks as though the quadratic residuosity of the polynomial $f(x)$ behaves randomly. We have something looking random but that, in fact, is not random at all. Obviously, this is welcome grist for the discrepancy method's mill.

9.2 Pairwise Independence

As a warmup, let us consider a weak but exceedingly simple way of achieving amplification without using too many random bits. Given a **BPP** algorithm that uses m random bits per run, we show how to reduce the probability of failure to $1/k$ by using only $O(m)$ random bits and $O(k)$ runs.

We identify strings of m bits with elements in the Galois field $\mathbf{F}_{2^m}$ in a natural way. This means that we map an m-bit string $(a_0, \ldots, a_{m-1})$ to the polynomial $\sum_i a_i x^i$ with coefficients in $\mathbf{F}_2$. Recall that field operations translate into polynomial operations modulo some fixed irreducible polynomial of degree m in $\mathbf{F}_2[X]$. Choose two random "seeds" a, b in $\mathbf{F}_{2^m}$. Pick t distinct field elements, $x_1, \ldots, x_t$ (which, of course, assumes that $t \leq 2^m$). Our pseudorandom strings are

$$R_i = ax_i + b,$$

---

[4]This was proved by Hasse and later generalized to arbitrary curves by Weil. The reader unfamiliar with these notions should read our discussion in §2.6.

for $1 \leq i \leq t$. Given any pair of distinct i, j, the two vectors $(x_i, 1)$ and $(x_j, 1)$ are independent over $\mathbf{F}_{2^m}$, and so any system of equations of the form $\{R_i = x$ and $R_j = y\}$ has a unique solution. It follows that the random variables R_i are pairwise independent, since the probability that $R_i = x$ and $R_j = y$ is exactly $1/4^m$, ie, the product of the probabilities of each event.

Run the algorithm $t = 8k$ times, and take a majority vote. A mistake means that at least $4k$ R_i's were bad. Let $v_i = 1$ if the i-th run was a mistake and 0 otherwise, and let $V = \sum_i v_i$. The expectation of V is at most $8k/3$. Pairwise independence implies that

$$\mathbf{var}\, V = \sum \mathbf{var}\, v_i \leq \frac{16k}{9}.$$

By Chebyshev's inequality,

$$\mathrm{Prob}[\,|V - \mathbf{E}V| \geq \delta\,] \leq \frac{\mathbf{var}\, V}{\delta^2} \leq \frac{16k}{9\delta^2} = \frac{1}{k},$$

for $\delta = 4k/3$. Thus,

$$\mathrm{Prob}[\,V \geq 4k\,] \leq \frac{1}{k}.$$

This shows that running the **BPP** algorithm $8k$ times and taking a majority vote produces the right answer with probability at least $1 - 1/k$. The number of random bits is $2m$.

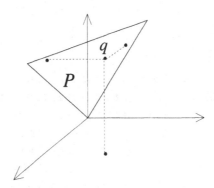

Fig. 9.1. The coordinates of a random point q in the plane P are pairwise independent.

The method for generating pairwise independent random strings has a simple geometric interpretation (Fig. 9.1). We choose a "generic" plane P in the vector space $\mathbf{F}_{2^m}^t$, ie, one whose projection onto any plane spanned by any two of the coordinate axes is itself a plane. We now sample P (or, rather, its parameter space) uniformly, which takes relatively few random bits, ie, a and b: The t coordinates of such a random point form the desired pseudorandom strings. Of course, restricted to any two coordinate axes, the projection is just as random as the point itself; hence the pairwise independence. Extending the construction to d-wise independence can be done by using linear codes, polynomials, and other algebraic tools.

9.3 Universal Hash Functions

Pseudorandom number generation is the operation of taking a random string on m bits and stretching it into an M-bit string ($M > m$). The longer string should look random enough to the "eyes" of a polynomially bounded computer. The previous section showed that linear codes can be used to ensure pairwise independence. We follow a different tack here.

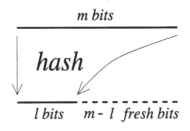

Fig. 9.2. From one pseudorandom string to the next.

We start with a random m-bit string R_1. To obtain the next string R_2, first we compress R_1 into a shorter ℓ-bit string $h(R_1)$, where $\ell < m$. Then, we append to it $m - \ell$ truly random new bits so as to restore the string to its previous length m. We iterate on this compress/expand process $k - 1$ times to produce, in the end, a sequence of k pseudorandom strings, $R_1, \ldots, R_k$, each of length m. Of course, the success of such a scheme is entirely in the hands of the function h. It is easy to come up with a dismal choice for h, eg, truncate a fixed-length suffix of the string. Fortunately, it is just as easy to find a good one. Intuitively, h should be "random" enough. A truly random function h would use up too many random bits. By sacrificing true randomness, we can limit the number of random bits needed to describe such a function. This leads to the key notion of a family of universal hash functions.

Let H be a family of functions from $\{0,1\}^m$ to $\{0,1\}^\ell$, where $\ell \leq m$. The set H is said to be a family of *universal hash functions* if, given any $x \neq y \in \{0,1\}^m$, a random $h \in H$ (chosen uniformly) satisfies:[5]

$$\mathrm{Prob}[\,h(x) = h(y)\,] \leq 2^{-\ell} + 2^{-m}.$$

In other words, the probability of a collision is bounded by (roughly) what it would be if $h(x)$ and $h(y)$ were random and independent. The question is, of course: Are there sparse families of universal hash functions? The answer is yes. We provide two examples.

1. Take H to be the set of all linear functions from the vector space $\{0,1\}^m$ to $\{0,1\}^\ell$, with coordinates in $\mathbf{F}_2$. If $x \neq y$, then $h(x) = h(y)$ implies that $h(z) = 0$, where $z = x - y \neq 0$. The i-th bit of $h(z)$ is the inner product of a random vector in $\{0,1\}^m$ with the nonzero vector z. The probability of its being 0 is therefore $1/2$. Over all i's, these events are mutually independent, so the probability that $h(x) = h(y)$ is $1/2^\ell$, and the family H is universal. What is the size of H? Any linear function can be represented by an ℓ-by-m matrix over $\mathbf{F}_2$, so the number of random bits needed to pick a random $h \in H$ is ℓm.

2. Fix a prime p whose binary representation has length $m+1$ (which is always possible since there are primes between 2^k and 2^{k+1}). Now, choose random integers a, b modulo p, and consider $h = h_{a,b}$, where

$$h(x) = ((ax + b) \bmod p) \bmod 2^\ell.$$

Intuitively, we form a standard linear congruential, but we truncate it by keeping only the ℓ lowest order bits. Fix $x \neq y$. How many pairs (a,b) can yield the same last ℓ bits in the number $(ax + b) \bmod p$? First let us ask, how many numbers modulo p can have a given ℓ-bit suffix? Obviously, at most $\lceil p/2^\ell \rceil$. For each such number, there are exactly p pairs (a,b) producing it (choosing a specifies b as well). Thus, the total number of pairs (a,b) producing the same suffix is at most $p\lceil p/2^\ell \rceil$. There are p^2 choices of (a,b), so this represents a probability bounded by $p\lceil p/2^\ell \rceil/p^2 \leq 1/2^\ell + 1/p$, and hence by $1/2^\ell + 1/2^m$. The family, therefore, is universal. To sample from it requires $2(m+1)$ truly random bits, which is much less than the ℓm needed in the previous example.

---

[5]The standard definition has an upper bound of $2^{-\ell}$, and our choice is often referred to as producing almost-universal hash functions.

We now describe our new pseudorandom number generator for **BPP**. Choose a parameter k and let H denote the family above for $\ell = m - 4k$. We define k pseudorandom m-bit strings $R_1, \ldots, R_k$ as follows:

(1) Pick a random hash function $h \in H$.
(2) Choose a random m-bit string $R_1 \in \{0,1\}^m$.
(3) Take $k - 1$ random *seeds*, ie, strings $S_1, \ldots, S_{k-1} \in \{0,1\}^{m-\ell}$.

The output is the sequence of strings: R_1 and, for $1 < i \leq k$,

$$R_i = h(R_{i-1}) \cdot S_{i-1},$$

where the dot sign denotes string concatenation. The strings $R_1, \ldots, R_k$ are the pseudorandom strings that we use in running the **BPP** algorithm k times. As usual, we conclude with a majority vote. We prove a result that is slightly weaker than Theorem 9.1 (page 318).

Lemma 9.2 *Given a* **BPP** *algorithm that uses m random bits per run, the probability of failure can be reduced to $1/2^k$ by using only $O(m + k^2)$ random bits and $O(k)$ runs.*

The connection with discrepancy theory is now made explicit. Consider a probability distribution $\pi = (\pi_1, \ldots, \pi_N)^T$; we use the notation $u = (1/N, \ldots, 1/N)^T$ for the uniform distribution. The ability of π to "simulate" u is to be measured here by the L^1-norm discrepancy

$$D(\pi) \stackrel{\text{def}}{=} \|\pi - u\|_1.$$

Lemma 9.3 *Let π be the probability distribution formed in $H \times \{0,1\}^\ell$ by $(h, h(R))$, where h and R are chosen randomly and uniformly in, respectively, H and some $X \subseteq \{0,1\}^m$. Then,*

$$D(\pi) \leq \sqrt{\frac{2^\ell}{|X|} + 2^{\ell - m}}.$$

This result, commonly known as the *leftover hash lemma*, tells the whole story in a nutshell. The input distribution (h, R) might be tremendously biased, because R is chosen only within a subset X. And yet the output distribution π is very close to uniform. Predictably, the smaller ℓ is, ie, the more compressive the hash function is, and the larger X is, ie, the less bias we force on R, the more uniform π is.

Proof: Given two random $h, h' \in H$ and $R, R' \in X$, the probability that $(h, h(R)) = (h', h'(R'))$ is exactly $\|\pi\|_2^2$. It follows that

$$
\begin{aligned}
\|\pi\|_2^2 &= \text{Prob}[h = h'] \times \text{Prob}[\, h(R) = h'(R') \mid h = h'\,] \\
&= \text{Prob}[h = h'] \times \text{Prob}[\, h(R) = h(R')\,] \\
&\leq \text{Prob}[h = h'] \times (\text{Prob}[\, R = R'\,] + \text{Prob}[\, h(R) = h(R') \mid R \neq R'\,]) \\
&\leq \frac{1}{|H|} \left(\frac{1}{|X|} + 2^{-\ell} + 2^{-m} \right).
\end{aligned}
$$

The distribution π is defined over a set of size $N = 2^\ell |H|$. By Cauchy-Schwarz and $\|\pi\|_1 = 1$,

$$
\begin{aligned}
D(\pi)^2 &= \|\pi - u\|_1^2 \leq N\|\pi - u\|_2^2 \leq N(\|\pi\|_2^2 + \|u\|_2^2 - 2\pi^T u) \\
&\leq N \left[\frac{1}{|H|} \left(\frac{1}{|X|} + 2^{-\ell} + 2^{-m} \right) + \frac{1}{N} - \frac{2}{N} \|\pi\|_1 \right] \leq \frac{2^\ell}{|X|} + 2^{\ell - m}.
\end{aligned}
$$

$\square$

It is now easy to show that the strings $R_1, \ldots, R_k$ behave much like uniform, independent random variables. Let $\pi^{(k)}$ be the distribution of that sequence of k strings. Fix an input x and let F denote the set of m-bit strings R that cause the **BPP** algorithm to fail on input x. For simplicity, we consider a specific example that illustrates the whole argument. Suppose that $k = 3$, and let p be the probability of both R_1 and R_3 being in F and R_2 being outside:

$$
\begin{aligned}
p &= \text{Prob}[\, R_1 \in F;\ R_2 \notin F;\ R_3 \in F\,] \\
&= \text{Prob}[\, R_1 \in F\,] \times \text{Prob}[\, R_2 \notin F;\ R_3 \in F \mid R_1 \in F\,] \\
&= \frac{|F|}{2^m} \cdot \text{Prob}[\, (h(R_1) \cdot S_1) \notin F;\ R_3 \in F \mid R_1 \in F\,],
\end{aligned}
$$

because R_2 is defined as the concatenation of $h(R_1)$ and S_1. Let Π be the distribution induced by

$$
\Big(h, (h(R_1) \cdot S_1), S_2, \ldots, S_{k-1} \Big),
$$

where R_1 is random in F and $h, S_1, \ldots, S_{k-1}$ are random as before. In our example above, where $k = 3$, the underlying universe is $\mathcal{U} = H \times \{0,1\}^m \times \{0,1\}^{m-\ell}$. So, $\Pi = (\Pi_1, \ldots, \Pi_{|\mathcal{U}|})^T$, where Π_u is the fraction of

$$
(h, R_1, S_1, S_2) \in H \times F \times \{0,1\}^{m-\ell} \times \{0,1\}^{m-\ell}
$$

producing the vector $u = (h, (h(R_1) \cdot S_1), S_2) \in \mathcal{U}$. Interpret any $u \in \mathcal{U}$ as a triplet (h, R_2^*, S_2), where R_2^* is a string in $\{0,1\}^m$, and define

$$\xi_u = \begin{cases} 1 & \text{if } R_2^* \notin F \text{ and } R_3 \in F, \text{ where } R_3 = h(R_2^*) \cdot S_2 , \\ 0 & \text{else.} \end{cases}$$

Given a random (h, R_1, S_1, S_2), it is clear that

$$\text{Prob}[(h(R_1) \cdot S_1) \notin F; R_3 \in F \mid R_1 \in F] = \sum_{u \in \mathcal{U}} \xi_u \Pi_u ,$$

while for a random (h, R_2^*, S_2)

$$\text{Prob}[R_2^* \notin F; R_3 \in F] = \sum_{u \in \mathcal{U}} \xi_u / |\mathcal{U}| ,$$

by the definition of ξ_u. Since

$$D(\Pi) = \sum_{u \in \mathcal{U}} |\Pi_u - 1/|\mathcal{U}||,$$

we have

$$\text{Prob}[(h(R_1) \cdot S_1) \notin F; R_3 \in F \mid R_1 \in F]$$
$$\leq \text{Prob}[R_2^* \notin F; R_3 \in F] + D(\Pi). \tag{9.3}$$

Let π be the distribution induced by $(h, h(R_1))$, where h (resp. R_1) is random in H (resp. F), and let u be the uniform distribution over $H \times \{0,1\}^\ell$. By the independence of $(h, h(R_1))$ and $(S_1, S_2, \ldots, S_{k-1})$, it is clear that[6] $\|\Pi - U\|_1 = \|\pi - u\|_1$, ie, $D(\Pi) = D(\pi)$. It follows that, by setting $X = F$, the leftover hash lemma (Lemma 9.3) yields

$$D(\Pi) \leq \sqrt{\frac{2^\ell}{|F|} + 2^{\ell-m}} .$$

By the definition of **BPP**, we have $|F| \leq 2^m/3$. By constant-factor amplification, we can strengthen this to $|F| \leq 2^m/5$. We can also easily assume

---

[6]There is a subtlety here. Let $\nu = |H|2^\ell$; each term $|\pi_i - 1/\nu|$ in $D(\pi)$ is in bijection with a sum within $D(\Pi)$ of the form

$$\left|\Pi_{i_1} - 1/|\mathcal{U}|\right| + \cdots + \left|\Pi_{i_r} - 1/|\mathcal{U}|\right|,$$

where $\pi_i = \sum_j \Pi_{i_j}$ and $r = |\mathcal{U}|/\nu$. What makes the identity true is that the Π_{i_j}'s are all equal because of independence, and therefore each term is equal to the sum in bijection with it.

that $|F|$ is not too small; in fact, we might as well assume that $|F| = 2^m/5$. By (9.3) and $\ell = m - 4k$, we find that

$$
\begin{aligned}
p &\leq \tfrac{1}{5}\mathrm{Prob}[\,(h(R_1) \cdot S_1) \notin F;\ R_3 \in F \mid R_1 \in F\,] \\
&\leq \tfrac{1}{5}\mathrm{Prob}[\,R_2^* \notin F;\ R_3 \in F\,] + D(\Pi) \\
&\leq \tfrac{1}{5}\mathrm{Prob}[\,R_2^* \notin F;\ R_3 \in F\,] + 2^{2-2k}.
\end{aligned}
$$

We can repeat essentially the same derivations with respect to

$$\mathrm{Prob}[\,R_2^* \notin F;\ R_3 \in F\,];$$

the only difference is that F is now replaced by its complement, which is of size $2^{m+2}/5$. The upper bound becomes (conservatively)

$$\mathrm{Prob}[\,R_2^* \notin F;\ R_3 \in F\,] \leq \mathrm{Prob}[\,R_3^* \in F\,] + 2^{2-2k}.$$

Finally, we find that

$$p \leq \frac{1}{5^2} + k2^{2-2k},$$

or, more generally,

$$p \leq \frac{1}{5^b} + k2^{2-2k},$$

where b is the number of bad strings R_i's. So, given a fixed sequence (good, good, bad, good, bad, etc) of length k with at least $k/2$ bads, the probability that the strings $R_1, \ldots, R_k$ fit that sequence is at most $(1/5)^{k/2} + k2^{2-2k}$. Thus, the probability that, out of $R_1, \ldots, R_k$, at least $k/2$ strings are bad, ie, the probability that the **BPP** algorithm fails, is at most $(2/\sqrt{5})^k + k2^{1-k}$, ie, $O(c^{-k})$, for some fixed $c > 1$. Going from k runs to Ck runs, for a large enough constant C, reduces the error probability to 2^{-k}, which proves Lemma 9.2. $\square$

9.4 Random Walk on an Expander

The time has come for us to bring out our most powerful artillery and prove Theorem 9.1 (page 318). We begin with a proof sketch. Let us model the process of choosing k random m-bit numbers $R_1, \ldots, R_k$ as a random walk in an undirected graph. Define K to be the complete graph on $n = 2^m$ nodes, labeled $0, \ldots, n-1$. In addition, we provide each node with a self-loop, and we define a Markov chain where each edge is assigned the probability $1/n$. Starting from node 0, we perform a random walk of length

k. Obviously, the labels of the first, second, etc, k-th node provide random m-bit numbers $R_1, \ldots, R_k$, as desired. Each step in the walk requires m random bits, so the total number of random bits is $k \log n$.

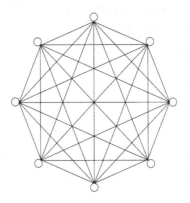

Fig. 9.3. A random walk in K_8 with self-loops produces random numbers but at a cost of $\log 8$ random bits per step.

Why this trivial analogy? Because it begs the question: Might we be able to sparse out the graph K by removing edges while still generating random enough R_i's? In particular, if we could limit our random walk to a subgraph G of constant degree, then we would expand only a constant number of random bits per step. We show below that choosing an expander does the trick.

Let G be a connected graph on $n = 2^m$ nodes. For simplicity, we assume that G has the following structure: (i) It is derived from a bipartite graph by attaching a self-loop to each node; (ii) each node is connected to d other nodes, not counting itself. We consider the Markov chain formed by assigning the probability $1/2$ to each self-loop and $1/2d$ to each of the other edges. The transition matrix $P = (P_{ij})$ specifies the probability of going from node i to node j. Because P is symmetric, it is diagonalizable and its eigenvalues $\lambda_1 \geq \cdots \geq \lambda_n$ are real. Furthermore,

Lemma 9.4

$$1 = \lambda_1 > \lambda_2 \geq \cdots \geq \lambda_n \geq 0.$$

Proof: Observe that $P = (I + Q)/2$, where I is the identity and Q is the transition matrix of the Markov chain formed by the bipartite graph (minus the self-loops) with all edge probabilities equal to $1/d$. Obviously, Q has the same eigenspaces as P, and its eigenvalues are $\mu_i = 2\lambda_i - 1$. Because the graph for Q is regular, and Q is doubly stochastic,[7] $|\mu_i| \le 1$. This shows that all of the λ_i's are between 0 and 1. Obviously, $(1/n, \ldots, 1/n)^T$ is an eigenvector for $\lambda_1 = 1$.

To complete the proof we show that, because G is connected, $\mu_1 > \mu_2$, and hence $\lambda_1 > \lambda_2$. We could invoke the Perron-Frobenius theorem, but a direct proof is easy. Because Q is diagonalizable, it suffices to show that the eigenspace for the eigenvalue 1 has dimension 1. In any nonzero vector $x = (x_1, \ldots, x_n)^T$ such that $Qx = x$, consider the largest entry x_i in absolute value. Obviously, $x_i = (1/d) \sum_j x_j$, where the sum extends over all of the nodes (of G minus the self-loops) adjacent to the i-th node. By the maximality of x_i, it follows that the x_j's of adjacent nodes are all equal to x_i. This gives us a starting base. Now, using the fact that the value at each node is the average of the neighboring values, we immediately derive, by connectivity, that all of the x_i's are equal. Therefore, the eigenspace for 1 is generated by $(1/n, \ldots, 1/n)^T$. □

Let $\pi^{(0)}$ be an initial probability distribution on the nodes of G (ie, the states of the Markov chain), represented as a vector column. The vector $\pi^{(1)} = P\pi^{(0)}$ gives the probability distribution after one step of the walk. We shall soon see that, as the number of steps k increases, the distribution $\pi^{(k)} = P^k \pi^{(0)}$ converges towards the uniform distribution

$$u = (1/n, \ldots, 1/n)^T,$$

and the convergence rate depends on λ_2. This implies that a random walk eventually leads to a uniform distribution among the nodes of G, regardless of the starting node. The intuition is clear. Powers of P bring all the eigenvalues down to 0 except for 1; in the limit we have the spectral distribution of the complete graph with self-loops. The second largest eigenvalue is thus a measure of how close to the complete graph our graph G really is.

Lemma 9.5

$$\|\pi^{(k)} - u\|_2 \le \lambda_2^k \|\pi^{(0)}\|_2 \le \lambda_2^k.$$

---

[7]A matrix is doubly stochastic if its elements are nonnegative and each row (and column) sums up to 1. It is immediately seen that such a matrix preserves the L^1 norm of any nonnegative vector and so, in particular, it maps a probability distribution to another one. No eigenvalue can exceed 1 in absolute value. To see why, assume that one did. Then any associated eigenvector would grow to unbounded lengths by repeated applications of the map Q. This would mean that large powers of Q would have to contain some very large matrix elements: But such a matrix cannot possibly map probability distributions to probability distributions, and we would have a contradiction.

Proof: Let $e_1, \ldots, e_n$ be an orthonormal eigenbasis, where λ_i is the eigenvalue associated with e_i. The vector $u = (1/n, \ldots, 1/n)^T$ is an eigenvector for the eigenvalue 1, and so $e_1 = \sqrt{n}\, u$. If

$$\pi^{(0)} = c_1 e_1 + \cdots + c_n e_n,$$

then

$$\pi^{(k)} = c_1 \lambda_1^k e_1 + \cdots + c_n \lambda_n^k e_n$$

and

$$\|\pi^{(k)} - c_1 e_1\|_2^2 \ \leq \ \sum_{i>1} c_i^2 \lambda_i^{2k} \leq \lambda_2^{2k} \sum_{i>1} c_i^2$$

$$\leq \ \lambda_2^{2k} \|\pi^{(0)}\|_2^2.$$

It follows that

$$\|\pi^{(k)} - c_1 e_1\|_2 \leq \lambda_2^k \|\pi^{(0)}\|_2 \leq \lambda_2^k \|\pi^{(0)}\|_1 \leq \lambda_2^k.$$

By Lemma 9.4, we have $0 \leq \lambda_2 < 1$, and so, as k goes to infinity, the distribution $\pi^{(k)}$ converges toward $c_1 e_1$. This means that $c_1 e_1$ is itself a probability distribution, and therefore it is equal to u. $\square$

Spectral Properties of Expanders

Lemma 9.5 indicates that, for a random walk to be "rapidly mixing," ie, to converge toward the uniform distribution fast, it suffices to ensure that λ_2 is small. Expander graphs have this property. An (n, d, c)-expander H is a connected, d-regular[8] bipartite graph $(X \cup Y, E)$, with $|X| = |Y| = n/2$. Furthermore, for each subset $W \subseteq X$, the number of nodes adjacent to at least one node in W is at least

$$\left(1 + c \left(1 - \frac{2|W|}{n} \right) \right) |W|.$$

It is easy to prove the existence of expanders (for suitable n, d, c) by a probabilistic argument: Choose a random graph from some appropriate distribution. This approach has two shortcomings. To be random is one of them; checking whether a given graph is an expander is in general co-NP complete (thus, likely intractable). Another weakness is that they must be specified in full. Since in many applications the number of nodes is

---

[8] A graph is d-regular if each node has degree d. In general, an expander need not be bipartite; but only such graphs will be considered here.

actually exponential in the problem size, it is preferable to have an implicit description requiring only logarithmic space.

Such graphs do exist. Here is one: The number of nodes is of the form $n = 2p^2$, for some integer p. The graph $H = (X \cup Y, E)$ is bipartite. There is a bijection between X and the pairs $(a, b) \in (\mathbf{Z}/p\mathbf{Z})^2$, and the same is true with Y. A node (a, b) is adjacent to any node of the form

$$(a, b), \ (a, 2a + b), \ (a, 2a + b + 1), \ (a, 2a + b + 2),$$

$$(a + 2b, b), \ (a + 2b + 1, b), \ \text{or} \ (a + 2b + 2, b),$$

with addition defined modulo p. It can be shown that the graph is an $(n, 7, \alpha)$-expander, where $\alpha = 1 - \sqrt{3}/2$. The n-by-n adjacency matrix[9] of H is symmetric, and so it has a full set of real eigenvalues, $\mu_1 \geq \cdots \geq \mu_n$, where obviously $\mu_1 = -\mu_n = d$ (for the first eigenvector, try a vector of 1's; for the second, switch the sign of the coordinates in the first half). The relation with the spectrum of the graph's adjacency matrix is made explicit in the following (we omit the proof).

Lemma 9.6 *If μ is the second largest distinct eigenvalue (in absolute value) of the adjacency matrix of an (n, d, c)-expander, then*

$$|\mu| \leq d - \frac{c^2}{2^{10} + 2c^2}.$$

Expanders behave like quasi-random graphs. The difference in behavior between them and random graphs can be expressed by a discrepancy measure. Given two subsets of nodes, $A \subseteq X$ and $B \subseteq Y$, let $E(A, B)$ be the set of edges joining A and B. The *discrepancy* for (A, B) is defined by

$$D(A, B) = \left| |E(A, B)| - \frac{2d|A||B|}{n} \right|.$$

The discrepancy can be bounded as a function of μ. The following result holds for any d-regular bipartite graph (no need for it to be an expander).

Theorem 9.7

$$D(A, B) \leq |\mu| \sqrt{|A||B|}.$$

Proof: Let M be the n-by-n adjacency matrix of H, where $n = |X \cup Y|$. Let $e_1, \ldots, e_n$ be an orthonormal eigenbasis, where μ_i is the eigenvalue

---

[9]This is the matrix whose element at row i and column j is 1 if (i, j) is an edge of the graph, and 0 otherwise.

associated with e_i. If ξ_A (resp. ξ_B) denotes the characteristic (column) vector of A (resp. B), then $\xi_A^T M \xi_B$ counts each edge between A and B once:

$$|E(A, B)| = \xi_A^T M \xi_B . \tag{9.4}$$

Expressing the characteristic vectors in the eigenbasis gives $\xi_A = \sum_i \alpha_i e_i$ and $\xi_B = \sum_i \beta_i e_i$. Because $e_1 = (1, \ldots, 1)^T / \sqrt{n}$,

$$\alpha_1 = \xi_A^T e_1 = \frac{|A|}{\sqrt{n}} .$$

Similarly, $\beta_1 = |B|/\sqrt{n}$. Now, using the fact that

$$e_n = (1, \ldots, 1, -1, \ldots, -1)^T / \sqrt{n} ,$$

we find that $\alpha_n = |A|/\sqrt{n}$ and $\beta_n = -|B|/\sqrt{n}$. Since $\mu_1 = -\mu_n = d$, it follows from (9.4) that

$$
\begin{aligned}
|E(A, B)| &= \left(\sum_i \alpha_i e_i \right)^T \left(\sum_i \mu_i \beta_i e_i \right) = \sum_i \mu_i \alpha_i \beta_i \\
&= \frac{2d|A||B|}{n} + \sum_{1 < i < n} \mu_i \alpha_i \beta_i
\end{aligned}
$$

and, by Cauchy-Schwarz, that

$$
\begin{aligned}
\left| |E(A, B)| - \frac{2d|A||B|}{n} \right| &= \sum_{1 < i < n} \mu_i \alpha_i \beta_i \leq |\mu| \, \|\xi_A\|_2 \, \|\xi_B\|_2 \\
&\leq |\mu| \sqrt{|A||B|} .
\end{aligned}
$$

$\square$

Recycling Random Bits

Returning to **BPP**, let F be the set of m-bit strings that cause the algorithm to fail. Because of constant-factor amplification of the type discussed earlier, we can assume that $|F| < n/100$, where $n = 2^m$. By abuse of notation, let F also denote the n-by-n diagonal matrix, where $F_{ij} = 1$ if $i = j$ and i (written in binary) is in F, and $F_{ij} = 0$ if not. Let G denote the graph derived from a bipartite (n, d, c)-expander by adding self-loops to each node. As before, we consider the Markov chain formed by assigning the probability $1/2$ to each self-loop and $1/2d$ to each of the other edges, and we let P denote its transition matrix. If π is a probability distribution on the nodes of G (or, equivalently, on the set of m-bit strings, by labeling the nodes between 0 and $n-1$ in binary), then $\|F\pi\|_1$ is the probability that a random string chosen from that distribution is in F. This generalizes easily. Let t be large enough that

$$\lambda_2^t < \tfrac{1}{10}. \tag{9.5}$$

Consider the strings $R_1, \ldots, R_k$ obtained in the following fashion: R_1 is picked by choosing a node of G at random (its label delivering the m bits of R_1). Next, we perform a random walk of t steps, which lands us at a node providing R_2; a subsequent t-step walk gives R_3, and so on, until we produce R_k. It is easy to keep track of the probability of each R_i being a bad or a good string, ie, being in or outside F.

For example, suppose that $k=4$; then the probability that R_1 and R_3 are in F, but R_2 and R_4 are outside, is exactly

$$\left\| (I-F)P^t F P^t (I-F) P^t F \pi^{(0)} \right\|_1. \tag{9.6}$$

We can bound such an expression by using two simple facts:

Lemma 9.8 *Given any nonnegative vector $x \in \mathbf{R}^n$,*

$$\|FP^t x\|_2 \le \tfrac{1}{5} \|x\|_2 \quad and \quad \|(I-F)P^t x\|_2 \le \|x\|_2.$$

Proof: For the first inequality, trivially we may assume that $x \ne 0$. It follows that $x/\|x\|_1$ is a probability distribution. By applying Lemma 9.5 for $k=t$, we get

$$\|F(P^t(x/\|x\|_1) - u)\|_2 \le \|P^t(x/\|x\|_1) - u\|_2 \le \lambda_2^t \frac{\|x\|_2}{\|x\|_1},$$

where $u = (1/n, \ldots, 1/n)^T$, and by (9.5)

$$\|F(P^t x - u\|x\|_1)\|_2 \le \frac{\|x\|_2}{10}.$$

By the triangular inequality and Cauchy-Schwarz,

$$\|FP^t x\|_2 \le \|x\|_1 \|Fu\|_2 + \frac{\|x\|_2}{10} \le \frac{\|x\|_1}{10\sqrt{n}} + \frac{\|x\|_2}{10} \le \frac{\|x\|_2}{5}.$$

The second inequality is even easier to prove. Obviously,

$$\|(I-F)P^t x\|_2 \le \|P^t x\|_2.$$

No eigenvalue of P^t exceeds 1 in absolute value, so as a linear map it cannot increase the L^2 norm of any vector, ie, $\|P^t x\|_2 \le \|x\|_2$. $\square$

The lemma shows that the probability in (9.6) is bounded by

$$\sqrt{n}\left\|(I-F)P^t F P^t (I-F) P^t F \pi^{(0)}\right\|_2 \le \frac{\sqrt{n}}{5}\|F\pi^{(0)}\|_2 .$$

Because the first string R_1 is obtained by picking a node at random, clearly, $\pi^{(0)} = u$ and $\|F\pi^{(0)}\|_2 \le 1/(10\sqrt{n})$; so, for a conservative upper bound on the probability in question, we can use $(1/5)^2$. More generally, given a fixed sequence (good, good, bad, good, bad, etc) of length k with at least $k/2$ bads, the probability that the strings $R_1, \ldots, R_k$ match that sequence is at most $(1/5)^{k/2}$. Thus, the probability that, out of $R_1, \ldots, R_k$, at least $k/2$ strings are bad is at most $2^k (1/5)^{k/2}$.

To summarize, we need m random bits to get R_1, and then only $O(k)$ random bits to obtain $R_2, \ldots, R_k$. The probability that a majority vote results in an error is at most $(2/\sqrt{5})^k$. We can increase k by a constant factor to bring the error probability below $1/2^k$ and still use only $O(m+k)$ random bits. This proves Theorem 9.1 (page 318). $\square$

9.5 Low Bias from Quadratic Residues

Instead of generating several perfectly random m-bit strings, hoping for some sort of independence among them, we consider a related but different problem: How to produce a single pseudorandom m-bit string, using much fewer than m truly random bits. Our criterion for pseudorandomness is based on linear tests. This has an immediate Fourier analysis interpretation, which we explain below. The connection between pseudorandomness, low discrepancy, and small Fourier coefficients is further exploited in §9.6.

We begin with a piece of terminology. The *bias* of a random variable X in $\{0,1\}$ is defined as $|\mathbf{E}(-1)^X|$, which is also

$$\left| \mathrm{Prob}[X=0] - \mathrm{Prob}[X=1] \right|.$$

We can generalize this notion to multivariate distributions. Let S be a subset of the hypercube $\{0,1\}^m$ with a distribution on it. For a given $t = (t_0, \ldots, t_{m-1}) \in \{0,1\}^m$, let X_t be the random variable $\sum t_i b_i \in \mathbf{F}_2$, where $(b_0, \ldots, b_{m-1})$ is chosen randomly in S. The bias of S is the maximum bias of X_t for all $t \ne 0$.

For any odd prime $p > m$, we show an easy construction of a set S of size p and bias $O(m/\sqrt{p})$. For any integer x, consider the vector:

$$s(x) \stackrel{\text{def}}{=} \Big(b_0(x), b_1(x), \ldots, b_{m-1}(x)\Big),$$

where $b_i(x)$ is the 0/1 function defined by the recipe:

$$b_i(x) = \begin{cases} \frac{1}{2}(1 - \chi_p(x + i)) & \text{if } x + i \not\equiv 0 \pmod{p}, \\ 1 & \text{else.} \end{cases}$$

The subset S is defined as $\{ s(x) \,|\, x \in \mathbf{F}_p \}$. To prove that its bias is small, we show that its discrepancy is low in the Fourier sense. What we mean is that the Fourier coefficients of its characteristic function are small: This is the function that is 1 at points of S and 0 elsewhere. The Fourier transform is defined over the abelian group $(\mathbf{Z}/2\mathbf{Z})^m$. By definition, the Fourier coefficient at the frequency $t = (t_1, \ldots, t_n) \in \{0, 1\}^m$ is (see Appendix B)

$$\widehat{f}(t) = \sum_{x \in \mathbf{F}_p} (-1)^{t_0 b_0(x) + \cdots + t_{m-1} b_{m-1}(x)}.$$

Obviously,

$$\text{bias}\,(S) = \max_{t \neq 0} | \mathbf{E}\,(-1)^{X_t} | = \frac{1}{p} \max_{t \neq 0} |\widehat{f}(t)|. \tag{9.7}$$

From now on, we may assume that $t \neq 0$. Observe that, for any fixed $x \in \mathbf{F}_p$,

$$\left| (-1)^{\sum_{i=0}^{m-1} t_i b_i(x)} - \prod_{i=0}^{m-1} \chi_p(x+i)^{t_i} \right| = \begin{cases} 1 & \text{if } \prod_{i=0}^{m-1}(x+i) \equiv 0 \pmod{p}. \\ 0 & \text{else.} \end{cases}$$

The first case happens for m values of x, so

$$\begin{aligned} |\widehat{f}(t)| &\leq \left| \sum_{x \in \mathbf{F}_p} \prod_{i=0}^{m-1} \chi_p(x+i)^{t_i} \right| + m \\ &\leq \left| \sum_{x \in \mathbf{F}_p} \chi_p \Big(\prod_{i=0}^{m-1} (x+i)^{t_i} \Big) \right| + m. \end{aligned}$$

The polynomial $\prod_{i=0}^{m-1}(x+i)^{t_i}$ is not a square[10] and it has exactly m distinct zeros; so, by Weil's bound (9.2) on page 321,

$$|\widehat{f}(t)| \leq (m-1)\sqrt{p} + m \leq 2m\sqrt{p}.$$

By (9.7), this shows that the bias of S is bounded by $2m/\sqrt{p}$. The savings is in the number of random bits. Instead of m of them we need only $\lfloor \log p \rfloor + 1$ truly random bits.

---

[10]Because $m < p$, the only square it could be is the constant polynomial 1, but this is impossible because $t \neq 0$.

9.6 Polynomial Interpolation

It is well known that a univariate real polynomial of degree $n - 1$ with rational coefficients can be reconstructed entirely by knowing only its value at n points. Furthermore, a judicious choice of n points (ie, roots of unity) helps to make this process extremely fast. A natural question arises: If we know the value of the polynomial only at a small, say polylogarithmic, number of points, can we reconstruct the polynomial, or some sufficiently close approximation of it? In this section, we show that if the polynomial is sufficiently sparse, the answer is yes. The idea is to pick roots of unity that form a low-discrepancy set with respect to arithmetic progressions along the unit circle. Such an approximate interpolation technique can be especially useful when n is very large.

Low-Discrepancy Arithmetic Progressions

Our goal is to produce a small subset of integers in $\{0, \ldots, n - 1\}$ with respect to which arithmetic progressions modulo n have low discrepancy. We review a few definitions. Let n be a large enough prime, and let $\mathbf{Z}_n$ be the set of integers modulo n. Given $A \subseteq \mathbf{Z}_n$, let $xA \subseteq \mathbf{Z}_n$ $(x \neq 0)$ denote the set $\{ xa \mid a \in A \}$. The discrepancy of an interval R (taken modulo n, ie, with wrap-around) in $\mathbf{Z}_n$ is defined as

$$D(A, R, x) \stackrel{\text{def}}{=} \left| \frac{|R|}{n} - \frac{|(xA) \cap R|}{|A|} \right|.$$

The maximum value over all R, x defines the discrepancy with respect to A,

$$D(A) \stackrel{\text{def}}{=} \max_{R, x} D(A, R, x).$$

An important remark: By perfect analogy with ε-approximations, the discrepancy is defined as a difference between a probability and a conditional probability. Note that

$$D(A, R, x) = \left| \frac{|R_x|}{n} - \frac{|A \cap R_x|}{|A|} \right|,$$

where $R_x = (x^{-1})R$ is an arbitrary arithmetic progression. Thus, $D(A)$ measures the maximum discrepancy of any arithmetic progression modulo n with respect to A (the set A playing the role of the coloring here). By Theorem 4.4 (page 174), we can construct a set A of size $O(\varepsilon^{-2} \log n)$ such that $D(A) \leq \varepsilon$. The following result gives a weaker bound, but the advantage is that A is given by a formula, so there is no need to look at the set system itself.

Theorem 9.9 *Fix a large prime n and a small enough constant $\alpha > 0$. Given any ε such that $n^{-\alpha} < \varepsilon < 1$, there exists $A \subset \mathbf{Z}_n$ of size $(\varepsilon^{-1} \log n)^{O(1)}$ such that $D(A) \leq \varepsilon$.*

Proof: Obviously, we can assume that ε is less than a suitably small positive constant. Choose some large enough constant c, and define $N = (\varepsilon^{-1} \log n)^c$. Because n is prime, $\mathbf{Z}_n$ is a field. The reciprocal of p is denoted by p^{-1}. We define

$$A = \{\, sp^{-1} \in \mathbf{Z}_n \mid 0 \leq s \leq \varepsilon N \text{ and } N \leq \text{ prime } p \leq 2N \,\}.$$

We easily verify that $|A| = (\varepsilon^{-1} \log n)^{O(1)}$. Fix a nonzero integer $x < n$ and an interval R in $\mathbf{Z}_n$. To facilitate the notation, the term random s, p refers to a random $0 \leq s \leq \varepsilon N$ and a random prime p such that $N \leq p \leq 2N$ (both uniformly distributed). Since ε can always be rescaled by a constant factor, it suffices to prove that

$$\left| \operatorname{Prob}[\, xsp^{-1} \bmod n \in R\,] - \frac{|R|}{n} \right| = O(\varepsilon).$$

By scaling down R to an interval $I \subseteq [0, 1)$ modulo 1, we have the equivalent bound

$$\left| \operatorname{Prob}\left[\frac{xsp^{-1} \bmod n}{n} \in I \right] - |I| \right| = O(\varepsilon). \tag{9.8}$$

Unless specified otherwise, the arithmetic below is performed over $\mathbf{Z}$ (ie, not modulo anything). By the Chinese remainder theorem,

$$p(p^{-1} \bmod n) + n(n^{-1} \bmod p) = pn + 1.$$

(Of course, n^{-1} is here the reciprocal of n modulo p, which exists since $p \neq n$.) Let $\{z\}$ denote the fractional part of z in $[0, 1)$, ie, $z \bmod 1$. We derive

$$\left\{ \frac{xs(p^{-1} \bmod n)}{n} + \frac{xs(n^{-1} \bmod p)}{p} \right\} = \left\{ xs + \frac{xs}{pn} \right\} = \frac{xs}{pn} \leq \varepsilon.$$

It follows that, for some interval I' of length at most $|I| + \varepsilon$,

$$\operatorname{Prob}\left[\left\{ \frac{xs(n^{-1} \bmod p)}{p} \right\} \in I \right] \leq \operatorname{Prob}\left[\left\{ \frac{xs(p^{-1} \bmod n)}{n} \right\} \in I' \right]$$

and, conversely,

$$\text{Prob}\left[\left\{\frac{xs(p^{-1}\bmod n)}{n}\right\}\in I\right]\le\text{Prob}\left[\left\{\frac{xs(n^{-1}\bmod p)}{p}\right\}\in I'\right].$$

As a result, (9.8) is a consequence of the bound

$$\left|\text{Prob}\left[\left\{\frac{xs(n^{-1}\bmod p)}{p}\right\}\in I\right]-|I|\right|=O(\varepsilon),\qquad(9.9)$$

where I is an arbitrary interval in $[0,1)$. We may assume that $|I|\ge\varepsilon$, since otherwise we can always prove (9.9) for an enlarged $|I|=\varepsilon$ and then shrink I back to its original size. We now say that p is *favorable* if, for all $s\le 1/\varepsilon$,

$$\left\{\frac{xs(n^{-1}\bmod p)}{p}\right\}^*>\frac{2}{\varepsilon^3 N},$$

where

$$\{z\}^*=\begin{cases}\{z\}&\text{if }\{z\}<1/2,\\1-\{z\}&\text{else.}\end{cases}$$

It remains for us to show that (9.9) holds for any favorable p, and that a random p is favorable with probability at least $1-O(\varepsilon)$. We begin with the former.

Assume that p is favorable. The pigeonhole principle ensures the existence of two integers, $0\le s<s'\le 1/\varepsilon$, whose difference $s_0=s'-s$ satisfies $s_0\le 1/\varepsilon$ and

$$y\stackrel{\text{def}}{=}\left\{\frac{xs_0(n^{-1}\bmod p)}{p}\right\}^*\le\varepsilon.$$

We can prove (9.9) separately for each residue class $i<s_0$, ie,

$$\left|\text{Prob}[P_r\in I]-|I|\right|=O(\varepsilon),\qquad(9.10)$$

where

$$P_r=\left\{\frac{x(rs_0+i)(n^{-1}\bmod p)}{p}\right\}.$$

By p being favorable, we know that $y\ge 2/\varepsilon^3 N$. This shows that, by exhausting all the values of r, P_r cycles around the interval $[0,1)$ at least $1/\varepsilon$ times. Indeed, r can step through about $\varepsilon N/s_0$ values. Each step shifts P_r by y (or by $-y$), wrapping around the unit interval when necessary. The total amount of shift is at least (about) $(\varepsilon N/s_0)(2/\varepsilon^3 N)\ge 2/\varepsilon$, and so P_r cycles around the unit interval at least $1/\varepsilon$ times. At each cycle, P_r visits I a certain number of times, creating a discrepancy of $O(y)=O(\varepsilon)$.

Discrepancy is defined here as a relative error, so by summing over all cycles around $[0, 1)$ and all residue classes i, we still find the discrepancy to be $O(\varepsilon)$, as desired.

We now show that a random p is favorable with probability at least $1 - O(\varepsilon)$. If p is not favorable, then

$$xs(n^{-1} \bmod p) \equiv b \pmod{p},$$

for some $s \leq 1/\varepsilon$ and[11] $|b| \ll p/(N\varepsilon^3) \ll 1/\varepsilon^3$. It follows that

$$
\begin{aligned}
bn &\equiv xsn(n^{-1} \bmod p) \pmod{p} \\
&\equiv xs \pmod{p},
\end{aligned}
$$

and so p divides $xs - bn$. Obviously, $xs \neq bn$, since otherwise, being prime, n would divide x or s, both of which are less than n. Because $|xs - bn| \leq n^2$, by a trivial count on the number of prime divisors of $xs - bn$, we find that each pair (s, b) gives rise to only $O(\log n)$ unfavorable p's. Since $s \leq 1/\varepsilon$ and $|b| \ll 1/\varepsilon^3$, this leads to a total of $O((\log n)/\varepsilon^4)$ unfavorable p's. The number of prime p is at least on the order of $N/\log N \gg (\varepsilon^{-1} \log n)^c / \log(\varepsilon^{-1} \log n)$, and so, with c large enough, a random prime p is favorable with probability at least $1 - O(\varepsilon)$. $\square$

Small Fourier Coefficients

In this section, n is a large prime and, as usual, $\mathbf{Z}_n$ denotes the set of integers modulo n. Given $A \subseteq \mathbf{Z}_n$, the discrete Fourier transform (DFT) of A is, by abuse of terminology, the DFT of its characteristic vector, ie,

$$f_A(t) = \sum_{x \in A} e^{2\pi i x t / n}.$$

Note the absence of a minus sign in the exponent; we use the conjugate of the standard definition for notational convenience. Intuitively, a low-discrepancy set should produce evenly distributed points on the unit circle (in the complex plane), which mutually cancel to produce a small Fourier coefficient (Fig. 9.4). It is easy to flesh out this intuition. Recall that $D(A)$ is the maximum value, over all x and intervals R in $\mathbf{Z}_n$, of

$$\left| \frac{|R|}{n} - \frac{|(xA) \cap R|}{|A|} \right|.$$

---

[11] Recall that $\ll$ and $\gg$ denote $O()$ and $\Omega()$, respectively.

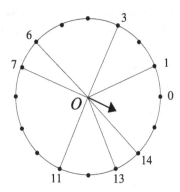

Fig. 9.4. The elements of $A \subset \mathbf{Z}_{16}$ are indicated by radial axes. Push the origin toward each point of A with a force equal to 1: The resulting force is the Fourier coefficient for $t = 1$. If the points of A are "uniformly" spread out around the circle, one expects the forces to cancel out and result in near-equilibrium, ie, a short resulting force vector. For $t > 1$, the t-th coefficient has a similar interpretation: Simply move the points around the circle by multiplying their radial angle by t.

Lemma 9.10 *Given any $A \subseteq \mathbf{Z}_n$ and $0 < t < n$,*

$$|f_A(t)| \leq 2\pi |A|\, D(A).$$

Proof: Present $\mathbf{Z}_n$ as $\{1, 2, \ldots, n\}$ and, with $m = |A|$, let $ta_1, \ldots, ta_m$ be the numbers in tA (all of which are distinct) in increasing order with respect to that presentation. Because $|e^{ix} - e^{iy}| \leq |x - y|$ and $|ta_j/n - j/m| \leq D(A)$,

$$\left| e^{2\pi i t a_j / n} - e^{2\pi i j / m} \right| \leq 2\pi D(A),$$

and therefore

$$\left| \sum_{j=1}^{m} \left(e^{2\pi i t a_j / n} - e^{2\pi i j / m} \right) \right| \leq 2\pi m D(A).$$

The lemma follows now from the fact that $\sum_{1 \leq j \leq m} e^{2\pi i j / m} = 0$. □

Interpolating a Sparse Polynomial

Let n be a prime and let $P = \sum_{0 \leq k < n} a_k x^k$ be a sparse polynomial in $\mathbf{Q}[x]$ over the reals. Suppose that the coefficients of P are unknown, but that an oracle is available for giving us the value of $P(y)$, given any real y. The

DFT is the standard vehicle for reconstructing P. As is well known, the coefficients of P satisfy:

$$a_k = n^{-1} \sum_{j=0}^{n-1} P(e^{2\pi i j/n}) e^{-2\pi i jk/n}. \tag{9.11}$$

Given $\varepsilon > 0$, let A be the set of Theorem 9.9 (page 339). Now, instead of computing the average in (9.11) over all n terms, let us compute the estimate

$$a_k^* = |A|^{-1} \sum_{j\in A} P(e^{2\pi i j/n}) e^{-2\pi i jk/n}.$$

How big an error are we making by pretending that the a_k^*'s are the true coefficients of P? The coefficient a_k is a sum of terms of the form

$$a_l n^{-1} \sum_{j=0}^{n-1} e^{2\pi i j(l-k)/n}.$$

By Lemma 9.10, $|f_A(t)| \le 2\pi|A|\varepsilon$ for any $0 < t < n$, so computing the average over A instead creates an additive error of

$$
\begin{aligned}
|a_k - a_k^*| &\le \sum_l |a_l| \left| n^{-1} \sum_{j=0}^{n-1} e^{2\pi i j(l-k)/n} - |A|^{-1} \sum_{j\in A} e^{2\pi i j(l-k)/n} \right| \\
&\le \sum_{l\ne k} |a_l| \, |A|^{-1} \, |f_A(l-k)| \\
&\le 2\pi(|a_1| + \cdots + |a_n|)\varepsilon.
\end{aligned}
$$

We conclude that interpolation can be done with a small additive error, ie, $O(\varepsilon L^1(P))$ per coefficient by evaluating the polynomial at $(\varepsilon^{-1}\log n)^{O(1)}$ places. This might be a good approximation method if P is sparse but of high degree.

9.7 Bibliographical Notes

Section 9.2: The construction of pairwise independent distributions was suggested by Joffe [170]. Alon, Babai, and Itai [13] generalized the construction to achieve k-wise independence by using standard linear codes (say, BCH codes). A nice exposition can also be found in Alon and Spencer's book [20]. In an influential paper, Luby [203] pioneered the use of limited independence for derandomization. The relevance to probability amplification was shown by Chor and Goldreich [90]. To achieve k-wise independence for large values of k is too expensive (provably so), which motivates studying weaker models of independence. Such an investigation was initiated by Naor and Naor [238] and pursued further by Alon et al. [15]. For general information, good sources are the books by Alon and Spencer [20], Luby [204], Motwani and Raghavan [236], and the monograph of Luby and Wigderson [205]. For an introduction to probabilistic algorithms, one will consult [179, 236, 253].

Section 9.3: Carter and Wegman [60] discovered universal hash functions. Impagliazzo and Zuckerman [166] designed the pseudorandom number generator given in the text. The leftover hash lemma is due to Impagliazzo, Levin, and Luby [163]. Our presentation differs somewhat from the original.

Section 9.4: For an introduction to algebraic graph theory, see the texts by Biggs [45] and Bollobás [53]. The co-NP completeness of checking whether a graph is an expander was established by Blum et al. [49]. The probabilistic construction of expanders was observed by Pinsker [249]. Gabber and Galil [139] discovered the bipartite expander described in the text. Other constructions were given by Lubotzky, Phillips, and Sarnak [202] and, independently, Margulis [208, 209]. The spectral properties of expanders expressed in Lemma 9.6 (page 333) and Theorem 9.7 (page 333) were established by Alon [12]; see also [18, 108, 302].

Ajtai, Komlós, and Szemerédi [9] exploited the rapid mixing property of expanders. Their use for probability amplification goes back to Sipser [290]. Extensions of the method were given by Cohen and Wigderson [98] and Impagliazzo and Zuckerman [166]. See also [241, 244] and the monograph by Luby and Wigderson [205] for other examples of pseudorandom number generators. Luby's book [204] offers a comprehensive treatment of the relation between pseudorandomness and one-way functions. Minimizing both the amount of randomness and the number of sampling rounds is investigated by Bellare, Goldreich, and Goldwasser [43] and Zuckerman [337]. The subject has been very active recently, with a flourish of results relating pseudorandom number generation and hardness (roughly, functions that are provably hard to compute can be used to generate pseudorandom numbers good for derandomization) together with work on functions extracting true randomness out of weak random sources [21, 22, 23, 50, 64, 127, 128, 143, 164, 165, 197, 243, 244, 245, 250, 255, 256, 266, 267, 296, 300, 310, 322, 330, 336]. To view randomness through the prism of computational complexity is a major, but fairly recent, development. The classical viewpoint, based on statistical tests, is discussed by Knuth [183].

Section 9.5: The quadratic residue construction is due to Alon et al. [15], and so is the lower bound on its size.

Section 9.6: The low-discrepancy set for arithmetic progressions was exhibited by Razborov, Szemerédi, and Wigderson [257], who proved Theorem 9.9 (page 339). The application to polynomial interpolation was observed by Alon and Mansour [17], whose paper investigates mostly the multivariate case.

10

Communication Complexity

 n the last two decades, communication complexity has emerged as an important methodology in the study of computational complexity. As opposed to, say, information theory, which seeks to model and analyze real-life problems, communication complexity is an abstraction aimed at isolating the computational bottlenecks of certain problems and providing tools for resolving their complexity. The discrepancy method has played—directly or indirectly—a pivotal role in the development of that theory.

We provide two examples, one very simple, the other less so. Both illustrate the underlying theme of this chapter: *Low discrepancy implies high complexity*. Informally, what happens is this. Two people play a game by taking turns. Each player has some distribution associated with him/her, which evolves over time at each move. As long as the players keep the two distributions "similar" enough, they can continue playing. This is a dynamic version of a discrepancy game, where a lower bound on the number of rounds depends on the players' collective ability to maintain low discrepancy as long as possible. Paradoxically, this phenomenon is the opposite of the one described in Chapter 6, where high discrepancy implied high complexity.

If all of this talk sounds a little surrealistic, the two examples treated in this chapter will surely clarify matters. In §10.1, we discuss the communication complexity of the inner product function. In §§10.2–10.4 we devise a communication complexity game and use it as a vehicle for deriving lower bounds for searching in a bounded universe.

10.1 Inner Product Modulo Two

Suppose that two parties, Alice and Bob, wish to compute

$$\langle x, y \rangle \stackrel{\text{def}}{=} \sum_i x_i y_i \quad (\text{mod } 2),$$

where x and y are $\{0,1\}^n$. Alice knows x but not y, and Bob is in the opposite situation, ie, he knows y but not x. How many bits must they exchange before both of them can know $\langle x, y \rangle$ with full confidence? Certainly $n+1$ is enough, since Alice can always hand over all of her bits to Bob, let him compute the inner product, and then have Bob send back the answer. Can they do better? Note that our measure of cost is the *communication complexity* of the problem, ie, the number of bits exchanged between the parties. How much internal computation either one does internally is immaterial.

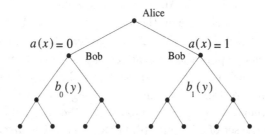

Fig. 10.1. The tree modeling the protocol between Bob and Alice.

In our model, Bob and Alice exchange bits one at a time. A protocol is a binary tree that specifies the rules of exchange (Fig. 10.1): An internal node is labeled Bob or Alice, depending on whom is sending a bit at that round. Assume that the root is labeled Alice (meaning that she talks first). She evaluates a function $a : \{0,1\}^n \mapsto \{0,1\}$ associated with the root and outputs $a(x)$ to Bob. The left (resp. right) branching from the root corresponds to the outcome $a(x) = 0$ (resp. $a(x) = 1$). Suppose that the two children of the root "belong" to Bob. With each one is associated a function $b_i \mapsto \{0,1\}$ ($i = 0,1$), which Bob uses to send his next bit. For example, if Alice sent 0 first, Bob would then send $b_0(y)$ back to her. In general, he sends her $b_{a(x)}(y)$. Thus, each node labeled Alice (resp. Bob) is associated with a particular function from x (resp. y) to $\{0,1\}$. Given x, y, the computation follows a unique path down the tree. Leaves are labeled 0 or 1, and correctness means that, for any x, y, the computation path reaches a leaf labeled $\langle x, y \rangle$. The *cost* of the protocol is the length of its longest path.

Distributional Communication Complexity

Variants of the model exist where random bits are allowed into the computation. To establish lower bounds for randomized protocols, it is useful to define the following notion of *distributional communication complexity*. Consider the possibility that a protocol is flawed and that correctness is not always guaranteed with certainty. Fix some $\varepsilon > 0$ and assume that, given a random pair x, y drawn independently and uniformly in $\{0, 1\}^n$, the probability that the protocol fails to compute the right answer is at most $1/2 - \varepsilon$. Let $C_\varepsilon(n)$ be the minimum cost of all the protocols with that property. This is, by definition, the distributional communication complexity of the function that Bob and Alice want to compute.

The relation with discrepancy theory is easy to see. A (combinatorial) *rectangle* is any Cartesian product $X \times Y$, where X and Y are subsets of $\{0, 1\}^n$. Each node splits the current set of x's (or y's) into two subsets (those causing a left or right turn), so all of the (x, y)'s leading to the same leaf ℓ form a rectangle R_ℓ; this has nothing to do with the particular function that they are trying to compute. Furthermore, the collection of R_ℓ's partitions the set of all inputs $\{0, 1\}^n \times \{0, 1\}^n$. The *discrepancy* $D(n)$ of the protocol is the maximum discrepancy $D(R_\ell)$, over all leaves ℓ, where

$$D(X \times Y) = \left| \sum \{ H(x, y) \mid x \in X \text{ and } y \in Y \} \right|,$$

with

$$H(x, y) = \begin{cases} 1 & \text{if } \langle x, y \rangle = 1, \\ -1 & \text{if } \langle x, y \rangle = 0. \end{cases}$$

The discrepancy $D(R_\ell)$ measures the difference between the number of right and wrong answers at leaf ℓ.

Lemma 10.1 *For any* $0 < \varepsilon \le 1/2$, $C_\varepsilon(n) \ge 2n + \log(2\varepsilon) - \log D(n)$.

Proof: Recall that, within the rectangle R_ℓ, the protocol produces the same answer regardless of the input pair. So, the difference between the number of pairs where the protocol succeeds and the number of those where it fails is bounded (in absolute value) by $D(n)$. Since over all input pairs this difference is at least $(1/2 + \varepsilon) - (1/2 - \varepsilon)$ times the number of possible inputs, ie, $2\varepsilon(2^n \times 2^n)$, it follows from summing up over all leaves that

$$\varepsilon 2^{2n+1} \leq D(n) \times \text{ number of leaves } \leq D(n) \cdot 2^{C_\varepsilon(n)}.$$

$\square$

By a simple discrepancy analysis we derive our main result on the distributional communication complexity of the inner product function.

Theorem 10.2 *For any $0 < \varepsilon \leq 1/2$, the distributional communication complexity of the inner product modulo 2 satisfies*

$$C_\varepsilon(n) \geq \frac{n}{2} + 1 - \log\frac{1}{\varepsilon}.$$

Proof: The 2^n-by-2^n matrix $H = (H(x,y))$ is the Hadamard matrix, and $H^T H = 2^n I$, where I is the identity (see Appendix B.1); implicit in this definition is the ordering of the rows and columns, which we choose as the lexicographical ordering of $\{0,1\}^n$. It follows that, for any vector v,

$$\|Hv\|_2 = 2^{n/2}\|v\|_2.$$

Given a rectangle $X \times Y$, let v (resp. w) be the characteristic vector in $\{0,1\}^{2^n}$ of the set X (resp. Y). By Cauchy-Schwarz,

$$D(X \times Y) = |v^T Hw| \leq \|v\|_2 \|Hw\|_2 = 2^{n/2}\|v\|_2 \|w\|_2 \leq 2^{n/2}\sqrt{2^n} \cdot \sqrt{2^n},$$

and therefore $D(n) \leq 2^{3n/2}$. By Lemma 10.1,

$$C_\varepsilon(n) \geq 2n + \log(2\varepsilon) - 3n/2,$$

which proves the theorem. $\square$

The Matrix Rank Bound

Having gone thus far, it would be a pity to leave out any discussion of the deterministic complexity of the inner product function. Consider the minimum-cost protocol for computing $\langle x, y \rangle$, and let $C(n)$ be its cost, ie, the length of its longest path. Let M be the 2^n-by-2^n matrix, where $M(x,y) = \langle x, y \rangle$. Each leaf of the protocol is associated with a rectangle of maximum discrepancy, ie, with only 0's or only 1's (any mix would cause errors at that leaf). Let ℓ be a 1-leaf, ie, one associated with a rectangle full of 1's, and let M_ℓ be the matrix obtained by zeroing out every element of M outside that rectangle. As we remarked earlier, the rectangles at the leaves partition the set of all input pairs (x, y), and so $M = \sum_\ell M_\ell$. Each M_ℓ has rank one,[1] and from

$$\text{rank}(A + B) \leq \text{rank}(A) + \text{rank}(B)$$

it follows that

$$\text{rank}(M) \leq \sum_\ell \text{rank}(M_\ell) \leq \ \text{number of 1-leaves.}$$

Of course, if we change the matrix M by switching all 0's and 1's, we obtain a new matrix M', and the same argument yields

$$\text{rank}(M') \leq \ \text{number of 0-leaves.}$$

Because $M + M' = U$, where U is the rank-1 matrix with 1's everywhere,

$$\text{rank}(M) \leq \text{rank}(-M') + \text{rank}(U),$$

and hence $\text{rank}(M') \geq \text{rank}(M) - 1$. Since a protocol is a binary tree, its cost is at least the logarithm of its number of leaves, and so

$$C(n) \geq \log(2 \cdot \text{rank}(M) - 1). \tag{10.1}$$

To find out the rank of M is easy. Consider the matrix $P = M^2$, where

$$P(x, y) = \sum_z \langle x, z \rangle \cdot \langle z, y \rangle.$$

- If either x or y is $(0, \ldots, 0)$, then $P(x, y) = 0$.
- Suppose that $x = y \neq (0, \ldots, 0)$. A random choice of z gives $\langle x, z \rangle = 1$ with probability $1/2$, and so $P(x, y) = 2^{n-1}$.
- Neither x nor y is $(0, \ldots, 0)$, and $x \neq y$. Then the two vectors x and y are independent, and so the n-by-2 matrix (x, y) is of rank 2 over $\mathbf{F}_2$. It follows that, for random z, the two random variables $\langle x, z \rangle$ and $\langle z, y \rangle$ are independent, and therefore the probability that both are 1 is equal to $1/4$. It follows that $P(x, y) = 2^{n-2}$.

Thus, the matrix derived from P by stripping its first row and its first column has 2^{n-1} on the diagonal and 2^{n-2} everywhere else. It is a circulant matrix whose determinant can thus be easily computed. We find that it is nonzero. The stripped matrix is nonsingular, and so the rank of P is

---

[1]Matrix rank is to be understood here over the reals, not over the integers mod 2. Indeed, over $\mathbf{F}_2$, the Galois field with two elements, the matrix M is of the form $u^T u$, and thus has rank 1: not a hopeful start for a rank-based argument. (See §9.2 on page 319 for background material on finite fields.)

$2^n - 1$. In turn, this implies that rank(M) $\geq 2^n - 1$. By (10.1), this shows that $C(n) \geq \log(2^{n+1} - 3)$. Obviously, $C(n) \leq n + 1$; therefore,

Theorem 10.3 *For any $n \geq 2$, the communication complexity of the inner product of two n-bit vectors modulo 2 is exactly $n + 1$.*

10.2 Searching in a Finite Universe

From a lower bound perspective, searching a database can be usefully modeled as a game between Alice (the querier) and Bob (the responder). Alice chooses a set $\mathcal{L}$ of potential queries and Bob a collection $\mathcal{P}$ of potential key-sets. Any pair $(\ell, P) \in \mathcal{L} \times \mathcal{P}$ defines a problem instance, with one or several solutions. Here are some examples:

- PREDECESSOR SEARCHING: The query ℓ is an integer and the key-set P is a collection of integers. The solution is the largest element in P that does not exceed ℓ, or $-\infty$ if there is no such thing.

- POINT SEPARATION: The query ℓ is a line in the plane $\mathbf{R}^2$ and the key-set P is a collection of points in the plane. The solution is 0 if all the points of P lie on the same side of ℓ and 1 otherwise. This is also known as *halfplane range detection*.

- APPROXIMATE NEAREST NEIGHBOR ON THE HAMMING CUBE: The query ℓ is a point in the cube $\{0, 1\}^d$, while the key-set P is a collection of points in that cube. Fix some approximation factor $\delta \geq 1$. The solution is any point of P whose L^1 distance to ℓ does not exceed δ times the shortest distance between ℓ and any point of P.

We investigate these problems in the *cell probe model*. The data structure consists of a table T of n^c cells of w bits each, where $n = |P|$ is the size of the key-set, c is a constant, and the word size w is a problem-dependent parameter. To strengthen our lower bounds, we assume that c is fixed but arbitrarily large. An algorithm consists of two parts:

(i) A table assignment strategy that indicates how, given P, the table T should be filled.

(ii) An infinite sequence of functions f_1, f_2, etc.

Presented with a query ℓ, the algorithm evaluates the index $f_1(\ell)$ and looks up the table entry $T[f_1(\ell)]$. If $T[f_1(\ell)]$ provides a solution, the algorithm terminates; otherwise, it moves on to evaluate $f_2(\ell, T[f_1(\ell)])$ and then looks up the entry $T[f_2(\ell, T[f_1(\ell)])]$, iterating in this fashion until a cell probe finally reveals a satisfactory solution.

To prove a lower bound for a given problem, we fix an algorithm once and for all. Then, we let Bob and Alice cooperate to exhibit a hard problem. Alice starts out with a set $\mathcal{L}_1$ of candidate queries and Bob with a collection $\mathcal{P}_1$ of key-sets. Their task is then to exhibit a hard problem instance $(\ell, P) \in \mathcal{L}_1 \times \mathcal{P}_1$, which they do by playing a communication complexity game.

The n^c possible values of the index $f_1(\ell)$ partition $\mathcal{L}_1$ into as many equivalence classes. Alice chooses one of them and sends to Bob the unique value of $f_1(\ell)$ corresponding to it. The assignment of T depends only on the key-set P that Bob has in mind. Of all the possible 2^w assignments of the entry $T[f_1(\ell)]$, Bob chooses one of them and narrows down his candidate set $\mathcal{P}_1$ to the set $\mathcal{P}_2$ of key-sets leading to that chosen value of $T[f_1(\ell)]$. Bob sends back to Alice his choice of $T[f_1(\ell)]$. Knowing $f_1(\ell)$ and $T[f_1(\ell)]$, Alice settles on a choice of a value for $f_2(\ell, T[f_1(\ell)])$, which she communicates to Bob, etc.

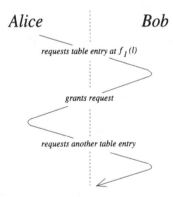

Fig. 10.2. Bob and Alice—the lower bound prover hopes—have a lot to say to each other.

Each round k produces a new pair $(\mathcal{L}_{k+1}, \mathcal{P}_{k+1})$ with the characteristic property that, for all queries in $\mathcal{L}_{k+1}$ and all key-sets in $\mathcal{P}_{k+1}$, Bob and Alice exchange the same information during the first k rounds. In other words, these first k rounds are unable to distinguish among any of the problem instances in $\mathcal{L}_{k+1} \times \mathcal{P}_{k+1}$. To alleviate notation, we say that a query (resp. key-set) is *active* at the beginning of round k to indicate that it belongs to $\mathcal{L}_k$ (resp. $\mathcal{P}_k$). The set $\mathcal{L}_k \times \mathcal{P}_k$ is called *unresolved* if it contains at least two problem instances (ℓ, P) and (ℓ', P') that have no common solution. In such a case, Bob and Alice need to proceed with round k, perhaps with more, and the cost of the protocol (ie, the minimum number of rounds necessary) is at least k.

Note that in one round Alice sends only $\log(n^c)$ bits to Bob (a table index), who then sends back w bits to Alice (a table entry). If, say, Alice, were to send the query ℓ explicitly, then Bob could conclude with no additional round. The same is true if Bob were to send Alice the key-set. Counting rounds can only, if anything, underestimate the true cost of any real algorithm, which is what we expect of a general lower bound model. Bob and Alice produce a nested sequence of unresolved sets

$$\mathcal{L}_1 \times \mathcal{P}_1 \supseteq \cdots \supseteq \mathcal{L}_t \times \mathcal{P}_t,$$

and so t constitutes a lower bound on the number of cell probes necessary to compute a solution to any problem instance in $\mathcal{L}_t \times \mathcal{P}_t$. We summarize the results below. All three of them are derived by specializing a *master argument*—developed in the next section—to the corresponding problem.

Theorem 10.4 *Given any algorithm for predecessor searching, there exist a key-set and a query that require $\Omega(\log b / \log \log b)$ probes to answer, where b is the number of bits needed to encode the integers used to specify the query and the keys. This holds for any word size $w = b^{O(1)}$.*

Theorem 10.5 *Given any algorithm for point separation, there exist a key-set and a query line that require $\Omega(\log b / \log \log b)$ probes to answer, where b is the number of bits needed to encode the (rational) coefficients of the line and the coordinates in the point set. This holds for any word size $w = b^{O(1)}$.*

Theorem 10.6 *Given any algorithm for approximate nearest neighbor searching on the Hamming cube, there exist a key-set and a query that require $\Omega(\log \log d / \log \log \log d)$ probes to answer, where d is the dimension of the cube. This holds for any approximation factor $\delta < 2^{(\log d)^{1-\varepsilon}}$, with any fixed $0 < \varepsilon < 1$, and for any word size $w = d^{O(1)}$.*

10.3 The Master Argument

We begin with a little notation: t refers to the lower bound sought on the number of rounds; queries are chosen from a universe of size at most 2^q; the word size w is defined as q^c. Note that it is appropriate to use the same exponent c for the word size q^c and table size n^c, since it is assumed to be arbitrarily large. Finally, for notational convenience, we write $h = \lfloor t \log^2 t \rfloor$. Throughout our discussion, it is understood that q and t exceed any constant that might show up during the proof.

Keys and queries are chosen in the universe. Keeping track of queries is relatively straightforward since, by and large, all that matters is size. We just need to ensure that intervals in the universe contain enough keys. Choosing key-sets is more difficult, however, because simply having many of them is not particularly useful. We need to maintain an entire spectrum of key-sets ranging over the whole universe, some concentrated in small intervals, others spread across the universe. Ideally, for any sequence of disjoint intervals (narrow or wide) we would like to have at least one key-set intersecting many intervals in the sequence. This translates into low discrepancy of the one-way sort; it is one-way because one must worry only about having too few hits, not too many.

Without pushing the analogy too far, one can think of the set of key-sets as a "signal" with a full spectrum in the Fourier transform sense. The analogue of a Fourier basis is here a set of "canonical intervals" defined hierarchically. Indeed, we build a large balanced tree over the universe, and associate leaves with universe elements and nodes with intervals (induced by the leaves below it). By contracting the tree in various ways we can define more trees and more intervals. The construction involves a double induction. One type of induction allows us to build each tree level by level; the other one defines the family of trees. This second type of induction is of length equal to the number of rounds.

We construct key-sets iteratively by a top-down process that selects nodes in trees and then recurses within subtrees rooted at these nodes. The only reason the construction is a little involved is that we want to achieve several objectives at once: One is that key-sets should cover a large variety of ranges; another is that the construction should be highly structured. In particular, for technical reasons, the family of valid key-sets should be defined as Cartesian products of smaller families (products ensure that projections have nice properties). We embed our key-sets in a probabilistic space to produce a highly nonuniform distribution of key-sets. This nonuniformity is what makes the lower bound proof unusual. If the reader can bear with a few paragraphs of formal definitions, it will all become crystal clear in the next section.

A Hierarchy of Tree Contractions

Let $\mathcal{T}_1$ denote the perfect γ-ary tree of depth h^t. The branching factor $\gamma \geq 2$ is a parameter whose setting depends on the application problem. Our choices of t and γ will always ensure that

$$q^{c^2} < \gamma^h \qquad \text{and} \qquad \gamma^{h^t} < 2^q, \qquad (10.2)$$

so that, in particular, the total number 2^q of queries available to Alice at the start exceeds the number γ^{h^t} of leaves in $\mathcal{T}_1$. By performing a sequence of edge contractions we define a sequence of auxiliary trees that guide Bob and Alice's strategy. Contract *all* the edges of $\mathcal{T}_1$ except those whose lower node is of depth divisible by h^{t-1} (depth of root being zero). This transforms the tree $\mathcal{T}_1$ into a smaller one, denoted $\mathcal{U}_1$, of depth h. Given a node v of the tree $\mathcal{T}_1$, let $\mathcal{T}_1(v)$ denote its subtree of depth h^{t-1} rooted at v. The depth-1 subtree formed by an internal node v of $\mathcal{U}_1$ and its $\gamma^{h^{t-1}}$ children forms a contraction of the tree $\mathcal{T}_1(v)$. Note that, in Figure 10.3, $\mathcal{T}_1(v)$ contains leaves of $\mathcal{T}_1$, but this need not be the case in general.

Repeating this process leads to the construction of $\mathcal{U}_k$ for $1 < k \leq t$. Figure 10.4 tells the whole story, but for completeness here are the details. Given an internal node v of $\mathcal{U}_{k-1}$, the depth-1 tree formed by v and its children is associated with the subtree $\mathcal{T}_{k-1}(v)$, which now plays the role of $\mathcal{T}_1$ earlier, and so can be renamed $\mathcal{T}_k$ (with v understood). Note that $\mathcal{T}_k$ is of depth h^{t-k+1}. For any node $u \in \mathcal{T}_k$ of depth divisible by h^{t-k} but distinct from h^{t-k+1}, let $\mathcal{T}_k(u)$ denote the subtree of $\mathcal{T}_k$ of depth h^{t-k} rooted at u: As before, turn the leaves of $\mathcal{T}_k(u)$ into the children of u by contracting the relevant edges. This transforms $\mathcal{T}_k$ into the desired tree $\mathcal{U}_k$ of depth h.

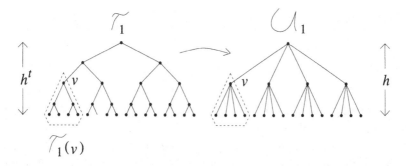

Fig. 10.3. The tree $\mathcal{T}_1$ and its contraction into $\mathcal{U}_1$: The tree $\mathcal{T}_2$ is defined, non-deterministically, as $\mathcal{T}_1(v)$, for some $v \in \mathcal{U}_1$; the branching factor γ is set to 2 in this example.

The contraction process is the same for all $k < t$. The final case, $k = t$, is a little different, however. We simply make all the leaves of $\mathcal{T}_t$ into the children of the root and remove the other internal nodes. This contraction produces a depth-1 tree $\mathcal{T}_t$ with γ^h leaves. We should always remember that a specific $\mathcal{T}_k$ implies the choice of various intermediate nodes in $\mathcal{U}_1, \ldots, \mathcal{U}_{k-1}$. Although $\mathcal{T}_k$ is defined nondeterministically, it is always a perfectly balanced γ-ary tree of depth h^{t-k+1}. It follows immediately from the construction that:

Lemma 10.7 *Any internal node of any $\mathcal{U}_k$ has exactly $\gamma^{h^{t-k}}$ children if $k < t$, and γ^h children if $k = t$.*

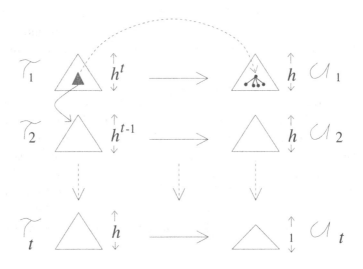

Fig. 10.4. The hierarchy of trees.

A Product Space Construction

Bob's starting collection $\mathcal{P}_1$ of active key-sets is defined implicitly by introducing a nonuniform probability distribution $\mathcal{D}_1$ over the universe of key-sets. Similarly, we define any $\mathcal{P}_k$ by means of a distribution $\mathcal{D}_k$. While every key-set with nonzero probability in $\mathcal{D}_1$ is active during the first round, this is not the case with subsequent distributions. So, we specify a lower bound on the probability that a random key-set P_k drawn from $\mathcal{D}_k$ is active prior to round k, ie, belongs to $\mathcal{P}_k$.

• *Distribution* $\mathcal{D}_1$: A random key-set P_1 is defined by two parameters, I_1 and σ_1, drawn independently of each other: The *index list* I_1 is a sequence of integers, while the *signature* σ_1 is a function assigning each element of I_1 to either 0 or 1. In the case of predecessor searching and approximate nearest neighbor searching, the signature function is constant and, in effect, plays no role at all. For point separation, however, σ_1 assigns 0 or 1 randomly to each element of I_1, uniformly and independently of the choice of I_1. Regardless of the distribution on σ_1 (trivial or uniform), in all cases we may say that a random P_1 is defined by a random index list I_1 and a random signature σ_1. For minor technical reasons, we require that P_1 should not be much larger than I_1:

$$n \stackrel{\text{def}}{=} |P_1| = O(|I_1|). \tag{10.3}$$

So, to define $\mathcal{D}_1$, we now need only explain the meaning of a random I_1. The index list I_1 is defined recursively in terms of I_2, which itself depends on $I_3, \ldots, I_t$. Each I_k is defined with respect to a certain tree $\mathcal{U}_k$ (the word "certain" being a reminder that $\mathcal{U}_k$ is not unique but depends on the choice of nodes in $\mathcal{U}_i$, for $i < k$). Number the leaves of $\mathcal{T}_1$ from left to right, $0, \ldots, \gamma^{h^t} - 1$, and with each node of the tree associate the interval formed by the integers at the leaves descending from it. Any node v of $\mathcal{U}_k$ inherits the interval of the corresponding node in $\mathcal{T}_1$. The *key* of v refers to the smallest integer in that interval. We define a random index list I_1 by setting $k = 1$ in the procedure below:

Random Index List I_k

• For $k = t$, a random I_k (within some $\mathcal{T}_k$) is formed by the keys of w^5 nodes selected at random, uniformly without replacement, among the leaves of the depth-1 tree $\mathcal{U}_k$.

• For $k < t$, a random I_k (within some $\mathcal{T}_k$) is defined in two stages:

  [1] For each $j = 1, 2, \ldots, h-1$, choose w^5 nodes of $\mathcal{U}_k$ of depth j at random, uniformly without replacement, among the nodes of depth j that are not descendants of nodes chosen at lower depth ($< j$). The $(h - 1)w^5$ nodes selected are said to be *picked by* I_k.

  [2] For each node v picked by I_k, recursively choose a random I_{k+1} within $\mathcal{T}_{k+1} = \mathcal{T}_k(v)$. The union of these $(h - 1)w^5$ disjoint sets I_{k+1} forms a random I_k within $\mathcal{T}_k$.

Note that the degree of any internal node of $\mathcal{U}_k$ ($k < t$) is at least

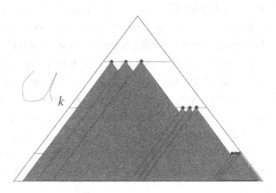

Fig. 10.5. Picking w^5 nodes at each level outside the dark subtrees rooted higher up.

γ^h (Lemma 10.7), which by (10.2) greatly exceeds $(h-1)w^5$. Thus, we never run out of nodes in step [1]. Since we need not be concerned with the signature σ_1 yet, this essentially completes the definition of the distribution $\mathcal{D}_1$. We see by induction that a random I_k consists of $(h-1)^{t-k}w^{5(t-k+1)}$ integers. Setting $k = 1$, we have the identity

$$|I_1| = (h-1)^{t-1}w^{5t}. \tag{10.4}$$

- *Distribution $\mathcal{D}_k$:* During round $k > 1$, our only control over active key-sets is through one number: the probability that a random key-set drawn from $\mathcal{D}_k$ is active (ie, belongs to $\mathcal{P}_k$). A random key-set drawn from $\mathcal{D}_1$ is active with probability 1, since no information has yet been exchanged between Bob and Alice. But this is no longer true of $\mathcal{D}_k$ for $k > 1$, and we need to enforce an invariant to keep the "activity" level high enough:

     - KEY-SET INVARIANT: For any $1 \leq k \leq t$, a random P_k from $\mathcal{D}_k$ is active with probability at least 2^{-w^2}.

But what exactly is $\mathcal{D}_k$? One can think of the distributions $\mathcal{D}_1, \mathcal{D}_2$, etc, as assigning probabilities to all possible key-sets parameterized by (I, σ). By abuse of terminology, we say that a key-set belongs to $\mathcal{D}_k$ if sampling from $\mathcal{D}_k$ produces it with nonzero probability. With this notation, we can write $\mathcal{P}_k \subseteq \mathcal{D}_k$. An important feature is that, once the probability of a key-set is 0 in some $\mathcal{D}_k$, it remains so in all subsequent distributions $\mathcal{D}_j$ $(j > k)$

or, put differently,

$$\mathcal{D}_1 \supseteq \cdots \supseteq \mathcal{D}_t. \tag{10.5}$$

To go from $\mathcal{D}_k$ to $\mathcal{D}_{k+1}$ entails choosing specific nodes of $\mathcal{U}_k$ and look-ing at the projections of key-sets of these nodes. We define this notion of projection next. Let P_1 be an input key-set in $\mathcal{D}_1$. In the recursive con-struction of I_1, if v is a node of $\mathcal{U}_k$ picked by I_k in step [1], let $I_{|v}$ denote the set I_{k+1} defined recursively within $\mathcal{T}_{k+1} = \mathcal{T}_k(v)$. Similarly, let $\sigma_{|v}$ be the restriction of the signature function σ_1 to $I_{|v}$. The key-set parameter-ized by $(I_{|v}, \sigma_{|v})$, denoted by $P_{|v}$, is called the v-projection of P_1. A few remarks:

(i) The notion of v-projection extends to any key-set in any $\mathcal{D}_j$ ($j \leq k$), since by (10.5) such a key-set also belongs to $\mathcal{D}_1$.

(ii) One should not confuse the key-sets in $\mathcal{D}_k$, such as P_1, all of which have exactly the same size n, with the v-projections $P_{|v}$, which are much smaller and are not part of Bob's arsenal of key-sets.

(iii) A v-projection might seem to depend on the choice of some P_1, but obviously one can speak of a random $P_{|v}$ (with $v \in \mathcal{U}_k$ fixed) independently of any P_1 as the key-set formed by a random I_{k+1} defined within $\mathcal{T}_{k+1} = \mathcal{T}_k(v)$ together with the restriction to it of a random σ_1. It is this distribution that will be understood in any further reference to a "random $P_{|v}$."

Assume that we have already defined $\mathcal{D}_k$, for $k < t$. Just like I_k, a distribution $\mathcal{D}_k$ is associated with a specific tree $\mathcal{T}_k$. So, to define $\mathcal{D}_{k+1}$, the first order of business is to choose a node v in $\mathcal{U}_k$ and make $\mathcal{T}_{k+1} = \mathcal{T}_k(v)$ our reference tree for $\mathcal{D}_{k+1}$. Any key-set of $\mathcal{D}_k$ whose probability is not explicitly set below is assigned probability 0 under $\mathcal{D}_{k+1}$. Consider each possible set K of the form $P_{|v}$. Note that this enumeration is independent of $\mathcal{D}_k$, and recall that v is fixed. For each K considered, apply the following rule:

- If K is the v-projection of at least one key-set in $\mathcal{P}_k$ (ie, active at the beginning of round k), then take one of them (any will do) and assign its probability under $\mathcal{D}_{k+1}$ to be the probability that a random $P_{|v}$ is equal to K.
- Otherwise, take one key-set in $\mathcal{D}_k$ whose v-projection is K (of course, it is inactive) and again assign its probability under $\mathcal{D}_{k+1}$ to be the probability that a random $P_{|v}$ is equal to K.

This completes the definition of a random P_{k+1} drawn from $\mathcal{D}_{k+1}$. The distribution is inherited directly from $P_{|v}$ (not from $\mathcal{D}_k$!), but the choice of which key-sets of $\mathcal{D}_k$ make it into $\mathcal{D}_{k+1}$ depends on $\mathcal{P}_k$, ie, on the active key-sets. Note that $\mathcal{D}_{k+1}$ is already fully specified *before* round k. During that round, Bob chooses the contents of a table entry pointed to by Alice, and this in effect reduces the number of active key-sets in $\mathcal{D}_{k+1}$: The surviving ones constitute $\mathcal{P}_{k+1}$. To summarize, a random key-set in $\mathcal{D}_{k+1}$ is defined with reference to a specific tree $\mathcal{T}_{k+1}$, and the distribution $\mathcal{D}_{k+1}$ is isomorphic to that of a random $P_{|v}$, for fixed $v \in \mathcal{U}_k$.

Candidate Queries

Alice starts out with the set $\mathcal{L}_1$ of queries formed by the keys of the leaves of $\mathcal{T}_1$. As the game progresses, this set decreases in size to produce the nested sequence $\mathcal{L}_1 \supseteq \cdots \supseteq \mathcal{L}_t$. For $k > 0$, the set $\mathcal{L}_{k+1}$ is obtained by identifying a special node v in $\mathcal{U}_k$ as well as a certain equivalence class in the partition of $\mathcal{L}_k$ induced by Alice's choice of a table index in round k: The set $\mathcal{L}_{k+1}$ consists of all the queries in that class that lie in the interval of v. (Recall that each node of $\mathcal{U}_k$ is associated with a unique interval in $\{0, \gamma^{h^t} - 1\}$.) Together with the invariant on key-sets, we impose a constraint on the active queries. Of course, prior to round k, the currently active query set $\mathcal{L}_k$ is defined with respect to *the same* reference tree $\mathcal{T}_k$ used to specify a random P_k.

- QUERY INVARIANT: For any $1 \leq k \leq t$, the fraction of the leaves in $\mathcal{U}_k$ whose intervals intersect $\mathcal{L}_k$ is at least $1/q$.

Naturally, $\mathcal{L}_1$ and $\mathcal{P}_1$ trivially satisfy their query and key-set invariants, respectively. The challenge for Bob and Alice is to coordinate their strategy to maintain these invariants throughout the t rounds of the game. In each application of the lower bound technique we must prove that, as long as the invariants hold, the game must go on.

Lemma 10.8 *If $\mathcal{L}_t$ and $\mathcal{P}_t$ satisfy their respective invariant, then $\mathcal{L}_t \times \mathcal{P}_t$ is unresolved.*

The lemma says that if both key-set and query invariants hold at the beginning of round t, then the protocol must proceed through round t if it is to solve all problem instances in $\mathcal{L}_t \times \mathcal{P}_t$ correctly. This shows that, indeed, t is a lower bound on the number of rounds necessary. Of course, the lemma cannot be proven in the abstract, and we must postpone the proofs until we have specific problems in hand (eg, predecessor searching, point separation, approximate nearest neighbor searching).

During the k-th round, Alice chooses an index in Bob's table. As we discussed earlier, the set of n^c possible choices partitions her current query set $\mathcal{L}_k$ into as many equivalence classes. To define $\mathcal{L}_{k+1}$, Alice first must choose a "special" node of $\mathcal{U}_k$ and then an equivalence class. For the node she will limit her choice to heavy ones: A node v of $\mathcal{U}_k$ is called *heavy* if

   (i) it is not a root or a leaf, and
   (ii) at least one of the equivalence classes intersects the intervals of a fraction $\geq 1/q$ of the children of v.

The levels of $\mathcal{U}_k$ span a whole spectrum of widely different interval widths. We use this variety to argue that some levels must have a certain number of heavy nodes.

Lemma 10.9 *The union of the intervals associated with the heavy nodes of $\mathcal{U}_k$ covers at least a fraction $1/2q$ of the leaves' intervals.*

Proof: Recall that q plays two roles: The total number of queries is at most 2^q, and the word size w is equal to q^c. To prove the lemma, let us fix an equivalence class C and color the nodes of $\mathcal{U}_k$ whose intervals intersect it. Mark every colored node that is heavy with respect to class C. Once this is done, mark every descendant in $\mathcal{U}_k$ of a marked node. Let N denote the number of leaves in $\mathcal{U}_k$, and let N_j be the number of leaves of $\mathcal{U}_k$ whose depth-j ancestors in $\mathcal{U}_k$ are colored and unmarked; in this terminology, a node is included among its ancestors. For $j > 1$, an unmarked, colored, depth-j node is the child of an unmarked, colored, depth-$(j-1)$ node that is not heavy for class C, and so $N_j < N_{j-1}/q$. Since, obviously, $N_1 \leq N$, it follows that, for any $j > 0$,

$$N_j \leq \frac{N}{q^{j-1}}.$$

If we repeat this line of reasoning for all of the other equivalence classes, we find that all the unmarked, colored nodes (at a fixed depth $j > 0$) are ancestors of at most $n^c N/q^{j-1}$ leaves. This implies that the number of unmarked, colored leaves is at most $n^c N/q^{h-1}$, which by (10.2–10.4) is less than $N/2q$. By the query invariant, at least N/q leaves of $\mathcal{U}_k$ are colored, and so at least half as many are both colored and marked. Walking up the tree from these leaves shows that the marked nodes whose parents are unmarked are themselves ancestors of at least $N/2q$ distinct leaves. All of these nodes are heavy, and the lemma is proven. □

Alice's strategy is to keep her active queries as "entangled" as possible with Bob's key-sets. Put differently, the active queries should form, in a weak sense, a one-way low-discrepancy approximation of the set system formed by the key-sets. The next result hints at such a close interaction between queries and key-sets. On at least one level of $\mathcal{U}_k$ many heavy nodes—the ones needed for maintaining the query invariant at the next round—end up being picked by a random I_k.

Lemma 10.10 *For any $0 < k < t$, there is a depth j in $\mathcal{U}_k$ $(0 < j < h)$, such that, with probability at least 2^{-w^2-1}, a random P_k from $\mathcal{D}_k$ is active and its index list I_k picks at least w^3 heavy, depth-j nodes in $\mathcal{U}_k$.*

Proof: We use our previous observation that $\mathcal{D}_k$ is isomorphic to a random (I_k, σ_k), where σ_k is the restriction to I_k of a random signature chosen independently of I_k. Fix σ_k once and for all. The heavy nodes of $\mathcal{U}_k$ are, together, ancestors of at least a fraction $1/2q$ of the leaves (Lemma 10.9). It follows that, for some $0 < j < h$, at least a fraction $1/2qh$ of the nodes of depth j are heavy. Among these, I_k may pick only those that are not picked further up the tree. This caveat rules out fewer than hw^5 candidate nodes, which, by Lemma 10.7, represents a fraction at most hw^5/γ^h of all the nodes of depth j. So, it appears that, among the allowable depth-j nodes (ie, those that may be picked by I_k), the fraction α of heavy ones satisfies (conservatively)

$$\alpha \geq \frac{1}{2hq} - \frac{hw^5}{\gamma^h},$$

where, as we mentioned earlier, $w = q^c$ and $h = \lfloor t \log^2 t \rfloor$. By (10.2), it follows that $\alpha > 1/3hq$. Recall that I_k picks w^5 depth-j nodes of $\mathcal{U}_k$ at random with no replacement. By Hoeffding's bounds,[2] the probability that the number of heavy ones picked exceeds the lemma's target of w^3 is at least[3]

$$1 - 2e^{-2w^5(\alpha-1/w^2)^2} > 1 - 2^{-w^3}.$$

It follows from the key-set invariant and the independence of I_k and σ_k that, with probability at least $2^{-w^2} - 2^{-w^3}$, a random P_k is active and its

---

[2] See Appendix A.

[3] To be fully rigorous, this bound holds when we condition upon a given set of allowable depth-j nodes. But since we get the same bound for all such sets, we can decondition right away.

index list I_k picks at least w^3 heavy depth-j nodes in its associated tree $\mathcal{U}_k$. $\square$

Probability Amplification by Projection

During the k-th round, Bob sends to Alice the contents of the cell

$$T[f_k(\ell, T[f_1(\ell)], \ldots)].$$

Prior to doing so, Bob must settle on one of the at most 2^w possible values of the cell. His best strategy is not to follow blind greed (ie, not necessarily to select the one leading to the largest number of active key-sets) but to maintain low discrepancy between active queries and key-sets. The 2^w choices partition the current collection $\mathcal{P}_k$ of active key-sets into as many equivalence classes. This is such a large number that one needs to build in a probability amplification mechanism to avoid an inevitable violation of the key-set invariant. We exploit the product nature of the distribution $\mathcal{D}_k$ and project it on one of its factors. This is a now-classical technique in combinatorics that has a straightforward geometric interpretation. In the plane, it says that, given a measurable region, one of its two axis-projections has extent at least the square root of its area (Fig. 10.6).

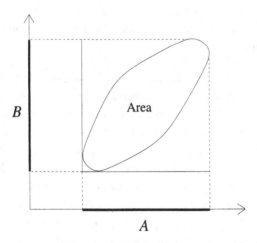

Fig. 10.6. Obviously, the larger of A or B is at least the square root of the area.

Lemma 10.11 *For any* $0 < k < t$, *there exists a heavy node* v *of* $\mathcal{U}_k$ *such that, with probability at least* $1/2$, *a random* P_{k+1} *drawn from the distribution* $\mathcal{D}_{k+1}$ *associated with* $\mathcal{T}_{k+1} = \mathcal{T}_k(v)$ *belongs to* $\mathcal{P}_k$.

Proof: The lemma claims the existence of a heavy node v "at which," with high enough probability, a random P_{k+1} is active at the beginning of round k. Considering how low the probability of "activity" specified by the key-set invariant is, this bears witness to truly impressive probability amplification. Of course, we should note that most such P_{k+1}'s might lose their active status during round k and thus fail to make it into $\mathcal{P}_{k+1}$.

To prove the claim, we refer to the depth j in Lemma 10.10. Let $p_{|S}$ denote the conditional probability that a random P_k from $\mathcal{D}_k$ belongs to $\mathcal{P}_k$, given that S is exactly the set of heavy nodes of depth j picked by I_k. Summing over all subsets S of heavy depth-j nodes of size at least w^3, we find that

$$\sum_S \text{Prob}\,[\,S = \text{set of heavy depth-}j\text{ nodes picked by } I_k\,] \cdot p_{|S} \qquad (10.6)$$

is the sum, over all S, of the probability that $P_k \in \mathcal{P}_k$ and that S is precisely the set of heavy nodes of depth j picked by its index list I_k. By Lemma 10.10, this sum is at least 2^{-w^2-1}, and therefore, by (10.6),

$$p_{|S^*} \geq 2^{-w^2-1}, \qquad (10.7)$$

for some set S^* of at least w^3 heavy nodes of depth j. The crux of the proof lies in the following two remarks, which together point to the desired projection node:

1. Fix some node v in some $\mathcal{U}_k$. Following our observations on page 360, the v-projection $P_{|v}$ of a random P_k from $\mathcal{D}_k$ is distributed according to $\mathcal{D}_{k+1}$. The same is true if the distribution on P_k is conditioned upon having S^* as the set of heavy depth-j nodes picked in that $\mathcal{U}_k$ by the index list of P_k. Furthermore, the projections on the nodes of S^* on the one hand, and the rest of P_k on the other hand, form $|S^*| + 1$ mutually independent random variables.

2. If P_k belongs to $\mathcal{P}_k$, then its v-projection maps to a unique set $P_{k+1} \in \mathcal{D}_{k+1}$ also in $\mathcal{P}_k$. So, in particular, given any $P_k \in \mathcal{P}_k$ claiming S^* has its set of picked heavy nodes of depth j, each of its v-projections maps to a unique $P_{k+1} \in \mathcal{D}_{k+1} \cap \mathcal{P}_k$.

Let $p_{|v}$ denote the probability that a random P_{k+1} drawn from the distribution $\mathcal{D}_{k+1}$ associated with $\mathcal{T}_{k+1} = \mathcal{T}_k(v)$ belongs to $\mathcal{P}_k$. From the two observations above, we find that

$$p_{|S^*} \leq \prod_{v \in S^*} p_{|v} \, .$$

Since $|S^*| \geq w^3$, it follows from (10.7) that

$$p_{|v} \geq \left(2^{-w^2-1}\right)^{1/|S^*|} \geq \frac{1}{2} \, ,$$

for some $v \in S^*$. $\square$

With the ground rules in place, Bob and Alice have no difficulty maintaining their invariants from round to round. Note that both query and key-set invariants are trivially satisfied at the outset. Assume now that they hold at the opening of round $k < t$. Let v denote the node of $\mathcal{U}_k$ referred to in Lemma 10.11. The n^c possible ways of indexing into the table T partition Alice's query set $\mathcal{L}_k$ into as many equivalence classes. Because v is heavy, the intervals associated with at least a fraction $1/q$ of its children intersect a particular equivalence class. Alice chooses such a class: The queries in it that lie in the interval associated with v constitute the new query set $\mathcal{L}_{k+1}$. The tree $\mathcal{U}_{k+1}$ is naturally derived from $\mathcal{T}_{k+1} = \mathcal{T}_k(v)$. By this choice, the fraction of the leaves of $\mathcal{U}_{k+1}$ whose intervals intersect $\mathcal{L}_{k+1}$ is at least $1/q$, and the query invariant is satisfied at the beginning of round $k + 1$.

Upon receiving the index from Alice, Bob must choose the contents of the table entry while staying consistent with all past choices. By Lemma 10.11, a random P_{k+1} from $\mathcal{D}_{k+1}$ (distribution associated with $\mathcal{T}_{k+1}$) is active at the beginning of round k with probability at least a half. There are at most 2^w choices for the table entry, and so for at least one of them, with probability at least $(1/2)2^{-w} > 2^{-w^2}$, a random key-set from $\mathcal{D}_{k+1}$ is active at the beginning of round k and produces a table with that specific entry value. These key-sets constitute the newly active collection P_{k+1}, and the key-set invariant still holds at the beginning of round $k + 1$.

To show that t rounds are needed, we must prove that $\mathcal{L}_k \times P_k$ is unresolved, for any $k \leq t$. In fact, because of the nesting structure of these products, it suffices to show that $\mathcal{L}_t \times P_t$ is unresolved, which follows from Lemma 10.8. In the applications that follow, we need to set t appropriately to make the lemma true.

10.4 Applications

We use the *master argument* to prove Theorems 10.4–10.6 (page 353). We begin with the simplest application.

Predecessor Searching

The starting set $\mathcal{L}_1$ of queries consists of all integers between 0 and $2^b - 1$. A key-set is specified entirely by its index list. The signature function is irrelevant and can be chosen to be constant. We set:

$$\begin{cases} q & = & b, \\ t & = & \lfloor \log b / 2 \log \log b \rfloor, \\ \gamma & = & 2. \end{cases}$$

We easily verify that conditions (10.2) are satisfied. To prove Theorem 10.4, we need only establish Lemma 10.8, ie, prove that if $\mathcal{L}_t$ and $\mathcal{P}_t$ satisfy their respective invariant, then $\mathcal{L}_t \times \mathcal{P}_t$ is unresolved. By contradiction, suppose that $\mathcal{L}_t \times \mathcal{P}_t$ is not unresolved. Then, all the key-sets of $\mathcal{P}_t$ must lie either to the left or to the right of all the queries. By the query invariant, the leaves of $\mathcal{U}_t$ whose interval contains a query of $\mathcal{L}_t$ constitute a fraction of at least $1/q$ of all the leaves. Within $\mathcal{D}_t$, the probability that the projections of a random $\mathcal{P}_t$ lie to the left or to the right of the leaves in question is at most (recall that $w = q^c$)

$$\left(1 - \frac{1}{q}\right)^{w^5} < 2^{-w^2},$$

which contradicts the key-set invariant. (We cheated a little because the left-most and right-most intervals containing a query are not necessarily forbidden, but the error is insignificant. Also note that, since $\mathcal{D}_t$ is hypergeometric, allowing replacement—as we did in the calculation—gives a valid upper bound.) This concludes the proof of Theorem 10.4. □

Point Separation

This problem is also known as halfplane range detection. We begin our discussion with the construction of point sets. We then turn our attention to the query lines. Let p_i denote the point (i, i^2) and, given $i < j$, let $A_{ij} = \frac{1}{2}(i + j, i^2 + j^2)$ and $B_{ij} = ((i + j)/2, ij)$. Note that A_{ij} ("A" for above) is the midpoint of $p_i p_j$ and lies above the parabola $y = x^2$. On the other hand, B_{ij} ("B" for below) is the point of intersection between the lines through p_i and p_j tangent to that parabola, and hence below it (Fig. 10.7). Any of Bob's n-point sets P is of the form

$$P = \left\{ p_{i_1}, X_{i_1 i_2}, p_{i_2}, X_{i_2 i_3}, \ldots, p_{i_{s-1}}, X_{i_{s-1} i_s}, p_{i_s} \right\},$$

for some $i_1 < \cdots < i_s$, where $n = 2s - 1$ and X denotes the symbol A or B (not necessarily the same one throughout the sequence). Thus, P can be specified by the index list $I = I(P) = \{i_1, \ldots, i_s\}$ consisting of s distinct b-bit integers (the abscissae of the points) and a signature function mapping each i_j to A or B. Note that the signature bit of i_s is irrelevant. For technical reasons, we require that all of the integers of the index list I be even. Going back to the *master argument*, a random P_1 is defined by forming a random I_1 over the set of even integers in $[0, 2^b)$, and for the signature of each index in I_1, choosing a random assignment, uniformly and independently, in $\{A, B\}$.

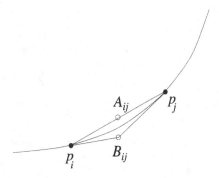

Fig. 10.7. The two points A_{ij} and B_{ij} above and below the parabola $y = x^2$.

The starting query set $\mathcal{L}_1$ consists of the lines of the form, $y = 2kx - k^2$, for all odd b-bit integers k. This is the equation of the line tangent to the parabola $y = x^2$ at p_k. From now on, we identify each line of $\mathcal{L}_1$ with the coefficient k in its equation.[4] The number of bits needed to encode any

---

[4] A minor technical point: The master argument does not allow us to restrict the index list to even integers or the query set to odd ones. To be rigorous, we should form the index list I_1 first and then multiply each index by two to produce the sequence $i_1, \ldots, i_s$. Similarly, the coefficient k of a query line should be of the form $2j + 1$, where, as required by the master argument, j is a key of a leaf of $\mathcal{T}_1$. We ignore these fine, but essentially unimportant, distinctions in our discussion below. So, when there is no ambiguity, we might say "key $2i$ and query $2j + 1$ are in the same leaf interval," instead of the correct statement, "key $2i$ and query $2j + 1$ are such that i and j belong to the same leaf interval."

point coordinate or line coefficient is $2b$ (and not b, as we claimed earlier, but this is of no consequence while it simplifies the notation). Observe that the problem does not become suddenly easier with other representations such as $\alpha x + \beta y = 1$, and that for the purposes of our lower bound, all such representations are essentially equivalent.

Our requirement that $i_j \in I$ be even and k be odd is meant to prevent any line of $\mathcal{L}_1$ from passing through any point of $\mathcal{P}_1$. The following lemma motivates our construction of points and lines. It is an immediate consequence of the fact that P is always in convex position, and we may leave out the proof.

Lemma 10.12 *Let p_{i_j} and $p_{i_{j+1}}$ be two points of P, and let ℓ be the line $y = 2kx - k^2$, where $i_j < k < i_{j+1}$. The line ℓ separates the point set P if and only if the symbol X in $X_{i_j i_{j+1}}$ is of type B.*

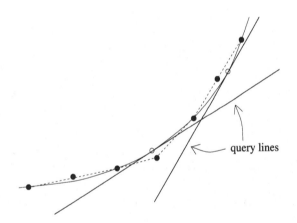

query lines

Fig. 10.8. A set P with $n = 7$ points and two queries with different answers.

We use the same parameter setting that we used for predecessor searching:

$$\left\{ \begin{array}{rcl} q & = & b\,, \\ t & = & \lfloor \log b / 2 \log \log b \rfloor\,, \\ \gamma & = & 2\,. \end{array} \right.$$

As we already mentioned, conditions (10.2), required by the construction of $\mathcal{T}_1$, do indeed hold. To prove Theorem 10.5, it now suffices to prove Lemma 10.8. Recall that the lemma states that if $\mathcal{L}_t$ and $\mathcal{P}_t$ satisfy their respective invariant, then $\mathcal{L}_t \times \mathcal{P}_t$ is unresolved. We assume that $\mathcal{L}_t$ satisfies the query invariant and $\mathcal{L}_t \times \mathcal{P}_t$ is not unresolved, and we show that $\mathcal{P}_t$ must then violate the key-set invariant. For each leaf of $\mathcal{U}_t$ whose interval intersects $\mathcal{L}_t$, select one $j_i \in \mathcal{L}_t$ in that interval. By Lemma 10.7 and the query invariant, this gives us an odd-integer sequence $j_1 < \cdots < j_m$ of length

$$m \geq \frac{2^h}{q}. \tag{10.8}$$

Given $P_t \in \mathcal{D}_t$, we define the *spread* of P_t, denoted $\texttt{spread}(P_t)$, to be the number of intervals of the form $[j_i, j_{i+1}]$ ($0 \leq i \leq m$) that intersect the index list $I(P_t)$ (Fig. 10.9); for consistency, we write $j_0 = -\infty$ and $j_{m+1} = \infty$. Informally, this is the number of intervals spanning consecutive queries that "enclose" at least one p_{i_j}. We argue that if the spread is small (resp. not small), then the points p_{i_j} (resp. $X_{i_j i_{j+1}}$) of an active P_t are too heavily constrained for the key-set invariant to hold.

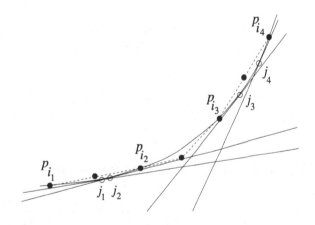

Fig. 10.9. A spread of 3 determined by $[j_0, j_1], [j_2, j_3], [j_4, j_5]$.

Fix a set S of intervals, where $|S| < w^4$, and consider the probability that S defines the spread of P_t. Of the $m+1$ intervals $[j_i, j_{i+1}]$, a random I_t must then avoid $m + 1 - |S|$ of them. Such an interval J may not always enclose a whole interval associated with a leaf of $\mathcal{U}_t$. By definition of the j_i's, however, the finite endpoints of J are odd integers in distinct leaf intervals, and therefore J contains at least one key (ie, the smallest even integer in the leaf interval). These keys are candidates in the construction of I_t and thus must be avoided. This limits the choice of I_t to at most $2^h - m - 1 + |S|$ leaves of $\mathcal{U}_t$. It follows that the probability that the spread is defined by S is bounded by

$$\binom{2^h + |S| - m - 1}{w^5} \Big/ \binom{2^h}{w^5} \le \left(1 - \frac{m - |S|}{2^h}\right)^{w^5}.$$

Recall that $w = q^c$ and $h = \lfloor t \log^2 t \rfloor$. By (10.2) and (10.8), we find that

$$1 - \frac{m - |S|}{2^h} \le 1 - \frac{1}{2q}.$$

Summing over all S's of size less than w^4 and using the fact that $m \le 2^h$ and c is large enough,

$$\text{Prob}\left[\text{spread}\,(P_t) < w^4\right] \le \sum_{k < w^4} \binom{m+1}{k}\left(1 - \frac{1}{2q}\right)^{w^5} \le 2^{-w^4}. \quad (10.9)$$

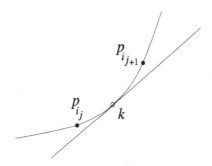

Fig. 10.10. Pairing the point p_{i_j} with the line $y = 2kx - k^2$.

Suppose now that the spread is at least w^4. Then, the point set

$$P_t = \Big\{ p_{i_1}, X_{i_1 i_2}, p_{i_2}, X_{i_2 i_3}, \ldots, p_{i_{s-1}}, X_{i_{s-1} i_s}, p_{i_s} \Big\}$$

includes a subset P^* of at least $w^4 - 1$ points p_{i_j}, every one of which can be paired with a line $y = 2kx - k^2$ of $\mathcal{L}_t$, where $i_j < k < i_{j+1}$ (Fig. 10.10). In Figure 10.9, for example, P^* can be chosen to be p_{i_1}, p_{i_3} and be paired with the lines j_1 and j_3, respectively. Draw a random P_t from $\mathcal{D}_t$, and let Ξ denote the event: "All queries from $\mathcal{L}_t$ give the same yes-or-no answer with respect to the point set P_t." By Lemma 10.12, the $X_{i_j, i_{j+1}}$'s corresponding to the p_{i_j}'s of P^* are either all of the form $A_{i_j, i_{j+1}}$ or all of the form $B_{i_j, i_{j+1}}$ (no mix). As we observed earlier, $\mathcal{D}_t$ is isomorphic to the distribution of a random I_t together with a string of w^5 bits (drawn uniformly, independently). The constraint on the X's reduces the choice of a random P_t by a factor of at least $2^{w^4-1}/2$, and hence

$$\text{Prob}\left[\Xi \mid \texttt{spread}(P_t) \geq w^4\right] \leq 2^{2-w^4}. \tag{10.10}$$

Putting together (10.9) and (10.10), we find that

$$\begin{aligned}
\text{Prob}[\Xi] &= \text{Prob}[\Xi \mid \texttt{spread}(P_t) < w^4] \cdot \text{Prob}[\texttt{spread}(P_t) < w^4] + \\
&\quad \text{Prob}[\Xi \mid \texttt{spread}(P_t) \geq w^4] \cdot \text{Prob}[\texttt{spread}(P_t) \geq w^4] \\
&\leq 2^{-w^4} + 2^{2-w^4} < 2^{-w^2},
\end{aligned}$$

which violates the key-set invariant. $\square$

This completes the proof of Theorem 10.5 (page 353). $\square$

Note that it is easy to rederive the lower bound on predecessor searching directly from this result.

Approximate Nearest Neighbors

We consider the problem of preprocessing a set P of n points in $\{0,1\}^d$, so that, given a query point $\ell \in \{0,1\}^d$, one can quickly find a δ-*approximate nearest neighbor* of ℓ in P, that is, any $p \in P$ such that $\|\ell - p\|_1 \leq \delta \|\ell - q\|_1$, for any $q \in P$. Setting $\delta = 1$ makes p the actual nearest neighbor. We fix the approximation factor $\delta = 2^{(\log d)^{1-\varepsilon}}$ once and for all, where ε is an arbitrary constant $(0 < \varepsilon < 1)$. Note that this is a very generous factor, leading therefore to an equally general lower bound. The data structure consists of a table T whose entries hold d^c bits each, where c is an arbitrarily large constant. This means that a point can be read in constant time. This assumption might be unrealistic when d is large, but it can only strengthen the lower bound result of Theorem 10.6 (page 353), which we now prove. For convenience we introduce the quantity,

$$\beta \stackrel{\text{def}}{=} 2^{4 + \lceil (\log d)^{1-\varepsilon} \rceil}. \tag{10.11}$$

Throughout our discussion, we assume, without loss of generality, that d is a large enough power of 2. In this way, we can divide d by powers of β and still get an integer. The term distance refers to the L^1 norm. A ball of radius r is the set of points in $\{0,1\}^d$ whose distance to a given point, the center, is at most r. We begin with a technical result that allows us to build the hierarchy of trees.

Lemma 10.13 *Let $B \subseteq \{0,1\}^d$ be a ball of radius $k \leq d$ large enough. There exists a collection of at least $2^{k/\beta}$ balls within B, each of radius $\lfloor k/\beta \rfloor$, such that the distance between any two points in distinct balls exceeds the distance between any two points in the same ball by a factor of at least $\beta/16$.*

Proof: Let V_r be the number of points in $\{0,1\}^d$ in a ball of radius r centered at a point in $\{0,1\}^d$. Obviously, this number does not depend on the center. In fact,

$$V_r = \sum_{i=0}^{\lfloor r \rfloor} \binom{d}{i}.$$

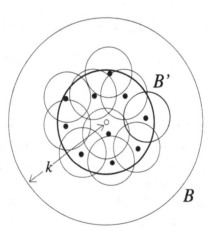

Fig. 10.11. Choosing a family of balls far apart.

Consider the ball B', concentric with B and of radius $\lfloor k/2 \rfloor$, and initially call its points *unmarked*. Next, mark the points of B' as follows. While there is an unmarked point left in B', select one and mark all of the points at distance at most $k/4$ from that point. The number N of points of B' selected in this manner satisfies

$$N \geq V_{\lfloor k/2 \rfloor}/V_{\lfloor k/4 \rfloor} \geq \binom{d}{\lfloor k/2 \rfloor} \Big/ \sum_{i=0}^{\lfloor k/4 \rfloor} \binom{d}{i}.$$

The denominator is dominated by a geometric series of ratio $1/3$; therefore,

$$\sum_{i=0}^{\lfloor k/4 \rfloor} \binom{d}{i} \leq \frac{3}{2} \binom{d}{\lfloor k/4 \rfloor}.$$

It follows easily that $N \geq 2^{k/\beta}$. To form the desired family, we take balls of radius $\lfloor k/\beta \rfloor$ centered at the N points chosen above. These balls lie within B since their centers are in B' and their common radius is much less than $k/2$. To verify the distance condition, first notice that any two points in a ball are at most $2k/\beta$ apart. So, suppose, by contradiction, that two points p and q in balls centered at distinct points p_0 and q_0 lie within $(\beta/16)(2k/\beta) = k/8$ of each other. Then,

$$\begin{aligned} \text{dist}(p_0, q_0) &\leq \text{dist}(p_0, p) + \text{dist}(p, q) + \text{dist}(q, q_0) \\ &\leq k/\beta + k/8 + k/\beta \leq k/4, \end{aligned}$$

which contradicts the fact that, by construction, the centers p_0 and q_0 are more than $k/4$ apart. $\square$

Again, to prove Theorem 10.6 it suffices to establish Lemma 10.8, ie, that if $\mathcal{L}_t$ and $\mathcal{P}_t$ satisfy their respective invariants, then $\mathcal{L}_t \times \mathcal{P}_t$ is unresolved. We set

$$\begin{cases} q &= d, \\ t &= \lfloor \varepsilon \log \log d / 2 \log \log \log d \rfloor, \\ \gamma &= 2^{d/\beta^{h^t}}. \end{cases}$$

Recall that $h = \lfloor t \log^2 t \rfloor$ and $\beta = 2^{4 + \lceil (\log d)^{1-\varepsilon} \rceil}$. We easily verify that (10.2) is satisfied. The γ children of the root of $\mathcal{T}_1$ are each associated with one of the $2^{d/\beta}$ balls in Lemma 10.13; note that this comes nowhere near exhausting all of the balls made available by the lemma. Repeated application of the lemma leads us to associate any node of $\mathcal{T}_1$ of depth $k \leq h^t$ with a ball of radius d/β^k. This works because γ, the number of children of an internal node, never exceeds the number of balls provided by the lemma, which is $2^{d/\beta^{k+1}}$ for a node of depth k. In particular, the balls associated with the nodes of $\mathcal{T}_1$ right above the leaves are of radius d/β^{h^t-1}, which, by our choice of t, is large enough for the application of Lemma 10.13. In fact, this is what, by and large, determines the lower bound t.

Recall that, by the *master argument*, any of Bob's n-point sets is specified by a pair (I, σ). In this case, the signature is the constant function mapping any element of I to zero; in other words, the signature plays no role and can be ignored. The index list I is a set of keys. Any key corresponds to a leaf of $\mathcal{T}_1$, and hence to a distinct ball. The collection of centers of these balls constitutes the key-set parameterized by I. The set $\mathcal{L}_1$ of queries consists of all the centers of the balls associated with the leaves of $\mathcal{T}_1$.

To establish Theorem 10.6, it suffices to establish Lemma 10.8, ie, that if $\mathcal{L}_t$ and $\mathcal{P}_t$ satisfy their respective invariant, then $\mathcal{L}_t \times \mathcal{P}_t$ is unresolved.

Each node of $\mathcal{T}_1$ is associated with a certain ball. Of course, the same is true of the nodes of any $\mathcal{U}_k$, since they originate from $\mathcal{T}_k$ and, hence, from $\mathcal{T}_1$. Below we prove the existence of a key-set P in $\mathcal{P}_t$ whose index list picks two distinct leaves of the tree $\mathcal{U}_t$, each of whose associated balls contains a query in $\mathcal{L}_t$. By Lemma 10.13, point distances between the balls and within them are in a ratio greater than $\beta/16 \geq 2^{(\log d)^{1-\varepsilon}}$. Within the approximation factor δ, the two queries thus cannot be answered the same way, and therefore $\mathcal{L}_t \times \mathcal{P}_t$ is unresolved.

We say that $P_t \in \mathcal{D}_t$ *hits* a leaf of $\mathcal{U}_t$ if its corresponding I_t picks it and the ball associated with the leaf in question intersects $\mathcal{L}_t$. Let χ denote the number of leaves hit by P_t. To prove the existence of P we proceed by contradiction. Suppose that no active P_t hits more than one leaf. Then, the probability π that a random P_t is active (ie, belongs to $\mathcal{P}_t$) satisfies

$$\pi \leq \mathrm{Prob}[\,\chi = 0\,] + \mathrm{Prob}[\,\chi = 1\,].$$

To form a random I_t, we choose w^5 leaves of $\mathcal{U}_t$ at random, uniformly. By the query invariant, at least $1/d$ of them are associated with a ball that contains at least one query in $\mathcal{L}_t$. It follows that

$$\pi < \left(1 - \frac{1}{d}\right)^{w^5} + 2^d \left(1 - \frac{1}{d}\right)^{w^5 - 1} < 2^{-w^2},$$

which contradicts the key-set invariant. This concludes the proof of Theorem 10.6 (page 353). $\square$

10.5 Bibliographical Notes

Communication complexity was created as a unifying field by Yao in the late seventies, who introduced the two-party model [328]. The field is nicely surveyed in a book by Kushilevitz and Nisan [191].

Section 10.1: The distributional communication complexity of the inner product was investigated by Chor and Goldreich [89], who derived Theorem 10.2 (page 349). The same result was also obtained by Babai, Frankl, and Simon [26]. The matrix rank bound is due to Mehlhorn and Schmidt [226].

Sections 10.3, 10.4: The cell probe model was invented by Yao [329]. The connection between searching in the cell probe model and communication complexity was developed by Miltersen [228, 229]. The *master argument* has its roots in the pioneering work of Ajtai [7] on predecessor searching.

(Lemma 10.9, in particular, is due to him.) Ajtai proved that constant search time cannot be achieved with polynomial storage. The lower bound is as weak as can be, but the remarkably original ideas behind its proof played a major influence in subsequent work on this and other problems.

The master argument goes beyond Ajtai's method in some important ways, however. For one thing, the calibration of the recursion parameters is optimized. Also, the recursions do not create new problem instances but always keep subsets of the original ones. This is essential for proving the Hamming cube lower bound. Finally, the master argument is more general than Ajtai's, because it covers search and decision problems (obviously a necessity for the point separation lower bound). Although phrased differently, most of the ideas behind the master argument can be found, in fragments, in the works of Ajtai [7], Beame and Fich [29, 30], Chakrabarti et al. [63], and Chazelle [76].

The lower bound for predecessor searching (Theorem 10.4, page 353) was established independently by Xiao [326] and Beame and Fich [30]. The latter authors also prove that, rather surprisingly, the bound is optimal. Weaker lower bounds were obtained earlier by Miltersen [228] and Miltersen et al. [230].

The lower bound for point separation (Theorem 10.5 page 353) is due to Chazelle [76]. As was pointed out to the author by Fich, an alternative proof involves keeping the same geometric construction but reducing from membership in a class of languages called "strongly-indecisive regular." Lower bounds for these languages were proven by Beame and Fich [29].

The Hamming cube lower bound (Theorem 10.6, page 353) was proven by Chakrabarti et al. [63]. In the same model of computation, but with randomization allowed, an (almost) matching upper bound of $O(\log\log d)$ follows from [167, 192]. Borodin, Ostrovsky, and Rabani [55] proved a stronger lower bound, but for the exact version of the nearest neighbor problem on the Hamming cube.

Minimum Spanning Trees

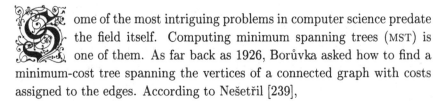ome of the most intriguing problems in computer science predate the field itself. Computing minimum spanning trees (MST) is one of them. As far back as 1926, Borůvka asked how to find a minimum-cost tree spanning the vertices of a connected graph with costs assigned to the edges. According to Nešetřil [239],

> This is a cornerstone problem of combinatorial optimization
> and in a sense its cradle.

Amazingly, after all of these years, the problem is still open. History aside, the minimum spanning tree problem is remarkable for several reasons. One of them is that it can be solved very quickly. Simple textbook solutions run in time $O(m \log m)$, where m is the number of edges. This is astonishing. Problems of that flavor are usually NP-complete or at least involve complicated polynomial-time procedures. But the minimum spanning tree problem is special. As a particular case of matroid optimization, it can be solved by a greedy approach. Crudely put, the hard part in finding the right answer is not *if* but *when*. Almost anything you try will eventually produce a minimum spanning tree. The question is, How long will it take?

Most methods based on divide-and-conquer split up the graph according to its combinatorial structure *or* to its distribution of edge costs. Random sampling allows us to do *both at once,* and by exploiting this fact, an optimal probabilistic solution was discovered. This sampling approach should be grist for the discrepancy method's mill. The challenge is to find, deterministically, a "low-discrepancy" subgraph whose own MST bears witness to the non-MST status of many edges.

By way of analogy, how should we go about finding the minimum element in a finite set of integers? Check all of the numbers one by one and keep track of the smallest. Perhaps a tad too dull and simplistic, no? Instead, extract a low-discrepancy subset, say the collection consisting of the highest number in each of the 100 percentiles of the set. All numbers outside the first percentile can be discarded. This leaves you with a set 1% the size of the original, to which we apply the procedure recursively. Using repeated linear selection for computing the sample of percentiles leads to a linear-time solution. Absurdly complicated? Yes, of course, but this low-discrepancy approach,[1] if not particularly simple, has the merit of being general. Such sampling can often be done even in the absence of a total ordering, and this will be the guiding theme of our discussion of the MST problem. In particular, it illustrates the nongreedy nature of our approach.

We need a form of low-discrepancy sampling that allows us to select edges that are both combinatorially (graph-wise) and numerically (cost-wise) representative of the whole. Furthermore, the selection must take place in a changing environment where insertions and deletions are allowed. The sampling is powered by an approximate priority queue, called a *soft heap*. We present its main features in §11.2 but postpone its implementation to the end (§11.4). The culmination of this chapter is the proof in §11.3 that the MST of a connected graph with n vertices and m edges can be computed deterministically in $O(m\alpha(m, n))$, where α is the classical inverse Ackermann function. This is currently the fastest deterministic algorithm for MST. If randomization is allowed, however, a solution of linear expected complexity is known.

11.1 Linear Selection as Low-Discrepancy Sampling

Linear selection is one of the earliest examples of the discrepancy method. Thirty years of age at the time of this writing, the method still stands as a model of algorithmic elegance and simplicity. Given a set X of n distinct numbers and an integer $1 \le k \le n$, linear selection returns the k-th smallest element of X by making only $O(n)$ comparisons. The pseudocode in the next figure explains how it all works. The algorithm is extremely simple: only a few lines of code. Yet it is hardly trivial: no fewer than three widely different recursive calls.

---

[1] As our example suggests, this notion of low discrepancy is one-way in the sense that the k smallest numbers should include enough sample points, but having too many would not hurt.

SELECT (X, k)

[1] If $|X| = O(1)$, sort X and return its k-th
    smallest element.

[2] Partition X into subsets X_i of size 5. Let Y
    be the set formed by SELECT $(X_i, 3)$, for all i.

[3] Compute $y = $ SELECT $(Y, \lceil n/10 \rceil)$, and form
    the subset Z of elements in X less than y.

[4] If $|Z| \geq k$ then return SELECT (Z, k) else
    return SELECT $(X \setminus Z, k - |Z|)$.

In Step 2, all (but at most one) of the X_i's have five elements. To compute the set Y of medians takes $O(n)$ time. Step 4 eliminates Z or its complement from contention. The issue is not correctness (which is trivial) but progress. The size of Z is between $n/4$ and $3n/4$, roughly, and so the time complexity $T(n)$ follows a recurrence of the form: $T(n) = O(1)$ if n is $O(1)$, else $T(n) = T(n/5) + T(3n/4) + O(n)$, which gives $T(n) = O(n)$.

We could easily tie up the loose ends of our explanation to produce industrial-strength code for linear selection. This is not our purpose here. The discrepancy method is the point of our discussion. Indeed, linear selection bears an uncanny resemblance with the computation of ε-approximations and, hence, low-discrepancy subsets. To see why, define the ε-percentiles ($0 < \varepsilon \leq 1/2$) of a set X to be its elements of ranks $i \lceil \varepsilon n \rceil$, for $i = 1, 2$, etc, where $n = |X|$. By iterating linear selection in trivial divide-and-conquer fashion, we derive an algorithm for computing all ε-percentiles in $O(n \log 1/\varepsilon)$ time. (We leave this as an easy warmup exercise.) These ε-percentiles constitute an optimal ε-approximation for the set system formed by the intervals of X (or, rather, the intervals of the sequence created by X sorted in increasing order).

Linear selection yields low-discrepancy subsets, which in turn set the grounds for divide-and-conquer. This pipeline, which can be found in most applications of linear selection, bears the marks of the discrepancy method. Also, in typical fashion, randomization can be used to simplify the proceedings greatly. For example, to select the k-th smallest element in X, pick random elements until one is found of rank between $n/4$ and $3n/4$; then recurse as in Step [4]. One expects to succeed after only a constant number of picks, and so the expected complexity of randomized selection is also linear.

Once again, we contemplate the same tradeoff encountered in many of the previous chapters, in fact, one of the running themes of this book: a deterministic solution of appealing algorithmic sophistication vs. a simpler, faster, but, shall we say, hopelessly banal, randomized algorithm. The algorithm designer might rush to the latter for practical applications, but it is toward the former that he or she will turn for algorithmic insight.

Linear selection has been gently maligned over the years as being too complicated. As outlandish as that particular charge surely is, the algorithm does have some weaknesses. One of them is its off-line nature. Numbers need to be known ahead of time; none can be added or deleted during the computation. Dynamic low-discrepancy maintenance has a number of applications, one of which—minimum spanning trees—is the focus of this chapter. We begin with a data structure, called a *soft heap*, that offers a dynamic alternative to linear selection. We apply this structure to computing the minimum spanning tree of a graph with arbitrary edge costs.

11.2 The Soft Heap: An Approximate Priority Queue

The soft heap is a data structure that supports the sort of dynamic sampling alluded to earlier. The data structure stores items with keys from a totally ordered universe. Like any full-fledged priority queue, it supports the following operations:

- create (S): Create an empty soft heap S.
- insert (S, x): Add a new item x to S.
- meld (S, S'): Form a new soft heap with the items stored in S and S' (assumed to be disjoint), and destroy S and S'.
- delete (S, x): Remove an item x from S.
- findmin (S): Return an item in S with the smallest key.

The catch is, the soft heap is allowed to make mistakes! Specifically, it may increase the value of certain keys at any time. Such keys, and by extension the corresponding items, are called *corrupted*. Keys cannot be decreased, so once corrupted always corrupted. Which item is corrupted when is entirely at the discretion of the data structure: The user has no control over it. The data structure remains heap-ordered with respect to *current* keys (corrupted or not), and findmin returns the minimum current key. The benefit is speed. During heap updates, items travel together in linked lists in a data structuring equivalent of "car pooling." To substitute mass transportation for individual travel cuts down on motion and, in our case, pointer updates. Similarly, by reducing the number of distinct keys, corruption cuts back on comparisons between keys.

The information theorist might call it entropy reduction, the coding theorist might name it lossy encoding, and the urban planner might invoke car pooling. Whatever one chooses to call it, the idea behind soft heaps is to make errors work to one's advantage. The amount of erring, the so-called *error rate*, is a parameter that we can adjust at will. It is the maximum allowed ratio between the number of corrupted items at any time and the total number of inserts so far.

Theorem 11.1 *Beginning with no prior data, consider a mixed sequence of operations that includes n inserts. For any $0 < \varepsilon \leq 1/2$, a soft heap with error rate ε supports each operation in constant amortized time, except for insert, which takes $O(\log 1/\varepsilon)$ time.[2] The data structure never contains more than εn corrupted items at any given time. In a comparison-based model, these bounds are optimal.*

In the worst case, every single item might end up being corrupted, and so the bound of εn applies only to a snapshot of the data structure at a given time. Despite this apparent weakness, the soft heap is optimal and—no mean feat—useful. Of course, by setting the error rate $\varepsilon = 1/2n$, we disallow corruption and the soft heap behaves like a regular heap with logarithmic insertion time.

To see the relation to linear selection, however, we must set ε to a higher value, say $\varepsilon = 1/4$: Insert n keys and then perform $\lfloor n/2 \rfloor$ findmins, each one followed by a delete. This takes $O(n)$ time. Among the keys deleted, the largest (original) one is at most $n/4$ away from the median of the n original keys. Like in the case of randomized linear selection, we can recurse through Step [4]. This gives us another linear selection algorithm, in fact one radically different from the classical deterministic solution.

The soft heap is built around a deceptively simple idea with powerful consequences. Most priority queues are represented as rooted trees with the

---

[2]This means that a particular operation may take longer than claimed, but that a sequence of m operations with n inserts runs in time $O(m + n \log 1/\varepsilon)$. Note that ε need not be a constant but can be chosen to be a function of n.

minimum key at the root and keys arranged in nondecreasing order along any path to a leaf. When the key at the root is deleted, the algorithm sifts items back up recursively to "refill" the now-empty root. Very roughly, the code might look something like this:

```
sift (tree)
  1. if tree has one node then exit;
  2. v = child (of root) with smallest key;
     move key (of v) to root;
  3. sift (subtree rooted at v);
```

The key novelty in the soft heap is to add a loop:

```
  4. if height of tree is odd then goto 1.
```

The conditional statement creates a branching process. Instead of looking like a path, the computation tree of `sift` is now truly a tree. Visually, items will no longer sift up one at a time, but they will collide and move together, creating the desired *car-pooling* effect. Of course, if we look too closely, most of what we just said is false: The loop condition is more than a parity test; lists must be formed to handle collisions, etc. From a distance, however, the picture we gave is the right one. A complete discussion of the soft heap with a full implementation in C is given in §11.4 (page 412).

11.3 Computing the MST

Let G be a connected, undirected graph with n vertices and m edges. The graph may have multiple edges but no self-loops (ie, no edge with the same endpoints). Each edge e is assigned a cost $c(e)$. We assume that these costs are all distinct real numbers: a nonrestrictive assumption. The MST of G is the tree spanning the vertices whose total edge cost is minimum. This definition involves the addition of costs, but remarkably one can compute MST (G) without ever adding any numbers. In fact, the beauty of the MST problem is that it is governed by a single *contraction rule*, from which everything else follows:

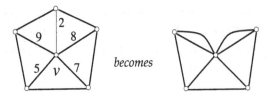

Given any vertex v of G, the cheapest edge incident to v belongs to the
MST *and can be contracted.*

"*To contract the edge*" means glueing together its two endpoints and
removing all the self-loops created in the process. By "*can be contracted*"
the rule indicates that it can be applied repeatedly anywhere, any time.
When in the end the graph G has shrunken to a single vertex, the collection
of contracted edges selected by the rule[3] forms precisely the MST of G.

Apply the contraction rule
to the vertex with the
filled circle. After five
iterations the MST is
identified: dashed edges in
bottom-right corner.

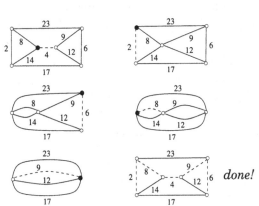

The contraction rule is about edges incident to a vertex. It implies a dual
rule, which governs the MST status of edges along a cycle. In most MST
algorithms the dual rule is seldom, if ever, used. The algorithm presented
in this book, however, features the dual rule in a leading role.

Given any cycle of G, its
most expensive edge does
not belong to the MST .

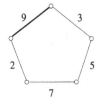

---

[3] Because of this rule, the MST problem can be redefined without mentioning the word
sum and edge costs can be taken in any totally ordered universe.

Obviously, the converse is true, too. Any non-MST edge forms a cycle with the MST and it is the most expensive. We have been speaking of the MST of G as though it were unique. In fact, it is. The dual rule and the distinctness of edge costs make sure of this.

There is usually a great deal of freedom in choosing contraction edges, and this high degree of nondeterminism is what explains the proliferation of different MST algorithms in the literature. All confront the same dilemma, however. As the graph gets contracted, the degrees of the vertices tend to grow, and so finding contraction edges becomes more and more difficult. A typical strategy is to maintain, for each vertex v, a heap H_v of incident edges. This data structure, also called a priority queue in the literature, maintains the edge costs under inserts, deletes, and melds (ie, forming the union of two heaps). Through an operation called findmin, the heap can return the minimum number in store. Usually, these operations take $O(\log k)$ time, where k is the number of edges present in the heap.

Heaps are often implemented as trees with items stored at the nodes. A rule of thumb says that the higher up an item is in the tree the more time-consuming its deletion is. Applying the contraction rule repeatedly tends to increase edge multiplicities and, hence, the number of edge deletions per contraction. One would like to store such deletion-prone edges low in the heap, but if their cost is low, how can we do that? This is the heart of the matter: The central question that has eluded 70 years of research on MST.

As contractions multiply, one should count on the degrees of the vertices to grow and on the heaps H_v to follow suit. Bigger heaps means slower heaps. Where this hurts is at deletion time. Indeed, as the average degree in the graph grows, for each MST edge discovered, more and more edges get deleted, which is increasingly costly.

A short digression: In some lucky cases, contracting G does not cause any growth in vertex degree. If the graph is planar, for example, it remains so under contraction; and since the average degree is less than 6, we are always able to apply the contraction rule to low-degree vertices. This leads to a straightforward linear-time algorithm for planar graphs. As a bonus we have this nice theorem from computational geometry: *Given the Voronoi diagram of a set of n points in* $\mathbf{E}^2$*, its Euclidean* MST *can be computed in linear time.* Proof: Form its Delaunay triangulation and use the classical fact that Euclidean-MST is a subgraph of Delaunay.

The contraction rule drives most MST algorithms. The difference between them has to do with the selection process for contraction edges, the organization of heaps, etc. To widen our range of action, we will relax the rule and contract edges that are not cost-minimum but only *nearly* so. The end result will be a spanning tree that is not minimum but—we hope —nearly so. The question will then be how to turn the nearly-so into the surely-so, using some form of bootstrapping.

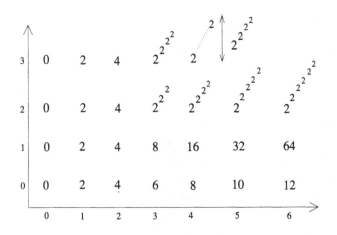

The first row is the function *two-times*(x), *the second one is derived by applying a feedback loop:* 2, *two-times*(2), *two-times*(*two-times*(2)), *etc. This yields the function* *two-to-the*(x)*-th. Another feedback loop gives the function* *tower-of*(x)*-twos. Then, we run out of names. Pretty soon, the rows seem to grow extremely fast. They do. But in fact the real story is in the columns. Past the column of fours, each one grows faster than any primitive recursive function. The diagonal grows at about the same speed as any column, which is much faster than any row!*

Before getting into the details, let us state the complexity of the algorithm. It involves Ackermann's function $A(i,j)$, which is defined recursively as follows:[4] For any integers $i,j \geq 0$,

$$\begin{cases} A(0,j) & = & 2j, & \text{for any } j \geq 0, \\ A(i,0) & = & 0 \quad \text{and} \quad A(i,1) = 2, & \text{for any } i \geq 1, \\ A(i,j) & = & A(i-1, A(i,j-1)), & \text{for any } i \geq 1, \ j \geq 2, \end{cases}$$

and for any $n, m > 0$,

$$\alpha(m,n) = \min\Big\{ i \geq 1 \, : \, A(i, 4\lceil m/n \rceil) > \log n \Big\}.$$

Theorem 11.2 *The* MST *of a connected graph with n vertices and m edges can be computed in $O(m\alpha(m,n))$ time.*

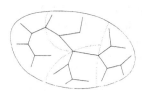

The graph G and its MST

Suppose we could, without knowing the MST *, partition the edges of G into subgraphs that each intersect the* MST *in a single connected component. Then we could work on each subgraph separately without worrying about inter-components edges. In other words, we could use genuine divide-and-conquer, something the classical methods do not allow.*

A Preview

The contraction rule tells us that the cheapest edge incident to a vertex is contractible. We generalize this notion by saying that a subgraph is *contractible* if its intersection with MST (G) is connected. To extend the contraction rule from single edges to entire subgraphs has an obvious benefit. Consider a contractible subgraph C, and let G' denote the graph G

---

[4]The definition is very robust and a number of variations can be found in the literature [99, 282, 303], all of them essentially equivalent.

after the contraction of C into a single vertex. The MST of G can be assembled directly from MST (C) and MST (G'), as illustrated in the figure below.

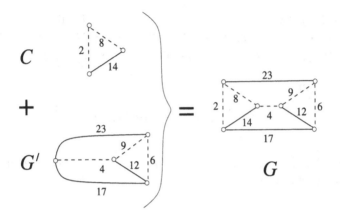

Suppose that an oracle tells us that the triangle C is contractible. Then, we can compute its MST , contract it, compute the MST of the remainder G', and put the the two trees together to form MST (G). What is the advantage? With standard methods, we would not know that the triangle is contractible until after its MST has been computed. By reversing this order, we are able to compute MST (C) without having to look at the edges labeled $4, 17, 23$: a small but important time savings.

Standard MST algorithms identify contractible subgraphs on the fly by computing their MST . The idea now is to reverse this process, ie, to certify the contractibility of C *before* computing its MST . The advantage is to speed up the computation of MST (C), since we can then do it unencumbered with edges having only one endpoint in C. This offers the possibility of effective divide-and-conquer. Of course, we need to be able to discover contractible subgraphs without computing their MST 's at the same time. That is where soft heaps come in. More on this shortly.

To compute MST (G), first, we decompose G into vertex-disjoint contractible subgraphs C_i of suitable size. Next, we contract each C_i into a single vertex and decompose the resulting minor[5] into another set of contractible subgraphs C_i'. We iterate on this process until G becomes a single

---

[5] A minor is a graph derived from a sequence of edge contractions and their implied vertex deletions.

vertex. This forms a hierarchy of contractible subgraphs, which we can model by a tree $\mathcal{T}$. The leaves correspond to the vertices of G; an internal node z with children $\{z_i\}$ is associated with a graph C_z whose vertices are the contractions of the graphs $\{C_{z_i}\}$. Assuming that each C_i has the same number of vertices at each level, $\mathcal{T}$ is perfectly balanced.

Each level of $\mathcal{T}$ represents a certain minor of G, and each C_z is a contractible subgraph of the minor associated with the level of its children. In this association, the leaf level corresponds to G while the root corresponds to the whole graph G contracted into a single vertex. Once $\mathcal{T}$ is available, we compute the MST of each C_z recursively. By the contractibility property, glueing together the trees MST (C_z) gives MST (G).

We have great freedom in choosing the height d of $\mathcal{T}$ and the number n_z of vertices of each C_z (which is also the number of children of z). As we shall see, by using soft heaps, we can compute the tree $\mathcal{T}$ in $O(m + d^3 n)$ time.

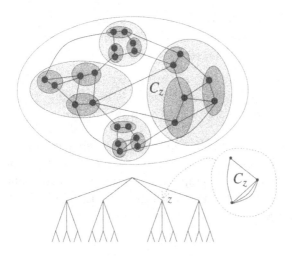

The bottom level $\ell = 0$ of the tree $\mathcal{T}$ corresponds to the graph G. Each level $\ell > 0$ corresponds to the graph obtained from level $\ell - 1$ by contracting C_z, for each z at level ℓ. The vertices of C_z are contractions of some C_{z_i} at level $\ell - 1$; typically, they are not vertices of G.

Obviously, there is a tradeoff: If d is chosen large, then the n_z's can be kept small; the recursive computation within each C_z is very fast, but building $\mathcal{T}$ is slow. Conversely, a small height speeds up the construction of

$\mathcal{T}$ but, by making the C_z's bigger, it makes the recursion more expensive. Ackermann's function provides the sort of Goldilocks height: not too big, not too small. For technical reasons, we define a slight variant of $A(i,j)$, which grows at about the same speed but is more in "synch" with the algorithm:

$$
\left\{
\begin{array}{llll}
S(1,j) & = & 2j, & \text{for any } j > 0, \\
S(i,1) & = & 2, & \text{for any } i > 0, \\
S(i,j) & = & S(i,j-1)S(i-1,S(i,j-1)), & \text{for any } i,j > 1.
\end{array}
\right.
$$

Let d_z denote the height of z in $\mathcal{T}$ (with the leaves at height zero). A judicious choice for the degree n_z of node z is

$$
n_z = \left\{
\begin{array}{ll}
S(t,1)^3 = 8 & \text{if } d_z = 1 \\
S(t-1,S(t,d_z-1))^3 & \text{if } d_z > 1.
\end{array}
\right.
\tag{11.1}
$$

The algorithm is built on a double recursion: One stepping through the levels of $\mathcal{T}$, from 0 to d; the other iterating within each C_z for a maximum recursion depth of t. For this reason we treat d and t as parameters, which we adjust at each recursive call. To compute the MST of G, we call the recursion for the first time by setting:

$$
\left\{
\begin{array}{lll}
d & = & c\lceil (m/n)^{1/3}\rceil, \text{ for some fixed } c \text{ large enough} \\
t & = & \text{smallest positive integer such that } n \le S(t,d)^3.
\end{array}
\right.
\tag{11.2}
$$

Given the graph G, the values of d and t are uniquely determined and, in turn, each n_z is fully specified. The graph C_z has n_z vertices, but what about its *expansion*, ie, the subgraph of G whose vertices are mapped into C_z? The number N_z of such vertices, which is also the number of leaves in the subtree rooted at z, satisfies

$$
N_z = S(t,d_z)^3,
\tag{11.3}
$$

as can be seen by induction using the identity

$$
S(t,d_z-1)^3 n_z = S(t,d_z-1)^3 S(t-1,S(t,d_z-1))^3 = S(t,d_z)^3.
$$

This might not quite be true at the root, as we might run out of vertices of G to provide the root with its prescribed number of children. At any rate, if z is the root, $S(t,d_z-1)^3 < n \le S(t,d_z)^3$ and so, by the monotonicity of $S(i,j)$, the height of $\mathcal{T}$ is at most d.

In this example,
$d_z = 2$ *and* $n_z = 3$.
The expansion of C_z
is a graph with six
vertices. The height
d *of* $\mathcal{T}$ *is 3.*

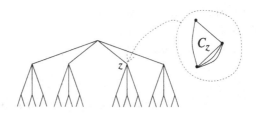

We prove by induction on t, d that if the number of vertices of G satisfies $n = S(t,d)^3$, then MST (G) can be computed in $bt(m + d^3 n)$ time, where b is a large enough constant.[6] We are not trying to give a rigorous proof at this point, so let us leave aside the basis case, $t = 1$. To apply the induction hypothesis on the time for computing MST (C_z), we observe that the number of vertices in C_z satisfies $n_z = S(t-1, S(t, d_z - 1))^3$, and so by visual inspection we see that, in the expression "$bt(m + d^3 n)$ time," t must be replaced by $t - 1$ and d by $S(t, d_z - 1)$. This gives a cost of $b(t-1)(m_z + S(t, d_z - 1)^3 n_z)$, where m_z is the number of edges in C_z. Summing up over all internal nodes $z \in \mathcal{T}$ and replacing n_z by its value in (11.1), we find that the cost of computing all the MST (C_z)'s is

$$b(t-1)\Big(m + \sum_z S(t, d_z - 1)^3 S(t-1, S(t, d_z - 1))^3\Big),$$

which is also, by (11.3)

$$b(t-1)\Big(m + \sum_z S(t, d_z)^3\Big) = b(t-1)\Big(m + \sum_z N_z\Big) = b(t-1)(m + dn).$$

The last derivation uses the fact that the tree is of height d and so each vertex is counted d times. Adding to this bound the time claimed earlier for computing $\mathcal{T}$ yields, for b large enough,

$$b(t-1)(m + dn) + O(m + d^3 n) \le bt(m + d^3 n),$$

which proves our claim. As we show later, our choice of t and d implies that $t = O(\alpha(m, n))$, and so the running time of the MST algorithm is $O(m\alpha(m, n))$. *Voilà!*

---

[6]The case where the equality $n = S(t,d)^3$ cannot be strictly enforced is a minor technicality we can overlook in this overview. Note that with this assumption, the height of $\mathcal{T}$ is precisely d.

This informal discussion leads to the heart of the matter: How to build $\mathcal{T}$ in $O(m + d^3 n)$ time. We have swept under the rug a number of peripheral difficulties. The end result is an algorithm far more subtle than the one we have outlined. Indeed, quite a few things can go wrong along the way; though none as serious as *edge corruption*, an unavoidable byproduct of soft heaps and our next topic of discussion.

The Effect of Corruption

In the course of computing $\mathcal{T}$, certain edges of G become *corrupted*, meaning that their costs are raised. (The costs of corrupted edges can be raised more than once but never decreased.) The reason for this has to do with the soft heap, the approximate priority queue that we use for identifying contractible subgraphs. One might think that all the corrupted edges need to be reprocessed from scratch some time later. This would spell doom, since in fact it could well happen that every single edge becomes corrupted. Fortunately, the problem is one of timing. Some corrupted edges cause trouble, while others do not. This depends on when the corruption occurs. To understand why, we must first discuss how the overall construction of $\mathcal{T}$ is scheduled.

It might be tempting to build $\mathcal{T}$ bottom-up level by level, but this won't work here. It is important to maintain a single connected structure at all times. Keeping around separate components would make the dual rule impossible to apply. So, we compute $\mathcal{T}$ in postorder, instead: *children first, parent last*. This is the order in which the C_z's are contracted.[7]

Let z be the current node visited in $\mathcal{T}$, and let $z_1, \ldots, z_k = z$ be the *active path*, ie, the path from the root z_1 to z. The subgraphs $C_{z_1}, \ldots, C_{z_k}$ are being currently assembled, and each one has at least one vertex. As soon as C_{z_k} is ready it is contracted into one vertex which is then added to $C_{z_{k-1}}$. Any edge of G with exactly one vertex in $C_{z_1} \cup \cdots \cup C_{z_k}$ is said to be of the *border* type. Naturally, the type of an edge changes over time. The example below should lift any remaining ambiguity.

---

[7]In our discussion, we actually refer to the order, first child first, parent next, other children last. In this way, at any time, the nodes visited so far form a tree, and so we can define paths among visited nodes and things of the sort. This is purely for notational purposes, however, and the sequence of contractions still follows the regular definition of postorder.

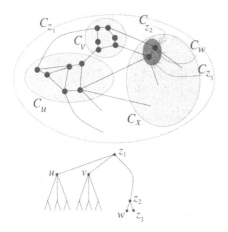

The subgraphs C_{z_1}, C_{z_2}, and C_{z_3} are under construction. Subgraphs C_u, C_v, and C_w have been completely processed already. Subgraph C_x is under construction. In accordance with our (modified) postorder visit of the nodes, x does not yet exist in $\mathcal{T}$. As soon as we are done with z_2, node x will be added. There are exactly five border edges.

Corruption can strike edges only while they are of the border type (since it is then that they are in soft heaps). At first, corruption might seem fatal, for if all edges become corrupted, aren't we solving an MST problem with entirely wrong edge costs? Fortunately, corruption causes harm only in one specific situation. A corrupted edge is called *bad* if it is still a border edge at the time its incident C_z is contracted into one vertex (ie, when z is popped off the postorder stack). Once bad, an edge remains bad always, but like any corrupted edge its cost can still rise. Remarkably, it can be shown that if no edges ever turned bad, the algorithm would behave as though no corruption ever occurred, regardless of how much actually took place. The recurrence of the previous section would be completely accurate, and our MST story would near the end. Our goal, thus, is to fight badness rather than corruption. We are able to limit the number of bad edges to within $m/2 + d^3 n$. The number of edges corrupted but never bad is irrelevant.

Once $\mathcal{T}$ is built, we restore all of the edge costs to their original values and remove all of the bad edges. We recurse within what is left of the C_z's to produce a spanning forest F. Finally, we throw back in the bad edges and recurse again to produce the minimum spanning tree of G. There are subtleties in these various recursions, which we explain in the next section.

The Algorithm

The previous section presented a road map of the algorithm. This one explores its twists and turns. First twist down the road: the *Borůvka phase*. This is the repeated application of the contraction rule at each vertex of G. It involves finding the cheapest edge incident to each vertex and contracting it. Once this is done, self-loops are removed from the resulting graph. It is easy to perform a Borůvka phase in $O(m)$ time in a single pass through the graph. The number of vertices drops by at least a factor of two.

To compute the MST of a connected graph G with n vertices and m edges, we call the function $\mathtt{msf}(G, t)$ for the parameter value

$$t = \min\{\, i > 0 \mid n \le S(i, d)^3 \,\}, \tag{11.4}$$

where $d = c\lceil (m/n)^{1/3} \rceil$. Throughout this chapter, c denotes a large enough integral constant. The function $\mathtt{msf}$ takes as input an integer $t \ge 1$ and a graph with distinct edge costs, and returns its minimum spanning forest (MSF). As the name suggests, we no longer assume that the input graph should be connected.[8] Note that t is a parameter. It is set in (11.4) at the beginning of the algorithm and is decremented during later recursive calls.

THE MST ALGORITHM

[1] If $t = 1$ or $n = O(1)$, return
MSF (G) by direct computation.
[2] Perform c consecutive Borůvka phases.
[3] Build $\mathcal{T}$ and form the graph B of bad edges.
[4] Set $F \leftarrow \bigcup_{z \in \mathcal{T}} \mathtt{msf}(C_z \setminus B, t - 1)$.
[5] Return $\mathtt{msf}(F \cup B, t) \cup \{\,\text{edges contracted in [2]}\,\}$.

Ackermann's double induction is visible in Steps [3] and [4]. The index i in the table $S(i, j)$ is the same throughout Step [3], and the tree $\mathcal{T}$ corresponds to the induction along the index j. Each new nesting in the recursion of Step [4] decreases the value of i (from t on down). So, a given $\mathcal{T}$ is built according to the numbers in a given row in the table of $S(i, j)$'s. Step [4] move to the next row below. Step [5] reprocesses the bad edges.

---

[8]This is a minor technicality required for the recursion invariants. As we shall see, the algorithm occasionally discards edges. Alternatively, we could keep everything connected at all times by adding dummy edges, but it is just as simple to allow the possibility of several connected components.

Steps [1] and [2]: Boundary Cases and Borůvka Phases

The case $t = 1$ is handled separately. We compute the MSF directly by performing as many Borůvka phases as it takes to contract G into a single vertex. If we keep the graph free of multiple edges by holding on to the cheapest edge in each group of same-endpoint edges, we can easily do all the work in $O(n^2 + (n/2)^2 + \cdots) = O(n^2)$ time. If $n = O(1)$, we trivially compute the MST in $O(m)$ time; note that we say $O(m)$ and not $O(1)$ because the graph may have multiple edges.

If the graph is not connected, we apply the algorithm to each connected component of G separately. To avoid repeating this last sentence incessantly, we assume connectivity throughout our description of the algorithm (but not in the complexity analysis). The purpose of Step [2] is to reduce the number of vertices. In $O(n + m)$ time, this transforms G into a graph G_0 with $n_0 \leq n/2^c$ vertices and $m_0 \leq m$ edges.

Step [3]: Building the Tree $\mathcal{T}$ of Contractible Subgraphs

With the argument $t > 1$ specified, so is the number n_z of vertices of each C_z, which is also the degree of z in $\mathcal{T}$. Indeed, $n_z = S(t - 1, S(t, d_{z_k} - 1))^3$, where d_{z_k} is the height of z. Actually, we should speak only of target size, because the algorithm sometimes fails to grow the contractible subgraphs to their intended vertex counts. Such failures are no mere technicality, but part of the fundamental structure of the algorithm. It stems from the occurrence of *fusions*, a phenomenon to be explained shortly.

To get things off the ground is a routine matter, and we may pick up the algorithm in midaction. Let $z_1, \ldots, z_k = z$ denote the active path. As shown below, the subgraphs $C_{z_1}, \ldots, C_{z_k}$ currently under construction are linked together by means of a cost-decreasing sequence of edges. The algorithm occasionally *discards*[9] edges from G_0. Each graph C_z includes all the nondiscarded edges of G_0 whose endpoints map into it. With this provision, a given C_z is entirely specified by its vertex set alone.

The invariants below hold at any time with respect to the current[10] G_0. We need to introduce the important concept of the *working cost*: At any time, the working cost of an edge is its current cost if the edge is bad, and

---

[9]The word "discarded" has a technical meaning and refers to edges removed from consideration only in the specific circumstances stated below.

[10]The term "current" refers here to the original G_0 minus all the edges discarded up to that point.

its original cost otherwise. Thus, we distinguish among three types of cost: original, current, and working.

INV1 For all $i < k$, we maintain an edge (called a *chain-link*) joining C_{z_i} to $C_{z_{i+1}}$ whose current cost is (i) at most that of any border edge incident to $C_{z_1} \cup \cdots \cup C_{z_i}$ and (ii) less than the working cost of any edge joining two distinct C_{z_j}'s ($j \leq i$). To enforce the latter condition efficiently, we keep a *min-link*, if it exists, for each pair $i < j$: This is an edge of minimum working cost joining C_{z_i} and C_{z_j}.

INV2 For all j, the border edges (u, v) with $u \in C_{z_j}$ are stored either in a soft heap, denoted $H(j)$, or in one, denoted $H(i, j)$, where $0 \leq i < j$. No edge appears in more than one heap. Besides the condition $u \in C_{z_j}$, membership in $H(j)$ implies that v is incident to at least one edge stored in some $H(i, j)$; membership in $H(i, j)$ implies that v is also adjacent to C_{z_i} but not to any C_{z_l} in between ($i < l < j$). We extend this to $i = 0$ to mean that v is incident to no C_{z_l} ($l < j$). All the soft heaps are chosen with error rate $1/c$.

The chain of subgraphs along the active path, with edge costs indicated by vertical height. Working costs are current for bad edges and original for others.

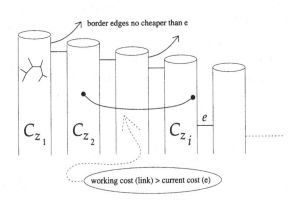

The main thrust of INV1 is to stipulate that the active path should correspond to a descending sequence of edges connecting various C_z's. This descending property is essential for ensuring contractibility. The chain-link between C_{z_i} and $C_{z_{i+1}}$ is the edge that contributes to $C_{z_{i+1}}$ its first vertex. Subsequently, as $C_{z_{i+1}}$ grows, lower cost edges might connect it to C_{z_i}

and so, in general, the chain-link is likely to be distinct from the min-link between C_{z_i} and $C_{z_{i+1}}$. The invariant INV2 specifies which edge should be stored in which heap.

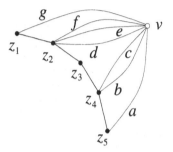

A possible assignment of edges to heaps:

$a \in H(4,5)$, $b \in H(4)$,

$c \in H(2,4)$, $d \in H(2)$,

$e \in H(1,2)$, $f \in H(1,2)$,

$g \in H(0,1)$.

From a distance, the algorithm exhibits the traits that we expect of an Ackermann-like complexity, in particular, a double induction.[11] A closer inspection reveals a more puzzling picture. Why does INV1 mix in current and working costs? Roughly, because using current costs alone would deny us the miraculous distinction between bad and corrupted, while using working costs alone would make soft heaps irrelevant. Why do we need so many heaps and not just one? The short answer is, to fight badness. The problem is that when bad edges are deleted from a soft heap, Theorem 11.1 allows the same amount of corruption to be produced anew. These newly corrupted edges might then turn bad and be deleted. Cycling through this process could have disastrous effects, perhaps even making every single edge bad. We use separate heaps to create a buffering effect meant to counter this process. This mechanism relies on structural properties of minimum spanning trees and a delicate interplay among heaps. This is rather subtle and can be fully explained only later.

The tree $\mathcal{T}$ is built in a postorder traversal driven by a stack whose two operations, pop and push, translate into, respectively, a *retraction* and an *extension* of the active path.

---

[11] Well, triple actually, but the last one is more bark than bite and plays no role in the Ackermann behavior.

• RETRACTION: This happens for $k \geq 2$ when the last subgraph C_{z_k} has attained its target size, ie, its number n_{z_k} of vertices has reached the value of $S(t-1, S(t, d_{z_k} - 1))^3$, where d_{z_k} is the height of z_k in $\mathcal{T}$. Recall that this is also the number of children of z_k in $\mathcal{T}$. In the particular case $d_{z_k} = 1$, the target size is set to $S(t, 1)^3 = 8$.

In a retraction, C_{z_k} contracts into a vertex and $C_{z_{k-1}}$ becomes the new end of the chain.

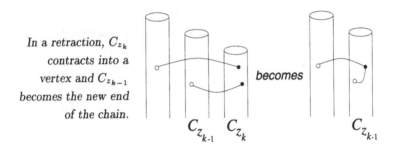

The subgraph C_{z_k} is contracted to become a new vertex of $C_{z_{k-1}}$, which thus gains one vertex and one or several new edges (including the chain-link); the end of the active path is now z_{k-1}. Maintaining INV1 in $O(k)$ time is straightforward; INV2 is less so. Here is what we do.

The heaps $H(k)$ and $H(k-1, k)$ are destroyed. All their corrupted edges are discarded. (These edges, if not bad already, become so now.) The remaining items from the two destroyed heaps are partitioned into subsets, called *clusters*, of edges that share the same endpoints outside the chain. For each cluster in turn, select the edge (r, s) of minimum current cost and discard the others (if any). Next, insert the selected edge into the heap implied by INV2. Specifically, if (r, s) comes from $H(k)$ and shares s with an edge in $H(k-1, k)$, or if it comes from $H(k-1, k)$, then by INV2 it also shares s with an edge in some $H(i, k-1)$ already, and so it can be inserted into $H(k-1)$. Otherwise, (r, s) comes from $H(k)$ and by INV2 it shares s with an edge in some $H(i, k)$, now with $i < k - 1$. The edge (r, s) should be inserted into $H(i, k)$.[12] Finally, for each $i < k - 1$, meld $H(i, k)$ into $H(i, k-1)$.

---

[12]Two remarks: (i) This insertion forces into $H(i, k)$ at least a second edge pointing to s; (ii) since the heap is then melded into $H(i, k-1)$, we could have inserted (r, s) into $H(k-1)$, instead of $H(i, k)$, and still maintain INV2. This would be a fatal mistake, for edges might then hop between $H(\star)$'s at each retraction, which could be prohibitively expensive. Intuitively, $H(j)$ is a buffer heap collecting edges while the action is below z_j. There is no bound on how many of these edges can share the same endpoint (what will be called "multiplicity"). For the $H(i, j)$'s, however, there is such a bound.

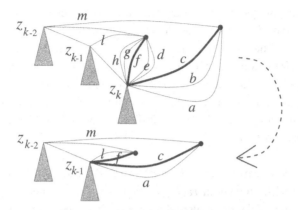

In this retraction, the thick edges are the cheapest in their clusters. Suppose that $a, c \in H(k)$ and $b \in H(k - 2, k)$: b remains in $H(k-2, k)$ while, in the cluster $\{a, c\}$, a is discarded and c is inserted into $H(k-2, k)$. This heap is melded into $H(k-2, k-1)$. In the cluster $\{d, e, f, g, h\}$, suppose that $d, e, f, g \in H(k)$ and $h \in H(k - 1, k)$. Note that $l \in H(k - 2, k - 1)$ is not in the cluster. Only f survives and moves into $H(k - 1)$.

• EXTENSION: Perform a `findmin` on *all* of the heaps and retrieve the border edge (u, v) of minimum current cost $c(u, v)$. We call it the *extension edge*. Of all the min-links of working cost at most $c(u, v)$, find the one (a, b) incident to the C_{z_i} of smallest index i. If such an edge does indeed exist, we perform a *fusion*: We contract the whole subchain $C_{z_{i+1}} \cup \cdots \cup C_{z_k}$ into a. It is best to think of this as a two-step process: First, we contract all of the edges with both endpoints in $C_{z_{i+1}} \cup \cdots \cup C_{z_k}$. By abuse of notation, call b the resulting vertex. We now contract the edge(s) joining a and b.

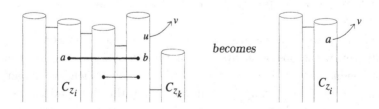

Fusion in action: All three subgraphs right of C_{z_i} collapse into b and then into a.

Next, we update all relevant min-links, which is easy to do in $O(k^2)$ time. To update heaps we generalize the retraction recipe in the obvious way. For the sake of completeness, here are the details. First, we extend the destruction of heaps to include not just $H(k)$ and $H(k-1,k)$, but $H(i+1), \ldots, H(k)$, and all $H(j,j')$, $i \leq j < j'$. Then, we discard all corrupted edges from those heaps since they are now bad. Next, we regroup the remaining edges into clusters and, for each one in turn, we reinsert the edge (r,s) of minimum current cost and discard the others. As before, we distinguish between two cases:

1. If (r,s) comes from some $H(j,j')$ or $H(j')$, but in the latter case shares s with an edge in $H(j,j')$, where $i \leq j < j'$, then by iterative application of Inv2 it also shares s with an edge (r',s) in some $H(h,l)$, with $h < i \leq l$. As we explain below, the edge (r',s) is to migrate into $H(h,i)$ through melding, if it is not there already, ie, if $l > i$; therefore, by Inv2 we can insert (r,s) into $H(i)$.

2. Otherwise, (r,s) comes from $H(j)$, where $i < j$, and it shares s with an edge in some $H(h,j)$, now with $h < i$. We insert the edge (r,s) into $H(h,j)$.

Finally, for each h,j with $h < i < j$, we meld $H(h,j)$ into $H(h,i)$. Observe that by Inv1(i) the vertex a, as defined, cannot be further down the chain than u. So, whether u originally belonged to the last C_z in the chain or not, it now does. In all cases (ie, fusion or not), we extend the chain by making v into the single-vertex C_{z_k} and the extension edge (u,v) into the chain-link incident to it. The end of the active path is now z_k, where $k \leftarrow i+1$ (fusion) and $k \leftarrow k+1$ (no fusion).

Old border edges incident to v cease to be of the border type. We delete them from their respective heaps, and we find among those that are not bad the min-link between v and each of $C_{z_1}, \ldots, C_{z_{k-1}}$. We insert the new border edges incident to v into the appropriate $H(i,k)$; in the case of multiple edges, we keep only the cheapest in each group and discard the others.

This completes our discussion of retractions and extensions. Let us briefly review the postorder-driven construction of $\mathcal{T}$. At any given node z_k of height at least 1, we perform extensions (and their accompanying fusions) as long as we can, stopping only when the size condition for retraction at that node has been met. There is no retraction condition for the root z_1, and so the algorithm stops only when border edges run out and extensions are no longer possible.

If no fusion ever takes place, then our previous argument leading to the identity (11.3) on the size of the expansion of C_z still holds, ie, $N_z = S(t, d_z)^3$. Fusions muddy the waters in two ways: They force contractions before the target size has been reached and they may cause arbitrarily large expansions. As we shall see, however, any C_z whose expansion exceeds its allowed size is naturally broken down into subgraphs of the right size, which then can be treated separately. In fact, those dreaded fusions turn out to be a blessing in disguise. We conclude this discussion of Step [3] with a few remarks:

▷ One should not mistake a fusion for a retraction into C_{z_i}. Because the edge (a, b) is contracted, too, it does not become an edge of C_{z_i}. Therefore, unlike a retraction, a fusion does not add new vertices to C_{z_i}, although it does increase its expansion.

▷ To be able to form B in Step [3], we need to keep track of the bad edges. Badness occurs to the corrupted border edges incident to C_{z_k} (in retraction) or $C_{z_{i+1}} \cup \cdots \cup C_{z_k}$ (in fusion). All such edges either are explicitly examined and, hence, discarded or belong to heaps that are being melded. A soft heap gives ready access to its corrupted items, so we can mark the relevant edges bad. It is routine to ensure that edges are marked only once.

STEP [4]: RECURSING IN SUBGRAPHS OF $\mathcal{T}$

Having built the tree $\mathcal{T}$, we consider each node z. Recall that C_z does not include any of the discarded edges. Let $C_z \setminus B$ denote the subgraph of C_z obtained by removing all of the bad edges.[13] We apply the algorithm recursively to $C_z \setminus B$ after resetting all edge costs to their original values and decrementing the parameter t by 1. With this new value of t, all the target sizes of the new tree $\mathcal{T}$ to be built are fully specified and the recursion has all that it needs to proceed.

The correspondence between the vertices of C_z and the children of z would be obvious, were it not for those pesky fusions. Consider the first fusion into vertex a of C_z (see figure on page 397). Prior to it, vertex a corresponded to some child z_0 of z, meaning that it was the contraction of C_{z_0}. What happens to it after the fusion? Nothing special: Step [4] recurses with respect to $C_{z_0} \setminus B$ as it would without any fusion. What happens to the part of $\mathcal{T}$ whose corresponding subchain $C_{z_{i+1}} \cup \cdots \cup C_{z_k}$ contracts into b? It is treated as a hierarchy of its own and handled like $\mathcal{T}$.

---

[13]Bear in mind that not all bad edges may have been discarded in Step [3]; in fact, bad edges can be selected as extension edges and play a direct role in building the C_z's.

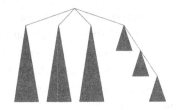

becomes

A fusion causes the pruning of a subtree constituting a suffix of the active path. Although "unfinished," this subtree is processed in Step [4] just like $\mathcal{T}$.

So, a and b are "handled" separately in Step [4]. Does this mean that no edge joining them finds its way into F? In fact, from the group of multiple edges joining a and (contracted) b, we retain in F the original[14] min-link (a, b) provided that it is not bad. By construction, this edge is the cheapest in the group (with respect to original costs, excluding bad edges). If the min-link is bad, we need not retain any edge of the group, since bad edges are all reprocessed in Step [5].

To summarize, if any fusion into a has taken the place, the expansion of a is treated as a collection of connected components, each of which is treated recursively in Step [4], possibly joined together by nonbad min-links which we include in F. Because each connected component stays within its mandated size, their potential proliferation does not increase the per-edge complexity of the algorithm. Fusions might blur the picture a little but, from a complexity viewpoint, the more the better.

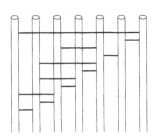

In this ultimate case, edges (resp. vertices) are horizontal (resp. vertical) segments with height indicating costs. In the absence of corruption, every extension gives rise to a fusion. The tree $\mathcal{T}$ remains of height 1 and Step [4] is trivial: a dream scenario.

---

[14] Recall that (a, b) was chosen as a min-link in the first place.

STEP [5]: THE FINAL RECURSION

In Step [3] we collected all of the bad edges created during the construction of $\mathcal{T}$ and formed the graph B. We now add to it the edges of F to assemble the subgraph $F \cup B$ of G_0. The output of $\mathrm{msf}(F \cup B, t)$ is now viewed as a set of edges with endpoints in G, not in G_0. Adding the edges contracted in Step [2] produces the MST of G.

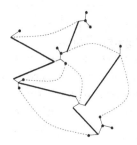

The recursion in Step [5] involves the bad edges (the dashed lines) and the edges of F computed in Step [4] (the thick lines). The output is a forest. Adding to it the edges contracted in Step [2] (the thin lines) gives us the MST of G.

Correctness

We prove by induction on t and n that $\mathrm{msf}(G, t)$ computes the minimum spanning forest of G. Since the algorithm iterates through the connected components separately, we can again assume that G is connected. Also, because of Step [1] we can obviously assume that $t > 1$ and that n is larger than a suitable constant. Borůvka phases contract only MST edges; so, by induction on the correctness of msf, the output of Step [5] is, indeed, MST (G), provided that any edge e of G_0 outside $F \cup B$ also lies outside MST (G_0). In other words, the proof of correctness of our MST algorithm hinges on the following:

Lemma 11.3 *If an edge of G_0 is not bad and lies outside F, then it is outside* MST (G_0) *as well.*

Note that in the lemma all costs are understood as original. Of course, this innocent-looking statement is the heart of the matter, since it pertains directly to the hierarchy $\mathcal{T}$. We omit the proof that invariant INV1 is maintained, which is straightforward. Note that the sole purpose of fusions is to maintain INV1(ii). Similarly, the carefully regulated updating of the heaps is meant to preserve INV2 and we need not revisit this territory. We now must show why maintaining these invariants produces contractible C_z's.

Contractibility is defined in terms of the MST, a global notion. Fortunately, we can certify by local means that the subgraph C of G_0 induced by a given subset of the vertices is contractible. Indeed, it suffices to check that C is *strongly contractible*, meaning that for every pair of edges e, f in G_0, each with exactly one vertex in C, there exists a path in C that connects e to f along which no edge exceeds the cost of both e and f. This implies contractibility (but not the other way around). Why is that so?

We argue by contradiction, assuming that C is not contractible. By definition, $C \cap \text{MST}(G_0)$ must have more than one connected component. Consider a shortest path π in $\text{MST}(G_0)$ that joins two distinct components (see next figure). The path has no edge in C (else it could be shortened), and it has more than one edge (else it would be in C since the graph contains all the edges induced by its vertices), so its end-edges e and f are distinct and each has exactly one endpoint in C. Any path in C joining e and f forms a cycle with π and, by elementary properties of MST, the most expensive cycle edge is outside $\text{MST}(G_0)$, ie, outside of π and, hence, in C. This contradicts the assumption and proves the claim.[15]

Strong contractibility implies contractibility. The advantage of the former notion is that it is a local condition, which involves only the subgraph and its neighborhood.

Lemma 11.4 *With respect to working costs, C_z is strongly contractible at the time of its contraction, and the same holds of every fusion edge (a, b).*

Proof: The lemma refers to the edges present in C_z and in its neighborhood at the time C_z is contracted; it does not include the edges of G_0 that have been discarded (in fact, the lemma is false otherwise). The graph C_z is formed by incrementally adding vertices via retractions. Occasionally, new neighboring edges are added by fusion into some $a \in C_z$. Because C_z does not contain border edges, edge discarding never takes place within it, and so it grows monotonically.

---

[15] Implicit to the proof was the assumption that all edge costs are distinct. In the presence of duplicate edge costs, strong contractibility implies contractibility with respect to at least one MST.

Assume for now that no fusion occurs. Each retraction has a unique chain-link (ie, an extension edge) associated with it, and together they form a spanning tree of C_z. Thus, given any two edges e, f, each with exactly one endpoint in C_z, the tree has a unique path π joining (but not including) them. Let $g = (u, v)$ be the edge of π of the highest current cost ever; break ties, if necessary, by choosing as g the last one selected for extension chronologically. As usual, we make the convention that v is the endpoint outside the chain at the time of extension. Along π, the vertex u lies between v and one of the two edges, say e. Throughout this proof the term "working" is to be understood at the time right after C_z is contracted, while "current" refers to the time when g is selected as a new chain-link (u, v) through extension. We claim that the working cost of e is at least the current cost of g. Since the working cost of no edge in π can ever exceed the current cost of g, the lemma follows.

Along the path π...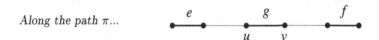

We prove the claim. If e currently joins C_z to some other $C_{z'}$, it follows from INV1(ii). Otherwise, let e' be the first (current) border edge encountered along the path from g to e. By INV1(i), its current cost is at least that of g, and so by our choice of g we have $e' \notin \pi$, and hence $e' = e$. Consequently, e currently is and still will be a border edge when C_z is contracted. If it is in a corrupted state, then it becomes bad after the contraction (were it not so already), and so, by definition, its working cost (right after C_z's contraction) is at least its current cost (right after g's extension); otherwise, both costs coincide with the original one. In both cases, the claim is true.

To deal with a fusion into C_z, we should think of it as a two-step process: (i) A sequence of retractions involving, successively, $C_{z_k}, C_{z_{k-1}}, \ldots, C_{z_{i+1}}$, where in this notation $C_z = C_{z_i}$; and (ii) the contraction of (a, b) into a. For the purpose of this discussion, let us run the algorithm right until the time C_z is contracted, while skipping Step (ii) in all fusions into C_z. Then, as far as C_z is concerned, its evolution is indistinguishable from the no-fusion case discussed earlier, and the same result applies. Executing all delayed applications of Step (ii) now results in contracting a number of edges already within C_z, which therefore keeps C_z strongly contractible. This proves the first part of the lemma.

Now, going back to the normal sequencing of the algorithm, consider the min-link (a, b) right before Step (ii) in a fusion into C_z. By construction, no other edge incident to (contracted) b is cheaper than (a, b) relative to working costs; remember that all corrupted border edges incident to b become bad, and so working and current costs agree. This shows that the edge (a, b) is strongly contractible. □

Proof of Lemma 11.3: The computation of $\mathcal{T}$ corresponds to a sequence of contractions of minors, which transforms G_0 into a single vertex. Denote these minors by $S_1, S_2, \ldots$ in chronological order of their contractions. Note that either S_i is of the form C_z or it consists of the multiple edges of some fusion edge (a, b).

Let G_0' be the graph G_0 minus all the edges discarded during Step [3]. It is easy to see that G_0' spans all the vertices of G_0 because no discarding takes place within any $H(0, j)$. Lemma 11.4 applies to C_z at the time of its contraction. The working costs of all edges within C_z are frozen once and for all. The current costs of edges with one endpoint in C_z might change, but the working costs can never decrease, so the lemma still applies relative to final working costs, ie, with each edge assigned its last working cost chronologically. Unless specified otherwise, such costs are understood throughout the remainder of our discussion.

Fix some S_i. A vertex of S_i is either a vertex of G_0 or the contraction of some S_j $(j < i)$. In turn, the vertices of S_j are either vertices of G_0 or contractions of S_k $(k < j)$, etc. By repeated applications of Lemma 11.4 (and again identifying graphs with their edge sets) it follows that the MST of G_0' is the union of all the MST (S_i)'s: We call this the *composition property*.[16]

In proving Lemma 11.3, we begin with the case where the edge e under consideration is never discarded, ie, belongs to G_0'. Consider the unique S_i that contains both endpoints of e among its vertices. By induction on the correctness of msf, the fact that e is not in F implies that it is not in MSF $(S_i \setminus B)$. Since it is not bad, the edge e is then outside MST (S_i) and, by the composition property, outside MST (G_0'). Recall that this holds relative to final working costs. Now, switch all edge costs to their original values. If changes occur, they can only be downward. The key observation now is that, by not being bad, the edge e witnesses no change and so still remains the most expensive edge on a cycle of G_0', with respect to original

---

[16] Keep in mind that the S_i's might include bad edges, and so the composition property does not necessarily hold for the graphs of the form $C_z \setminus B$. In fact, it is worth noticing that, for all their "badness," bad edges are useful for making contractibility statements.

costs. This shows that e is not in MST (G_0') and, hence, MST (G_0) relative to original costs.

Assume now that e is not in G_0'. Before being discarded, $e = (u, v)$ shared a common endpoint v with a cheaper uncorrupted edge $e' = (u', v)$. In the case of a retraction, u and u' coincide, while in a fusion both are merged together through the contraction of a subgraph. In both cases, u and u' end up in a vertex that, by repeated applications of Lemma 11.4, is thus seen to be the contraction of a contractible subgraph of G_0' relative to working costs. It follows that e is outside MST (G_0). Again, observe the usefulness of bad edges. Indeed, because e' might later become bad, we could not conclude that e is outside MST $(G_0 \setminus B)$. This completes the proof of Lemma 11.3. $\square$

FREQUENTLY ASKED QUESTIONS

▷ Why are current costs used in INV1(i) and not original ones? Because the errors caused by soft heaps make it impossible to enforce order relations among original costs.

▷ But, then, why are we comparing working costs against current ones in INV1(ii)? The weaker invariant obtained by comparing only current costs would force us to treat all corrupted edges as bad. Indeed, strong contractibility would then hold only relative to current costs, and we would be unable to restore any corrupted edge to its original cost without violating the contractibility conditions.

▷ Why does INV1(i) say "at most" and INV1(ii) say "less" ? A very minor point, indeed: Soft heaps tend to assign the same corrupted keys to many items; so if we want extensions to take place, INV1(i) should avoid strict inequalities. On the other hand, to say "less" in INV1(ii) is to favor fusions, which are highly desirable events from a complexity standpoint.

The Decay Lemma

We need to show that the creation of bad edges tapers off rapidly while iterating through Step [5]. To do that, we bound the size of B after Step [3].

Lemma 11.5 (Decay Lemma) *The total number of bad edges produced while building $\mathcal{T}$ is $|B| \leq m_0/2 + d^3 n_0$.*

We begin by bounding the total number of inserts. Recall that n_0 (resp. m_0) denotes the number of vertices (resp. edges) of G_0.

Lemma 11.6 *The total number of inserts in all the heaps is at most* $4m_0$.

Proof: The first insertion of a given edge into a heap occurs during an extension. In fact, all extensions witness exactly m_0 inserts. To bound the number of reinserts, we provide each edge with three credits when it is first inserted. At a currency rate of one credit per reinsert, we show that the credits injected cover the reinserts.

We maintain the following invariant: For any j, any edge in $H(j)$ has two credits; for any i, j and any vertex s outside the chain, the κ edges of $H(i, j)$ incident to s contain a total of $\kappa + 2$ credits (or, of course, 0 if $\kappa = 0$). With its three brand-new credits the first insertion of edge (r, s), which takes place in some $H(i, k)$, easily conforms with the invariants.

Consider the case of a reinsert of (r, s) through retraction. If (r, s) originates from $H(k - 1, k)$ or comes from $H(k)$ but shares s with an edge in $H(k - 1, k)$, then its cluster of edges pointing to s releases at least three credits, ie, $\kappa + 2 +$ (zero or more credits from $H(k)$), for $\kappa > 0$: One pays for the insert into $H(k-1)$, while the other two are enough to maintain the credit invariant for $H(k - 1)$. Otherwise, (r, s) originates from $H(k)$ and has two credits at its disposal, as it is inserted into some $H(i, k)$, where $i < k - 1$. After the insertion, the heap $H(i, k)$ contains more than one edge pointing to s, and so only one credit is needed for the heap as (r, s) moves into it. The remaining credit pays for the insert.

We just revisited the retraction procedure step-by-step and followed the movement of credits alongside. We can do exactly the same for a fusion and cover extensions in a similar fashion. We omit the details. $\square$

The time has come to explain why we need all those heaps. This is elementary but subtle. Intuitively, the difficulty is that deleting corrupted items does not necessarily imply a decrease in corruption. In fact, all edges of G_0 might end up being corrupted. To prevent all of them from becoming bad we must force them to enter into cycles with extension edges before their incident C_z's get contracted. At first this might seem impossible since we have no control over the cycle structure. The trick is to combine three ingredients: (i) Making $\mathcal{T}$ shallow enough; (ii) keeping the $H(\star, \star)$'s sparse; and (iii) using the $H(\star)$'s as overflow heaps. We explain with the following picture.

In the worst case, a typical vertex v ends up with about m/n border edges incident to it. This is much more than the height of $\mathcal{T}$, therefore many of these edges end up sharing the same C_z's. In such clusters most of the edges end up being stored in $H(\star)$. For them, deletions may cause added corruption but not added badness, since the latter can occur only at the very end of an $H(\star)$'s life.

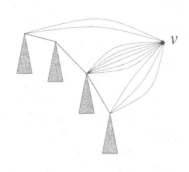

Furthermore, since the heap is entirely dismantled and bad edges are discarded, corruption does not propagate to other heaps the way it does with the $H(\star, \star)$'s through melding. In this regard, the $H(\star, \star)$'s are the true villains, which is why we keep them sparse. But why can't we do with $H(\star)$ only? Because dismantling would be too costly. This is why we make the $H(\star)$'s act as overflow heaps, whose dismantling can always be amortized against discarded edges. To summarize, we need a number of $H(\star)$'s to keep badness low and a number of $H(\star, \star)$'s to keep the running time low.

After the obligatory words of wisdom, the facts. Let $B(i,j)$ be the bad edges in the heap $H(i,j)$ at the time of its disappearance (either via melding or actual destruction). To avoid double-counting, we focus only on the edges of $B(i,j)$ that were not already bad while in $B(i',j')$, for any (i',j') lexicographically greater than (i,j). Actually, we can assume that $i = i'$, since for $i' > i$ all such edges are discarded before they can have a chance to appear in $B(i,j)$. We also have the bad edges from the heaps $H(\star)$. They are easy to handle because, unlike $H(\star, \star)$, these are never melded together: By Theorem 11.1 and Lemma 11.6, the total number of corrupted edges in all the $H(\star)$'s at the time of their destruction is at most $4m_0/c$. Thus, the total number $|B|$ of bad edges satisfies (by abuse of notation, i,j is a shorthand for all pairs (node, descendant) in $\mathcal{T}$):

$$|B| \le 4m_0/c + \sum_{i,j} \left| B(i,j) \setminus \bigcup_{j'>j} B(i,j') \right|. \tag{11.5}$$

We define the *multiplicity* of the heap $H(i,j)$ to be the maximum number of edges in it that share the same endpoint (outside the chain). Melding $H(i,j')$ into $H(i,j)$ does not increase its multiplicity; that is precisely the reason why we keep separate heaps for each pair i,j. An insert into some $H(i,j)$ during an extension sets the multiplicity to one. During a retraction, an insert can increment the multiplicity by at most one, but then the heap is immediately melded into $H(i,j-1)$. It follows that the multiplicity of any $H(\star,\star)$ is at most the height of $\mathcal{T}$.[17]

Any edge in $H(i,l)$ that ends up in $H(i,j)$ passes through all of the intermediate heaps $H(i,l')$ created for l' between i and l (because of fusions, not all l' might occur). So, with the summation sign ranging over the children j' of node j in $\mathcal{T}$, we find that the summand in (11.5) is equal to

$$|B(i,j)| - \sum_{j'} |B(i,j')| +$$

$$\sum_{j'} \# \text{ bad edges deleted from } H(i,j') \text{ during extensions.} \qquad (11.6)$$

The last additive term comes from the fact that the only deletes from $H(i,j')$ are caused by extensions. Indeed, deletes occur after findmins: All the edges sharing the same endpoint with the edge selected by findmin are deleted. As we just observed, there are at most d of them in each $B(i,j)$. There are at most $\binom{d+1}{2} \leq d^2$ heaps $H(\star,\star)$ at any time, so the total number of edges deleted from $H(i,j')$, for all i,j', is at most $d^3 n_0$. In view of (11.6), expanding (11.5) gives us a telescoping sum resulting in

$$|B| \leq 4m_0/c + d^3 n_0 + \sum_{i,i'} |B(i,i')|,$$

where i' denotes any child of node i. The inserts that caused corruption within the $H(i,i')$'s are all distinct, and so, again by Theorem 11.1 and Lemma 11.6, the $|B(i,i')|$'s sum up to at most $4m_0/c$. We conclude that $|B| \leq 8m_0/c + d^3 n_0$. (In fact, we are overcounting.) With c large enough, the decay lemma (Lemma 11.5) is now proven. $\square$

---

[17]Pause a minute to understand why repeated melds into $H(i,j-1)$ cannot keep bringing in more and more edges incident to a given endpoint.

Complexity Analysis

We show that executing $\mathtt{msf}(G, t)$ takes at most $bt(m + d^3(n - 1))$ time, where b is a constant large enough (but arbitrarily smaller than c) and d is any integer such that $n \leq S(t, d)^3$. We prove this by induction on t and n. The basis case $t = 1$ is easy. We have $n \leq S(1, d)^3 = 8d^3$ and the computation takes time $O(n^2) = O(d^3 n) \leq b(m + d^3(n-1))$. So we assume that $t > 1$ and, because of Step [1], that n is large enough.

The Borůvka phases in Step [2] transform G into a graph G_0 with $n_0 \leq n/2^c$ vertices and $m_0 \leq m$ edges. This transformation requires $O(n + m)$ time. The complexity of building $\mathcal{T}$ in Step [3] is dominated by the heap operations. By Lemma 11.6, there are $O(m_0)$ inserts and, hence, $O(m_0)$ deletes and melds. There are $n_0 - 1$ edge extensions, each of them calling up to $O(d^2)$ findmins. Each heap operation takes constant time so computing $\mathcal{T}$ takes $O(m_0 + d^2 n_0)$ time plus the bookkeeping costs of accessing the right heaps at the right time. Naturally, for each node z_j, we maintain a linked list giving access to $H(j)$ and to the $H(i, j)$'s, with the nodes z_i appearing in order of increasing height and pointers linking $H(i, j)$ to z_i. In addition, we need:

1. *v-lists* keeping the border edges incident to each v, sorted along the active path;

2. *z-lists* keeping the border edges incident to each C_z.

The only nontrivial bookkeeping operation is this one (needed in extension and retraction). Given a border edge (u, v) incident to some C_z, find the next z_i up the active path such that C_{z_i} is adjacent to v. This is easily done in constant time.[18]

There is only one difficulty. During a retraction (or fusion), two or more C_z's collapse together. But we cannot link the corresponding z-lists (else edge-to-z is no longer a constant-time operation) nor can we copy one into the other (too costly). So, we modify each z-list into a tree as follows: Partition it into groups of size d, with a remainder group of size $< d$. For each size-d group, create a node of height 1 whose children are the corresponding leaves. The root of the tree is the parent of these nodes and of the leaves associated with the remainder group. The updating cost per C_z is now $O(d)$ plus $1/d$ per border edge. The latter cost can be incurred at most d times by a given edge, since the height of its corresponding z

---

[18] A minor technicality: Since several edges in the v-list might join the same C_z, we might have to walk past them in the list to find z_i. A quick inspection shows that with a little care we can charge the extra cost to the discarded edges in the list.

increases by at least one at every retraction/fusion. This gives a per-edge cost of $O(1)$. There are at most d C_z's involved during a given retraction or extension; so, conservatively, the reconfiguration costs of $O(d)$ add up to $O(d^2 n_0)$. All together, this gives bookkeeping costs of $O(m_0 + d^2 n_0)$.

A z-list is a tree which provides constant-time edge-to-z access with efficient dynamic linking.

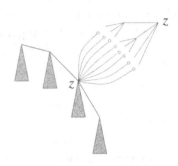

In sum, by choosing b large enough, we ensure that

- **time for Steps [2, 3]** $< \frac{b}{2}(n + m + d^2 n_0)$.

Turning now to Step [4], consider an internal node z of $\mathcal{T}$. If $d_z = 1$, we say that z is *full* if its number n_z of children (ie, # vertices in C_z) is equal to $S(t, 1)^3 = 8$. If $d_z > 1$, the node z is full if $n_z = S(t-1, S(t, d_z-1))^3$ and its children are also full; recall that the condition on n_z is precisely (11.1). As we saw in (11.3), given a full z, the expansion of C_z relative to G_0 has a number N_z of vertices equal to $S(t, d_z)^3$; we do not include the fusion subgraphs in the count. For z not to be full, the construction of C_z must have terminated prematurely, either because a fusion pruned a part of $\mathcal{T}$ including z or, more simply, because the algorithm finished. Therefore, either z is full or else all of its children but (the last) one are. This shows that $N_z \geq (n_z - 1)S(t, d_z - 1)^3$, for all $d_z > 1$. By construction, the number of vertices in $C_z \setminus B$ is at most $S(t-1, S(t, d_z - 1))^3$, and so we can apply the induction hypothesis and bound the time for $\mathtt{msf}(C_z \setminus B, t-1)$ by

$$b(t-1)\left(m_z + S(t, d_z - 1)^3(n_z - 1)\right) \leq b(t-1)(m_z + N_z), \qquad (11.7)$$

where m_z is the number of edges in $C_z \setminus B$. Accounting for fusions, recall that a vertex of C_z may not be the contraction of just one $C_{z'}$, for some child z' of z in $\mathcal{T}$, but also subgraphs of the form C_v, where v is a node pruned from the active path of $\mathcal{T}$ together with its "fusion tree" below. Fusion trees are treated separately, and so the inequality in (11.7) applies to any such v as well. Over all nodes of $\mathcal{T}$ (and all fusion trees), we have $\sum m_z \le m_0 - |B|$ and $\sum N_z \le dn_0$ (at most n_0 vertices per level), so the overall recursion time is bounded by $b(t-1)(m_0 - |B| + dn_0)$:

- **time for Step [4]** $< b(t-1)(m_0 - |B| + dn_0)$.

Finally, Step [5] recurses with respect to the graph $F \cup B$. Its number of vertices is $n_0 < n \le S(t,d)^3$ and F is cycle-free; so, by induction,

- **time for Step [5]** $< bt(n_0 - 1 + |B| + d^3(n_0 - 1))$.

Adding up all of these costs gives a running time at most

$$btm_0 + b\left(\frac{m}{2} - m_0 + |B|\right) + 2btd^3 n_0 + \frac{bn}{2}.$$

By the decay lemma (Lemma 11.5), this is no more than

$$btm - b(m - m_0)\left(t - \frac{1}{2}\right) + 3btd^3 n_0 + \frac{bn}{2}.$$

Finally, using the fact that $n_0 \le n/2^c$, we find that the complexity of $\mathtt{msf}(G,t)$ is bounded by $bt(m + d^3(n-1))$, which completes the proof by induction.

When we call the function $\mathtt{msf}$ for the first time, our choice of d ensures that $d^3 n = O(m)$. As shown below, this implies that $t = O(\alpha(m,n))$. This proves Theorem 11.2 (page 385). $\square$

Lemma 11.7 *If $d = c\lceil (m/n)^{1/3} \rceil$ and $t = \min\{\, i > 0 \mid n \le S(i,d)^3 \,\}$, then*

$$t = O(\alpha(m,n)).$$

Proof: It follows from the definition of Ackermann's function (page 385) that, for $i \ge 1$ and $j \ge 4$,

$$A(3i, j) = A(3i - 1, A(3i, j-1)) > 2^{A(3i, j-1)} = 2^{A(3i-1, A(3i, j-2))}.$$

By using the monotonicity of A, since $A(3i, j-2) \ge j$, we find that

$$A(3i, j) > 2^{A(i,j)}. \tag{11.8}$$

It is easily shown by induction that, for any $u \ge 2, v \ge 3$, $A(u,v) \ge 2^{v+1}$, and so

$$A(3i, j) = A(3i-1, A(3i, j-1)) \ge A(3i-1, 2^j) \ge A(i, 2^j). \tag{11.9}$$

Trivially, $A(u-1, v) \le S(u,v)$, for any $u, v \ge 1$, which implies that

$$S(9\alpha(m,n) + 1, d) \ge A(9\alpha(m,n), d).$$

Therefore, by (11.8) and (11.9) and with $d \geq 4$,

$$S(9\alpha(m,n)+1,d) \quad > \quad 2^{A(3\alpha(m,n),d)} \geq 2^{A(\alpha(m,n),2^d)}$$
$$\geq \quad 2^{A(\alpha(m,n),4\lceil m/n \rceil)} > n,$$

and therefore the smallest t such that $n \leq S(t,d)^3$ satisfies $t \leq 9\alpha(m,n)+1$.
□

11.4 The Soft Heap, Cont'd

We pursue our discussion of the soft heap begun in §11.2. We prove Theorem 11.1 (page 380) and provide a working implementation of the soft heap. A *binomial tree* of rank k is a rooted tree of 2^k nodes. It is formed by the combination of two binomial trees of rank $k-1$, where the root of one becomes the new child of the other root. A soft heap is a sequence of modified binomial trees of distinct ranks, called *soft queues*. The modifications come in two ways:

- A soft queue q is a binomial tree with possibly a few subtrees pruned. The original binomial tree from which q is derived is called its *master tree*. The *rank* of a node of q is the number of children of the corresponding node in the master tree. It is an upper bound on its actual number of children in q. We maintain the following *rank invariant*: The number of children of the root of q is at least $\lfloor \text{rank}(\text{root})/2 \rfloor$.

- A node v may store several items, in fact, a whole *item-list*. The *ckey* of v denotes the common value of all the current keys of the items in item-list(v). It is an upper bound on the original keys. The soft queue is heap-ordered with respect to ckeys, ie, a ckey of a node does not exceed the ckeys of any of its children. We fix an integer parameter $r = 2\lceil \log 1/\varepsilon \rceil + 2$, and we require that all corrupted items be stored at nodes of rank greater than r.

Two soft queues of rank 2 were combined to form one soft queue of rank 3. The light edges are missing, but the ranks of both of their upper nodes are 1 (not 0). The soft queue is heap-ordered with respect to ckeys (indicated in brackets), but not with respect to original keys.

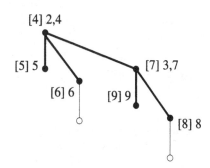

The actual code for soft heaps is quite short (about 100 lines of C). Including it helps resolve ambiguities that might arise from pseudocode. Readers allergic to C code should be able to get by with our plain-English explanations. An item-list is a singly-linked list of items with one field indicating the original value of the key:

```
typedef struct ILCELL
        { int key;
          struct ILCELL *next;
        } ilcell;
```

A node of a soft queue indicates its ckey and its rank in the master tree. Pointers **next** and **child** give access to the children. If there are none, the pointers are NULL. Otherwise, the node is the root of a soft queue of rank 1 less (pointed to by **next**) and the parent of the root of another one of the same type (pointed to by **child**). The figure below indicates the binary-tree representation of a soft queue of rank 2. Item-lists are stored in nodes marked by filled circles. We also include a pointer **il** to give access to the head of the item-list. To facilitate the concatenation of item-lists, we add a pointer **il_tail** to the tail of the list.

```
typedef struct NODE
        { int ckey, rank;
          struct NODE *next, *child;,
          struct ILCELL *il, *il_tail;
        } node;
```

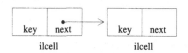

The top structure of the heap[19] consists of a doubly-linked list $h_1, \ldots, h_m$, called the *head-list*. Each *head* h_i has two extra pointers: One (queue) points to the root r_i of a distinct queue, and another (suffix_min) points to the root of minimum ckey among all r_j's ($j \geq i$). We require that rank(r_1) $< \cdots <$ rank(r_m). By extension, the rank of a queue (resp. heap) refers to the rank of its root (resp. r_m). It is stored in the head h_i as the integer variable rank.

```
typedef struct HEAD
        { struct NODE *queue;
          struct HEAD *next, *prev, *suffix_min;
          int rank;
        } head;
```

The head-list points to the roots of the queues and to the minimum ckeys ahead (indicated by the integers 2, 7, 1, 8).

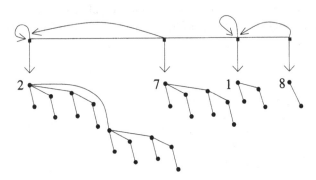

We initialize the soft heap by creating two dummy heads (global variables): header gives access to the head-list while tail, of infinite rank, represents the end of that list. The functions new_head and new_node create and initialize a new head and a new node in the obvious way. The third global variable is the parameter $r = r(\varepsilon)$.

```
head *header, *tail; int r;
header = new_head (); tail = new_head ();
tail->rank = INFTY; header->next = tail; tail->prev = header;
printf (''Enter r: ''); scanf (''%d'', &r);
```

---

[19]For brevity we drop the adjective "soft."

Implementing the Four Operations

To simplify our discussion we bypass the `create` operation and integrate it within `insert`. Similarly, `delete` warrants no discussion, since to delete an item we can simply remove it from its item-list, leaving ckeys untouched. Actual work is required only when `findmin` returns an empty item-list. For this reason, we ignore `findmin` and `delete` altogether and, instead, discuss `deletemin`, the operation that finds an item with minimum current key and deletes it. Our discussion includes an informal explanation of what happens, actual C code, and often a picture. The remainder of this section should basically read itself...

INSERT

To insert a new item, we create an uncorrupted one-node queue, and we meld it into the heap. The code is just a sequence of straightforward initializations:

```
insert (newkey)
int newkey;
{       node *q; ilcell *l;
        l = (ilcell *) malloc (sizeof (ilcell));
        l->key = newkey; l->next = NULL;
        q = new_node (); q->rank = 0; q->ckey = newkey;
        q->il = l; q->il_tail = l;
        meld (q);
}
```

MELD

We discuss the melding of two heaps S and S'. The idea is to break apart the heap of lesser rank, say S' (or either one if the ranks are equal), and meld each of its queues into S separately. To meld a queue of rank k into S, we look for the minimum index i such that $\text{rank}(r_i) \geq k$. (Note that the dummy head `tail` ensures that i always exists.) If $\text{rank}(r_i) > k$, we insert the head right before h_i, instead; otherwise, a conflict arises because all ranks must be distinct. So, we meld the two queues into one of rank $k + 1$, which we do by making the root with the larger key a new child of the other root. If $\text{rank}(r_{i+1}) = k + 1$, a new conflict arises. We repeat the process as long as necessary, like a carry propagation in binary addition. Finally, we update the `suffix_min` pointers between h_1 and the last head visited. When melding not a single queue but a whole heap, the last step can be done at the very end in just one pass through S.

Melding a queue is much like adding numbers in binary notation. Ranks are indicated next to the roots.

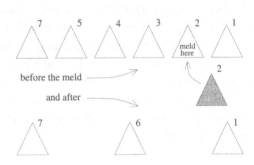

And now the C code: Let q be a pointer (**node *q**) to the soft queue to be melded into the soft heap. First, we scan the head-list until we reach the point at which the melding proper can begin. This leads us to the first head of rank at least that of q, which is denoted by **tohead**. To facilitate the insertion of the new queue, we also remember the preceding head, called **prevhead**.

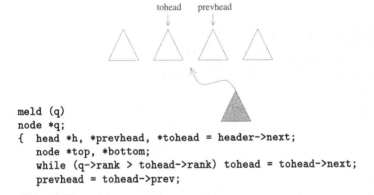

```
meld (q)
node *q;
{  head *h, *prevhead, *tohead = header->next;
   node *top, *bottom;
   while (q->rank > tohead->rank) tohead = tohead->next;
   prevhead = tohead->prev;
```

If there is already a queue of the same rank as q, we perform the carry propagation, as discussed earlier. When merging two queues, we use the variables top and bottom to specify which of the two queues end up at/below the root. We create a new node q pointing to top and bottom. Its item-list is inherited from top, and its rank is 1 plus that of top (ie, top->rank +1). Finally, we update tohead to point to the next element down the head-list.

```
while (q->rank == tohead->rank)
  { if (tohead->queue->ckey > q->ckey)
      { top = q; bottom = tohead->queue; }
    else
      { top = tohead->queue; bottom = q; }
    q = new_node ();
    q->ckey = top->ckey; q->rank = top->rank +1;
    q->child = bottom; q->next = top;
    q->il = top->il; q->il_tail = top->il_tail;
    tohead = tohead->next;
  }
```

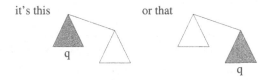

it's this   or that

q   q

Next, we insert the new queue in the list of heads. We use a little hack: If a carry has actually taken place, then the head pointed to by `prevhead->next` is now unused and can be recycled as the head of the new queue; otherwise, we create a new head h. We insert h between `prevhead` and `tohead`. Finally, we call a function, called `fix_minlist(h)`, to update the `suffix_min` pointers.

```
  if (prevhead == tohead->prev) h = new_head ();
  else  h = prevhead->next;
  h->queue = q;  h->rank = q->rank;
  h->prev = prevhead; h->next = tohead;
  prevhead->next = h; tohead->prev= h;
  fix_minlist (h);
}  /* end of meld */
```

```
                    FIX_MINLIST
```

Before calling `fix_minlist(h)`, all `suffix_min` pointers are assumed correct except for those between `header` and h. We update them by walking from h back to `header`, as shown below.

```
fix_minlist (h)
head *h;
{       head *tmpmin;
        if (h->next == tail) tmpmin = h;
        else tmpmin = h->next->suffix_min;
        while (h != header)
          { if (h->queue->ckey < tmpmin->queue->ckey)
                tmpmin = h;
            h->suffix_min = tmpmin;
            h = h->prev;
          }
}
```

SIFT

Our discussion up to this point is unlikely to have drawn from the reader more than a yawn. After all, we have been busy doing mostly the obvious. With `deletemin`, the attention level should perk up. The `suffix_min` pointer at the beginning of the head-list points to the head h with the minimum `ckey` (corrupted or not). The trouble is, the item-list at that node might be empty. Should this happen, we refill the item-list with items taken lower down in the queue pointed to by h: The call `sift(h->queue, h->rank)` attempts to do that by replacing the empty item-list with another one.

Of course, there is no reason why the new item-list might not itself be empty, in which case we repeat the same calls until good things happen. The function `sift` is at the heart of the soft heap, and this is where we focus our discussion. The argument v is the node at which the sifting takes place. We begin with some pseudocode:

---

sift(v)

item-list(v) $\leftarrow T \leftarrow \emptyset$;
if v has no child
      then set ckey(v) to ∞ and return;
1. `sift(v->next)`;
if ckey(v->next) > ckey(v->child)
      then exchange v->next and v->child;
$T \leftarrow T\cup$ item-list(v->next);
if loop condition holds **then** goto 1;
item-list(v) $\leftarrow T$.

The "loop condition" is what makes soft heaps special. Without it, sift *would be indistinguishable from the standard deletemin operation of a binomial tree. This loop condition holds if and only if (i) the goto has not yet been executed during this invocation of* sift *(ie, branching is at most binary), and (ii) the rank of v exceeds the threshold* r *and either it is odd or it exceeds the rank of the highest ranked child of v by at least two.*

The rank condition ensures that no corruption takes places too low in the queue; the parity condition is there to keep branching from occurring too often; and, finally, the last condition guarantees that branching does occur frequently enough. The variable T implements the car-pooling in the concatenation $T \leftarrow T \cup$ item-list(v->next). The cleanup is intended to prune the tree of nodes that have lost their item-lists to ancestors and whose ckeys have been set to ∞.

On the left, the computation tree without a goto. On the right, the goto creates branching.

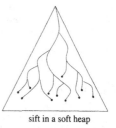

sift in a binomial tree   sift in a soft heap

We now give and explain the code for sift. The item-list at v is worthless and it is effectively emptied out at the beginning. We test whether the node v is a leaf. If so, we bottom out by setting its ckey to infinity (ie, a large integer), which will cause the node to stay at the bottom of the queue. If v is not a leaf, then neither v->next nor v->child is NULL. In fact, this is a general invariant: Both are null or neither one is. This might change temporarily within a call to sift, but it is restored before the call ends.

```
node *sift (v)
node *v;
{  node *tmp;
   v->il = NULL; v->il_tail = NULL;
   if (v->next == NULL && v->child == NULL)
     { v->ckey = INFTY; return v; }
   v->next = sift (v->next);
```

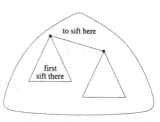

to sift here

first
sift there

The new item-list at `v->next` now might have a large `ckey` that violates the heap ordering. If so, we exchange the children `v->next` and `v->child`, an operation that we call a *rotation*.

```
if (v->next->ckey > v->child->ckey)
    { tmp = v->child;
      v->child = v->next;
      v->next = tmp;
    }
```

Once the children of `v` are in place, we update the various pointers from `v`. In particular, the item-list of `v->next` is passed on to `v`, and so is its `ckey`. Recall that `v->child` is truly a child of `v` in the soft queue, but the node `v->next` is a child of `v` only in the binary-tree implementation of the queue.

```
v->il =  v->next->il;
v->il_tail  =  v->next->il_tail;
v->ckey = v->next->ckey;
```

Next in line, the most distinctive feature of soft heaps: the possibility of sifting twice, ie, of creating a branching process in the recursion tree for `sift`. If the loop condition is satisfied (rank of v is $> r$ and either odd or 2 over the ranks over its children), we sift again. As a result of the sifting, another rotation might be needed to restore the heap ordering.

```
if (v->rank > r &&
        (v->rank % 2 == 1 || v->child->rank < v->rank-1))
    { v->next = sift (v->next);
      if (v->next->ckey > v->child->ckey)
          { tmp = v->child;
            v->child = v->next;
            v->next = tmp;
          }
    }
```

The item-list at `v->next` should now be concatenated with the one at `v`, unless, of course, it is empty or no longer defined. The latter case occurs when `ckey` is infinite at both `v->child` and `v->next`. One can easily verify that this cannot happen after the first sift but only after the second one.

```
    if (v->next->ckey != INFTY && v->next->il != NULL)
       { v->next->il_tail->next = v->il;
         v->il = v->next->il;
         if (v->il_tail == NULL)
            v->il_tail = v->next->il_tail;
         v->ckey = v->next->ckey;
       }
 }  /* end of 2nd sift */
```

is concatenated
to item-list
at root
to
give

We clean up the queue by removing the nodes with infinite `ckeys`. We do *not* update `v->rank` since rank is defined with respect to the master tree. This is where the rank and the number of children can be made to differ. In fact, we ensure that for any node v the ranks of its children (in the binary tree) are always equal, ie, `v->next->rank = v->child->rank`.

```
    if (v->child->ckey == INFTY)
       { if (v->next->ckey == INFTY)
            { v->child = NULL;   v->next = NULL; }
         else
            { v->child = v->next->child;
              v->next = v->next->next; }
       }
    return v;
 }  /* end of sift */
```

rank does not decrease despite pruning

DELETEMIN

The function `deletemin` finds an item with minimum `ckey` and then removes it from the soft heap. The first `suffix_min` pointer takes us to the smallest `ckey`, which is where we want to go, unless of course the corresponding item-list is empty. In that case, we call `sift`—perhaps more than once—to bring items back to the root. Calling `sift` may prune the tree and cause a violation of the rank invariant. (Recall that this invariant stipulates that # children of root $\geq \lfloor \text{rank(root)}/2 \rfloor$.) We count the children of the root; alternatively, we could add a field to keep track of this number.

```
deletemin ()
{  node *sift (), *tmp;
   int min, childcount;  head *h = header->next->suffix_min;
   while (h->queue->il == NULL)      /* outer while loop */
     { tmp = h->queue;  childcount = 0;
       while (tmp->next != NULL)
         { tmp = tmp->next; childcount ++; }
```

The advantage in spotting a rank invariant violation so late in the game is that to fix it is much easier since the root's item-list is empty (else, what would we do with it?) If the rank invariant is violated (ie, childcount $<$ $\lfloor h \rightarrow \text{rank}/2 \rfloor$), then we remove the queue and update the head-list and suffix_min pointers. Then, we *dismantle* the root by remelding back its children.

```
if (childcount < h->rank/2)
    { h->prev->next = h->next;
      h->next->prev = h->prev;
      fix_minlist (h->prev);
      tmp = h->queue;
      while (tmp->next != NULL)
         { meld (tmp->child);
           tmp = tmp->next; }
    }
```

meld back

If the rank invariant holds, we are ready to refill the item-list at the root by calling sift.

```
else
    { h->queue = sift (h->queue);
      if (h->queue->ckey == INFTY)
         { h->prev->next = h->next;
           h->next->prev = h->prev;
           h = h->prev; }
      fix_minlist (h);
    }
h = header->next->suffix_min;
}  /* end of outer while loop */
```

sift

& fix heads

We are now in a position to delete the minimum-key item.

```
      min = h->queue->il->key;
      h->queue->il = h->queue->il->next;
      if (h->queue->il == NULL)
            h->queue->il_tail = NULL;
      return min;
}     /* end of deletemin */
```

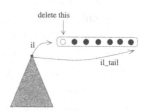

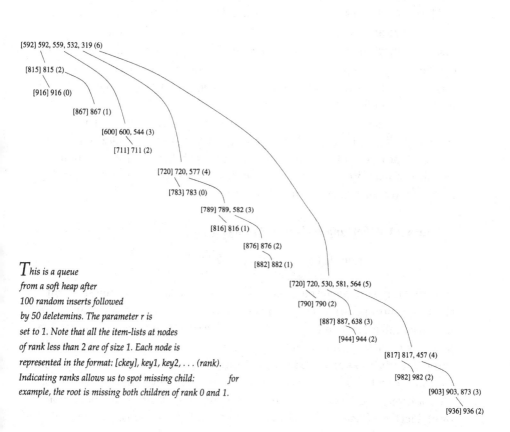

[592] 592, 559, 532, 319 (6)

[815] 815 (2)

[916] 916 (0)

[867] 867 (1)

[600] 600, 544 (3)

[711] 711 (2)

[720] 720, 577 (4)

[783] 783 (0)

[789] 789, 582 (3)

[816] 816 (1)

[876] 876 (2)

[882] 882 (1)

[720] 720, 530, 581, 564 (5)

[790] 790 (2)

[887] 887, 638 (3)

[944] 944 (2)

[817] 817, 457 (4)

[982] 982 (2)

[903] 903, 873 (3)

[936] 936 (2)

This is a queue from a soft heap after 100 random inserts followed by 50 deletemins. The parameter r is set to 1. Note that all the item-lists at nodes of rank less than 2 are of size 1. Each node is represented in the format: [ckey], key1, key2, . . . (rank). Indicating ranks allows us to spot missing child: for *example, the root is missing both children of rank 0 and 1.*

The Error Rate

To prove that the soft heap meets its claims is surprisingly easy. The first point to clarify is the correspondence between a soft queue and its master tree. When no deletion takes place the equivalence is obvious, and it is trivially preserved through inserts and melds. During a sift, we are careful to keep v → next → rank and v → child → rank identical. As we mentioned earlier, it is the enforcement of this equality that causes discrepancies between ranks and numbers of children. But it allows us to think of a rotation as an exchange between soft queues of the same rank (albeit with perhaps missing subtrees). The corresponding master trees, having the same rank, are isomorphic and a rotation thus has no effect on the correspondence. Similarly, the cleanup prunes away subtrees, which in this regard is of no consequence either.

The interesting aspect of the correspondence is that the leaves of the master tree that are missing from the soft queue correspond to items that have migrated upward to join item-lists of nodes of positive rank. Such items can never again appear in leaves of any soft queue. Note that dismantling a node by remelding its children does not contradict this statement, since it merely reconfigures the soft heap.

A sift branches at (roughly) every other recursive call until the rank reaches r. Thus, its execution is modeled by a perfectly balanced binary tree of depth $(k - r)/2$. Going bottom-up, each level in the tree doubles the size of each item-list, so the item-list at v should be of size roughly $2^{(k-r)/2}$. We formalize this intuition below.

Lemma 11.8 *For any node v in the soft heap,*

$$\Big| \text{item-list}\,(v) \Big| \leq \max\Big\{ 1,\, 2^{\lceil \text{rank}(v)/2 \rceil - r/2} \Big\}.$$

Proof: Only sifts have the power to increase the size of an item-list. Furthermore, no operation can by itself cause a violation of the lemma (in fact, some like meld can only, if anything, strengthen the inequality). We prove by induction on their number that sifts cannot violate the lemma. If, through sift(v→next), sift(v) calls itself recursively exactly once, then the loop condition is not satisfied and the item-list of v→next (after possible rotation) migrates to a higher ranking node by itself: The lemma holds by induction. Otherwise, the item-list at v becomes the concatenation of the two item-lists associated with v→next after each call sift(v→next). For this to happen, v→rank must exceed r and one of two conditions must hold: Either v → rank is odd or it exceeds v → child → rank + 1. In the first case, after either recursive call, the rank of v → next is strictly less than v → rank, and by induction the size of either one of the item-lists of v → next is at most

$$\max\left\{ 1 , 2^{\lceil (\mathrm{rank}(v)-1)/2 \rceil - r/2} \right\} = 2^{\lceil (\mathrm{rank}(v)-1)/2 \rceil - r/2} = 2^{\lceil \mathrm{rank}(v)/2 \rceil - 1 - r/2}.$$

Note that the max disappears because $\mathrm{rank}(v) > r$. In the other case, ie, v → rank > v → child → rank + 1, the size of either one of the item-lists of v → next is at most

$$\max\left\{ 1 , 2^{\lceil (\mathrm{rank}(v)-2)/2 \rceil - r/2} \right\} = 2^{\lceil \mathrm{rank}(v)/2 \rceil - 1 - r/2}.$$

This time, the max disappears because r and the rank of v are both even, and so $\mathrm{rank}(v) \geq r + 2$. We have shown that the concatenated item-list is of size at most $2 \times 2^{\lceil \mathrm{rank}(v)/2 \rceil - 1 - r/2}$, and so the lemma holds. □

Lemma 11.9 *During a mixed sequence of operations that includes n inserts, the soft heap contains at most $n/2^{r-3}$ corrupted items at any given time.*

Proof: We begin with a simple observation. Let S be the node set of a binomial tree. It is immediate to prove by induction that

$$\sum_{v \in S} 2^{\mathrm{rank}(v)/2} \leq 2^{k+2} - 3 \cdot 2^{k/2},$$

where k is the rank of the binomial tree. Since S consists of 2^k nodes, it follows that

$$\sum_{v \in S} 2^{\mathrm{rank}(v)/2} \leq 4|S|. \tag{11.10}$$

The ranks of the nodes of a queue q are derived from the corresponding nodes in its master tree q'. So, the set R (resp. R') of nodes of rank greater than r in q (resp. q') is such that $|R| \leq |R'|$. Within q', the nodes of R' account for a fraction at most $1/2^r$ of all the leaves. Summing over all master trees, we find that

$$\sum_{q'} |R'| \leq \frac{n}{2^r}. \tag{11.11}$$

There is no corrupted item at any rank $\leq r$, and so by Lemma 11.8 their total number does not exceed

$$\sum_{q'} \sum_{v \in R'} 2^{(\mathrm{rank}(v)+1-r)/2} = 2 \sum_{q'} \sum_{v \in R'} 2^{(\mathrm{rank}(v)-r-1)/2}. \tag{11.12}$$

Each R' forms a binomial tree by itself, where the rank of node v becomes $\text{rank}(v) - r - 1$. So, by (11.10) and (11.11), the sum in (11.12) is at most

$$\sum_{q'} 8|R'| \leq \frac{n}{2^{r-3}}.$$

□

The Running Time

All operations take constant time, except for `meld` and `sift`, which can take longer. Melds are easily disposed of: Assign one credit per queue. Every time two queues of the same rank are combined into one, one credit can be released to pay for the work involved. Updating `suffix_min` pointers can take time, however. Specifically, carries aside, the cost of melding two soft heaps S and S' is at most the smaller rank of the two (up to a constant factor). The entire sequence of soft heap melds can be modeled as a binary tree $\mathcal{M}$. A leaf z denotes a one-item heap (its cost is 1). An internal node z indicates the melding of two heaps. Since heaps can grow only through melds, the added size of the master trees in the soft heap at z is proportional to the number $N(z)$ of descendants in $\mathcal{M}$. Since the rank of a soft queue is exactly the logarithm of the number of nodes in its master tree, the cost of the meld associated with node z is at most $1 + \log \min\{ N(x), N(y) \}$, where x and y are the left and right children of z. It is a simple exercise to show that adding together all of these costs gives a total melding cost linear in the size of $\mathcal{M}$ or, in other words, $O(n)$.

This analysis implicitly assumes the absence of any node dismantling. We can easily amend it, however, to cover the general case. For the purpose of the analysis, let us regard the remelding caused by dismantling not as heap melds but as queue melds. The benefit is to leave the tree $\mathcal{M}$ unchanged. The dismantle-induced melds associated with a node z of $\mathcal{M}$ reconfigure the soft heap at z by removing some of its nodes and restoring the rank invariant. This can only decrease the value of $N(z)$, so the previous analysis remains correct.

Of course, the queue melds associated with node z now must be accounted for. Dismantling node v causes at most $\text{rank}(v)$ queue melds. From the violation of the rank invariant, we conclude that the node v has at least one missing child of rank $\geq \lceil \text{rank}(v)/2 \rceil$. In the master tree there are at least $2^{\lceil \text{rank}(v)/2 \rceil - 1}$ leaves at or below that child, all of which are gone from the soft queue. We charge the dismantle-induced melds against these leaves and thus observe that melding still takes $O(n)$ time.

Last but not least, we show that the cost of all calls to sift is $O(rn)$. Consider any decreasing sequence of integers. We call an integer m *noticeable* if it is odd or if its successor is less than $m-1$. Clearly, any subsequence of size 2 contains at least one noticeable integer. Now, consider the computation tree corresponding to an execution of sift(v). Noticeable ranks higher than r coincide with nodes where the loop condition holds. So, along any path of size greater than r, at least one branching must occur; in fact, more than that at ranks higher than r. It follows that, excluding the updating of suffix_min pointers, the running time is $O(rC)$, where C is the number of times the loop condition succeeds.

We omit the easy proof by induction that if v is the root of a subtree with fewer than two finite ckeys in the subtree below, the computation tree of sift(v) is of constant size. Conversely, if the subtree contains at least two finite ckeys at distinct nodes, and if the loop condition is satisfied at v, then both calls sift(v→next) bring finite ckeys to the root, which triggers the concatenation of two nonempty item-lists. There can be at most $n-1$ such concatenations, so $C \leq n$ and, as claimed, calls to sift cost a total of $O(rn)$ time.

We have accounted for the costs of updating suffix_min pointers (via fix_minlist(h)) during melds, but not during deletemins. Maintaining the rank invariant makes the cost of suffix_min updating negligible. Indeed, each update takes time proportional to the rank of the queue: (i) If the rank invariant holds, then the updating time is dominated by the cost of the call to sift that precedes fix_minlist(h) (which has already been accounted for); (ii) otherwise, the root v is dismantled, which, as we just saw, releases $2^{\lceil \text{rank}(v)/2 \rceil - 1}$ leaves, against which we can charge the updating cost. By Lemma 11.9, the total number of corrupted items is bounded by $n/2^{r-3}$. Setting $r = 2\lceil \log(1/\varepsilon) \rceil + 2$ proves Theorem 11.1 (page 380). We omit the proof of optimality.[20] □

---

[20] Space optimality requires a few simple modifications.

11.5 Bibliographical Notes

The classical linear selection algorithm mentioned at the beginning was discovered by Blum et al. [48]. The soft heap was invented by Chazelle [72], where Theorem 11.1 (including optimality) is also established. The binomial queue was designed by Vuillemin [318]. Fredman and Tarjan [136] created the Fibonacci heap, which differs from the binomial queue mostly by its ability to support a constant-time decrease-key operation.

The first page of Borůvka's 1926 minimum spanning tree paper [56]. [Courtesy J. Nešetřil]

PRÁCE
MORAVSKÉ PŘÍRODOVĚDECKÉ SPOLEČNOSTI
SVAZEK III., SPIS 3.   1926   SIGNATURA: F 23
BRNO, ČESKOSLOVENSKO.

ACTA SOCIETATIS SCIENTIARUM NATURALIUM MORAVICAE.
TOMUS III., FASCICULUS 3.; SIGNATURA: F 23; BRNO, CECHOSLOVAKIA; 1926.

Dr. OTAKAR BORŮVKA:

O jistém problému minimálním.

V tomto článku podávám řešení následujícího problému:

Budiž dána matice M čísel $r_{\alpha\beta}$ ($\alpha, \beta = 1, 2, \ldots n$; $n \geqslant 2$), až na podmínku $r_{\alpha\alpha} = 0$, $r_{\alpha\beta} = r_{\beta\alpha}$, kladných a vzájemně různých.

Jest vybrati z ní skupinu čísel vzájemně a od nuly různých takovou, aby

1° bylo možno, jsou-li p_1, p_2 libovolná od sebe různá přirozená čísla $\leqslant n$, vybrati z ní skupinu částečnou tvaru

$$r_{p_1 c_1}, \; r_{c_2 c_3}, \; r_{c_3 c_4}, \ldots \ldots r_{c_{q-3} c_{q-1}}, \; r_{c_{q-1} p_2}$$

2° součet jejich členů byl menší než součet členů kterékoliv jiné skupiny čísel vzájemně a od nuly různých, hovíci podmínce 1°.[*]

Řešení. Budiž f_0 libovolné z čísel α a budiž $[f_0 f_1]$ nejmenší z čísel $[f_0 \gamma_0]$ $[\gamma_0 \neq f_0]$. Množství čísel $[f_1 \gamma_1]$ ($\gamma_1 \neq f_0$, f_1) jest pak buď prázdné, anebo nikoliv. V případě prvním položme

$$F = [f_0 f_1],$$

v případě druhém jest nejmenší z čísel $[f_1 \gamma_1]$ buď větší než $[f_0 f_1]$, anebo menší. Je-li větší, položme

$$F = [f_0 f_1],$$

je-li menší, budiž $[f_1 f_2]$ nejmenší z čísel $[f_1 \gamma_1]$. Množství čísel $[f_2 \gamma_2]$ ($\gamma_2 \neq f_0$, f_1, f_2) jest pak buď prázdné anebo nikoliv. V případě prvním položme

$$F \equiv [f_0 f_1], \; [f_1 f_2],$$

v případě druhém jest nejmenší z čísel $[f_2 \gamma_2]$ buď větší než $[f_1 f_2]$, anebo menší. Je-li větší, položme

[*] V dalším značím pro stručnost číslo $r_{\alpha\beta}$ symbolem $[\alpha\beta]$.

Computing a minimum spanning tree optimally is perhaps the oldest open problem in computer science. Borůvka inaugurated the pursuit for an optimal solution in 1926 [56, 147, 239]. The $O(m\alpha(m, n))$ algorithm was discovered by Chazelle [73]. It is the fastest deterministic solution to date. As befits one of the most central problems in combinatorial optimization, the literature on MST is vast. Textbooks usually discuss algorithms attributed to Prim (but also discovered by Jarník [169] in 1930) or Kruskal, both of which run in $O(m \log m)$ time. Improvements to $O(m \log \log m)$ were given independently by Yao [327] and Cheriton and Tarjan [87]. About ten years later, Fredman and Tarjan [136] brought the complexity down to $O(m\beta(m,n))$, where $\beta(m,n)$ is the number of log-iterations necessary to map n to a number less than m/n. In the worst case, $m = O(n)$ and the running time is $O(m \log^* m)$. Soon after, the complexity was further reduced to $O(m \log \beta(m,n))$ by Gabow et al. [140]. More recently, Karger, Klein, and Tarjan [177] discovered a randomized algorithm with linear expected complexity. The algorithm relies on a linear-time procedure for verifying whether a spanning tree is minimal. The ideas behind this procedure go back to Komlós [187] and were developed into a full-fledged algorithm by Dixon, Rauch, and Tarjan [110], which was later simplified by King [182]. With this verification procedure, one can apply a very general result of Levin [196] to produce an optimal algorithm of undetermined complexity. Basically, one enumerates all algorithms by size until the right one is found; see also [154, 172] and a variant based on soft heaps by Pettie and Ramachandran [248]. Whether any of these ideas can be used to obtain a linear-time algorithm for MST or even just to improve Chazelle's method [73] in any way is an intriguing open problem.

All of the algorithms mentioned so far make no assumption on the edge costs. In models where costs are small enough integers one can do better [137]. There is probably little to learn from such models, however, if one's goal is to resolve the MST question and settle what truly is one of the most remarkable open problems of computer science.

Appendix A

Probability Theory

 e review some simple probabilistic facts that are used throughout the text. In particular, we estimate the tails of common
probability distributions, and we discuss a general proof technique for deriving such results. We also mention basic properties of the
entropy of a probability distribution. Again, we remind the reader that all
logarithms are to the base two.

A.1 Common Distributions

Two integer random variables X and Y are said to be *independent* if, for
any k, l,

$$\text{Prob}[\, X = k \text{ and } Y = l \,] = \text{Prob}[X = k]\,\text{Prob}[Y = l]\,.$$

The *conditional probability* of $X = k$ given $Y = l$ is defined by

$$\text{Prob}[\, X = k \,|\, Y = l \,] = \frac{\text{Prob}[\, X = k \text{ and } Y = l \,]}{\text{Prob}[Y = l]}\,. \qquad (A.1)$$

Independence implies that the distribution of X is equal to its conditional
distribution for any value of Y. The lack of symmetry in (A.1) leads to
some interesting formulas, such as *Bayes' rule*:

$$\frac{\text{Prob}[\, X = k \,|\, Y = l \,]}{\text{Prob}[\, Y = l \,|\, X = k \,]} = \frac{\text{Prob}[\, X = k \,]}{\text{Prob}[Y = l]}\,.$$

Exercise: Use Bayes' rule to prove the well-known fact that testing positive for
a rare disease still leaves you unlikely to have the disease in question; this is
assuming, of course, that the test has a small but nonnegligible chance of being
faulty.

430

The notion of independence extends naturally to the case of several random variables: $X_1, \ldots, X_n$ are (mutually) independent if

$$\text{Prob}[X_1 = k_1, \ldots, X_n = k_n] = \text{Prob}[X_1 = k_1] \times \cdots \times \text{Prob}[X_n = k_n].$$

A more general notion is that of *t-wise independence*, meaning that any subset of t variables is mutually independent.

The *expectation* $\mathbf{E}X$ (also called the *mean*) of the random variable X is defined as

$$\sum_{k \geq 0} k \, \text{Prob}[X = k].$$

The equivalent formulation

$$\mathbf{E}X = \sum_{k \geq 1} \text{Prob}[X \geq k]$$

is often useful. Arguably the single most useful fact in probability theory, the expectation operator is linear: given two random variables X and Y, and two reals α, β, regardless of any dependency between the variables,

$$\mathbf{E}[\alpha X + \beta Y] = \alpha \, \mathbf{E}X + \beta \, \mathbf{E}Y.$$

Given two random variables X and Y, the *conditional expectation* of X *given* Y is defined as

$$\mathbf{E}[X \mid Y] = \sum_{k \geq 0} k \, \text{Prob}[X = k \mid Y].$$

Note that it is itself a random variable. It satisfies the identity

$$\mathbf{E}X = \mathbf{E}_Y \, \mathbf{E}_X[X \mid Y],$$

where the subscripts indicate which random variable is being averaged over.

The *variance* $\mathbf{var}X$ of X is defined as $\mathbf{E}[(X - \mathbf{E}X)^2]$. It measures the average square deviation of X from its mean. It is immediate that

$$\mathbf{var}X = \mathbf{E}X^2 - (\mathbf{E}X)^2,$$

which is often a more useful formulation of the variance. If X and Y are independent, it is easy to see that $\mathbf{E}[XY] = (\mathbf{E}X)(\mathbf{E}Y)$, from which it immediately follows that

$$\mathbf{var}[X + Y] = \mathbf{var}X + \mathbf{var}Y.$$

Notice that this is usually *false* if X and Y are not independent; for example, $\mathbf{var}[X + X] = 4 \, \mathbf{var}X$. The *standard deviation* is defined as the square root of the variance. Next, we discuss a few important classes of probability distributions.

Uniform Distribution. Toss a coin in the air. If it comes down without landing on its side, we have what is known as a *Bernoulli trial*. Not assuming that the coin is necessarily fair, the outcome is heads (resp. tails) with some probability p (resp. $1 - p$). If the coin is fair, ie, $p = 1/2$, the distribution is called *uniform*. If we define a random variable X equal to 1 if the outcome is heads and 0 otherwise, we can easily verify that $\mathbf{E}X = p$ and $\mathbf{var}X = p(1 - p)$. It is worth noticing that the assignment $p = 1/2$ corresponding to the uniform distribution maximizes the variance.

Geometric Distribution. Keep tossing the coin as long as the outcome is tails. The number X of tosses (excluding the final "heads" toss) follows a *geometric distribution:* $\text{Prob}[X = k] = p(1 - p)^k$, for $k \geq 0$. The expectation is $\mathbf{E}X = (1 - p)/p$ and the variance is $\mathbf{var}X = (1 - p)/p^2$. If we consider heads to be a success and tails a failure, then the geometric distribution describes the time that it takes to get a first success.

Binomial Distribution. For our next experiment, we toss the coin n times. The number X of heads follows a *binomial distribution*, denoted by $B(n, p)$. Clearly,

$$\text{Prob}[X = k] = \binom{n}{k} p^k (1 - p)^{n-k}.$$

We can write X as a sum $X_1 + \cdots + X_n$ of mutually independent 0/1 variables, so it immediately follows that $\mathbf{E}X = np$ and $\mathbf{var}X = np(1 - p)$. It is often useful to have an upper bound on the higher moments of the distribution.

Lemma A.1 *If X is distributed in $B(n, p)$, then, for any integer $c \geq 0$, $\mathbf{E}X^c \leq (c + np)^c$.*

Proof: To prove the lemma, we establish the slightly stronger bound: $\mathbf{E}(X + j)^c \leq (c + np + j)^c$, for any integer $j \geq 0$. Proceeding by induction on c, we begin with the observation that the cases $c = 0, 1$ are obvious. If the bound holds for some $c \geq 0$ and all $n > 0$ and $j \geq 0$, we find that

$$\begin{aligned}\mathbf{E}(X + j)^{c+1} &= j\,\mathbf{E}(X + j)^c + \sum_{i=1}^{n} p\,\mathbf{E}[\,(X + j)^c \,|\, X_i = 1\,] \\ &\leq np\,\mathbf{E}(Y + j + 1)^c + j\,\mathbf{E}(X + j)^c,\end{aligned}$$

where Y is distributed in $B(n-1, p)$. Thus,

$$\mathbf{E}(X+j)^{c+1} \leq np(c+(n-1)p+j+1)^c + j(c+np+j)^c \leq (c+1+np+j)^{c+1}.$$

$\square$

Depending on the magnitude of the expectation, denoted by μ, the binomial distribution resembles a Poisson distribution (μ small) or a normal distribution (μ large). What does this mean?

Poisson Distribution. Suppose that $\mu = pn$ is at most a constant. Then, $\binom{n}{k} p^k (1-p)^{n-k}$ is dominant for small values of k. It can be approximated as

$$\left(\frac{n^k}{k!}\right) p^k e^{-pn} = e^{-\mu} \frac{\mu^k}{k!}.$$

The distribution $X \in \{0, 1, \dots\}$ such that

$$\mathrm{Prob}[X = k] = e^{-\mu} \frac{\mu^k}{k!}$$

is called *Poisson*. Notice the crucial fact that the sum of the probabilities is, indeed, 1.

Another way to derive these probabilities is via the generating function of the binomial distribution,

$$G(z) \stackrel{\text{def}}{=} \sum_{k \geq 0} \mathrm{Prob}[X = k] z^k = \mathbf{E}\, z^X.$$

Generating functions (or Fourier transforms) usually shine when dealing with sums of independent random variables (because convolutions become products). By independence of the variables,

$$G(z) = (\mathbf{E}\, z^{X_1})^n = (1 + p(z-1))^n,$$

which can be approximated by $P(z) = e^{\mu(z-1)}$. Expanding in Taylor series gives us

$$P(z) = e^{-\mu} \sum_{k \geq 0} \frac{\mu^k}{k!} z^k,$$

which is the generating function of the Poisson distribution. Because $\mathbf{E}\, X = \sum_{k \geq 0} k \mathrm{Prob}[X = k] = P'(1)$, the expectation is μ. We also find that the variance is μ.

Normal Distribution. At the other end of the spectrum, suppose now that p is a constant, and therefore μ is a fixed fraction of n (as in tossing a coin). The (standard) *normal* distribution with mean 0 and variance 1, denoted by $N(0,1)$, is the continuous distribution defined over the entire real line by

$$\text{Prob}[\, X \le t \,] = \frac{1}{\sqrt{2\pi}} \int_{-\infty}^{t} e^{-x^2/2} \, dx.$$

What makes the normal distribution so natural and compelling is that "typically" the average of any set of independent random variables converges towards a normal distribution as n goes to infinity. This applies, in particular, to the binomial distribution when p does not depend on n. More generally, let $y_1, \ldots, y_n$ be n i.i.d. (independent, identically distributed) random variables with mean μ and variance σ^2. Define the normalized average

$$Y_n = \frac{1}{\sigma\sqrt{n}} \sum_{i=1}^{n} (y_i - \mu).$$

By the Central Limit Theorem, as n goes to infinity, $\text{Prob}[\, Y_n \le t \,]$ tends to

$$\frac{1}{\sqrt{2\pi}} \int_{-\infty}^{t} e^{-x^2/2} \, dx.$$

Hypergeometric Distribution. We close this laundry list by briefly mentioning the *hypergeometric* distribution $H(N,n,p)$. Suppose that N urns contain red and blue balls. Each urn contains exactly one ball and a fraction p of the balls are blue. Pick n balls at random without replacement, that is, without refilling the emptied urns. If X is the number of blue balls, we have

$$\text{Prob}[\, X = k \,] = \binom{pN}{k} \binom{N(1-p)}{n-k} \Big/ \binom{N}{n}.$$

The mean and variance of the distribution are, respectively, $\mathbf{E}X = np$ and $\mathbf{var}X = np(1-p)(N-n)(N-1)$.

A.2 Tail Estimates

In the following we assume that X denotes a nonnegative integer random variable. It is often needed to have some estimation of the probability that X exceeds a given quantity. Exact bounds usually do not come in closed form, so we seek asymptotic approximations. Naturally, the more information we have about the distribution, especially in terms of its higher moments, the sharper tail estimates tend to be. Thus, there is a trade-off between sharpness and generality. The weakest but most general tail estimate is *Markov's* inequality: For any $t > 0$,

$$\text{Prob}[\,X \geq t\,] \leq \frac{\mathbf{E}X}{t},$$

which is obtained by straightforward truncation. (Note the importance of the assumption $X \geq 0$.) Observe that

$$\text{Prob}[\,|X - \mathbf{E}X| \geq t\,] = \text{Prob}[(X - \mathbf{E}X)^2 \geq t^2].$$

So, by applying Markov's inequality to $(X - \mathbf{E}X)^2$, we derive *Chebyshev's* inequality,

$$\text{Prob}[\,|X - \mathbf{E}X| \geq t\,] \leq \frac{\text{var}\,X}{t^2}.$$

We can iterate on this idea and generate k-moment tail estimates, which give us a vanishing rate of the order of $1/t^k$: For k even,

$$\text{Prob}[\,|X - \mathbf{E}X| \geq t\,] \leq \frac{\mathbf{E}(X - \mathbf{E}X)^k}{t^k}.$$

One limitation of this technique is that it is often difficult to obtain good estimates on the higher moments of the distribution (not to mention the fact that these moments might not even exist). A useful property, however, is that if X is a sum of random variables, then such tail estimates require only bounded independence.

Lemma A.2 *Let $x_1, \ldots, x_n$ be 2k-wise independent 0/1 random variables, each equal to 1 with probability p. There is an absolute constant β (independent of n, k) such that, given any constant $\alpha > 0$,*

$$\text{Prob}[\,|X - np| \geq \alpha np\,] \leq \left(\frac{\beta}{\alpha^2 np}\right)^k,$$

where $X = x_1 + \cdots + x_n$.

Proof: Setting $t = \alpha np$, we obtain

$$\text{Prob}[\,|X - np| \geq \alpha np\,] \leq \frac{\mathbf{E}(X - \mathbf{E}X)^{2k}}{(\alpha np)^{2k}}.$$

Because expectation is linear, we can expand the numerator to obtain a sum of terms of the form

$$\mathbf{E}\prod_{i \in I}(x_i - p),$$

where $|I| \leq 2k$. By regrouping identical factors together, this becomes $\mathbf{E}\prod_i(x_i - p)^{c_i}$, where the i's are distinct and the c_i's sum up to $2k$. Because there are at most $2k$ factors and the x_i's are $2k$-wise independent, expectation and product commute, and so we obtain $\prod_i \mathbf{E}(x_i - p)^{c_i}$. Note that if $c_i = 1$ for some i, then the product vanishes. Thus, the numerator is the sum of at most $O(n^k)$ nonvanishing terms. Furthermore, every term is the product of at most k factors of the form $\mathbf{E}(x_i - p)^{c_i}$. A simple calculation shows that each such factor is at most $2p$. Thus, the numerator is in $O(pn)^k$ and the proof is complete. $\square$

The Chernoff and Hoeffding Bounds

In the case of the binomial distribution, we can push the method one step further and consider all the moments of the distribution at once to obtain tail estimates that vanish exponentially fast. Assume now that $X = \sum_{i=1}^n x_i$ obeys the binomial distribution $B(n,p)$. *Chernoff's bounds* are tail estimates obtained by applying Markov's inequality to the moment generating function $G(z) = \mathbf{E}\, e^{zX}$.

Lemma A.3 *Let $x_1, \ldots, x_n$ be mutually independent 0/1 random variables, each equal to 1 with probability p. If $X = \sum_{i=1}^n x_i$, then, for any $0 < \delta < 1/2$,*

$$\text{Prob}[\, X < (1 - \delta)pn \,] < e^{-\delta^2 pn/2},$$

and

$$\text{Prob}[\, X > (1 + \delta)pn \,] < e^{-\delta^2 pn/4}.$$

Another useful version of Chernoff's bounds brings upper and lower tails together:

Lemma A.4 *Given $p_1, \ldots, p_n \in [0,1]$, let $X = \sum_{i=1}^n x_i$, where $x_1, \ldots, x_n$ are mutually independent random variables, and for each x_i, $\text{Prob}[x_i = p_i - 1] = p_i$ and $\text{Prob}[x_i = p_i] = 1 - p_i$. For any $\Delta > 0$,*

$$\text{Prob}[\, |X| > \Delta \,] < 2e^{-2\Delta^2/n}.$$

The technique used to prove these lemmas is useful to know, and so we illustrate it by establishing a slightly simpler tail estimate. In the next section, we show that it is part of a general scheme that can be used to derive other interesting bounds.

Let $X = \sum_{i=1}^{n} x_i$ be the sum of n mutually independent random variables $x_i \in \{-1, 1\}$, where

$$\text{Prob}[\, x_i = 1 \,] = \text{Prob}[\, x_i = -1 \,] = \tfrac{1}{2}.$$

Note that the expectation $\mathbf{E}X$ is 0. The standard deviation is $\sqrt{n}$, so by Chebyshev we know that 99% of the distribution is concentrated within $[-10\sqrt{n}, 10\sqrt{n}]$, but a Chernoff-like derivation allows us to tighten the deviation from the mean much more sharply.

Lemma A.5 *If $X = \sum_{i=1}^{n} x_i$ is the sum of n mutually independent random variables x_i uniformly distributed in $\{-1, 1\}$, then, for any $\Delta > 0$,*

$$\text{Prob}[X \geq \Delta] < e^{-\Delta^2/2n}.$$

Proof: By Markov's inequality,

$$\text{Prob}[X \geq \Delta] = \text{Prob}[e^{\lambda X} \geq e^{\lambda \Delta}] \leq e^{-\lambda \Delta} \mathbf{E}\, e^{\lambda X}.$$

Because of the independence among the x_i's,

$$\mathbf{E}\, e^{\lambda X} = (\mathbf{E}\, e^{\lambda x_1})^n = \cosh^n \lambda,$$

with $\cosh \lambda = \tfrac{1}{2}(e^\lambda + e^{-\lambda})$. Expanding $\cosh \lambda$ in Taylor series at 0 gives, for any $\lambda > 0$,

$$\cosh \lambda = \sum_{i \geq 0} \frac{\lambda^{2i}}{(2i)!} < \sum_{i \geq 0} \frac{\lambda^{2i}}{2^i i!} = e^{\lambda^2/2}.$$

Thus, by choosing $\lambda = \Delta/n$, we find

$$\text{Prob}[X \geq \Delta] < e^{n\lambda^2/2 - \lambda \Delta} = e^{-\Delta^2/2n}.$$

$\square$

Note that by symmetry the same upper bound holds for $\text{Prob}[X \leq \Delta]$, if $\Delta \leq 0$.

Finally, we consider the case of the hypergeometric distribution. Recall that X belongs to $H(N, n, p)$ means that X is the number of blue balls found among n (blue or red) balls picked at random without replacement from N urns containing a total of pN blue balls. Interestingly, the following result, known as *Hoeffding's bound*, holds even if we allow replacement.

Lemma A.6 *If $X \in H(N, n, p)$, then, for any $0 \leq t \leq 1$,*

$$\text{Prob}[\, |X/n - p| \geq t \,] \leq 2e^{-2nt^2}.$$

A Unifying View of Tail Estimation

Although the proof techniques above might seem different at first, they all follow the same general pattern. For a change, we consider integrable functions that are not necessarily discrete, but, as shall soon be obvious, the same technique applies to discrete sums almost verbatim. Let $f : \mathbf{R} \mapsto \mathbf{R}^+$ be a function whose tail $\int_{y_0}^{\infty} f(y)\,dy$ we wish to upper-bound. The strategy is to choose an appropriate kernel, ie, a function $g(x,y) \geq 0$, nondecreasing in y, such that the corresponding integral

$$f^*(x) = \int_{-\infty}^{\infty} g(x,y)f(y)\,dy$$

has a *known* upper bound. Since

$$f^*(x) \geq \int_{y_0}^{\infty} g(x,y_0)f(y)\,dy,$$

we have

$$\int_{y_0}^{\infty} f(y)\,dy \leq \min_x \frac{f^*(x)}{g(x,y_0)}\,.$$

It is easy to see that every tail estimate we have obtained so far was derived through this process. For example, in the proof of Lemma A.5, we have $f(y) = \text{Prob}[\,X = \lfloor y \rfloor\,]$ and $g(x,y) = e^{x\lfloor y \rfloor}$.

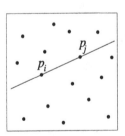

Fig. A.1. $n(i,j) = 8$.

Example: *k-Sets.* Let $p_1,\ldots,p_n$ be a collection of points in E^2. For simplicity, we assume that no two points are vertically aligned and that no three points are collinear. Given $1 \leq i < j \leq n$, let $n(i,j)$ be the number of points lying strictly below the line passing through p_i and p_j, and let f_k denote the number of pairs (i,j) such that $n(i,j) = k$ (Fig. A.1). In the figure, for example, $f_0 = 4$. We apply the previous technique to derive an upper bound on the prefix sum $f_{\leq k} = f_0 + \cdots + f_k$. We define the function g by conducting a randomized experiment. Pick each point p_i with probability r/n (independently), and denote by X the number of edges on the *lower hull* of the chosen points. The lower hull is the portion of the convex hull visible from $y = -\infty$. Obviously,

$$\mathbf{E}X \leq r. \tag{A.2}$$

On the other hand, $\mathbf{E}X$ is the sum of all $\mathbf{E}\chi_{ij}$ ($i < j$), where χ_{ij} is 1 if edge $p_i p_j$ is on the lower hull and 0 otherwise. For χ_{ij} to be 1, both p_i and p_j must be picked (each event has probability r/n) and none of the $n(i,j)$ points below the line passing through $p_i p_j$ must be chosen (each event with probability $1 - r/n$). So,

$$\mathbf{E}X = \sum_{i<j} \left(\frac{r}{n}\right)^2 \left(1 - \frac{r}{n}\right)^{n(i,j)} = \sum_{k \geq 0} f_k g_k,$$

where

$$g_k = \left(\frac{r}{n}\right)^2 \left(1 - \frac{r}{n}\right)^k.$$

Here, $g_k = g(r,k)$ is nonincreasing in k (as opposed to nondecreasing), so the order of truncation must be reversed:

$$\mathbf{E}X \geq \left(\frac{r}{n}\right)^2 \left(1 - \frac{r}{n}\right)^k f_{\leq k}.$$

From the upper bound (A.2), setting $r = n/k$ yields

$$f_{\leq k} = O(kn).$$

Remark: It is easy to see that the upper bound $f_{\leq k}$ is asymptotically tight (take n points on a semicircle).

A.3 Entropy

We mention useful facts about the entropy of a distribution. The *entropy* of a probability distribution $\mathcal{R} = (p_1, \ldots, p_n)$ is defined as

$$H(\mathcal{R}) = \sum_{k=1}^{n} p_k \log \frac{1}{p_k}.$$

In information theory, the motivation behind the notion of entropy is that if a memoryless source generates random symbols with the distribution $\mathcal{R}$, then $H(\mathcal{R})$ is the average number of bits per source symbol needed by the "best possible" data compaction code. More relevant to the purposes of this book is fact that the entropy measures the randomness of $\mathcal{R}$, that is, how many bits are required on average to sample from the distribution. For example, if $\mathcal{R}$ is uniform, its entropy is $\log n$. Skewing the distribution can only lower the entropy.

Lemma A.7 *If $\mathcal{R}$ is a probability distribution of size n, then $H(\mathcal{R}) \leq \log n$, with equality occurring if and only if the distribution is uniform.*

This follows easily from the fact that, given two distributions $p_1, \ldots, p_n$ and $q_1, \ldots, q_n$,

$$\sum_k p_k \log \frac{1}{p_k} \leq \sum_k p_k \log \frac{1}{q_k},$$

with equality if and only if the two distributions are identical. It is also immediate to see that if $p_i \leq p_j < 1$, then decreasing p_i by $\varepsilon > 0$ while increasing p_j by the same amount can only decrease the entropy. The following is a simple corollary of that fact.

Lemma A.8 *Let X be a random variable over a universe of size n. If the probability of no single value of X exceeds 2^{-t}, for some $0 \leq t \leq \log n$, then $H(X) \geq t$.*

We follow common practice here by using $H(X)$ to designate the entropy of the underlying distribution. Let X, Y be two (possibly correlated) random variables. Given a fixed value of $Y = y$, let H_y be the entropy of the distribution $\text{Prob}[X \mid Y = y]$. The *conditional entropy* of X given Y, denoted by $H(X \mid Y)$, is defined as the expectation of the random variable H_Y. It is expressed explicitly as,

$$\sum_y \text{Prob}[Y = y] \sum_x \text{Prob}[X = x \mid Y = y] \log \frac{1}{\text{Prob}[X = x \mid Y = y]}.$$

We have the standard identity,

$$H(X, Y) = H(X) + H(Y \mid X), \tag{A.3}$$

from which it immediately follows that joining distributions cannot decrease the entropy:

$$H(X) \leq H(X, Y). \tag{A.4}$$

A less trivial fact that is that conditioning can never increase the entropy:

$$H(X \mid Y) \leq H(X). \tag{A.5}$$

This implies that the entropy function is *subadditive* in the following sense: If $X_1, \ldots, X_p$ are random variables and $Z = (X_1, \ldots, X_p)$ is the random variable with the joint distribution, then

$$H(Z) \leq H(X_1) + \cdots + H(X_p).$$

Equality holds if the X_i's are mutually independent. Finally, we mention the useful inequality,

$$H(X \mid Y) \leq H(X \mid Z) + H(Z \mid Y). \tag{A.6}$$

A.4 Bibliographical Notes

Section A.1: The reader should consult Feller [130] for an introduction to probability theory. The bound on the higher moments of the binomial distribution (Lemma A.1) is taken from Brönnimann, Chazelle, and Matoušek [57].

Section A.2: The tail estimate on the sum of random variables with bounded independence given in Lemma A.2 is due to Reif and Sen [258]. Chernoff's bounds are named after their originator [88]. The bounds given in Lemmas A.3 and A.4 are borrowed (with some minor modifications) from Alon and Spencer [20]. Lemma A.6 is due to Hoeffding [162]. The proof technique used to bound the number of k-sets is due to Clarkson and Shor [97].

Section A.3: We left out the proofs, most of which are quite elementary and can be found in standard texts on information theory, eg, Blahut [47].

Appendix B

Harmonic Analysis

 e give a quick review of Fourier transforms and Fourier series and discuss the properties that are used in this text. A good introduction to the subject can be found in Dym and McKean [117].

B.1 Fourier Transforms

We begin with the classical setting for Fourier transforms, ie, real (or complex-valued) functions with bounded L^2 norm. Then, we discuss the Fourier transform over abelian groups, restricting ourselves to the two cases, $\mathbf{Z}$ and $(\mathbf{Z}/p\mathbf{Z})^n$. Finally, we briefly review the discrete Fourier transform. We skip over spherical harmonics, which are discussed in the text itself.

Functions in $L^2(\mathbf{R}^d)$

Let $f : \mathbf{R}^d \mapsto \mathbf{C}$ be a function in $L^2(\mathbf{R}^d)$, meaning that $\|f\|_2 < \infty$. Recall that the L^p norm of a function f is defined by

$$\|f\|_p \overset{\text{def}}{=} \left(\int |f(x)|^p \, dx \right)^{1/p}.$$

Throughout this section, the integration domain is assumed to be $\mathbf{R}^d$. The *Fourier transform* of f is defined as

$$\widehat{f}(t) \overset{\text{def}}{=} \int f(x)e^{-2\pi i \langle x,t \rangle} \, dx,$$

where $\langle x, t \rangle = x_1 t_1 + \cdots + x_d t_d$ is the inner product of $x = (x_1, \ldots, x_d) \in \mathbf{R}^d$ and $t = (t_1, \ldots, t_d) \in \mathbf{R}^d$. By the inversion formula we can recover f from its Fourier transform:[1]

$$f(x) = \int \widehat{f}(t) e^{2\pi i \langle x, t \rangle} \, dt.$$

The *Parseval-Plancherel identity* expresses the fact that the Fourier operator preserves the L^2 norm of the function f:

$$\int |f(x)|^2 \, dx = \int |\widehat{f}(t)|^2 \, dt.$$

The *convolution* of f and g is the function

$$(f \star g)(x) \stackrel{\text{def}}{=} \int f(y) g(x - y) \, dy.$$

Note that convolution is commutative. More interesting,

$$\widehat{f \star g} = \widehat{f} \times \widehat{g}.$$

We use the notation δ for the Dirac delta function. Many fascinating things can be said about that function, beginning with the fact that it is not a function, but for us only its behavior on other, smooth functions is what matters; in particular, its integral over an open interval containing the origin is 1 (and zero over the outside) while, more generally, the integral of $f(x)\delta(x)$ is $f(0)$. The Fourier transform of δ is $\widehat{\delta}(t) = 1$. Another useful function is

$$f(x) = \begin{cases} 1 & \text{if } x \in [-r/2, r/2]^d, \\ 0 & \text{else,} \end{cases}$$

whose Fourier transform is

$$\widehat{f}(t) = \prod_{i=1}^{d} \frac{\sin(\pi r t_i)}{\pi t_i}.$$

Abelian Groups

As is well known, every finitely generated abelian group G is isomorphic to a finite direct sum of a free group $\bigoplus^k \mathbf{Z}$ and a number of cyclic groups of the form $\mathbf{Z}_{p_i} = \mathbf{Z}/p_i \mathbf{Z}$, where p_1 divides p_2, which itself divides p_3, etc. The rank k is called the *Betti number* of the group. The finite groups are the *torsion subgroups* and the p_i's are the *torsion coefficients*. We examine the two cases $G = \mathbf{Z}$ and $G = \mathbf{Z}_p^n$. We omit all discussion of characters,

---

[1]Equality of functions is to be understood here in the sense of the Hilbert space $L^2(\mathbf{R}^d)$, ie, $f = g$ if and only if $\|f - g\|_2 = 0$.

invariant subspaces, eigenfunctions, etc, and refer the reader to [117] for all of the necessary mathematical justifications.

Fourier Transform over Z. Let f be a function in $L^2(\mathbf{Z})$. We define its Fourier transform by

$$\widehat{f}(t) = \sum_{n \in \mathbf{Z}} f(n) e^{-2\pi int},$$

for $0 \le t \le 1$. The inversion formula,

$$f(n) = \int_0^1 \widehat{f}(t) e^{2\pi int}\, dt,$$

follows directly from the fact that

$$\int_0^1 e^{2\pi i(n-m)t}\, dt = \begin{cases} 1 & \text{if } n = m, \\ 0 & \text{else.} \end{cases}$$

The Parseval-Plancherel identity becomes

$$\sum_{n \in \mathbf{Z}} |f(n)|^2 = \int_0^1 |\widehat{f}(t)|^2\, dt.$$

As usual, the convolution theorem says that if

$$(f \star g)(n) \overset{\text{def}}{=} \sum_{m \in \mathbf{Z}} f(m) g(n - m),$$

then

$$\widehat{f \star g} = \widehat{f} \times \widehat{g}.$$

Fourier Transform over $\mathbf{Z}_p^n$. Let p be a positive integer and let $\mathbf{Z}_p = \mathbf{Z}/p\mathbf{Z}$ denote the set of integers modulo p. Given $x = (x_1 \ldots, x_n)$ and $y = (y_1 \ldots, y_n)$ in $\mathbf{Z}_p^n$, we use $\langle x, y \rangle$ to denote the inner product $x_1 y_1 + \cdots + x_n y_n$ over $\mathbf{Z}$. Given any function $f : \mathbf{Z}_p^n \mapsto \mathbf{C}$, its Fourier transform is defined by

$$\widehat{f}(t) = \sum_{x \in \mathbf{Z}_p^n} f(x) e^{-2\pi i \langle t, x \rangle / p},$$

where $t \in \mathbf{Z}_p^n$. The inversion formula is

$$f(x) = \frac{1}{p^n} \sum_{t \in \mathbf{Z}_p^n} \widehat{f}(t) e^{2\pi i \langle x, t \rangle / p}.$$

The case $p = 2$ is particularly important for us: $\langle t, x \rangle$ indicates the number of 1-bits common to t and x, and we have

$$\widehat{f}(t) = \sum_{t \in \{0,1\}^n} f(x)(-1)^{\langle t,x \rangle}.$$

Viewing f as a vector in $\mathbf{C}^{2^n}$ (whose coordinates are $f(0), \ldots, f(2^n - 1)$), the transform is simply a linear mapping in finite-dimensional space: Its associated matrix $H = (h_{ij})$ is called a *Hadamard matrix* of order 2^n. Note that h_{ij} is 1 or -1, depending on the parity of the number of common 1's in the binary expansion of i and j. This leads to the following recursive definition of this type of Hadamard matrix[2] of order 2^n, denoted by $H^{(n)}$. Beginning with

$$H^{(1)} = \begin{pmatrix} 1 & 1 \\ 1 & -1 \end{pmatrix},$$

we define $H^{(n)}$ as a tensor product. We take the so-called Kronecker product of $H^{(n-1)}$ with $H^{(1)}$, ie,

$$H^{(n)} = \begin{pmatrix} H^{(n-1)} & H^{(n-1)} \\ H^{(n-1)} & -H^{(n-1)} \end{pmatrix}.$$

For example, we have

$$H^{(3)} = \begin{pmatrix} 1 & 1 & 1 & 1 & 1 & 1 & 1 & 1 \\ 1 & -1 & 1 & -1 & 1 & -1 & 1 & -1 \\ 1 & 1 & -1 & -1 & 1 & 1 & -1 & -1 \\ 1 & -1 & -1 & 1 & 1 & -1 & -1 & 1 \\ 1 & 1 & 1 & 1 & -1 & -1 & -1 & -1 \\ 1 & -1 & 1 & -1 & -1 & 1 & -1 & 1 \\ 1 & 1 & -1 & -1 & -1 & -1 & 1 & 1 \\ 1 & -1 & -1 & 1 & -1 & 1 & 1 & -1 \end{pmatrix}.$$

The Discrete Fourier Transform. Given a vector $x = (x_0 \ldots, x_{n-1})^T$ in $\mathbf{C}^n$, its discrete Fourier transform is its image Fx under the linear map defined by the Fourier matrix F of order n. If ζ denotes $e^{-2\pi i/n}$, then F is of the form:

---

[2]Generally speaking, a Hadamard matrix of order m is any m-by-m matrix with ± 1-elements, whose row vectors are mutually orthogonal; of course, so are its column vectors.

$$F = \begin{pmatrix} 1 & 1 & 1 & \cdots & 1 \\ 1 & \zeta & \zeta^2 & \cdots & \zeta^{n-1} \\ 1 & \zeta^2 & \zeta^4 & \cdots & \zeta^{2(n-1)} \\ \vdots & \vdots & \vdots & \ddots & \vdots \\ 1 & \zeta^{n-1} & \zeta^{2(n-1)} & \cdots & \zeta^{(n-1)^2} \end{pmatrix}.$$

One will recognize the same coefficients used in the Fourier transform over $\mathbf{Z}_n$ defined earlier. From the fact that, for any k,

$$0 = 1 - \zeta^{kn} = (1 - \zeta^k)(1 + \zeta^k + \zeta^{2k} + \cdots + \zeta^{(n-1)k}),$$

it easily follows that the inverse of F is its Hermitian transpose (up to a factor of n):

$$F^{-1} = \frac{1}{n} \begin{pmatrix} 1 & 1 & 1 & \cdots & 1 \\ 1 & \zeta^{-1} & \zeta^{-2} & \cdots & \zeta^{-(n-1)} \\ 1 & \zeta^{-2} & \zeta^{-4} & \cdots & \zeta^{-2(n-1)} \\ \vdots & \vdots & \vdots & \ddots & \vdots \\ 1 & \zeta^{-(n-1)} & \zeta^{-2(n-1)} & \cdots & \zeta^{-(n-1)^2} \end{pmatrix}$$

This implies the Parseval-Plancherel identity, $\|Fx\|_2 = \sqrt{n}\,\|x\|_2$. The convolution theorem says that if $x = (x_0, \ldots, x_{n-1})$ and $y = (y_0, \ldots, y_{n-1})$, then $\widehat{x \star y}$ is the coordinate-wise product $(\widehat{x}_0 \widehat{y}_0, \ldots, \widehat{x}_{n-1} \widehat{y}_{n-1})$, where $x \star y$ is the vector whose i-th coordinate is

$$\sum_{j=0}^{n-1} x_j y_{(i-j) \bmod n} \cdot$$

B.2 Fourier Series

We have defined the Fourier transform for functions whose L^2 norms converge. Thus the theory does not seem to apply to periodic functions, even bounded ones. This would seem an unfortunate technical flaw since such functions are nothing more than infinite copies of functions for which Fourier transforms exist. To salvage the theory we just need to tweak it a little and introduce *Fourier series*. Let $f : \mathbf{R} \mapsto \mathbf{C}$ be a function in $L^2(S^1)$, ie, periodic over $[0, 1]$ and such that $\int_0^1 |f(x)|^2 \, dx < \infty$. Setting

$$\widehat{f}(n) = \int_0^1 f(x) e^{-2\pi i n x} \, dx,$$

for any $n \in \mathbf{Z}$, we define the partial sum

$$S_n = \sum_{|k| \leq n} \widehat{f}(k) e^{2\pi i k x},$$

which is a function of x (note that the sum is over negative *and* positive integers). By the orthogonality of the function basis, we have the *Parseval-Plancherel identity* for Fourier series:

$$\int_0^1 |f(x)|^2 \, dx = \sum_{n \in \mathbf{Z}} |\widehat{f}(n)|^2.$$

It can be shown that the function S_n converges to f in the L^2 sense, ie, $\int_0^1 |S_n(x) - f(x)|^2 \, dx$ tends to 0 as $n \to \infty$. If f is continuously differentiable p times, then the convergence is uniform and, furthermore,[3]

$$\|S_n - f\|_\infty \ll \frac{\sqrt{n}}{n^p}.$$

Thus, high smoothness (a local property) translates into fast uniform convergence (a global property). This is a typical phenomenon in harmonic analysis: Local properties of a function translate into global properties of its Fourier series. As shown below, however, *Gibbs' phenomenon* dashes all hopes of uniform convergence in the presence of discontinuities. We illustrate this difficulty by introducing a function of wide use in discrepancy theory.

Example: *The sawtooth function.* Let $f(x) = \{x\}$, where $\{x\} \stackrel{\text{def}}{=} x \bmod 1$ (Fig. B.1). If $n \neq 0$, integration by parts shows that $\widehat{f}(n) = -1/(2\pi i n)$, while trivially $\widehat{f}(0) = \int_0^1 x \, dx = 1/2$. For nonintegral x, we have

$$\{x\} = \frac{1}{2} - \sum_{n \neq 0} \frac{e^{2\pi i n x}}{2\pi i n} = \frac{1}{2} - \sum_{n > 0} \frac{\sin 2\pi n x}{\pi n}. \tag{B.1}$$

1. Note that the Fourier series estimation for $\{0\}$ is $1/2$, which is the average of the two limits around 0. Outside of integral values of x, however, S_n converges to f pointwise.

2. In the vicinity of any integer x, the convergence is not uniform: Gibbs' phenomenon says that, for any value of n, there is some place near x at which $S_n - f$ deviates from 0 by at least a fixed amount.

---

[3] Recall that $\ll$ and $\gg$ denote $O()$ and $\Omega()$, respectively.

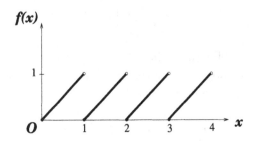

Fig. B.1. The sawtooth function.

3. The sawtooth function yields unexpected results: From the Parseval-Plancherel identity we find that $\int_0^1 f(x)^2\, dx = 1/3$ is equal to

$$\frac{1}{4} + \sum_{n>0} \frac{1}{2\pi^2 n^2},$$

and therefore the sum $\zeta(2) \overset{\text{def}}{=} \sum_{n>0} 1/n^2$ is equal to $\pi^2/6$, which is a nice, nontrivial result.

Even though S_n might fail us at bad points, we should mention that, as long as f is continuous (and periodic over the unit interval), the average $\frac{1}{n} \sum_{k=0}^{n-1} S_k$ converges uniformly to f (Fejér's theorem). We close this brief review by introducing the exponential sum

$$D_n(x) \overset{\text{def}}{=} \sum_{|k| \le n} e^{2\pi i k x},$$

known as the *Dirichlet kernel*. It is a geometric series, so we can write it in closed form:

$$D_n(x) = \frac{e^{\pi i(2n+1)x} - e^{-\pi i(2n+1)x}}{e^{\pi i x} - e^{-\pi i x}},$$

and hence

$$D_n(x) = \frac{\sin \pi(2n+1)x}{\sin \pi x}. \tag{B.2}$$

Note that if x is not a multiple of π, then the denominator is nonzero and the exponential sum $D_n(x)$ remains bounded as $n \to \infty$.

Appendix C

Convex Geometry

 e review basic facts about polytopes, cell complexes, Voronoi diagrams, and duality. These topics are treated in detail in the texts [118, 335] and the collection of surveys [152]. We assume that the reader is familiar with the notion of a linear subspace V of $\mathbf{R}^d$ and its affine version, a flat, ie, $x + V$ ($x \in \mathbf{R}^d$). The affine span of a set is the lowest dimensional flat enclosing it.

C.1 Polytopes

A *convex polyhedron* in $\mathbf{R}^d$ is the intersection of a finite number of closed halfspaces, ie, sets of the form $\{\, x \in \mathbf{R}^d \mid a \cdot x \leq b \,\}$, for $a, b \in \mathbf{R}^d$, where $a \neq 0$ and $a \cdot x$ denotes the inner product of a and x. A *polytope* is a bounded convex polyhedron. Equivalently, it is the convex hull of a finite point set. A *face* of a polytope $P \subset \mathbf{R}^d$ is the relative interior[1] of the intersection of P with a supporting hyperplane.[2] The dimension of a face is that of its affine span. A face of dimension 0 (resp. 1 or $d-1$) is called a *vertex* (resp. *edge* or *facet*). The collection of all faces, ordered by inclusion of their closures, forms a *cell complex* with a lattice structure. It is convenient to represent it by a *facial graph*. Each node denotes a face, and an arc connects two incident faces whose dimensions differ by exactly one. The cell complex and facial graph of an arrangement of hyperplanes can be defined in much the same way. The faces of an arrangement are sometimes called *cells*, or j-cells to specify their dimension.

---

[1] The term "relative" refers to the topology of the affine span of the set in question.
[2] A hyperplane is supporting if it intersects the polytope but not its relative interior.

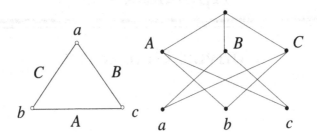

Fig. C.1. The facial graph of the triangle abc.

Given a point $x \in \mathbf{R}^d$ distinct from the origin O, we define its *polar hyperplane* as

$$\{\, y \in \mathbf{R}^d \,|\, x \cdot y = 1 \,\}.$$

This is the hyperplane normal to Ox at distance $1/\|x\|_2$ from the origin and on the same side of O as x. Conversely, we can define the polar point of any hyperplane that does not pass through the origin. This transformation, called *polarity*, creates a duality between points and hyperplanes, in the sense that it preserves incidence relations.

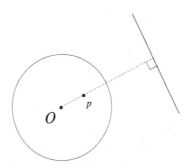

Fig. C.2. The polar hyperplane of a point p: The circle is of unit radius.

We mention in passing another popular dual transform (this one is not involutory, but it is still commonly referred to as a dual transform):

point $(p_1, \ldots, p_d) \mapsto$ *hyperplane* $x_d = p_1 x_1 + \cdots + p_{d-1} x_{d-1} + p_d$

and

$$x_d = p_1 x_1 + \cdots + p_{d-1} x_{d-1} + p_d \mapsto (-p_1, \ldots, -p_{d-1}, p_d).$$

This maps any point different from the origin to a hyperplane not parallel to the x_d-axis. It is easily verified that the transform preserves "above/below" relationships. In other words, a point p above a hyperplane h (in the sense of x_d) maps to a hyperplane above the image point of h.

Returning to the polarity, suppose that the polytope $P \subset \mathbf{R}^d$ contains the origin O in its interior. Then its polar polytope P^*, which is defined as the region containing the origin that is bounded by the polar hyperplanes of the vertices of P, is dual to P. This means that there exists an isomorphism between the facial lattices of P and P^* that maps k-faces (ie, faces of dimension k) to $(d - k - 1)$-faces (Fig. C.3). Algorithmically, this implies that to compute the convex hull of n points ($n > d$), it suffices to place the origin inside the hull (say, inside the simplex formed by any $d + 1$ of the points) and then compute the polytope formed by the intersection of the n halfspaces (containing the origin) derived from the points by polarity.

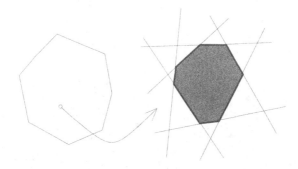

Fig. C.3. Isomorphism between a convex hull and its polar polytope.

If we place n distinct points on the moment curve, their convex hull P forms what is called a *cyclic polytope*. (The moment curve consists of the points $(t, t^2, \ldots, t^d)$, for any $t \in \mathbf{R}$.) The number of $(j - 1)$-faces of P is exactly $\binom{n}{j}$, for any $j \le d/2$. It follows that the combinatorial complexity of the convex hull of n points can be as high as $\Omega(n^{\lfloor d/2 \rfloor})$. By McMullen's Upper Bound Theorem, the cyclic polytope is, in a strong sense, the convex hull of n points with the most faces. The asymptotic version of the theorem is the one of interest to us. It states that the convex hull of n points in $\mathbf{R}^d$ has $O(n^{\lfloor d/2 \rfloor})$ faces and that, for some point configurations, this bound is tight.

C.2 Voronoi Diagrams

Let S be a finite set of points in $\mathbf{R}^d$. For each point $p \in S$, we define its *Voronoi cell* as the region of points closer (in the Euclidean sense) to p than to any other $q \in S$. The collection of cells forms the d-cells of a cell complex.[3] We assume that no $d+1$ points lie in a common hyperplane and no $d+2$ points lie on a common $(d-1)$-sphere. Then, the cell complex is *simple*, meaning that each vertex is incident to exactly $d+1$ edges. A vertex belongs to $d+1$ Voronoi d-cells, and obviously the points of S giving rise to these cells lie on a common $(d-1)$-sphere centered at the vertex, with no point of S strictly inside it. The $d+1$ points on such a sphere form a *Delaunay* simplex. The collection of all Delaunay simplices triangulates the convex hull of S and forms the *Delaunay triangulation*. From our discussion, there is obviously a dual relationship between the Voronoi diagram and the Delaunay triangulation. Computing one provides the other, and vice versa.

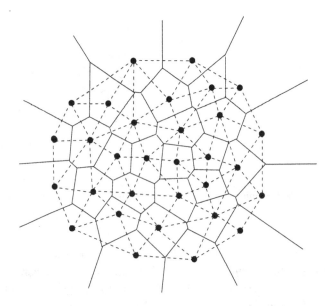

Fig. C.4. A Voronoi diagram and its Delaunay triangulation.

---

[3]For convenience, a cell complex is defined in this book as a collection of nonempty, disjoint, relatively open, polyhedral cells whose closures C_1, C_2, etc, satisfy the usual conditions: (i) The boundary of any C_i is the union of some C_j's; and (ii) any intersection $C_i \cap C_j$ is empty or some C_k. A k-cell is a cell whose affine span has dimension k.

There are several ways to see why Voronoi diagrams (and, hence, Delaunay triangulations) are basically special cases of convex hulls in one dimension higher. A relationship between Voronoi diagrams and polytopes in one dimension higher was exhibited by Brown [59]. Another, discussed below, is due to Edelsbrunner and Seidel [121].

Choose a system of coordinates $(O; x_1, \ldots, x_{d+1})$ for $\mathbf{R}^{d+1}$; given a point $p = (p_1, \ldots, p_d)$ in the hyperplane spanned by the first d axes, map p to the hyperplane $h(p)$ in $\mathbf{R}^{d+1}$ whose equation is:

$$x_{d+1} = 2p_1 x_1 + \cdots + 2p_d x_d - (p_1^2 + \cdots + p_d^2).$$

Geometrically, $h(p)$ is obtained by lifting the point p toward the paraboloid $x_{d+1} = x_1^2 + \cdots + x_d^2$ in the direction x_{d+1}, and taking the tangent hyperplane at that point. Let $h(p)^+$ denote the halfspace bounded by $h(p)$ that contains the origin, ie, lies "above" $h(p)$. Given a set S of n points in the hyperplane $x_{d+1} = 0$, it is easy to show that the faces of the convex polyhedron $\bigcap \{ h(p)^+ \mid p \in S \}$ project normally to the hyperplane $x_{d+1} = 0$ into the faces of the Voronoi diagram of S. Intuitively, this is because the squared distance from p to any point q in the hyperplane $x_{d+1} = 0$ is equal to the vertical drop between $h(p)$ and the vertical projection of q on the paraboloid, ie,

$$(q_1^2 + \cdots + q_d^2) - \left(2p_1 q_1 + \cdots + 2p_d q_d - (p_1^2 + \cdots + p_d^2) \right) = \sum_{1 \le i \le d} (p_i - q_i)^2.$$

We conclude that any algorithm for computing the intersection of halfspaces (or, equivalently, convex hulls) can be automatically converted into one for computing Voronoi diagrams (or, equivalently, Delaunay triangulations) in one dimension lower. The converse is not true.

Bibliography

[1] van Aardenne-Ehrenfest, T. Proof of the impossibility of a just distribution of an infinite sequence of points over an interval, *Proc. Kon. Nederl. Akad. Wetensch.* 48 (1945), 266–271.

[2] Abramowitz, M., Stegun, I.A. *Handbook of Mathematical Functions*, Dover Publications, Inc., 1972.

[3] Agarwal, P.K. Partitioning arrangements of lines I: An efficient deterministic algorithm, *Disc. Comput. Geom.* 5 (1990), 449–483.

[4] Agarwal, P.K. Partitioning arrangements of lines II: Applications, *Disc. Comput. Geom.* 5 (1990), 533–573.

[5] Agarwal, P.K. Geometric partitioning and its applications, in *Computational Geometry: Papers from the DIMACS Special Year*, eds., Goodman, J.E., Pollack, R., Steiger, W., Amer. Math. Soc., 1991.

[6] Agarwal, P.K., Erickson, J. Geometric range searching and its relatives, in *Advances in Discrete and Computational Geometry*, eds. Chazelle, B., Goodman, J.E., Pollack, R., *Contemporary Mathematics* 223, Amer. Math. Soc., 1999, pp. 1–56.

[7] Ajtai, M. A lower bound for finding predecessors in Yao's cell probe model, *Combinatorica* 8 (1988), 235–247.

[8] Ajtai, M., Fredman, M., Komlós, J. Hash functions for priority queues, *Information and Control* 63 (1984), 217–225.

[9] Ajtai, M., Komlós, J., Szemerédi, E. Deterministic simulation in LOGSPACE, *Proc. 19th Annual ACM Symp. Theory Comput.* (1987), 132–140.

[10] Alexander, R. Geometric methods in the study of irregularities of distribution, *Combinatorica* 10 (1990), 115–136.

[11] Alexander, R. Principles of a new method in the study of irregularities of distribution, *Invent. Math.* 103 (1991), 279–296.

[12] Alon, N. Eigenvalues and expanders, *Combinatorica* 6 (1986), 83–96.

[13] Alon, N., Babai, L., Itai, A. A fast and simple randomized parallel algorithm for the maximal independent set problem, *J. Algorithms* 7 (1986), 567–583.

[14] Alon, N., Bárány, I., Füredi, Z., Kleitman, D. Point selections and weak ε-nets for convex hulls, *Combinatorics, Probability and Computing* 3 (1992), 189–200.

[15] Alon, N., Goldreich, O., Hastad, J., Peralta, R. Simple constructions of almost k-wise independent random variables, *J. Random Structures and Algorithms* 3 (1992), 289–304.

[16] Alon, N., Karchmer, M., Wigderson, A. Linear circuits over GF(2), *SIAM J. Comput.* 19 (1990), 1064–1067.

[17] Alon, N., Mansour, Y. ε-discrepancy sets and their application for interpolation of sparse polynomials, *Inform. Process. Lett.* 54 (1995), 337–342.

[18] Alon, N., Milman, V.D. Eigenvalues, expanders and superconcentrators, *Proc. 25th Annual IEEE Symp. Found. Comput. Sci.* (1984), 320–322.

[19] Alon, N., Rónyai, L., Szabó, L. Norm-graphs: variations and applications, *J. Combinatorial Theory B* 76 (1999), 280–290.

[20] Alon, N., Spencer, J.H. *The Probabilistic Method*, Wiley-Interscience, 1992.

[21] Andreev, A.E., Clementi, A.E.F., Rolim, J.D.P. A new general derandomization method, *J. ACM* 45 (1998), 179–213.

[22] Andreev, A.E., Clementi, A.E.F., Rolim, J.D.P., Trevisan, L. Weak random sources, hitting sets, and BPP simulations, *SIAM J. Comput.* 28 (1999), 2103–2116.

[23] Armoni, R., Saks, M., Wigderson, A., Zhou, S. Discrepancy sets and pseudorandom generators for combinatorial rectangles, *Proc. 37th Annual IEEE Symp. Found. Comput. Sci.* (1996), 412–421.

[24] Artin, M. *Algebra*, Prentice Hall, 1991.

[25] Assouad, P. Densité et dimension, *Annales de l'Institut Fourier* 33 (1983), 233–282.

[26] Babai, L., Frankl, P., Simon, J. Complexity classes in communication complexity theory, *Proc. 27th Annual IEEE Symp. Found. Comput. Sci.* (1986), 337–347.

[27] Baker, R.C. On irregularities of distribution, *Bull. London Math. Soc.* 10 (1978), 289–296.

[28] Barret, W.W., Jarvis, T.J. Spectral properties of a matrix of Redheffer, *Linear Algebra Appl.* 162-164 (1992), 673–683.

[29] Beame, P.W., Fich, F.E. On searching sorted Lists: a near-optimal lower bound, *Tech. Rep. UW-CSE-97-09-02*, Univ. Washington, 1997.

[30] Beame, P.W., Fich, F.E. Optimal bounds for the predecessor problem, *Proc. 31st Annual ACM Symp. Theory Comput.* (1999), 295–304.

[31] Beck, J. Roth's estimate of the discrepancy of integer sequences is nearly sharp, *Combinatorica* 1 (1981), 319–325.

[32] Beck, J. van der Waerden and Ramsey-type games, *Combinatorica* 2 (1982), 103–116.

[33] Beck, J. On a problem of K.F. Roth concerning irregularities of point distribution, *Invent. Math.* 74 (1983), 477–487.

[34] Beck, J. Sums of distances between points on a sphere – an application of the theory of irregularities of distribution to discrete geometry, *Mathematika* 31 (1984), 33–41.

[35] Beck, J. Irregularities of distribution, I, *Acta Math.* 159 (1987), 1–49.

[36] Beck, J. Quasi-random 2-colorings of point sets, *Random Structures and Algorithms* 2 (1991), 289–302.

[37] Beck, J., Chen, W.W.L. *Irregularities of Distribution*, Cambridge Tracts in Mathematics, 89, Cambridge University Press, 1987.

[38] Beck, J., Fiala, T. "Integer-making" theorems, *Discrete Applied Mathematics* 3 (1981), 1–8.

[39] Beck, J., Sós, V.T. Discrepancy theory, in *Handbook of Combinatorics*, Chap. 26, eds., Graham, R.L., Grötschel, M., Lovász, L., North-Holland, 1995, pp. 1405–1446.

[40] Bednarchak, D., Helm, M. A note on the Beck-Fiala theorem, *Combinatorica* 17 (1997), 147–149.

[41] Behnke, H. Über die Verteilung von Irrationalitäten mod 1, *Abh. Math. Semin. Univ. Hamburg* 1 (1922), 252–267.

[42] Behnke, H. Zur Theorie der diophantischen Approximationen I., *Abh. Math. Semin. Univ. Hamburg* 3 (1924), 261–318.

[43] Bellare, M., Goldreich, O., Goldwasser, S. Randomness in interactive proofs, *Computational Complexity* 3 (1993), 319–351.

[44] Berger, M. *Geometry II*, Springer-Verlag, 1987.

[45] Biggs, N. *Algebraic Graph Theory*, Cambridge Mathematical Library, Cambridge University Press, 2nd Ed., 1993.

[46] Biggs, N., Anthony, M.H.G. *Computational Learning Theory: An Introduction*, Cambridge Tracts in Theoretical Computer Science, No. 30, Cambridge University Press, 1992.

[47] Blahut, R.E. *Principles and Practice of Information Theory*, Addison-Wesley, 1987.

[48] Blum, M., Floyd, R.W., Pratt, V., Rivest, R.L., Tarjan, R.E. Time bounds for selection, *Journal of Computer and System Sciences* 7 (1973), 448–461.

[49] Blum, M, Karp, R.M., Vornberger, O., Papadimitriou, C.H., Yannakakis, M. The complexity of testing whether a graph is a superconcentrator, *Inform. Process. Lett.* 13 (1981), 164–167.

[50] Blum, M., Micali, S. How to generate cryptographically strong sequences of pseudo-random bits, *SIAM J. Comput.* 13 (1984), 850–864.

[51] Blumer, A., Ehrenfeucht, A., Haussler, D., Warmuth, M. Learnability and the Vapnik-Chervonenkis dimension, *J. ACM* 36 (1989), 929–965.

[52] Boissonnat, J.-D, Yvinec, M. *Algorithmic Geometry*, Cambridge University Press, 1998.

[53] Bollobás, B. *Modern Graph Theory*, Graduate Texts in Mathematics, 184, Springer-Verlag, 1998.

[54] Bollobás, B., Chung, F.R.K., Diaconis, P. (eds), *Probabilistic Combinatorics and Its Applications* Proc. Symposia in Applied Mathematics, Vol 44, Amer. Math. Soc., 1992.

[55] Borodin, A., Ostrovsky, R., Rabani, Y. Lower Bounds for High Dimensional Nearest Neighbor Search and Related Problems, *Proc. 31st Annual ACM Symp. Theory Comput.* (1999), 312–321.

[56] Borůvka, O. O jistém problému minimálním, *Práce mor. přírodověd. spol. v Brně* III, 3 (1926), 37–58. (In Czech.) English Translation in: "Otakar Borůvka on Minimum Spanning Tree Problem", by J. Nešetřil, E. Milková, H. Nešetřilová, to appear in *DIMATIA Surveys*, North Holland 2000.

[57] Brönnimann, H., Chazelle, B., Matoušek, J. Product range spaces, sensitive sampling, and derandomization, *SIAM J. Comput.* 28 (1999), 1552–1575.

[58] Brönnimann, H., Chazelle, B., Pach, J. How hard is halfspace range searching, *Disc. Comput. Geom.* 10 (1993), 143–155.

[59] Brown, K.Q. Voronoi diagrams from convex hulls, *Inform. Process. Lett.* 9 (1979), 223–228.

[60] Carter, J.L., Wegman, M.N. Universal classes of hash functions, *J. Comput. and System Sciences* 18 (1979), 143–154.

[61] Capoyleas, V. An almost linear upper bound for weak ε-nets of points in convex position, manuscript, 1992.

[62] Cassels, J.W.S. *Lectures on Elliptic Curves*, Cambridge University Press, 1991.

[63] Chakrabarti, A., Chazelle, B., Gum, B., Lvov, A. A lower bound on the complexity of approximate nearest-neighbor searching on the Hamming cube, *Proc. 31st Annual ACM Symp. Theory Comput.* (1999), 305–311.

[64] Chari, S., Rohatgi, P., Srinivasan, A. Improved algorithms via approximation of probability distributions, *Proc. 26th Annual ACM Symp. Theory Comput.* (1994), 584–592. Also, TR97-01, DIMACS, 1997.

[65] Chazelle, B. Lower bounds on the complexity of polytope range searching, *J. Amer. Math. Soc.* 2 (1989), 637–666.

[66] Chazelle, B. Lower bounds for orthogonal range searching: I. The reporting case, *J. ACM* 37 (1990), 200–212.

[67] Chazelle, B. Lower bounds for orthogonal range searching: II. The arithmetic model, *J. ACM* 37 (1990), 439–463.

[68] Chazelle, B. An unbiased greedy algorithm for low discrepancy, manuscript, 1992.

[69] Chazelle, B. Cutting hyperplanes for divide-and-conquer, *Disc. Comput. Geom.* 9 (1993), 145–158.

[70] Chazelle, B. An optimal convex hull algorithm in any fixed dimension, *Disc. Comput. Geom.* 10 (1993), 377–409.

[71] Chazelle, B. Lower bounds for off-line range searching, *Disc. Comput. Geom.* 17 (1997), 53–65.

[72] Chazelle, B. The soft heap: an approximate priority queue with optimal error rate, *J. ACM* 47 (2000), 1012–1027,

[73] Chazelle, B. A minimum spanning tree algorithm with inverse-Ackermann type complexity, *J. ACM* 47 (2000), 1028–1047.

[74] Chazelle, B. A spectral approach to lower bounds with applications to geometric searching, *SIAM J. Comput.* 27 (1998), 545–556.

[75] Chazelle, B. Discrepancy bounds for geometric set systems with square incidence matrices, in *Advances in Discrete and Computational Geometry*, eds. Chazelle, B., Goodman, J.E., Pollack, R., *Contemporary Mathematics* 223, Amer. Math. Soc., 1999, pp. 103–107.

[76] Chazelle, B. Geometric searching over the rationals, *Proc. 7th Annual Euro. Symp. Algo.* (1999), 354–365.

[77] Chazelle, B., Edelsbrunner, H., Grigni, M., Guibas, L.J., Sharir, M., Welzl, E. Improved bounds on weak ε-nets for convex sets, *Disc. Comput. Geom.* 13 (1995), 1–15.

[78] Chazelle, B., Friedman, J. A deterministic view of random sampling and its use in geometry, *Combinatorica* 10 (1990), 229–249.

[79] Chazelle, B., Lvov, A. A trace bound for the hereditary discrepancy, *Disc. Comput. Geom.* (2001), to appear. Prelim. version in *Proc. 16th Annual ACM Symp. Comput. Geom.* (2000), 64–69.

[80] Chazelle, B., Lvov, A. The discrepancy of boxes in higher dimension, *Disc. Comput. Geom.* 25 (2001), 519–524.

[81] Chazelle, B., Matoušek, J. On linear-time deterministic algorithms for optimization problems in fixed dimension, *J. Algorithms* 21 (1996), 579–597.

[82] Chazelle, B., Matoušek, J., Sharir, M. An elementary approach to lower bounds in geometric discrepancy, *Disc. Comput. Geom.* 13 (1995), 363–381.

[83] Chazelle, B., Rosenberg, B. The complexity of computing partial sums off-line, *Int. J. Comput. Geom. and Appl.* 1 (1991), 33–45.

[84] Chazelle, B., Rosenberg, B. Simplex range reporting on a pointer machine, *Comput. Geom.: Theory and Appl.* 5 (1996), 237–247.

[85] Chazelle, B., Sharir, M., Welzl, E. Quasi-optimal upper bounds for simplex range searching and new zone theorems, *Algorithmica* 8 (1992), 407–429.

[86] Chazelle, B., Welzl, E. Quasi-optimal range searching in spaces of finite VC-dimension, *Disc. Comput. Geom.* 4 (1989), 467–489.

[87] Cheriton, D., Tarjan, R.E. Finding minimum spanning trees, *SIAM J. Comput.* 5 (1976), 724–742.

[88] Chernoff, H. A measure of asymptotic efficiency for tests of a hypothesis based on the sum of observations, *Annals of Mathematical Statistics* 23 (1952), 493–507.

[89] Chor, B., Goldreich, O. Unbiased bits from sources of weak randomness and probabilistic communication complexity, *SIAM J. Comput.* 17 (1988), 230–261.

[90] Chor, B., Goldreich, O. On the power of two-point sampling, *J. of Complexity* 5 (1989), 96–106.

[91] Chung, F.R.K. The Laplacian of a hypergraph, in *Expanding Graphs*, ed. Friedman, J., Dimacs Series 1, Amer. Math. Soc., 1993, pp. 21–36.

[92] Chung, F.R.K. Spectral Graph Theory, CBMS Regional Conference Series in Mathematics, Vol. 92, Amer. Math. Soc., Providence, 1997.

[93] Clarkson, K.L. Linear programming in $O(n \times 3^{d^2})$ time, *Inform. Process. Lett.* 22 (1986), 21–24.

[94] Clarkson, K.L. New applications of random sampling in computational geometry, *Disc. Comput. Geom.* 2 (1987), 195–222.

[95] Clarkson, K.L. Randomized geometric algorithms, in *Computing in Euclidean Geometry*, eds., Du, D.-Z., Kwang, F.K., Lecture Notes Series on Comp. 1, 1992, World Scientific, pp. 117–162.

[96] Clarkson, K.L. Las Vegas algorithms for linear and integer programming when the dimension is small, *J. ACM* 42 (1995), 488–499.

[97] Clarkson, K.L., Shor, P.W. Applications of random sampling in computational geometry, II, *Disc. Comput. Geom.* 4 (1989), 387–421.

[98] Cohen, A., Wigderson, A. Dispersers, deterministic amplification, and weak random sources, *Proc. 30th Annual IEEE Symp. Found. Comput. Sci.* (1989), 14–19.

[99] Cormen, T.H., Leiserson, C.E., Rivest, R.L. *Introduction to Algorithms*, MIT Press/McGraw-Hill, 1990.

[100] Cornell, G., Silverman, J.H., Stevens, G. (eds.). *Modular Forms and Fermat's Last Theorem*, Springer, 1997.

[101] Coxeter, H.M.S. *Non-Euclidean Geometry*, University of Toronto Press, 1942.

[102] Dantzig, D.B. *Linear Programming and Extensions*, Princeton University Press, 1963.

[103] Danzer, L., Laugwitz, D., Lenz, H. Über das Löwnersche Ellipsoid und seine Anlagen unter dem einem Eikörper einbeschriebenen Ellipsoid, *Archiv der Mathematik* 8 (1957), 214–219.

[104] Daubechies, I. *Ten Lectures on Wavelets*, CBMS-NSF, No. 61, 1992.

[105] Davenport, H. Note on irregularities of distribution, *Mathematika* 3 (1956), 131–135.

[106] Deligne, P. Formes modulaires et représentations ℓ-adiques, Séminaire Bourbaki, 355 (1968/69), Lecture Notes in Mathematics, 179, Springer-Verlag, Berlin, 1971, pp. 139–172.

[107] Deligne, P. La conjecture de Weil I, *Publications de l'I.H.E.S.* 43 (1974), 273–307.

[108] Diaconis, P., Stroock, D. Geometric bounds for eigenvalues of Markov chains, *Ann. Appl. Prob.* 1 (1991), 36–61.

[109] Dickson, L.E. Arithmetic of quaternions, *Proc. London Mathematical Soc.* 2 (1922), 225–232.

[110] Dixon, B., Rauch, M., Tarjan, R.E. Verification and sensitivity analysis of minimum spanning trees in linear time, *SIAM J. Comput.* 21 (1992), 1184–1192.

[111] Drmota, M.,Tichy, R.F., *Sequences, Discrepancies and Applications*, Lecture Notes in Mathematics, Vol. 1651, Springer, 1997.

[112] Dudley, R.M. Central limit theorems for empirical measures, *Ann. Probability* 6 (1978), 899–929.

[113] Dunford, N., Schwartz, J.T. *Linear Operators*, Interscience, I, II, III, 1958, 1963, 1971.

[114] Dyer, M.E. On a multidimensional search technique and its application to the Euclidean one-centre problem, *SIAM J. Comput.* 15 (1986), 725–738.

[115] Dyer, M.E. A class of convex programs with applications to computational geometry, *Proc. 8th Annual ACM Symp. Comput. Geom.* (1992), 9–15.

[116] Dyer, M.E., Frieze, A.M. A randomized algorithm for fixed-dimensional linear programming, *Mathematical Programming* 44 (1989), 203–212.

[117] Dym, H., McKean, H.P. *Fourier Series and Integrals*, Probability and Mathematical Statistics, Vol. 14, Academic Press, 1972.

[118] Edelsbrunner, H. *Algorithms in Combinatorial Geometry*, Springer, 1987.

[119] Edelsbrunner, H., Guibas, L.J., Herschberger, J., Seidel, R., Sharir, M., Snoeyink, J., Welzl, E. Implicitly representing arrangements of lines or segments, *Disc. Comput. Geom.* 4 (1989), 433–466.

[120] Edelsbrunner, H., Mücke, E. Simulation of simplicity: a technique to cope with degenerate cases in geometric algorithms, *ACM Trans. Graphics* 9 (1990), 66–104.

[121] Edelsbrunner, H., Seidel, R. Voronoi diagrams and arrangements, *Disc. Comput. Geom.* 1 (1986), 25–44.

[122] Edelsbrunner, H., Seidel, R., Sharir, M. On the zone theorem for hyperplane arrangements, *SIAM J. Comput.* 22 (1993), 418–429.

[123] Edelsbrunner, H., Welzl, E. Halfplanar range search in linear space and $O(n^{0.695})$ query time, *Inform. Process. Lett.* 23 (1986), 289–293.

[124] Erdős, P., Selfridge, J.L. On a combinatorial game, *J. Combinatorial Theory B* 14 (1973), 298–301.

[125] Erickson, J. New lower bounds for Hopcroft's problem, *Disc. Comput. Geom.* 16 (1996), 389–418.

[126] Erickson, J. New lower bounds for halfspace emptiness, *Proc. 37th Annual IEEE Symp. Found. Comput. Sci.* (1996), 472–481.

[127] Even, G., Goldreich, O., Luby, M., Nisan, N., Veličković, B. Approximations of general independent distributions, *Proc. 24th Annual ACM Symp. Theory Comput.* (1992), 10–16.

[128] Even, G., Goldreich, O., Luby, M., Nisan, N., Veličković, B. Efficient approximation of product distributions, *Random Structures & Algorithms* 13 (1998), 1–16.

[129] Fang, K.-T., Wang, Y. *Number-Theoretic Methods in Statistics*, Chapman and Hall, London, 1994.

[130] Feller, W. *An introduction to Probability Theory and Its Applications*, Vol.1, John Wiley & Sons, 3rd Ed., 1968.

[131] Fenchel, W. *Elementary Geometry in Hyperbolic Space*, de Gruyter Studies in Mathematics, 11, 1989.

[132] Fortune, S. A sweepline algorithm for Voronoi diagrams, *Algorithmica* 2 (1987), 153–174.

[133] Fredman, M.L. A lower bound on the complexity of orthogonal range queries, *J. ACM* 28 (1981), 696–705.

[134] Fredman, M.L. *Lower bounds on the complexity of some optimal data structures*, *SIAM J. Comput.* 10 (1981), 1–10.

[135] Fredman, M.L., Komlós, J., Szemerédi, E. Storing a sparse table with $O(1)$ worst case access time, *J. ACM* 31 (1984), 538–544.

[136] Fredman, M.L., Tarjan, R.E. Fibonacci heaps and their uses in improved network optimization algorithms, *J. ACM* 34 (1987), 596–615.

[137] Fredman, M.L., Willard, D.E. Trans-dichotomous algorithms for minimum spanning trees and shortest paths, *J. Comput. and System Sci.* 48 (1993), 424–436.

[138] Friedman, A. *Foundations of Modern Analysis*, Dover Publications, Inc., 1982.

[139] Gabber, O., Galil, Z. Explicit constructions of linear-sized superconcentrators, *J. Comput. and System Sci.* 22 (1981), 407–420.

[140] Gabow, H.N., Galil, Z., Spencer, T., Tarjan, R.E. Efficient algorithms for finding minimum spanning trees in undirected and directed graphs, *Combinatorica* 6 (1986), 109–122.

[141] Gärtner, B. A subexponential algorithm for abstract optimization problems, *SIAM J. Comput.* 24 (1995), 1018–1035.

[142] Gärtner, B., Welzl, E. Linear programming – randomization and abstract frameworks, *Proc. 13th Annual Symp. Theoret. Aspects Comput. Sci.* LNCS, 1046, Springer-Verlag, 1996, pp. 669–688.

[143] Goldreich, O. *Modern Cryptography, Probabilistic Proofs and Pseudorandomness*, Algorithms and Combinatorics, Vol 17, Springer-Verlag, 1998.

[144] Goldwasser, S., Kilian, J. Almost all primes can be quickly certified, *Proc. 18th Annual ACM Symp. Theory Comput.* (1986), 316–329.

[145] Graham, R.L. An efficient algorithm for determining the convex hull of a planar point set, *Inform. Process. Lett.* 1 (1972), 132–133.

[146] Graham, R.L., Grötschel, M., Lovász, L. (eds.). *Handbook of Combinatorics*, North-Holland, 1995.

[147] Graham, R.L., Hell, P. On the history of the minimum spanning tree problem, *Ann. Hist. Comput.* 7 (1985), 43–57.

[148] Graham, R.L., Patashnik, O., Knuth, D.E. *Concrete Mathematics : A Foundation for Computer Science*, Addison-Wesley, 1994.

[149] Graham, R.L., Rothschild, B.L., Spencer, J.H. *Ramsey Theory*, Wiley-Interscience Series in Discrete Mathematics and Optimization, John Wiley & Sons, Inc., 2nd Ed., 1990.

[150] Griffiths, P., Harris, J. *Principles of Algebraic Geometry*, Wiley-Interscience, John Wiley & Sons, 1994.

[151] Gritzmann, P., Klee, V. Mathematical programming and convex geometry, in *Handbook of Convex Geometry*, eds., Gruber, P.M., Wills, J.M., North-Holland, 1993, pp. 627–674.

[152] Gruber, P.M, Wills, J.M. (eds.). *Handbook of Convex Geometry*, Vol. A, North-Holland, 1993.

[153] Guralnik, G., Zemach, C., Warnock, T. An algorithm for uniform random sampling of points in and on a hypersphere, *Inf. Process. Lett.* 21 (1985), 17–21.

[154] Gurevitch, Y. Kolmogorov machines and related issues: the column on logic in computer science, Bull. EATCS 35 (1988), 71–82.

[155] Halász, G. On Roth's Method in the Theory of Irregularities of Point Distributions, in *Recent Progress in Analytic Number Theory*, Vol. 2, Academic Press, 1981, pp. 79–94.

[156] Halton, J.H. On the efficiency of certain quasi-random sequences of points in evaluating multi-dimensional integrals, *Numer. Math.* 2 (1960), 84–90.

[157] Hammersley, J.M. Monte Carlo methods for solving multivariable problems, *Ann. New York Acad. Sci.* 86 (1960), 844-874.

[158] Hardy, G., Wright, E. *The Theory of Numbers*, Oxford University Press, 4th. Ed., 1965.

[159] Hartshorne, R. *Algebraic Geometry*, Springer, New York, 1977.

[160] Haussler, D., Welzl, E. ε-nets and simplex range queries, *Disc. Comput. Geom.* 2 (1987), 127–151.

[161] Hlawka, E. Funktionen von beschränkter Variation in der Theorie der Gleichverteilung, *Ann. Mat. Pura Appl.* 54 (1961), 325–333.

[162] Hoeffding, W. Probability inequalities for sums of bounded random variables, *J. Amer. Statistical Association* 58 (1963), 13–30.

[163] Impagliazzo, R., Levin, L., Luby, M. Pseudo-random generation from one-way functions, *Proc. 21st Annual ACM Symp. Theory Comput.* (1989), 12–24.

[164] Impagliazzo, R., Shaltiel, R., Wigderson, A. Near-optimal conversion of hardness into pseudo-randomness, *Proc. 40th Annual IEEE Symp. Found. Comput. Sci.* (1999), 181–190.

[165] Impagliazzo, R., Wigderson, A. P = BPP if E requires exponential circuits: Derandomizing the XOR lemma, *Proc. 29th Annual ACM Symp. Theory Comput.* (1997), 220–229.

[166] Impagliazzo, R., Zuckerman, D. How to recycle random bits, *Proc. 30th Annual IEEE Symp. Found. Comput. Sci.* (1989), 248–253.

[167] Indyk, P., Motwani, R. Approximate nearest neighbors: Towards removing the curse of dimensionality, *Proc. 30th Annual ACM Symp. Theory Comput.* (1998), 604–613.

[168] Ireland, K. Rosen, M. *A Classical Introduction to Modern Number Theory*, Graduate Texts in Mathematics, 84, Springer-Verlag, 2nd Ed., 1991.

[169] Jarník, V. O jistém problému minimálním, *Práce mor. přírodověd. spol. v Brně* VI, 4 (1930), 57–63. (In Czech.)

[170] Joffe, A. On a set of almost deterministic *k*-independent random variables, *Ann. Probability* 2 (1974), 161–162.

[171] Johnson, D.S. Approximation algorithms for combinatorial problems, *J. Comput. Syst. Sci.* 9 (1974), 256–278.

[172] Jones, N.D. *Computability and Complexity: From a Programming Perspective*, Foundations of Computing, The MIT Press, Cambridge, Mass., USA, 1997.

[173] Jones, M.N. *Spherical Harmonics and Tensors for Classical Field Theory*, Research Studies Press, John Wiley & Sons, 1985.

[174] Juhnke, F. Volumenminimale Ellipsoidüberdeckungen, *Beiträge zur Algebra und Geometrie* 30 (1990), 143–153.

[175] Kalai, G. A subexponential randomized simplex algorithm, *Proc. 24th Annual ACM Symp. Theory Comput.* (1992), 475–482.

[176] Kalai, G. Linear programming, the simplex algorithm and simple polytopes, *Math. Prog. B* 79 (1997), 217-234.

[177] Karger, D.R., Klein, P.N., Tarjan, R.E. A randomized linear-time algorithm to find minimum spanning trees, *J. ACM* 42 (1995), 321-328.

[178] Karmarkar, N. A new polynomial-time algorithm for linear programming, *Combinatorica* 4 (1984), 373-395.

[179] Karp, R.M. An introduction to randomized algorithms, *Discrete Applied Mathematics* 34 (1991), 165–201.

[180] Karp, R.M., Upfal, E., Wigderson, A. Constructing a maximum matching is in Random NC, *Combinatorica* 6 (1986), 35-48.

[181] Khachiyan, L.G. Polynomial algorithm in linear programming, *U.S.S.R. Comput. Math. and Math. Phys.* 20 (1980), 53–72.

[182] King, V. A simpler minimum spanning tree verification algorithm, *Algorithmica* 18 (1997), 263-270.

[183] Knuth, D.E. *The Art of Computer Programming, Vol. 2: Seminumerical Algorithms*, Addison-Wesley, 3rd Ed., 1997.

[184] Koblitz, N. *Introduction to Elliptic Curves and Modular Forms*, Graduate Texts in Mathematics, 97, Springer-Verlag, 2nd Ed., 1993.

[185] Koblitz, N. *A Course in Number Theory and Cryptography*, Graduate Texts in Mathematics, 114, Springer-Verlag, 2nd Ed., 1994.

[186] Koksma, J.F. Een algemeene stelling uit de theorie der gelijkmatige verdeeling modulo 1, *Mathematica B (Zutphen)* 11 (1942/43), 7–11. (English treatment in [240].)

[187] Komlós, J. Linear verification for spanning trees, *Combinatorica* 5 (1985), 57–65.

[188] Komlós, J., Pach, J., Woeginger, G. Almost tight bounds for ε-nets, *Disc. Comput. Geom.* 7 (1992), 163–173.

[189] Komlós, J., Pintz, J., Szemerédi, E., On Heilbronn's triangle problem, *J. London Math. Soc.* 2, 24 (1981), 385–396.

[190] Komlós, J., Pintz, J., Szemerédi, E., A lower bound for Heilbronn's problem, *J. London Math. Soc.* 2, 25 (1982), 13–24.

[191] Kushilevitz, E., Nisan, N. *Communication Complexity*, Cambridge University Press, 1997.

[192] Kushilevitz, E., Ostrovsky, R., Rabani, Y. Efficient search for approximate nearest neighbor in high-dimensional spaces, *Proc. 30th Annual ACM Symp. Theory Comput.* (1998), 614–623.

[193] Lancaster, P., Tismenetsky, M. *The Theory of Matrices*, Academic Press, 2nd Ed., 1985.

[194] Lang, S. *Elliptic Functions*, Graduate Texts in Mathematics, 112, Springer-Verlag, 2nd Ed., 1987.

[195] Lenstra, H.W. Factoring integers with elliptic curves, *Ann. Math.* 126 (1987), 649–673.

[196] Levin, L.A. *Computational complexity of functions*, Tech. Rep., Boston University, BUCS-TR-85-005, 1985. (Translation of a 1974 article by author in Russian.)

[197] Linial, N., Luby, M., Saks, M., Zuckerman, D. Efficient construction of a small hitting set for combinatorial rectangles in high dimension, *Combinatorica* 17 (1997), 215–234.

[198] Lovász, L. On the ratio of optimal integral and fractional covers, *Disc. Math.* 13 (1975), 383–390.

[199] Lovász, L., Spencer, J., Vesztergombi, K. Discrepancy of set systems and matrices, *European J. Combinatorics* 7 (1986), 151–160.

[200] Lubotzky, A., Phillips, R., Sarnak, P. Hecke operators and distributing points on the sphere, I, *Commun. Pure and Appl. Math.* 39 (1986), S149–S186.

[201] Lubotzky, A., Phillips, R., Sarnak, P. Hecke operators and distributing points on S^2, II, *Commun. Pure and Appl. Math.* 40 (1987), 401–420.

[202] Lubotzky, A., Phillips, R., Sarnak, P. Explicit expanders and the Ramanujan conjectures, *Combinatorica* 8 (1988), 261–277.

[203] Luby, M. A simple parallel algorithm for the maximal independent set problem, *SIAM J. Comput.* 15 (1986), 1036–1053.

[204] Luby, M. *Pseudorandomness and Cryptographic Applications*, Princeton Computer Science Notes, Princeton University Press, 1996.

[205] Luby, M., Wigderson, A. Pairwise Independence and Derandomization, *Tech. Rep. UC Berkeley* UCB/CSD-95-880, 1995.

[206] McKean, H.P., Moll, V. *Elliptic Curves: Function Theory, Geometry, Arithmetic*, Cambridge University Press, 1997.

[207] Magnus, W. *Noneuclidean Tesselations and Their Groups*, Academic Press, 1974.

[208] Margulis, G.A. Explicit constructions of concentrators, *Problems of Information Transmission* (translation) 9 (1973), 325–332.

[209] Margulis, G.A. Explicit group-theoretical constructions of combinatorial schemes and their application to the design of expanders and superconcentrators, *Problems of Information Transmission* (translation) 24 (1988), 39–46.

[210] Matoušek, J. Construction of ε-nets, *Disc. Comput. Geom.* 5 (1990), 427–448.

[211] Matoušek, J. Cutting hyperplane arrangements, *Disc. Comput. Geom.* 6 (1991), 385–406.

[212] Matoušek, J. Efficient partition trees, *Disc. Comput. Geom.* 8 (1992), 315–334.

[213] Matoušek, J. Range searching with efficient hierarchical cuttings, *Disc. Comput. Geom.* 10 (1993), 157–182.

[214] Matoušek, J. Geometric range searching, *ACM Comput. Surv.* 26 (1994), 421–461.

[215] Matoušek, J. Tight upper bounds for the discrepancy of halfspaces, *Disc. Comput. Geom.* 13 (1995), 593–601.

[216] Matoušek, J. Approximations and optimal geometric divide-and-conquer, *J. Comput. Syst. Sci.* 50 (1995), 203–208.

[217] Matoušek, J. Derandomization in computational geometry, *J. Algorithms* 20 (1996), 545–580.

[218] Matoušek, J. On discrepancy bounds via dual shatter functions, *Mathematika* 44 (1997), 42–49

[219] Matoušek, J. *Geometric Discrepancy: An Illustrated Guide*, Algorithms and Combinatorics, 18, Springer, 1999.

[220] Matoušek, J., Sharir, M., Welzl, E. A subexponential bound for linear programming, *Algorithmica* 16 (1996), 498–516.

[221] Matoušek, J., Spencer, J. Discrepancy in arithmetic progressions, *J. Amer. Math. Soc.* 9 (1996), 195–204.

[222] Matoušek, J., Welzl, E., Wernisch, L. Discrepancy and ε-approximations for bounded VC-dimension, *Combinatorica* 13 (1993), 455–466.

[223] Mazur, B. Number theory as gadfly, *Amer. Math. Monthly* 98 (1991), 593–610.

[224] Megiddo, N. Linear-time algorithms for linear programming in R^3 and related problems, *SIAM J. Comput.* 12 (1983), 759–776.

[225] Megiddo, N. Linear programming in linear time when the dimension is fixed, *J. ACM* 31 (1984), 114–127.

[226] Mehlhorn, K., Schmidt, E. Las Vegas is better than determinism in VLSI and distributed computing, *Proc. 14th Annual ACM Symp. Theory Comput.* (1982), 330–337.

[227] Milnor, J. Hyperbolic geometry: the first 150 years, *Bull. Amer. Math. Soc.* 6 (1982), 9–24.

[228] Miltersen, P.B. Lower bounds for union-split-find related problems on random access machines, *Proc. 26th Annual ACM Symp. Theory Comput.* (1994), 625–634.

[229] Miltersen, P.B. On the cell probe complexity of polynomial evaluation, *Theoret. Comput. Sci.* 143 (1995), 167–174.

[230] Miltersen, P.B., Nisan, N., Safra, S., Wigderson, A. On data structures and asymmetric communication complexity, *Proc. 27th Annual ACM Symp. Theory Comput.* (1995), 103–111.

[231] Mitchell, D.P. Spectrally optimal sampling for distribution ray tracing, *Computer Graphics* 25 (1991), 157–164.

[232] Montgomery, H.L. On irregularities of distribution, in *Congress of Number Theory* (Zarautz, 1984), Universidad del País Vasco, Bilbao, 1989, pp. 11–27.

[233] Montgomery, H.L. *Ten Lectures on the Interface Between Analytic Number Theory and Harmonic Analysis*, CBMS Regional Conference Series in Mathematics, No. 84, Amer. Math. Soc., Providence, 1994.

[234] Morgenstern, J. Note on a lower bound of the linear complexity of the fast Fourier transform, *J. ACM* 20 (1973), 305–306.

[235] Moser, W.O.J. Problems on extremal properties of a finite set of points, in *Discrete Geometry and Convexity*, Ann. New York Acad. Sci. 440 (1985), 52–64.

[236] Motwani, R., Raghavan, P. *Randomized Algorithms*, Cambridge University Press, 1995.

[237] Mulmuley, K. *Computational Geometry: An Introduction Through Randomized Algorithms*, Prentice-Hall, 1994.

[238] Naor, J., Naor, M. Small-bias probability spaces: efficient constructions and applications, *SIAM J. Comput.* 22 (1993), 838–856.

[239] Nešetřil, J. A few remarks on the history of MST-problem, *Archivum Mathematicum, Brno* 33 (1997), 15–22. Prelim. version in *KAM Series*, Charles University, Prague, No. 97-338, 1997.

[240] Niederreiter, H. *Random Number Generation and Quasi-Monte Carlo Methods*, CBMS-NSF, SIAM, Philadelphia, PA, 1992.

[241] Nisan, N. $RL \subseteq SC$, *Proc. 24th Annual ACM Symp. Theory Comput.* (1992), 619–623.

[242] Nisan, N. Pseudorandom generators for space-bounded computation, *Combinatorica* 12 (1992), 449–461.

[243] Nisan, N. Extracting randomness: how and why, a survey, *IEEE Conf. Comput. Complexity* (1996), 44–58.

[244] Nisan, N., Zuckerman, D. More deterministic simulation in logspace, *Proc. 25th Annual ACM Symp. Theory Comput.* (1993), 235–244.

[245] Nisan, N., Zuckerman, D. Randomness is linear in space, *J. Comput. Syst. Sci.* 52 (1996), 43–52.

[246] Odlyzko, A.M. Discrete logarithms in finite fields and their cryptographic significance, in *Advances in Cryptology, Proc. Eurocrypt 84*, LNCS, 209, Springer-Verlag (1985), 224–314.

[247] Pach, J., Agarwal, P.K. *Combinatorial Geometry*, Wiley-Interscience Series in Discrete Mathematics and Optimization, John Wiley & Sons, Inc., 1995.

[248] Pettie, S., Ramachandran, V. An optimal minimum spanning tree algorithm, *Proc. 27th ICALP* (2000), 49–60.

[249] Pinsker, M. On the complexity of a concentrator, *Proc. 7th Int. Teletraffic Conf.* Stockholm (1973), 318/1–318/4.

[250] Pippenger, N. Superconcentrators, *SIAM J. Comput.* 6 (1977) 298–304.

[251] Post, M.J. Minimum spanning ellipsoids, *Proc. 16th Annual ACM Symp. Theory Comput.* (1984), 108–116.

[252] Preparata, F.P., Hong, S.J. Convex hulls of finite sets of points in two and three dimensions, *Comm. ACM* 20 (1977), 87–93.

[253] Rabin, M.O. Probabilistic algorithms, in *Algorithms and Complexity, Recent Results and New Directions*, Traub, J.F., ed., Academic Press, 1976, pp. 21–39.

[254] Raghavan, P. Probabilistic construction of deterministic algorithms: approximating packing integer programs, *J. Comput. Syst. Sci.* 37 (1988), 130–143.

[255] Raz, R., Reingold, O., Vadhan, S. Error reduction for extractors, *Proc. 40th Annual IEEE Symp. Found. Comput. Sci.* (1999), 191–201.

[256] Raz, R., Reingold, O., Vadhan, S. Extracting all the randomness and reducing the error in Trevisan's extractors, *Proc. 31st Annual ACM Symp. Theory Comput.* (1999), 149–158.

[257] Razborov, A., Szemerédi, Wigderson, A. Constructing small sets that are uniform in arithmetic progressions, *Combinatorics, Probability and Computing* 2 (1993), 513–518.

[258] Reif, J.H., Sen, S. Optimal randomized parallel algorithms for computational geometry, *Algorithmica* 7 (1992), 91–117.

[259] Renegar, J. On the computational complexity and geometry of the first order theory of the reals, *J. Symbolic Comput.* 13 (1992), 255–352.

[260] Roth, K.F. On a problem of Heilbronn, *J. London Math. Soc.* 26 (1951), 198–204.

[261] Roth, K.F. On irregularities of distribution, *Mathematika* 1 (1954), 73–79.

[262] Roth, K.F. Remark concerning integer sequences, *Acta Arithmetica* 9 (1964), 257–260.

[263] Roth, K.F. Developments in Heilbronn's triangle problem, *Advances in Mathematics* 22 (1976), 364–385.

[264] Roth, K.F. On irregularities of distribution, IV, *Acta Arithmetica* 37 (1980), 67–75.

[265] Santaló, L.A. Integral geometry and geometric probability, in *Encyclopedia of Mathematics and its Applications*, Vol. 1, ed. Rota, G.-C., Addison-Wesley, 1976.

[266] Saks, M., Srinivasan, A., Zhou, S. Explicit OR-dispersers with polylogarithmic degree, *J. ACM* 45 (1998), 123–154.

[267] Santha, M., Vazirani, U.V. Generating quasi-random sequences from semi-random sources, *J. Comput. Syst. Sci.* 33 (1986), 75–87.

[268] Sarnak, P. *Some Applications of Modular Forms*, Cambridge Tracts in Mathematics, 99, Cambridge University Press, 1990.

[269] Sauer, N. On the density of families of sets, *J. Combinatorial Theory A* 13 (1972), 145–147.

[270] Schmidt, W.M. Irregularities of distribution, IV, *Invent. Math.* 7 (1969), 55–82.

[271] Schmidt, W.M. On a problem of Heilbronn, *J. London Math. Soc.* 4 (1971/72), 545–550.

[272] Schmidt, W.M. Irregularities of distribution, III, *Pac. J. Math.* 29 (1969), 225–234.

[273] Schmidt, W.M. Irregularities of distribution, VII, *Acta Arithmetica* 21 (1972), 45–50.

[274] Schmidt, W.M. Irregularities of distribution, X, *Number Theory and Algebra*, Academic Press, 1977, pp. 311–329.

[275] Schmidt, W.M. *Equations over Finite Fields: An Elementary Approach*, Lecture Notes in Mathematics, 536, Springer-Verlag, 1976.

[276] Seidel, R. A convex hull algorithm optimal for point sets in even dimensions, *Tech. Rep. Univ. British Columbia* 81–14, 1981.

[277] Seidel, R. Constructing higher-dimensional convex hulls at logarithmic cost per face, *Proc. 18th Annual ACM Symp. Theory Comput.* (1986), 404–413.

[278] Seidel, R. Small-dimensional linear programming and convex hulls made easy, *Disc. Comput. Geom.* 6 (1991), 423–434.

[279] Serre, J.P. *A Course in Arithmetic*, Springer-Verlag, 1985.

[280] Shafarevich, I.R. *Basic Algebraic Geometry*, Springer-Verlag, 1977.

[281] Shamos, M.I., Hoey, D. Closest-point problems, *Proc. 16th Annual IEEE Symp. Found. Comput. Sci.* (1975), 151–162.

[282] Sharir, M., Agarwal, P.K. *Davenport-Schinzel sequences and their geometric applications*, Cambridge University Press, 1995.

[283] Sharir, M., Welzl, E. A combinatorial bound for linear programming and related problems, *Proc. 9th Annual Symp. Theoret. Aspects Comput. Sci.*, LNCS, 577, Springer-Verlag, 1992, pp. 569–579.

[284] Shelah, S. A combinatorial problem; stability and order for models and theories in infinitary languages, *Pac. J. Math.* 41 (1972), 247–261.

[285] Shimura, G. *Introduction to the Arithmetic Theory of Automorphic Functions*, Publications of the Mathematical Society of Japan, 11, Iwanami Shoten Publishers and Princeton University Press, 1971.

[286] Shirley, P. Discrepancy as a quality measure for sample distributions, *Proc. Eurographics'91*, (1991), 183–194.

[287] Silverman, J.H. *The Arithmetic of Elliptic Curves*, Graduate Texts in Mathematics, 106, Springer-Verlag, 1986.

[288] Silverman, J.H., Tate, J. *Rational Points on Elliptic Curves*, Undergraduate Texts in Mathematics, Springer-Verlag, 1992.

[289] Sinclair, A. *Algorithms for Random Generation and Counting: A Markov Chain Approach*, Progress in Theoretical Computer Science, Birkhauser, 1993.

[290] Sipser, M. Expanders, randomness, or time versus space, *J. Comput. Syst. Sci.* 36 (1988), 379–383.

[291] Sloane, N.J.A. Encrypting by random rotations, *Proc. Workshop on Cryptography, Burg Feuerstein, Germany*, LNCS, 149, Springer-Verlag, 1983, pp. 71–128.

[292] Spencer, J. Balancing games, *J. Combinatorial Theory B* 23 (1977), 68–74.

[293] Spencer, J. Six standard deviations suffice, *Trans. Amer. Math. Soc.* 289 (1985), 679–706.

[294] Spencer, J. *Ten Lectures on the Probabilistic Method*, CBMS-NSF, SIAM, 1987.

[295] Srinivasan, A. Improving the discrepancy bound for sparse matrices: better approximations for sparse lattice approximation problems, *Proc. 8th ACM-SIAM Symp. Discrete Algorithms* (1997), 692–701.

[296] Srinivasan, A., Zuckerman, D. Computing with very weak random sources, *Proc. 35th Annual IEEE Symp. Found. Comput. Sci.* (1994), 264–275.

[297] Stein, E.M., Weiss, G. *Introduction to Fourier Analysis on Euclidean Spaces*, Princeton University Press, 1971.

[298] Stolarsky, K.B. Sums of distances between points on a sphere. II, *Proc. Amer. Math. Soc.* 41 (1973), 575–582.

[299] Strang, G. *Introduction to Applied Mathematics*, Wellesley-Cambridge Press, 1986.

[300] Sudan, M., Trevisan, L., Vadhan, S. Pseudorandom generators without the XOR Lemma, *Proc. 31st Annual ACM Symp. Theory Comput.* (1999), 537–546.

[301] Szegö, G. *Orthogonal Polynomials*, Amer. Math. Soc., Coll. Publ. Vol. XXIII, 1939.

[302] Tanner, R.M. Explicit construction of concentrators from generalized n-gons, *SIAM J. on Algebraic and Discrete Methods* 5 (1984), 287–293.

[303] Tarjan, R.E. Efficiency of a good but not linear set-union algorithm, *J. ACM* 22 (1975), 215–225.

[304] Tarjan, R.E. Complexity of monotone networks for computing conjunctions, *Ann. Disc. Math.* 2 (1978), 121–133.

[305] Tarjan, R.E. *Data Structures and Network Algorithms*, SIAM, Philadelphia, PA, 1983.

[306] Taylor, R., Wiles, A. Ring-theoretic properties of certain Hecke algebras, *Ann. Math.*, 141 (1995), 553–572.

[307] Thurston, W.P., Levy, S. (ed.). *Three-Dimensional Geometry and Topology*, Princeton Mathematical Series, 35, Princeton University Press, 1997.

[308] Traub, J.F., Wasilkowski, G.W., Woźniakowski, H. *Information-Based Complexity*, Academic Press, 1988.

[309] Traub, J.F., Werschulz, A.G. *Complexity and Information*, Cambridge University Press, 1998.

[310] Trevisan, L. Construction of extractors using pseudo-random generators, *Proc. 31st Annual ACM Symp. Theory Comput.* (1999), 141–148.

[311] Vaidya, P.M., Space-time tradeoffs for orthogonal range queries, *SIAM J. Comput.* 18 (1989), 748–758.

[312] Valiant, L. Graph-theoretic arguments in low-level complexity, *Proc. 6th Math. Foundations Comp. Sci.* LNCS, 53, Springer-Verlag, 1977, pp. 162–176.

[313] van der Corput, J.G. Verteilungsfunktionen I. *Proc. Nederl. Akad. Wetensch.* 38 (1935), 813–821.

[314] van der Corput, J.G. Verteilungsfunktionen II. *Proc. Nederl. Akad. Wetensch.* 38 (1935), 1058–1066.

[315] van Emde Boas, P. Preserving order in a forest in less than logarithmic time and linear space, *Inform. Process. Lett.* 6 (1977), 80–82.

[316] van Emde Boas, P., Kaas, R., Zijlstra, E. Design and implementation of an efficient priority queue, *Math. Systems Theory* 10 (1977), 99–127.

[317] Vapnik, V.N., Chervonenkis, A.Ya. On the uniform convergence of relative frequencies of events to their probabilities, *Theory of Probability and its Applications* 16 (1971), 264–280.

[318] Vuillemin, J. A data structure for manipulating priority queues, *Commun. ACM* 21 (1978), 309–315.

[319] Welzl, E. Partition trees for triangle counting and other range searching problems, *Proc. 4th Annual ACM Symp. Comput. Geom.* (1988), 23–33.

[320] Welzl, E. Smallest enclosing disks (balls and ellipsoids), in *New Results and New Trends in Computer Science*, ed., Maurer, H., LNCS, 555, Springer-Verlag, 1991, pp. 359–370.

[321] Weyl, H. Über die Gleichverteilung von Zahlen mod. Eins. *Math. Ann.* 77 (1916), 111–147.

[322] Wigderson, A., Zuckerman, D. Expanders that beat the eigenvalue bound: Explicit construction and applications, *Proc. 25th Annual ACM Symp. Theory Comput.* (1993), 245–251.

[323] Wiles, A. Modular elliptic-curves and Fermat's Last Theorem, *Ann. Math.* 141 (1995), 443–551.

[324] Willard, D. Polygon retrieval, *SIAM J. Comput.* 11 (1982), 149–165.

[325] Willard, D.E. Log-logarithmic worst-case range queries are possible in space $\Theta(n)$, *Inform. Process. Lett.* 17 (1983), 81–84.

[326] Xiao, B. *New Bounds in Cell Probe Model*, PhD Thesis, UC San Diego, 1992.

[327] Yao, A.C. An $O(|E|\log\log|V|)$ algorithm for finding minimum spanning trees, *Inf. Process. Lett.* 4 (1975), 21–23.

[328] Yao, A.C. Some complexity questions related to distributed computing, *Proc. 11th Annual ACM Symp. Theory Comput.* (1979), 209–213.

[329] Yao, A.C. Should tables be sorted?, *J. ACM* 28 (1981), 615–628.

[330] Yao, A.C. Theory and applications of trapdoor functions, *Proc. 23rd Annual IEEE Symp. Found. Comput. Sci.* (1982), 80–91.

[331] Yao, A.C. On the complexity of maintaining partial sums, *SIAM J. Comput.* 14 (1985), 277–288.

[332] Yap, C.K. Symbolic treatment of geometric degeneracies, *J. Symbolic Comput.* 10 (1990), 349–370.

[333] Yap, C.K. A geometric consistency theorem for a symbolic perturbation scheme, *J. Comput. Sys. Sci.* 40 (1990), 2–18.

[334] Zagier, D. Introduction to modular forms, in *From Number Theory to Physics*, eds., Waldschmidt, M., et al., Springer-Verlag, 1992, pp. 238–291.

[335] Ziegler, G.M. *Lectures on Polytopes*, Graduate Texts in Mathematics, 152, Springer-Verlag, 1995.

[336] Zuckerman, D. Simulating BPP using a general weak random source, *Algorithmica* 16 (1996), 367–391.

[337] Zuckerman, D. Randomness-optimal oblivious sampling, *Random Structures and Algorithms* 11 (1997), 345–367.

Index